Conducting Polymers

Conducting Polymers

Chemistries, Properties and Biomedical Applications

Edited by

Ram K. Gupta

CRC Press
Taylor & Francis Group
Boca Raton London New York

CRC Press is an imprint of the
Taylor & Francis Group, an **informa** business

First edition published 2022
by CRC Press
6000 Broken Sound Parkway NW, Suite 300, Boca Raton, FL 33487-2742

and by CRC Press

4 Park Square, Milton Park, Abingdon, Oxon, OX14 4RN

CRC Press is an imprint of Taylor & Francis Group, LLC

© 2022 selection and editorial matter, Ram K. Gupta; individual chapters, the contributors

---

Library of Congress Cataloging-in-Publication Data

Names: Gupta, Ram K., editor.
Title: Conducting polymers : chemistries, properties and biomedical
  applications / edited by Ram K. Gupta.
Description: First edition. | Boca Raton : Taylor and Francis, 2022. |
  Includes bibliographical references and index.
Identifiers: LCCN 2021049179 | ISBN 9781032071091 (hardback) | ISBN
  9781032071138 (paperback) | ISBN 9781003205418 (ebook)
Subjects: LCSH: Conducting polymers.
Classification: LCC QD382.C66 C652 2022 | DDC
  620.1/9204297--dc23/eng/20211201
LC record available at https://lccn.loc.gov/2021049179

ISBN: 978-1-032-07109-1 (hbk)
ISBN: 978-1-032-07113-8 (pbk)
ISBN: 978-1-003-20541-8 (ebk)

DOI: 10.1201/9781003205418

Typeset in Times
by KnowledgeWorks Global Ltd.

I would like to dedicate this book to my parents, family members, friends, teachers, and students. Special thanks to my wife Rajani, my daughter Anjali, and my niece Payal Gupta for their love, motivation, and encouragement.

Contents

Preface

Conducting polymers are versatile materials that possess both the unique properties of polymeric materials (elastic behavior, reversible deformation, flexibility, etc.) and the ability to conduct electricity with bulk conductivities comparable to those of metals and semiconductors. Shirakawa, MacDiarmid, and Heeger received the 2000 Nobel Prize in Chemistry for the discovery of conducting polymers and since then conducting polymers find their wide applications in many areas such as energy, sensors, semiconducting devices due to their tunable properties. Conducting polymers could be attractive for various biomedical applications due to their intelligent response to electrical fields from different types of tissues, including muscle, connective tissue, epithelium, and nervous tissue. Biocompatibility and biodegradability are some of the main requirements for conducting polymers to be used in biomedical applications therefore chemical/physical modifications are needed to make them suitable for biomedical applications. Some examples of biomedical applications are the development of nerve conduits that can promote nerve regeneration which is required in organ transplant processes and post-trauma cases. Also, the fabrication of synthetic scaffolds to promote the growth of myocardial cells can be used as components for pacemakers. The combination of biocompatibility and electric conduction opens the possibility for complex brain surgery with the potential to cure Parkinson's disease, for instance. Applications for drug and gene delivery systems, synthetic skin, bone fixation, and biosensors for diagnostics are within reach. Hence, conducting polymers are pointed as a viable solution to tackle the current challenges in medicine and bring new materials that can greatly improve the life quality of patients.

The main purpose of this book is to provide current, state-of-the-art knowledge of conducting polymers and their composites for biomedical applications. This book covers the fundamentals of conducting polymers, strategies to modify the structure of conducting polymers to make them biocompatible, and their applications in various biomedical areas like drug/gene delivery, tissue engineering, antimicrobial activities, biosensors, etc. Prospects of biodegradable conducting polymers for medical implant devices are also covered. Many chapters are covered by experts in these areas around the world making this a suitable textbook for students and providing new guidelines to researchers and industries.

Ram K. Gupta

Editor

Ram K. Gupta, PhD, is Associate Professor at Pittsburg State University. Dr. Gupta's research focuses on conducting polymers and composites, green energy production and storage using biowastes and nanomaterials, optoelectronic and photovoltaic devices, organic-inorganic heterojunctions for sensors, bio-based polymers, flame-retardant polymers, biocompatible nanofibers for tissue regeneration, scaffold and antibacterial applications, corrosion inhibiting coatings, and biodegradable metallic implants. Dr. Gupta has published over 235 peer-reviewed articles, made over 300 national, international, and regional presentations, chaired many sessions at national and international meetings, edited many books, and written several book chapters. He has received over two and a half million dollars for research and educational activities from many funding agencies. He serves as Editor-in-Chief, Associate Editor, and editorial board member of numerous journals.

Contributors

Zahra Allahyari
Department of Biomedical Engineering
Rochester Institute of Technology
Rochester, New York

Seyed Mohammad Amini
Radiation Biology Research Center
Iran University of Medical Sciences (IUMS)
Tehran, Iran

Khairunnisa Amreen
Microfluidics and Nanoelectronics Lab
Department of Electrical and Electronics
    Engineering
Birla Institute of Technology and Science
Hyderabad, India

Edith Amuhaya
Department of Pharmaceutics and Pharmacy
    Practice
School of Pharmacy and Health Sciences
United States International University—Africa
Nairobi, Kenya

Salim Anisha
Biomaterials and Bioprocess Laboratory
Department of Microbial Biotechnology
Bharathiar University
Coimbatore, India

Ismail Bal
Vocational School of Health Services
Istanbul Okan University
Opticianry Programme
Istanbul, Turkey
and
Programme of Nanoscience and
    Nanoengineering
Institute of Nanotechnology
Gebze Technical University
Gebze, Turkey

Trishna Bal
Department of Pharmaceutical Sciences and
    Technology
Birla Institute of Technology
Mesra, Ranchi, India

Robert Becky
Biomaterials and Bioprocess
    Laboratory
Department of Microbial
    Biotechnology
Bharathiar University
Coimbatore, India

Nilay Bereli
Chemistry Department
Hacettepe University
Ankara, Turkey

Trinath Biswal
Department of Chemistry
VSS University of Technology
Burla, India

Hema Bora
Biomaterials and Tissue Engineering
    Laboratory
School of Medical Science and
    Technology
Indian Institute of Technology
Kharagpur, India

Emilio Bucio
Departamento de Química de Radiaciones y
    Radioquímica
Instituto de Ciencias Nucleares
Universidad Nacional Autónoma de
    México
Mexico City, México

Merve Çalışır
Chemistry Department
Hacettepe University
Ankara, Turkey

Luis A. Camacho-Cruz
Departamento de Química de Radiaciones y
    Radioquímica
Instituto de Ciencias Nucleares
Universidad Nacional Autónoma de
    México
Mexico City, México

Shuo Chen
State Key Laboratory for Modification of
  Chemical Fibers and Polymer Materials
College of Materials Science and Engineering
Donghua University
Shanghai, PR China

Bavatharani Chokkiah
Nano Electrochemistry Lab (NEL)
Department of Chemistry
National Institute of Technology
Puducherry, Karaikal, India

Soumen Das
Bio-MEMS Laboratory
School of Medical Science and Technology
Indian Institute of Technology
Kharagpur, India

Felipe de Souza
Kansas Polymer Research Center
Pittsburg State University
Pittsburg, Kansas

Adil Denizli
Chemistry Department
Hacettepe University
Ankara, Turkey

Katayoun Derakhshandeh
Medicinal Plants and Natural Products
  Research Center
Hamadan University of Medical Sciences
Hamadan, Iran

Hossein Derakhshankhah
Pharmaceutical Sciences Research Center,
  Health Institute
Kermanshah University of Medical Sciences
Kermanshah, Iran

Ragupathy Dhanusuraman
Nano Electrochemistry Lab
Department of Chemistry
National Institute of Technology
Puducherry, Karaikal, India

Santanu Dhara
Biomaterials and Tissue Engineering Laboratory
School of Medical Science and Technology
Indian Institute of Technology
Kharagpur, India

Chenthamara Dhrisya
Biomaterials and Bioprocess
  Laboratory
Department of Microbial Biotechnology
Bharathiar University
Coimbatore, India

Pawan K. Diwan
Department of Applied Science
Kurukshetra University
Kurukshetra, India

Krishna Dixit
Biomaterials and Tissue Engineering
  Laboratory
School of Medical Science and Technology
Indian Institute of Technology
Kharagpur, India

Girirajasekhar Dornadula
Department of Pharmacy Practice
Annamacharya College of Pharmacy
Rajampet, India

Muthusankar Eswaran
Nano Electrochemistry Lab (NEL)
Department of Chemistry
National Institute of Technology
Puducherry, Karaikal, India
and
Indian Institute of Technology Jammu
Jammu, India

Shah Z. Farooq
Department of Civil and Environmental
  Engineering
Korea Advanced Institute of Science and
  Technology (KAIST)
Daejeon, South Korea

Mbuso Faya
Discipline of Pharmaceutical Sciences
School of Health Sciences
University of KwaZulu-Natal
Durban, South Africa

Shayan Gholizadeh
Department of Biomedical Engineering
Rochester Institute of Technology
Rochester, New York

Sanket Goel
Microfluidics and Nanoelectronics Lab
Department of Electrical and Electronics
  Engineering
Birla Institute of Technology and
  Science
Hyderabad, India

Thirumala Govender
Discipline of Pharmaceutical Sciences
School of Health Sciences
University of KwaZulu-Natal
Durban, South Africa

Lakshmi Narasimha Gunturu
Scientimed Solutions Private Limited
Mumbai, India

Ram K. Gupta
Department of Chemistry
Pittsburg State University
Pittsburg, Kansas

Maimoona Ilyas
Sustainable Development Study Centre
Government College University
Lahore, Pakistan

Muntaha Ilyas
Department of Zoology, Wildlife and
  Fisheries
University of Agriculture Faisalabad
Faisalabad, Pakistan
and
International Society of Engineering Science
  and Technology
Nottingham Trent University
Nottingham, United Kingdom

Eman Abdallah Ismail
Discipline of Pharmaceutical Sciences
School of Health Sciences
University of KwaZulu-Natal
Durban, South Africa

Daeik Jang
Department of Civil and Environmental
  Engineering
Korea Advanced Institute of Science and
  Technology (KAIST)
Daejeon, South Korea

Dharmendra K. Jena
Department of Chemistry
Utkal University
Vani Vihar, Bhubaneswar, India

Mizan İbrahim Kahyaoğlu
Department of Chemistry
Faculty of Science and Arts
Ondokuz Mayis University
Samsun, Turkey

Shagun Kainth
Virginia Tech Center of Excellence in
  Emerging Materials
Thapar Institute of Engineering and
  Technology
Patiala, India

Pamayyagari Kalpana
Department of Pharmacy Practice
Annamacharya College of Pharmacy
Rajampet, India

Selcan Karakuş
Department of Chemistry
Faculty of Engineering
Istanbul University—Cerrahpasa
Istanbul, Turkey

N. Veni Keertheeswari
Nano Electrochemistry Lab
Department of Chemistry
National Institute of Technology
Puducherry, Karaikal, India

Hammad R. Khalid
Civil and Environmental Engineering
  Department
King Fahd University of Petroleum and
  Minerals (KFUPM)
Dhahran, Saudi Arabia

Israfil Kucuk
Programme of Nanoscience and
  Nanoengineering
Institute of Nanotechnology
Gebze Technical University
Gebze, Turkey

Gaurav Kulkarni
Biomaterials and Tissue Engineering
    Laboratory
School of Medical Science and Technology
Indian Institute of Technology
Kharagpur, India
and
Bio-MEMS Laboratory
School of Medical Science and Technology
Indian Institute of Technology
Kharagpur, India

Anuj Kumar
Department of Chemistry
GLA University
Mathura, India

H.K. Lee
Department of Civil and Environmental
    Engineering
Korea Advanced Institute of Science and
    Technology (KAIST)
Daejeon, South Korea

Umer Liaqat
Department of Zoology, Wildlife and
    Fisheries
University of Agriculture Faisalabad
Faisalabad, Pakistan

Eder C. Lima
Institute of Chemistry
Federal University of Rio Grande do Sul
    (UFRGS)
Porto Alegre, Rio Grande do Sul, Brazil

José C. Lugo-González
Facultad de Química
Universidad Nacional Autónoma de México
Mexico City, México

Suba Lakshmi Madaswamy
Nano Electrochemistry Lab
Department of Chemistry
National Institute of Technology
Puducherry, Karaikal, India

Palanichamy Nandhini
Biomaterials in Medicinal Chemistry
    Laboratory
Department of Natural Products Chemistry
School of Chemistry
Madurai Kamaraj University
Madurai, India

Nimbagal Raghavendra Naveen
Department of Pharmaceutics
Sri Adichunchanagiri College of Pharmacy
Adichunchanagiri University
Nagar, India

Calvin A. Omolo
Department of Pharmaceutics and Pharmacy
    Practice
School of Pharmacy and Health Sciences
United States International University—Africa
Nairobi, Kenya
and
Discipline of Pharmaceutical Sciences
School of Health Sciences
University of KwaZulu-Natal
Durban, South Africa

Cemal Özeroğlu
Department of Chemistry
Faculty of Engineering
Istanbul University—Cerrahpasa
Istanbul, Turkey

Erdoğan Özgür
Advanced Technologies Application and
    Research Center
Hacettepe University
Ankara, Turkey

Jhansi Lakshmi Parimi
Biomaterials and Tissue Engineering
    Laboratory
School of Medical Science and Technology
Indian Institute of Technology
Kharagpur, India

Paloma Patra
Department of Biomedical Engineering
Indian Institute of Technology Hyderabad
Hyderabad, India

Murugan Prasathkumar
Biomaterials and Bioprocess Laboratory
Department of Microbial Biotechnology
Bharathiar University
Coimbatore, India

Mariappan Rajan
Biomaterials in Medicinal Chemistry
    Laboratory
Department of Natural Products Chemistry
School of Chemistry
Madurai Kamaraj University
Madurai, India

Aditya Dev Rajora
Department of Pharmaceutical Sciences and
    Technology
Birla Institute of Technology
Mesra, Ranchi, India

Aravind Kumar Rengan
Department of Biomedical Engineering
Indian Institute of Technology Hyderabad
Hyderabad, India

Aram Rezaei
Nano Drug Delivery Research Center
Health Technology Institute
Kermanshah University of Medical Sciences
Kermanshah, Iran

Subramaniam Sadhasivam
Biomaterials and Bioprocess Laboratory
Department of Microbial Biotechnology
Bharathiar University
Coimbatore, India
and
Department of Extension and Career Guidance
Bharathiar University
Coimbatore, India

Baisakhee Saha
Biomaterials and Tissue Engineering
    Laboratory
School of Medical Science and Technology
Indian Institute of Technology
Kharagpur, India

Prafulla K. Sahoo
Department of Chemistry
Utkal University
Vani Vihar, Bhubaneswar, India
and
School of Applied Sciences
Centurion University of Technology and
    Management
Jatni, India

Hadi Samadian
Nano Drug Delivery Research Center
Health Technology Institute
Kermanshah University of Medical Sciences
Kermanshah, Iran

Srijita Sen
Department of Pharmaceutical Sciences and
    Technology
Birla Institute of Technology
Mesra, Ranchi, India

Neelima Sharma
Department of Pharmaceutical Sciences and
    Technology
Birla Institute of Technology
Mesra, Ranchi, India

Piyush Sharma
School of Physics and Materials Science
Thapar Institute of Engineering and
    Technology
Patiala, India

Shreya Sharma
Department of Pharmaceutical Sciences and
    Technology
Birla Institute of Technology
Mesra, Ranchi, India

Shubha Rani Sharma
Department of Bioengineering
Birla Institute of Technology
Mesra, Ranchi, India

Farooq Sher
Department of Engineering
School of Science and Technology
Nottingham Trent University
Nottingham, United Kingdom

Mika Sillanpää
Chemistry Department
College of Science
King Saud University
Riyadh, Saudi Arabia

Jasmina Sulejmanović
Department of Chemistry
Faculty of Science
University of Sarajevo
Sarajevo, Bosnia and Herzegovina

Aswathi Thomas
Department of Biomedical Engineering
Indian Institute of Technology Hyderabad
Hyderabad, India

Aykut Arif Topcu
Medical Laboratory Program
Vocational School of Health Service
Aksaray University
Aksaray, Turkey

İbrahim Vargel
Deparment of Plastic, Reconstructive and
    Aesthetic Surgery
Hacettepe University
Ankara, Turkey

Marlene A. Velazco-Medel
Departamento de Química de Radiaciones y
    Radioquímica
Instituto de Ciencias Nucleares
Universidad Nacional Autónoma de México
Mexico City, México

H.N. Yoon
Department of Civil and Environmental
    Engineering
Korea Advanced Institute of Science and
    Technology (KAIST)
Daejeon, South Korea

Zhengwei You
State Key Laboratory for Modification of
    Chemical Fibers and Polymer Materials
College of Materials Science and Engineering
Donghua University
Shanghai, PR China

Ayesha Zafar
International Society of Engineering Science
    and Technology
Nottingham Trent University
Nottingham, United Kingdom
and
Institute of Biochemistry and Biotechnology
Faculty of Biosciences
University of Veterinary and Animal Sciences
Lahore, Pakistan

1 Conducting Polymers
An Introduction

Luis A. Camacho-Cruz,[1] Marlene A. Velazco-Medel,[1]
José C. Lugo-González,[2] and Emilio Bucio[1]

[1]Departamento de Química de Radiaciones y Radioquímica, Instituto de Ciencias Nucleares, Universidad Nacional Autónoma de México, Circuito Exterior, Ciudad Universitaria, Mexico City, México

[2]Facultad de Química, Universidad Nacional Autónoma de México, Ciudad Universitaria, Mexico City, México

CONTENTS

1.1 INTRODUCTION

The development of electricity as a tool has been, throughout centuries of mankind, one of the most useful feats that humanity has achieved. Undoubtedly, our modern world would not be the same without electricity because we now have achieved technologies that people who lived merely 200 years ago would only be able to rationalize as magical or ineffable (e.g., computers, smartphones, videoconferences, instant messaging, self-driving cars) [1]. Since the end of the 18th century, the development of new theories for the electrical behavior of materials has allowed scientists to profit better from the characteristics of existing materials, and, naturally, develop materials that allowed for different applications. At the start of these studies, materials were classified as either electricity conductors (at that time, only metals, graphite, electrolytic solutions, and some organic tissues were known to conduct electricity) and insulators; therefore, the applications of these materials were limited to analog systems that explored the limits of what classical conductors/insulators were capable of doing [2]. In the 20th century, quantum theory and its applications in understanding charge mobility in metals and insulators allowed for the discovery of the third group of materials called semiconductors, which again revolutionized the applications of electricity, allowing for digital systems to be conceived [3].

Even though the discovery of materials in the 19th and 20th centuries was characterized by new electrical behaviors and applications of those, all these materials (conductors and semiconductors)

DOI: 10.1201/9781003205418-1

shared very similar characteristics. For instance, metals are shiny materials capable of conducting electricity and heat very effectively, and despite being malleable and ductile (mandatory characteristics for their use in electric equipment and electronics), they cannot change shape easily because they deform irreversibly, they are susceptible to ruptures, and they are not elastic. Additionally, to these disadvantages, metals which are practical for their use on electric equipment and electronics are limited to copper, silver, gold, and alloys of some other metals such as aluminum and iron. On the other hand, semiconductors are typically brittle solid-state materials that can only be incorporated with special handling onto electronic circuits [4].

Due to the characteristics of the materials comprising typical conductors and semiconductors, constructing a flexible conductive substrate was not a trivial task. Opposite from the characteristics of these materials, also in the 20th century, polymer chemistry revolutionized all aspects of human life because of the interesting characteristics of these macromolecular compounds. Because of this, in the last decades of the 20th century, there was a shift in which many materials were replaced with polymeric alternatives [5]. Typical polymeric systems are those that possess characteristics that the laypeople commonly associate with plastics, for example, elastic behavior, reversible deformations, ease of processing, cheap manufacturing, and ubiquity. However, polymer chemistry has also produced materials with many other characteristics, such as gigantic mechanical resistances, unprecedented thermal resistance for organic compounds, self-healing behaviors, responses to chemical and physical stimuli, among many others [6].

It is not surprising that polymers seemed like an attractive alternative for the development of electronics; nevertheless, since polymers are organic macromolecules, most are insulators for electrical current or heat. Despite this, by modulating the molecular structures of the monomers, tweaking polymerization steps, and adding chemical modifiers that allow for the movement of charges through the polymeric system (dopants), it has become possible to generate polymeric systems that conduct electricity with great effectiveness [7]. These systems, which have also been named synthetic metals or plastic metals, have gained a lot of popularity in several areas, and the research on these materials has increased substantially in recent years [8]. Conducting polymers have become important in areas such as the design of electronic components like organic light-emitting diodes (OLED), the development of new screens for mobile devices, and the design of solar cells. However, besides classical uses in electronics, one area in which these systems exhibit great potential is the biomedical since conducting polymers can also be crafted, so they are compatible with organic tissue while also performing goals such as drug delivery, tissue engineering, *in vivo* analyte sensing, artificial muscle development, neural probes, among others [9, 10].

In this introduction to the book "Conducting Polymers: Chemistries, Properties, and Biomedical Applications", a simplified overview on topics like the history of conducting polymers, the different ways in which conducting polymers transport electric current, and examples of conducting polymeric systems relevant to biomedical applications are presented. This section functions as a prelude to understanding the specific mechanisms used for the different applications in a variety of medical areas that will be presented in the following chapters. Because this topic is relevant not only to medicine but also to other areas of chemistry and physics of materials, some of the descriptions in this introduction (especially, those referring to molecular orbital theory and charge transport through non-crystalline solids) are simplified; nevertheless, we encourage the readers who may be interested in more technical explanations to review the existing literature about conducting polymers [7, 8, 11–13].

1.2 TECHNICALITIES

Up until now, we have treated the term *conducting polymers* as it was self-explanatory, namely, a polymer that can conduct electricity. Nevertheless, this term is not usually applied to all polymeric systems that may conduct electricity; therefore, it is important to define two types of systems [14]:

1. Polymers may be intrinsically electrically conductive because of particular structural characteristics (conjugated double bonds) and the addition of chemical dopants that alter the

structure of the base polymer (e.g., through redox processes, ionic reactions, photoreactions), allowing for the presence of charge carriers (ions, solitons, polarons, or bipolarons) that can conduct electric current through the bulk material. These materials are called *conductive polymers*.

2. Nonconductive polymers may acquire electrical conductivity through the addition of conducting particles (metallic powder, metal nanoparticles [NPs], carbon nanomaterials) by impregnation, covalent bonding, or dissolution. The electrical conduction on these materials occurs through these conducting particles and not through the polymer chains. These materials are called *conducting polymer composites*.

Although *conducting polymer composites* are tremendously important, the mechanism in which they conduct electricity depends on the identity and concentration of the added conductive particles and can be understood with classical band theory and percolation conduction [15]. In contrast, canonical *conductive polymers* are systems with greater complexity; therefore, the following sections intend to explain these polymers. An important thing to keep in mind is that even though these two technologies have clear differences, many authors use the term *conductive polymer* to describe both systems, and since the final goal of the systems is to have polymers that conduct electricity for a plethora of applications, it is always wise to know the differences before studying any specific implementation.

1.3   HISTORY

The history of conducting polymers is relatively young simply because synthetic polymers are also a recent invention. For instance, the first synthesis of polymers was explored at the end of the 19th century and the start of the 20th century with the discovery of Bakelite, and the fundamentals for understanding polymer chemistry were described until the 1920s by Staudinger. Polymer chemistry later exploded with the development of techniques for the polymerization of a great variety of monomers and the characterization of increasingly complex polymeric systems [5]. Relatively at the same time in which synthetic polymers were being discovered, in 1910, through solid-state chemistry techniques, a multiatom compound composed of shiny bluish crystals with a molecular formula of $(SN)_x$ was discovered and characterized by Burt [7]. At that time, Burt did not have the analytical tools to determine the exact properties of this material; furthermore, since polymer chemistry was still not completely understood, the material was not identified as a polymer until the 1950s. From the 1950s and onward, poly(sulfur nitride) or $(SN)_x$ was correctly identified as a polymeric system and its synthesis was perfected; through these developments finally in 1973 this material was demonstrated to exhibit semiconductor properties and superconductivity below 0.73 K. $(SN)_x$ was, from that moment, the first nonmetallic solid to exhibit metal-like properties; nevertheless, due to low conductivities, and its complicated synthesis, this system was not seen as a candidate for many applications (Figure 1.1a) [16].

The following years to the discovery of the conductivity of the inorganic polymer, $(SN)_x$, it was clear to many researchers that the electrical conductivity through organic polymers should profit from electron delocalization like in metals; for example, through the presence of conjugated double bonds, nevertheless, the ideal system was not trivial to found. Through these years and until 1977,

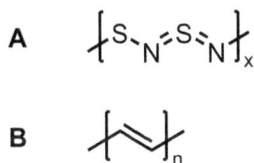

A $\left\{ S_{\diagdown N} \diagup S_{\diagdown N} \right\}_x$

B $\left\{ \diagup\!\!\!\diagdown\!\!\!\diagup \right\}_n$

FIGURE 1.1   Chemical structures of (a) poly(sulfur nitride) and (b) linear polyacetylene.

three researchers Alan G. MacDiarmid, Hideki Shirakawa, and Alan J. Heeger collaborated to find the first completely organic conducting polymer: Halogen doped polyacetylene (PA) [17].

PA (Figure 1.1b) is the product of the polymerization of its monomer, acetylene, the simplest alkyne. The typical radical polymerization of this monomer is not an effective route, because since PA contains conjugated double bonds, excessive crosslinking is unavoidable. It was in the 1950s, however, when Ziegler-Natta catalysis was introduced by Karl Ziegler and Giulio Natta (later Nobel laureates in 1963) as a means to produce ultrahigh molecular weight linear polymers of alkenes and alkynes [7, 18]. Linear PA, although needing specialized catalysis to be produced, by itself is a nonconductive black powder with little application that showed little promise as a useful material. However, in the 1970s, Hideki Shirakawa's group was experimenting with the Ziegler-Natta polymerization of acetylene at a high concentration of catalysts, which allowed them to obtain a silvery polymeric film. The properties of these products were later compared with those of $(SN)_x$ through the collaboration with Alan J. Heeger and Alan G. MacDiarmid. Eventually, these three collaborators discovered that PA became conductive when oxidized with halogens such as bromine and iodine, and thus, conductive organic polymers were born. For their work with these systems, and the further description of their characteristics, conductive mechanism, and further support for the development of many other conductive polymer substrates, Shirakawa, Heeger, and MacDiarmid received the Nobel Prize in chemistry in the year 2000. Following this prime discovery, the chemistry of conductive polymers expanded to what it is today, namely, a great array of applications and developments in many areas [7, 9, 19].

1.4 STRUCTURAL CHARACTERISTICS AND MECHANISM OF CONDUCTION

Upon discovery of conducting polymers, naturally, a model for the conduction of electric current on these types of systems was greatly desired. For instance, it was interesting to note that only polymers that were modified with other chemical substances, called dopants, were the only ones able to conduct electricity, while their unmodified counterparts were electric insulators; this behavior was somewhat similar with n- and p-type doping on classic semiconductors. Additionally, it was noted that double-bond conjugation (π-conjugation) was an essential characteristic of this system; nevertheless, since undoped π-conjugated polymers were insulators, this delocalization was not the only characteristic responsible for the electric conduction. Progressively, research on this topic allowed for the conclusion that the doping was not only essential for the presence of charge carriers, but that in most of the known systems proceeds either via a redox process. This redox process (Figure 1.2) could follow one of two scenarios [9]:

1. The base polymer is reduced during the redox reaction, which means that it received electrons from a reducing agent (negative charge doping or *n-doping*).
2. The base polymer is oxidized during the redox reaction, which means that it donated electrons to an oxidizing agent (positive charge doping or *p-doping*).

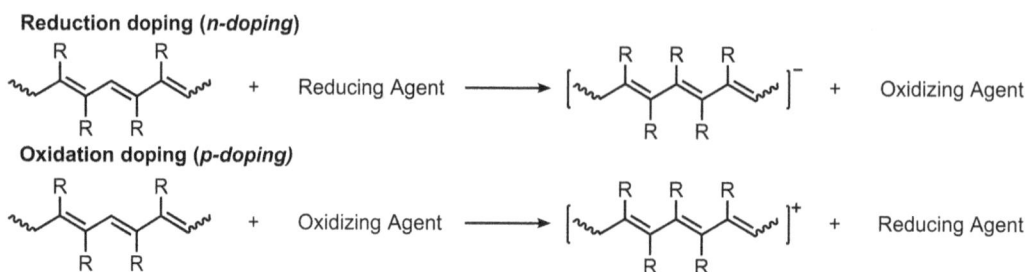

FIGURE 1.2 Redox doping of conjugated polymers.

Redox doping of conjugated polymers allows for the conduction of electricity within the polymeric bulk structure due to the introduction of electrical charges to the system. It is important to remark that all redox process occurs in pairs; therefore, the exact mechanism of the reduction or the oxidation reactions depends on the identity of both the polymer and the reduction agent. This dependence is due to specific factors such as the thermodynamic values for standard electrode potentials of both the polymer and the redox agent: E^0, the conditions of the reaction media: solvent, concentration of reagents, solubilities of the reagents, pH, presence of other chemical species, etc., and the structure of the polymer: degree of polymerization, branching, crosslinking, dispersity, etc. Independently of this, the general idea is that redox reactions introduce charges to the otherwise neutral polymer structure in the form of extra electrons, which occupy extra molecular orbitals, typically of higher energy states, within the polymer; or an electron deficiency, which allows for molecular orbitals of lower energy states to become unoccupied, within the polymer [19]. In some cases, it is also possible to dope polymers through other mechanisms to obtain conductive substrates; for instance, for conjugated polymers with acid groups, it is sometimes possible to protonate/deprotonate these groups to add/remove charges through the proton ion. For example, in polyaniline (PANI), the pH conditions in which the reaction is performed affect the final conductivity of the polymer because of different degrees of protonation [7].

One intuitive first approach to explain the conductivity of these substrates is that electrons have free mobility on the bulk material like in metals. However, this is not the case, and the explanation relies on the understanding of some basic concepts of molecular orbital theory and band theory in solids. According to band theory and molecular orbital theory, in any chemical compound, valence atomic orbitals from all the atoms in the compound (mathematical wave functions for the behavior of electrons) combine through linear superimposition to form an equal number of molecular orbitals (again, mathematical wave functions for the behavior of electrons). The different linear combinations of the original atomic orbitals naturally form molecular orbitals with different energy between each other, in small molecules, these orbitals are divided between *bonding* and *antibonding* orbitals (Figure 1.3). It is important to note that the generation of linear combinations to form the molecular

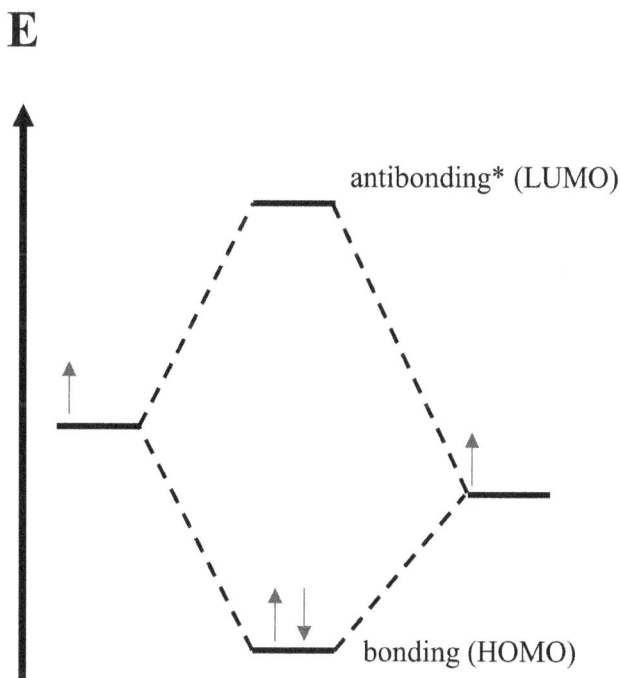

FIGURE 1.3 Representation of the formation of molecular orbitals on a small molecule.

Metal   Semiconductor   Insulator

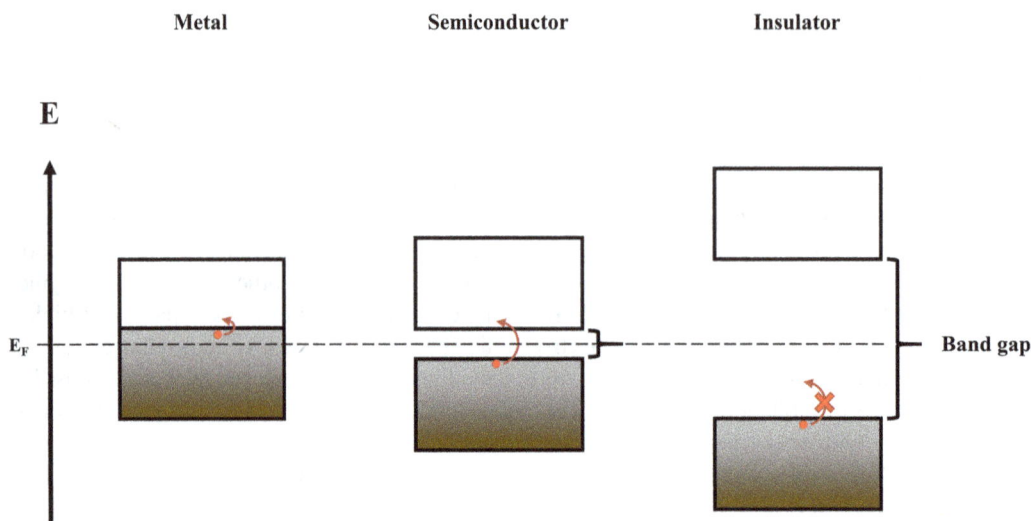

FIGURE 1.4 Representation of the band structure for metal, semiconductor, and insulator.

orbitals is governed by the symmetry of the system, and naturally that the combinations remain valid solutions to the constraints imposed by the Schrödinger equation for the system [20].

Materials that present much larger bonding interactions like metals and crystalline solids can be treated in the same way as small molecules; nevertheless, the complexity of these materials renders it impossible to model them through a plain linear combination of molecular orbitals. In contrast, for these crystalline systems, it is possible to profit from the periodicity of their structure, and the fact that electrons delocalize, to construct solutions to their Schrödinger equations considering that the aspect of each small group of atoms repeats infinitely through the crystal. These solutions are called Bloch functions, and these are vector equations that allow calculating the energy of different electronic states (orbitals) through the lattice, which in the case of crystalline materials form continuous bands of energy levels [20]. With this model, diagrams presenting the continuous energy levels like shown in Figure 1.4. In these diagrams, all the materials show two bands, the valence band that is filled with electrons and contains the highest occupied molecular orbital (HOMO) and the valence band that may contain the lowest unoccupied molecular orbital (LUMO). The difference between metal, a semiconductor, and an insulator is the energy distance between the conduction and valence bands, also called the bandgap. An insulator, for instance, has a large bandgap (>3 eV), so no electrons may jump to the conduction band. On the other hand, metals have no bandgap, and therefore, electrons may move freely between both valence and conduction bands, typical for all conductors. As an intermediate case are semiconductors that have a small bandgap, allowing for the flow of electrons under certain conditions [20].

According to molecular orbital theory and band theory, π-conjugated polymers are insulator systems because there exists a large bandgap between the valence band (bonding: π orbitals), and the conduction band (antibonding: π^* orbitals); however, in conducting polymers, extra conditions aide in the conduction of charged carriers (formed through redox processes as described before, for example). First, and less importantly, the energy gap between bonding and antibonding orbitals in π-conjugated polymers decreases in size when the polymerization degree increases; therefore, conduction is facilitated in polymers with high degrees of polymerization. Second, polymer doping not only allows for the presence of extra charges (more electrons in *n-doping* and more holes in *p-doping*), but this partial oxidation/reduction of the π-conjugated polymeric substrate may create intermediate energy levels within the bandgap with which the charge carriers may ultimately reach

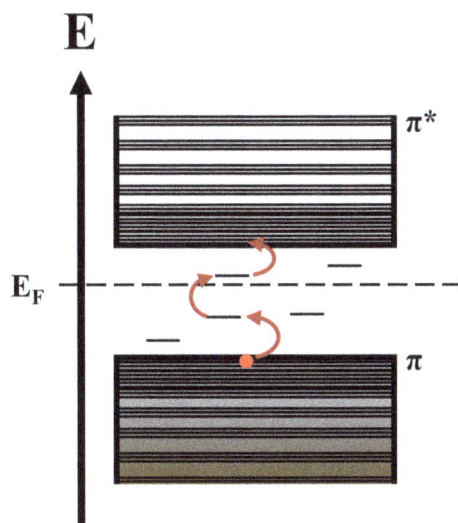

FIGURE 1.5 Representation of the band structure for a conducting polymer with intermediate states within the bandgap.

the conduction band and transmit electric current through the double-bond conjugation (Figure 1.5). Finally, the third condition that allows for the conduction is that charge carriers on π-conjugated polymers are not electrons, but other entities [7, 12, 21].

Despite the delocalization caused by the π-conjugation on these types of polymers, the electrons are still strongly bonded to the atoms of their bonds; this is different to metals in which electrons are considered to have more mobility [22]. This poses a problem in understanding the conduction mechanism because even when a surplus of charge is introduced to the system through doping, these electrons/holes cannot move easily and thus cannot carry electric current by themselves. Instead, conducting polymers rely on three types of charge carriers: solitons, polarons, and bipolarons, the behavior of which is governed by interactions between charged particles in the system (electrons/holes through doping), and the vibrations intrinsic to the crystal lattice (phonons) [23–25]. This type of conduction, although complicated to rationalize, implies that the electric charge transported on conducting polymers is dispersed through all the lattice of the material through interactions between surplus charges (electrons/holes from doping) and phonons. This can be effectively conceptualized and simplified as charge migration through hops through individual polymer chains and between polymer chains (Figure 1.6) [7, 26]. An important note is that when high enough doping is achieved in certain polymer systems, these polymers may start behaving like a metal in which a complete delocalization (Fermi Sea) of charges may be seen. However, in polymers, polarons and bipolarons from this sea of delocalization are the entities responsible for the electrical conduction [24].

Since this mechanism of conduction is preferred on these systems, any distortion to the polymer lattice may significantly alter the electrical conduction of any system. For instance, polymerization degree, crystallinity, dispersity, hybridization variations during polymerization, temperature, and degree of doping all influence the electrical conduction of any given system that can behave just like a metal, a semiconductor, or an insulator [7, 19, 24]. Although this may seem like a disadvantage, research has profited these characteristics to find both conducting and semiconducting polymers from a limited choice of monomers. To exemplify this, in the following sections, we will discuss examples of some of the most important conducting polymers and their potential applications.

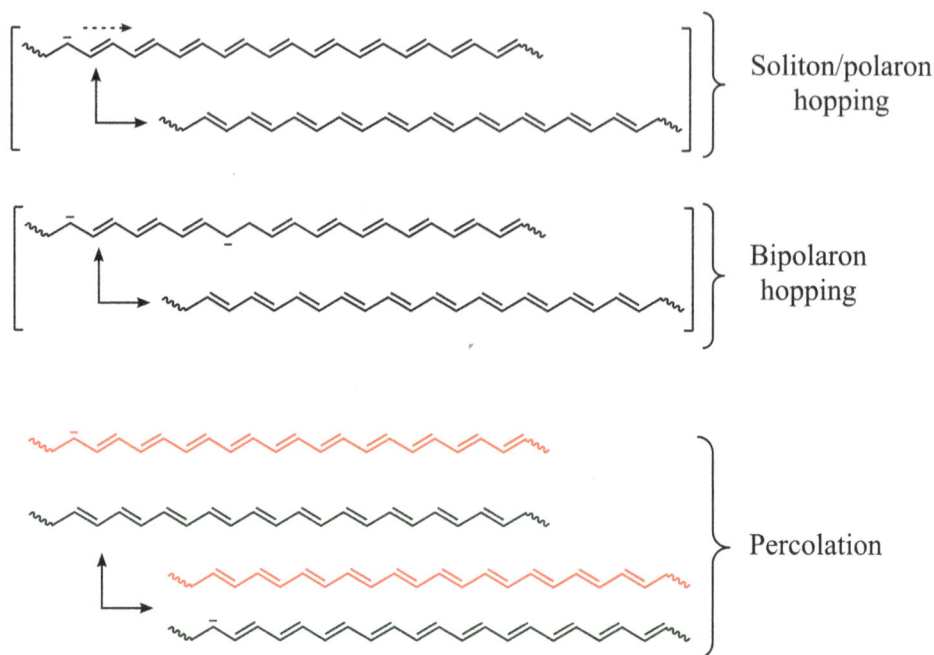

FIGURE 1.6 Hopping charge migration through conducting polymers. Solitons, polarons, and bipolarons are responsible for this charge migration [26].

1.5　EXAMPLES OF CONDUCTING POLYMERS

In previous sections, we reviewed the history, conducting mechanisms, and some properties of the conducting polymers, and for this section, we enlist one of the most used conducting polymers for biomedical applications, their chemical structure, and synthesis, as well as a list of their main uses in medicine field. Conducting polymers can be obtained by different procedures, which involve conventional chemistry or electrochemical reactions. Since all of these systems are studied extensively in many works, we refer the reader to the previous citations for more information about these polymers; additionally, we include a couple of references that include characteristics of these polymers [27–34].

1.5.1　POLYACETYLENE

PA is an organic conjugated polymer derived from the polymerization of acetylene, and it is the simplest polyolefin with conducting properties. It was the first conducting polymer discovered, the description of the conductivity in PA enriched the interest in this field, and since its discovery, it has been used for several applications, especially in electronics. PA is a linear flat molecule, and its structure presents alternate carbon-carbon double bonds with an angle of 120° between them. The alkene-type functionalization allows the existence of two spatial configurations: the *cis-* and *trans-*PA. It can be modified by extraction of hydrogens, which can be replaced by other functional groups or molecules, this grafting approach is highly used in the development of biosensors; nevertheless, the modification of the polymer backbone provokes deformations through the polymer chain, which can affect the intensity of the conductivity.

　　The synthesis of PA can be achieved by different techniques. The polymerization methods that involve gaseous acetylene include the use of (a) transition metals catalysts (e.g., Ziegler-Natta reaction) or (b) ionizing radiation, and with other monomers and precursors the main procedures, (c) the ring-opening metathesis polymerization (ROMP) with Grubbs catalysts, and (d) dehydrohalogenation

TABLE 1.1
Chemical Structure, Properties, and Applications of PA

Chemical Structure	Properties	Biomedical Uses
trans-Polyacetylene	• Lack of stability in air • Hard synthetic procedures • Conformational defects • Conductivity: undoped-PA 10^{-5} S/cm; doped-PA upper than 10^2 S/cm	• Biosensors • Bioelectrodes
cis-Polyacetylene		

reactions, e.g., from poly(vinyl chloride). The conductivity of PA can be increased by *p*-doping, with the addition of oxidizing agents to the polymer. The most common doping agents used to increase PA conductivity are bromine, dihexadecyl hydrogen phosphate, iodine, quaternized cellulose NPs, and metallic NPs. Table 1.1 shows the chemical structure, properties, and applications of PA.

1.5.2 POLYPHENYLENE DERIVATIVES

Poly(*p*-phenylene) (PPP) is an insoluble powder with high mechanical and chemical stability as well as electrical properties. It can be transformed from an electrical insulator into an electrical conductor by doping. It is widely used as a photocatalyst due to its energy transfer through its backbone and light-harvesting properties, which has been exploited in X/polyphenylene composites (X = Pd, ZnO, NPs). Poly-*para*-phenylene copolymer (also called self-reinforced polyphenylene) has high strength and modulus. The synthesis of pristine PPP (unsubstituted) by oxidative polymerizations remains a synthetic problem, although it has been obtained through 1,4-dimethoxycyclohexadienylene monomers polymerization, yielding the nonaromatic precursor polymer, and by additive-free thermal treatment, aromatization was induced. The product exclusively contains phenyl units having lengths of about 75. Poly(*p*-phenylenevinylene) (PPV) has extremely low conductivity that can be enhanced by doping with alkali metals or iodine. The synthesis can be achieved by Wittig couplings between the bis(ylide) and the dialdehyde (e.g., 1,4-benzenedialdehyde). Table 1.2 shows the chemical structure, properties, and applications of PPP and its derivative.

TABLE 1.2
Chemical Structure, Properties, and Applications of PPP and Its Derivative

Chemical Structure	Synthesis and Physicochemical Properties	Biomedical Uses
Poly(*p*-phenylene) (PPP)	• Glass transition temperature: ~150°C • Heat distortion temperature: 150–160°C • Flame resistance • Abrasion, solvent, and acid-base resistance • Thermal-oxidative stability • Highly fluorescent • Conductivity of PPP: 10^2–10^3 S/cm • Conductivity of PPV 3–5×10^3 S/cm	• 3D tissue engineering
Poly(*p*-phenylenevinylene) (PPV)		

1.5.3 POLYANILINE

PANI is one of the most studied conducting polymers, and its chemical structure arises from the polymerization of aniline molecules. PANI can be found in three oxidation states: (a) emeraldine (partially oxidized), (b) pernigraniline (oxidized), and (c) leucoemeraldine (reduced). The chemical structure of the oxidized form presents imine groups, forming an imino-quinoide structure, while emeraldine presents both states, it is the most stable and conductive of the three different states of PANI. Generally, PANI conductivity can be increased by the protonation of the amine groups; the neutral emeraldine (called emeraldine base) is commonly doped by the addition of acid to the polymer, to obtain the emeraldine salt, which is the conducting form. The common acidic dopants used in PANI are HBr, HCl, and peracids. The best conductivity is obtained when pH is around 1–3, and at these pH values, PANI behaves similar to metallic conductors. The mechanism of conductivity of PANI is by the formation of bipolaron in the doped form.

Chemical oxidation is the most preferred way to obtain PANI, and the synthetic route involves the mixture of (a) oxidizing initiator (e.g., ammonium persulfate and ceric salts), (b) aniline as a monomer, and (c) acid solution (e.g., hydrochloric acid). The reaction is highly exothermic and promotes the release of a high amount of heat. The solution changes its coloration indicating the formation of PANI, each oxidation state (or form) has a different color, and it depends directly on the pH: emeraldine salt is green, emeraldine base is blue, and pernigraniline base is violet. Due to the solubility of the reagents, the reaction usually is performed in two-phases mixtures or using the microemulsion technique, with the addition of surfactants, and further purification techniques are needed. In the synthetic process, the main product obtained is emeraldine, which is the most used. The synthesis of the other two PANI states can be reached by the treatment of emeraldine with strong acids or reducing agents. In addition, PANI can be obtained by electrochemical procedures, and this technique has allowed the electrodeposition of thin films onto different materials, including electrodes, to provide electrical conductivity. Table 1.3 shows the chemical structure, properties, and applications of PANI.

TABLE 1.3
Chemical Structure, Properties, and Applications of PANI

Chemical Structure	Properties	Biomedical Uses
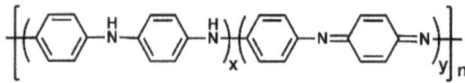 Emeraldine 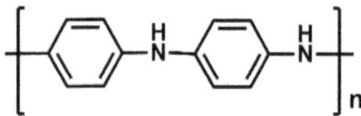 Leucoemeraldine 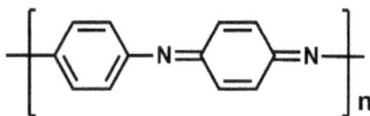 Pernigraniline	• High stability in air • Poor solubility in common organic solvents • Ease synthesis and processability • Lack of flexibility • Nonbiodegradable • Tunable optical properties • Conductivity can be controlled by protonation • Acid-doped emeraldine conductivity between 10^{-3} and 10^2 S/cm	• Tissue engineering • Biosensors • Actuators • Neural probes

1.5.4 POLYTHIOPHENE

Polythiophene (PT) derivatives are extensively used for biomedical applications, because they present high conductivity. These derivatives are stimuli-responsive polymers, they respond to changes in temperature or media, and change their color. The simplest derivative of this group is the PT, its structure consists of thiophene as a repeated unit (i.e., a five-membered ring containing sulfur) linked through carbons 2 and 5. Other derivatives, such as poly(alkyl thiophenes) present alkyl groups in carbons 3 and 4, alkyl groups or other functional groups (e.g., diether), for example, poly(3-methyl thiophene) (P3MT), poly(3-hexylthiophene) (P3HT), and poly(3,4-ethyl-enedioxythiophene) (PEDOT). The conductivity of PT derivatives depends on the oxidizing dopants (e.g., halogen anions: bromine and iodine), when they are treated with these agents, they convert into salt form, which is the conductive structure. Several small and large doping agents have been used to provide conductivity to PEDOT, one of the most employed is the sodium poly(4-styrene sulfonate) (PSS), (PEDOT:PSS); additionally, other strong acids are used as dopants, such as trifluoroacetic acid. The conducting mechanism for p-doping is followed by the formation of a bipolaron through the structure. PEDOT is one of the most explored derivatives, and it is synthesized by the oxidation polymerization of the monomer 3,4-ethylenedioxythiophene; it presents better stability and conductivity compared to polypyrrole (PPy). Different composites have been prepared by mixing PEDOT with other materials, such as polymers or carbon nanomaterials, for several applications.

PT derivatives can be synthesized by electrochemical reactions from the thiophene monomer, similar to the other conducting polymers, by modulating the electric current and redox potential, the polymer is deposited onto the anode surface and can be easily removed and purified. In the case of chemical synthesis, several techniques have been employed, as well as chemical reagents. PTs can be obtained from the polymerization of alkyl substituted-bromothiophenes by carbon-carbon coupling. Other polymerization techniques involve the use of the catalyst for atom transfer radical polymerization (ATRP), or strong oxidants, such as ferric (III) chloride. As well as the selected dopant, the polymerization technique influences on the conductivity too. Table 1.4 shows the chemical structure, properties, and applications of PT its derivatives.

TABLE 1.4

Chemical Structure, Properties, and Applications of PT and Its Derivatives

Chemical Structure	Properties	Biomedical Uses
 Polythiophene (PT) 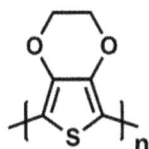 Poly(3,4-ethyl-enedioxythiophene) (PEDOT)	• Stability to oxidation in different media • Low redox potential • Easily processable • Electrical and thermal stability • PEDOT insoluble in water • Thin films: optical transparency • Conductivity from 10^{-5} to 10^3 S/cm	• PT: tissue engineering • PEDOT: neural probes and electrodes, biosensing, and drug delivery systems

TABLE 1.5
Chemical Structure, Properties, and Applications of PPy

Chemical Structure	Properties	Biomedical Uses
Polypyrrole (PPy)	• High stability in air and water • Amorphous • Insoluble in several solvents • *In vivo* and *in vitro* biocompatibility • Facile synthetic procedures • Poor flexible • Brittle • Conductivity in physiological environment • Conductivity around 10^{-3} to 10^3 S/cm	• Tissue engineering • Neural probes and implants • Drug delivery systems • Biosensors and actuators

1.5.5 POLYPYRROLE

PPy is one of the biocompatible conducting polymers and has been reported to allow cell adhesion and growth of different cells such as epithelial and neural cells, fibroblasts, among others. Additionally, PPy presents stimuli-responsive properties (to the electric field and pH), and this behavior allows the development of smart biomaterials. Moreover, due to the biocompatibility of PPy, it has been mixed with other polymers or compounds to form composites used for biomedical applications. Undoped PPy presents zero conductivity and insulating properties, but treatment with halogen anions such as chlorine, bromine, or iodine provides high conductive behavior. The conductivity in this polymer follows a *p*-type mechanism, governed by the formation of bipolaron; additionally, it can be enhanced by variations in the synthetic procedure (i.e., the solvent, oxidizing agent nature, redox potential, and reagents concentration), as well as it can be controlled varying the concentration and type of dopant. PPy is synthesized by the oxidation of pyrrole units, several oxidizing agents are used as catalysts, for example, iron (III) and copper (II) salts, and hydrogen peroxide. The synthetic procedure is flexible and can be achieved in different solvents including water. The electrosynthesis of PPy allows the obtention of highly conductive polymer, and the polymer is deposited onto the electrode surface. Table 1.5 shows the chemical structure, properties, and applications of PPy.

1.6 CONCLUSION

Advances in materials for the electrical field have been a crucial task for the last 200 years since electricity was first understood. In this regard, conducting polymers are interesting substrates because they integrate electrical conductivity to the unique characteristics of polymers, allowing for the creation of devices that were not conceivable even as close as 50 years ago; therefore, it is no surprise that the advances in this field have not only a Nobel Prize for their discovery in the late 1970s but also yielded a plethora of useful materials for many applications. In this introductory chapter, we outlined some of the most interesting characteristics of these systems, so our readers may more easily understand why they have become increasingly important in many areas. Currently, applications in the biomedical fields have the greatest attractiveness due to the big potential for substitution and enhancement of tissue with compatible and functional systems.

ACKNOWLEDGMENTS

The authors are thankful to Dirección General de Asuntos del Personal Académico, Universidad Nacional Autónoma de México under Grant IN202320. Thanks to CONACyT for the master's degree scholarship for Luis Alberto Camacho Cruz (916557), the doctoral scholarship provided for Marlene Alejandra Velazco Medel (696062/583700).

REFERENCES

1. N. Roztocki, P. Soja and H.R. Weistroffer, *The role of information and communication technologies in socioeconomic development: towards a multi-dimensional framework*, Inf. Technol. Dev. 25 (2019), pp. 171–183.
2. A. Corbin, *The Third Element: A Brief History of Electronics*, AuthorHouse, Bloomington, IN, 2006.
3. L. Hoddeson, G. Baym and M. Eckert, *The development of the quantum-mechanical electron theory of metals: 1928–1933*, Rev. Mod. Phys. 59 (1987), pp. 287–327.
4. R.I. Haasen and P. Jafee, eds., *Amorphous Metals and Semiconductors*, Elsevier, Coronado, CA, 1986.
5. D. Feldman, *Polymer history*, Des. Monomers Polym. 11 (2008), pp. 1–15.
6. M. Wei, Y. Gao, X. Li and M.J. Serpe, *Stimuli-responsive polymers and their applications*, Polym. Chem. 8 (2017), pp. 127–143.
7. G. Alan, *Introduction of Conducting Polymers*, in *Conducting Polymers with Micro or Nanometer Structure*, Springer Berlin Heidelberg, Berlin, Heidelberg, 2009, pp. 1–15.
8. K. Namsheer and C.S. Rout, *Conducting polymers: a comprehensive review on recent advances in synthesis, properties and applications*, RSC Adv. 11 (2021), pp. 5659–5697.
9. T. Nezakati, A. Seifalian, A. Tan and A.M. Seifalian, *Conductive polymers: opportunities and challenges in biomedical applications*, Chem. Rev. 118 (2018), pp. 6766–6843.
10. N.K. Guimard, N. Gomez and C.E. Schmidt, *Conducting polymers in biomedical engineering*, Prog. Polym. Sci. 32 (2007), pp. 876–921.
11. T.K. Das and S. Prusty, *Review on conducting polymers and their applications*, Polym. Plast. Technol. Eng. 51 (2012), pp. 1487–1500.
12. S.D. Kang and G.J. Snyder, *Charge-transport model for conducting polymers*, Nat. Mater. 16 (2017), pp. 252–257.
13. R. Siegmar and D. Carroll, *One-Dimensional Solid-State Physics*, in *One-Dimensional Metals*, Wiley-VCH Verlag GmbH & Co. KGaA, Weinheim, 2015, pp. 113–151.
14. V. Gold, ed., *The IUPAC Compendium of Chemical Terminology, International Union of Pure and Applied Chemistry (IUPAC)*, Research Triangle Park, North Carolina, 2019.
15. J. Yang, Y. Liu, S. Liu, L. Li, C. Zhang and T. Liu, *Conducting polymer composites: material synthesis and applications in electrochemical capacitive energy storage*, Mater. Chem. Front. 1 (2017), pp. 251–268.
16. M.M. Labes, P. Love and L.F. Nichols, *Polysulfur nitride – a metallic, superconducting polymer*, Chem. Rev. 79 (1979), pp. 1–15.
17. H. Shirakawa, E.J. Louis, A.G. MacDiarmid, C.K. Chiang and A.J. Heeger, *Synthesis of electrically conducting organic polymers: halogen derivatives of polyacetylene, (CH)x*, J. Chem. Soc. Chem. Commun. 16 (1977), pp. 578.
18. J.P. Claverie and F. Schaper, *Ziegler-Natta catalysis: 50 years after the Nobel Prize*, MRS Bull. 38 (2013), pp. 213–218.
19. N. Yi and M.R. Abidian, *Conducting Polymers and Their Biomedical Applications*, in *Biosynthetic Polymers for Medical Applications*, Elsevier Ltd., Amsterdam, 2016, pp. 243–276.
20. J.J. Quinn and K.-S. Yi, *Elements of Band Theory*, in *Solid State Physics: Principles and Modern Applications*, Springer Berlin Heidelberg, Berlin, Heidelberg, 2009, pp. 109–127.
21. K. Doblhofer and C. Zhong, *The mechanism of electrochemical charge – transfer reactions on conducting polymer films*, Synth. Met. 43 (1991), pp. 2865–2870.
22. J.J. Quinn and K.-S. Yi, *Free Electron Theory of Metals*, in *Solid State Physics: Principles and Modern Applications*, Springer Berlin Heidelberg, Berlin, Heidelberg, 2009, pp. 79–107.
23. R. Siegmar and D. Carroll, *Polarons, Solitons, Excitons, and Conducting Polymers*, in *Foundations of Solid State Physics*, Wiley-VCH Verlag GmbH & Co. KGaA, Weinheim, 2019, pp. 301–401.
24. Y. Lu, *Solitons and Polarons in Conducting Polymers*, World Scientific, 1988.
25. R. Siegmar and D. Carroll, *Conducting Polymers: Solitons and Polarons*, in *One-Dimensional Metals*, Wiley-VCH Verlag GmbH & Co. KGaA, Weinheim, 2004, pp. 85–112.
26. R. Siegmar and D. Carroll, *Conducting Polymers: Conductivity*, in *One-Dimensional Metals*, Wiley-VCH Verlag GmbH & Co. KGaA, Weinheim, 2004, pp. 113–151.
27. T. Pal, S. Banerjee, P.K. Manna and K.K. Kar, *Characteristics of Conducting Polymers*, in *Handbook of Nanocomposite Supercapacitor Materials I*, Springer, Cham, 2020, pp. 247–268.
28. K. Namsheer and C.S. Rout, *Conducting polymers: a comprehensive review on recent advances in synthesis, properties and applications*, RSC Adv. 11 (2021), pp. 5659–5697.
29. T.-H. Le, Y. Kim and H. Yoon, *Electrical and electrochemical properties of conducting polymers*, Polymers (Basel) 9 (2017), pp. 150.

30. L. Dai, *Conducting Polymers*, in *Intelligent Macromolecules for Smart Devices*, Springer-Verlag, London, 2004, pp. 41–80.
31. R. Prakash, *Electrochemistry of polyaniline: study of the pH effect and electrochromism*, J. Appl. Polym. Sci. 83 (2002), pp. 378–385.
32. R. Dirl and P. Weinberger, *Group Theory in Materials Science*, in *Applications, in Encyclopedia of Condensed Matter Physics*, F. Bassani, G.L. Liedl and P. Wyder, eds., Elsevier, Oxford, 2005, pp. 290–302.
33. S.A. Brazovskii, N.N. Kirova and S.I. Matveenko, *The Peierls effect in conducting polymers*, Zhurnal Eksp. i Teor. Fiz. 86 (1984), pp. 743–757.
34. R. Siegmar and D. Carroll, *Electron–Phonon Coupling, Peierls Transition*, in *One-Dimensional Metals*, Wiley-VCH Verlag GmbH & Co. KGaA, Weinheim 2004, pp. 77–83.

2 Conducting Polymers
Fundamentals to Biomedical Applications

Felipe de Souza,[1] *Anuj Kumar,*[2] *and Ram K. Gupta*[1,3]
[1]Kansas Polymer Research Center, Pittsburg State University, Pittsburg, Kansas, United States
[2]Department of Chemistry, GLA University, Mathura, India
[3]Department of Chemistry, Pittsburg State University, Pittsburg, Kansas, United States

CONTENTS

2.1 INTRODUCTION

The development and industrial application of polymers had a great impact on society as these materials are used virtually everywhere and because of that, the early 20th century has been named the "Age of Plastics" [1]. The widely known polymers such as polyethylene or polystyrene are insulators. However, it was demonstrated that some organic polymers can obtain properties like those of inorganic semiconductors even reaching values of conductivity close to metals. The amount of conduction in these materials depends on the level of conjugation (alternative π bonding) which allows the electrons to hop over the polymer chain. Conducting polymers (CPs) have been around since 1834 when polyaniline was first reported, followed by polypyrrole (PPy) (1915), polyphenylene (1949), polyacetylene (1955), polyphenylene vinylene (1969), and polythiophene (1980) [2]. Usually, CPs undergo redox processes that change their electronic and optical properties similar to inorganic materials while maintaining the properties of organic polymers such as low-cost, facile synthesis, and flexibility. The combination of such properties sparked the interest in research for the further development of conductive polymers. Through that, many new polymers have been synthesized. Further research on these polymers unraveled their applications for organic photovoltaic cells, field-effect transistors, supercapacitors, organic light-emitting diodes, electrochromic devices, sensors, biosensors, artificial organs, scaffolds, and others [3, 4]. In addition, CPs presented great suitability for biomedical applications like tissue engineering as they can harmonize with cells allowing cell adhesion, proliferation, and providing electric stimuli to control their activities [5]. Also, the soft interface of CPs allowed them to replace more rigid and costly materials used to make implants such as platinum, gold, titanium, etc. Because of that, several studies were performed on nerves, muscles, cardiac cells, bones, among others. However, CPs may often require some sort of chemical modifications since in some cases these materials can present relatively high brittleness, low processability in commonly used organic solvents, and poor interactions with cells. To address that, a common strategy relies on the development of composites that incorporate biocompatible as

DOI: 10.1201/9781003205418-2

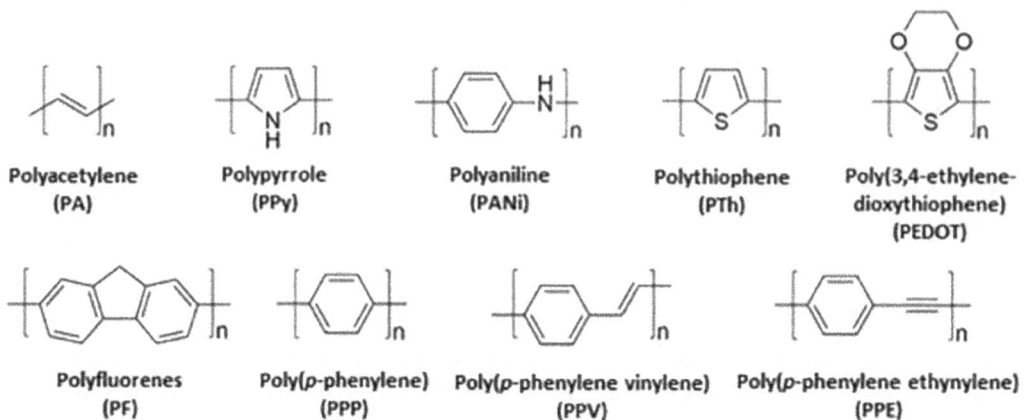

FIGURE 2.1 Mostly known CPs are widely used for biomedical applications. (Adapted with permission from Ref. [6], Copyright 2018, American Chemical Society.)

well as biodegradable polymeric segments which can be derived either from synthetic or natural sources. Monomers can be also functionalized with biomolecules or doped with counterions that can work as dopants [6].

Polymer architecture such as linear, hyperbranched, star-shaped, or cross-linked is another factor that plays a major role in properties such as thermal stability, mechanical strength, and degradability as more exposed sites may allow enzymes to decompose the polymeric material. The use of CPs in the niche of biomedical applications alone has many possibilities due to the number of materials available on the side of monomers for the synthesis of CPs as well as composites that can be incorporated, for example, carbon-based nanomaterials such as carbon nanotubes (CNT), graphene, fullerene, and nanodiamond. Natural components such as polymeric sugars i.e., chitosan- or protein-based biocomponents like collagen and fibroin, have been studied [7]. Accompanied to that there are several synthetic methods applicable for CPs such as chemical, electrochemical, metathesis, concentrated emulsion, inclusion method, solid-state, plasma polymerization, pyrolysis, chemical vapor deposition (CVD), and electrospinning. Based on that, one of the main strategies that have been adopted by scientists is the design of composites made of CPs and biocomponents to introduce conductivity, biocompatibility, and biodegradability, which are usually the main properties required for application in the biomedical field. The chemical structure of the most synthesized CPs is provided in Figure 2.1.

The potential applications for CPs within the medical and clinical areas can solve many complex health issues. Some examples of this bright scenario are the development of nerve conduits that can promote nerve regeneration that is a requirement in organ transplant processes and post-trauma cases. Also, the fabrication of synthetic scaffolds to promote the growth of myocardial cells can be used as components for pacemakers. The combination of biocompatibility and electric conduction opens the possibility for complex brain surgery with the potential to cure Parkinson's disease for instance. Metallic implants usually lack biocompatibility and proper arrangement around the tissues. On top of that, applications for drug and gene delivery systems, synthetic skin, bone fixation, and biosensors for diagnostics are within reach. Hence, CPs are pointed as a viable solution to tackle the current challenges in medicine and bring new materials that can greatly improve the life quality of patients.

2.2 SYNTHESIS OF CONDUCTING POLYMERS

One of the most attractive points in the research of CPs is the facile and diverse synthetic approaches that yield materials with appreciable properties. Chemical and electrochemical methods are the widely used approaches for the synthesis of CPs. Chemical polymerization generally consists of using an initiator and/or an oxidizing agent to form the polymer. It's an advantageous method due to bulky production and can be performed on several monomers through different procedures. However, it leads to thicker films as well as inorganic salts, which require an extra purification step to remove them. A well-known process for the synthesis of CP is the Ziegler-Natta synthesis of acetylene into polyacetylene, performed under the presence of $Ti(OBu)_4$ and $AlEt_3$ in toluene used as catalysts. Pressure is used to drive the polymerization of acetylene gas to form thick films that quickly appear at the reaction flask within an hour of reaction.

Chemical polymerization has been vastly explored for several monomers', aniline being a common example, which is an extremely versatile monomer that can be polymerized through many ways aside from chemically such as electrochemically, CVD, X-ray radiation, and so on. The same principle applies to aniline derivatives. To demonstrate that, an experiment was performed using the monomers: aniline, nitroaniline, chloroaniline, and methylaniline to understand the effect on properties when an electron-withdrawing group such as $-NO_2$ or $-Cl$ or an electron-donating group such as $-CH_3$ is attached to their respective polymers [8]. The same procedure was used to polymerize aniline and its substituted derivatives. The synthesis consisted of dissolution of the monomer in acid media, the addition of initiator, in this case, ammonium persulfate, and performing the reaction at low temperature under N_2 atmosphere. After that, purification and neutralization processes were made to obtain the polymer. It was noted that the pending group with electron-withdrawing effect presented longer reaction time, lower conductivity, and higher solubility compared with the others. This change in property is related to the decrease in charge density in the amino group that decreases reactivity. Also, polyaniline with no substituents presented higher electronic conductivity because the structure tends to be aligned in one place allowing high delocalization of π electrons. On the other hand, the substituent groups may disrupt the structure's alignment causing the conductivity to decrease.

Many polymers have been synthesized chemically through facile procedures that consists of dissolving the monomer in a polar solvent such as water, methanol, ethanol, or a mixture of them while using several types of initiators like iron salts: $Fe(NO_3)_3$, $Fe(ClO_4)_3$, $FeBr_3$, $Fe(BF_4)_3$, and $FeCl_3$. Variations of these procedures yielded satisfactory results for the chemical synthesis of PPy, polythiophene, and polyaniline [9]. The use of chemical initiators led to rough and amorphous morphologies due to the inherent lack of control that chemical polymerization provides. This condition can be controlled by diluting the initiator and decreasing the temperature. Through that, the reaction rate tends to decrease and therefore suppress side reactions that would lead to a loss in conjugation. A more organized and therefore crystalline structure tends to be formed. It was observed that $FeCl_2/FeCl_3$ ratio of $1/3.5$ for the polymerization of pyrrole over the surface of a polyurethane positively influenced the conductivity of the polymer [10]. Also, vapor phase polymerization was used to obtain a thin film over the substrate and control the morphology.

Other types of synthetical procedures employ a sacrificial template that can be soft or hard. These approaches aim at controlling the CPs morphology and dimension, allowing the growth of organized and thin films over a substrate. In that sense, soft templates such as liquid crystals, surfactants, or sugars like cyclodextrin can be used. These structures function as *in situ* nanoreactors by acting as surfactants that form a micellar dispersion with the monomer, which allows the polymerization to take place within or around the micelle. Through that, nanosized materials, from 10 to 100 nm, can be obtained which preferably take a cylindrical or spherical shape. CPs such as poly(3,4-ethylene dioxythiophene) (PEDOT), and PPy have been synthesized through this approach to obtain nanotubes or nanospheres [11, 12]. However, this technique requires careful reaction control due to the thermodynamic and kinetic instability. The hard template method is also widely used to synthesize nanofibers, nanotubes, or nanorods by using matrixes such as polycarbonates or anodic aluminum oxide. It is a convenient process as the hard template's pore size can lead to nanomaterials with a narrower size distribution. Template-free methods accompanied to CVD are also an option that can lead to unique nanostructures according to the versatility of the procedure [13, 14].

Electrochemical polymerization is the other main synthetic approach that is widely employed for the synthesis of CPs. It occurs by performing anodic oxidation or cathodic reduction of a monomer. The facile synthetic procedure of this method makes it attractive as it can quickly yield high-performance polymeric films in one reaction cell. Unlike chemical polymerization, the electrochemical process allows more control of the reaction system as the electric potential can be varied through a potentiostat as the polymerization occurs. Hence, the optimal condition for a certain monomer to polymerize can be quickly identified. Several monomers can also be used in this method such as aniline, pyrrole, thiophene, and their derivatives. Such a high control of the reaction system allows the formation of smooth and organized structures, which greatly improve the conductivity of the polymers. On top of that, flexible films can be obtained. However, this method has the drawback of low yield compared to bulk chemical polymerization. Finding the optimal electrical potential along with the temperature is one of the main driving forces to obtain polymers with high conductivity. For example, 3-methyl thiophene was electropolymerized at 1180 mV vs. saturated calomel electrode (SCE) at −20°C for 10 min [15]. The polymeric film's conductivity was 170 S/cm. The higher electrical potential required in comparison to thiophene to perform the reaction (500 mV) is related to the steric hindrance of the substituent groups.

Photochemical synthesis is another route that is facile, fast, and low-cost and can produce relatively larger quantities of CPs. PPy has been synthesized through this method by exposing the monomer to a 172-nm ultraviolet (UV) lamp [16]. It was reported that the longer exposure time provides a thinner polymeric film which could be due to surface etching because of high-frequency UV radiation. As a consequence of this, the polymeric film was observed to be more porous compared with the same materials synthesized through the chemical method. Another tested procedure is plasma polymerization, which has been performed to synthesize polyaniline [17]. The polymer tended to be cross-linked due to the excessive impact of electrons in the material's surface even leading to breakage of aromatic rings that also disrupt the conjugated system. Because of that, the reported conductivity was between in the range of 10^{-10} to 10^{-12} S/cm. A similar observation was found for PPy and polythiophene [18, 19].

In the realm of biomedical applications, there are some synthetic procedures such as freeze-drying, gas foaming, phase separation, solvent casting, self-assembly, particulate leaching, electrospinning, and deposition that have been preferred [20]. Within those, the latter two have been explored extensively. The electrospinning method is highly convenient due to its facile execution and relatively versatility in the processing of fibrous polymeric nanomaterials. It requires three basic components: a syringe with a blunt needle to pump the polymer, a source of high voltage (from 5 to 30 kV), and a collector. Through that, nanofibrous structures with high surface area concerning their volume and interconnected pores can be obtained. This type of material serves as an extracellular matrix (ECM) serving as a scaffold that can prompt cell growth, expansion of covered area by the cells, and differentiation. The electrospinning is carried out by a melted or diluted monomer inside the syringe. Then it is pumped out of the syringe at a set rate while it is exposed to a high electric voltage that is applied in between the needle and collector. The main parameters that need to take into consideration while performing electrospinning are the solution's viscosity, flow rate, solution charge density, and applied voltage. These parameters grant versatility to this process, allowing the same material to have considerably different properties. However, there are some challenges as many polymers do not dissolve in common solvents. Also, due to the high charge density and relative rigidity of the polymeric chains of CPs, the fibrous structure may be hard to obtain [21]. To counter that, other biocompatible and biodegradable polymers such as poly(L-lactide) (PLLA), poly(Ɛ-caprolactone) (PCL), poly(lactic-co-glycolic) acid (PLGA), silk or collagen gel can be blended with the CP monomers to make it more processable and biocompatible [22, 23]. Some examples of biodegradable polymers that can be used are given in Figure 2.2.

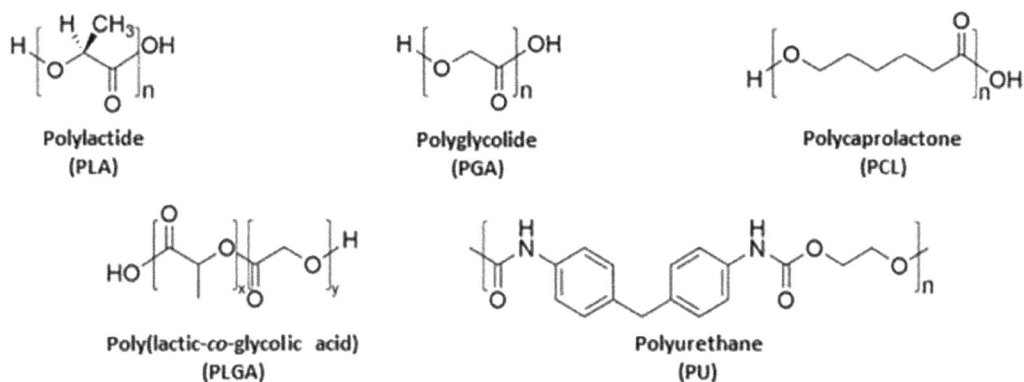

Polylactide
(PLA)

Polyglycolide
(PGA)

Polycaprolactone
(PCL)

Poly(lactic-*co*-glycolic acid)
(PLGA)

Polyurethane
(PU)

FIGURE 2.2   Biodegradable polymers suitable for biomedical applications. (Adapted with permission from Ref. [6], Copyright 2018, American Chemical Society.)

2.3 BIOMEDICAL APPLICATIONS OF CONDUCTING POLYMERS

The tunability of electron conduction, facile approaches, and different morphologies that can be achieved by varying the procedures for the synthesis of CPs grant them a valuable spot in the biomedical field. This concept has been demonstrated by Borriello and his team who synthesized polyaniline nanoneedles doped with camphor sulfonic acid over PCL substrate to be used as a scaffold for regeneration of heart tissue [24]. PCL was used to encapsulate polyaniline and to provide room for the cells to grow around the polyaniline due to its slow degradation. This was an important step to make CPs more biocompatible because polyaniline by itself can trigger inflammatory responses which could worsen the medical scenario. On the other hand, polyaniline's conductivity was the main characteristic that prompts the regeneration of cardiac muscle by providing electric stimuli to the neighbor cells. The chemical polymerization reaction of polyaniline was performed under high stirring and for a relatively short time of 2 h to purposely obtain short nanoneedles as these can properly blend with the PCL and still provide electrical conduction for cell stimuli. Also, the cells' survival rate was higher when polyaniline was present, suggesting that the electric stimuli positively influences cell migration and survivability. An often strategy to make CPs more biocompatible consists of blending them with biocompatible materials that are also biodegradable. This can prevent an inflammatory reaction and slow decomposition of the biodegradable material aids in the settlement of the CP into the cellular system and provide continuous electric stimuli for cellular growth.

PPy is another CP that has been studied for biomedical applications. PPy is convenient due to the solubility of monomer in water, low oxidation potential for electropolymerization, and high conductivity. When oxidized, PPy presents delocalized positive charges throughout the polymeric chain and therefore it can be neutralized by doping with negatively charged counter ions. The interest in PPy within biomedical applications has been noted through an *in vitro* study that showed an increase in neurite extension of neurons when CP was exposed to an external electric field suggesting potential application for brain degeneration diseases [25]. Even though PPy is not biodegradable, there have been approaches to make it biocompatible. This process has been performed by synthesizing a copolymer composed of PPy and polythiophene along with flexible and linear chains of ester linkages to increase the materials' biodegradability (Figure 2.3). This type of structure allows this material to be used for spinal cord regeneration, stimulation of muscle tissue, and bone repair applications.

Developing a proper interface between neural tissue and conductive materials is a challenge as it requires materials to be highly conducting to stimulate the tissue growth and to gather brain signals with low noise/signal ratio as the intensity of these signals is in order of 10^{-6} to 10^{-3} V. In addition,

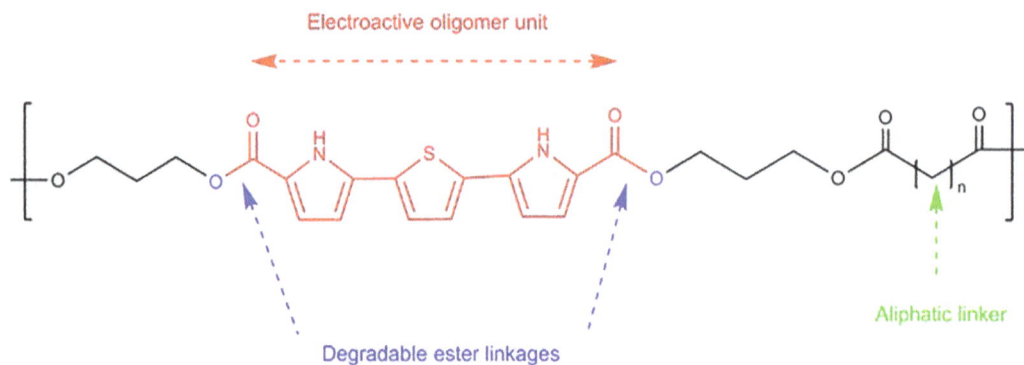

FIGURE 2.3 A biodegradable conducting copolymer of pyrrole-thiophene-pyrrole attached with aliphatic ester segments. (Adapted with permission from Ref. [26], Copyright 2013, The Authors, some rights reserved. Distributed under a Creative Commons Attribution License 4.0 (CC BY) https://creativecommons.org/licenses/by/4.0/.)

they should not cause an inflammatory reaction which can be prevented by designing a flexible composite with a soft chemical structure. A neural implant should have a small size to decrease the dissimilarity with the bio tissue, which may cause an increase in the electrode's resistance. This effect is related to the frequency of the induced potential. The higher the frequency smaller is the impedance due to the limited space that diffused species can transit from the electrode's surface. On the other hand, at low frequency, the diffused species can easily diffuse, distancing for the electrode and therefore increasing the impedance. Hence, a larger electrode would address this issue. However, it can damage the tissue. Because of that, a smaller electrode which highly sensible to an electrical signal is preferred [27].

Despite these requirements, the development of a neural probe with appreciable properties allows an in-depth study of the human nervous system as well as to help people with neural diseases such as Parkinson's disease, epilepsy, paralysis, among others [28, 29]. Studies on the recovery of tetraplegic patients have shown positive results as scientists were able to connect neural probes with an array of neurons that allowed patients to control prosthetic robotic limbs [30]. Hence, the development of neural probes has the potential to improve medical conditions that were considered irreversible. With these objectives, a research group fabricated a flexible neural probe composed of three parts: (a) an electrode array made of ZnO nanowires, with a thin Au shell wrapped around it, and PEDOT that was electrochemically deposited over them; (b) interconnected line for the signal transfer composed of a graphene and Au structure; and (c) the connector. Polyimide was used as a substrate due to its flexibility and soft structure. The schematic representation of the neural probe is shown in Figure 2.4 [27]. ZnO nanowires are low-cost materials that are synthesized through facile procedures and have a high surface area with the drawback of low conductivity. To address that, a thin layer of Au was deposited over it. After that, PEDOT was incorporated into the electrode's surface to further improve the surface area. Also, PEDOT leads to a decrease in impedance because it provides conduction through ionic or electronic routes. The composite electrode's (Au-ZnO-Au-PEDOT) testing was performed *in vivo* in a rat's brain and demonstrated an enhancement in signal reception in comparison to the bare Au electrode for the type of stimulus. Hence, a more sensitive electrode to track brain activity.

Promoting a fast wound-healing process is an effective way to address recurrent injuries. To address that, an injectable hydrogel copolymer was synthesized [31]. First, chitosan was quaternized and grafted with polyaniline. Second, sebacic acid, polyethylene glycol, and glycerol were polymerized through a polycondensation reaction, and it was further functionalized with 4-formylbenzoic acid (FA). These two compounds were chemically bonded leading to a hydrogel network. Increasing

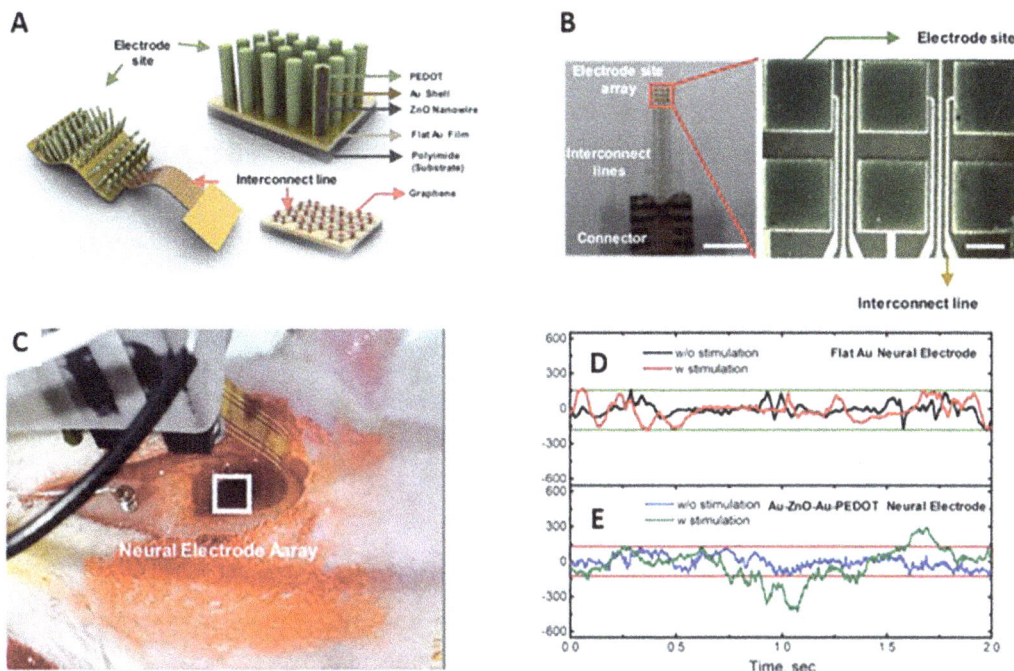

FIGURE 2.4 (a) Schematics for the composite electrode functionalized with ZnO nanowires wrapped with Au and PEDOT while using graphene as a conductive and flexible for the interconnected lines. (b) Electrode's array images under the microscope: 10 mm and 450 μm, left and right, respectively. (c) In vivo testing for the composite electrode in a rat's brain. Neural signal for 2s with and without whisker stimulus (d) bare Au and (e) Au-ZnO-Au-PEDOT electrodes. (Adapted with permission from Ref. [27], Copyright 2017, American Chemical Society.)

amounts of FA were used to increase the cross-linking density. The copolymeric hydrogel presented several properties such as self-healing, anti-oxidation, anti-infection, conductivity, free radical scavenging, skin adhesiveness, and biocompatibility. The wound-healing properties of the composite were measured by applying the hydrogel into the wound and tracking its diminishment for 15 days while comparing with a commercial film dressing (Tegaderm Film™), the hydrogel without polyaniline (QCS3/PEGS-1.5FA), and the hydrogel with polyaniline (QCSP3/PEGS-FA1.5). The addition of higher amounts of cross-linker (FA) promoted a predictable increase in adhesion strength due to a higher number of dynamic cross-linking bonds and more interaction with the tissue as shown in Figure 2.5a. Adhesion is important as it correlates with the improvement of hemostatic properties, hence stopping hemorrhaging sites as for the control around 2025.9 mg of blood were spilled over the filter paper while for the copolymeric hydrogel only around 214.7 mg were spilled as demonstrated in Figure 2.5b. A wound-healing test performed *in vivo* showed the effect that the polymeric hydrogel had on the injured tissue for over 15 days. It was notable that the hydrogel that contained polyaniline surpassed both commercial and hydrogel without polyaniline in terms of tissue regeneration. This effect was observed due to the simultaneous antibacterial activity of quaternary amine groups at chitosan and positively charged sessions over polyaniline that can neutralize the cellular wall of bacteria. On top of that, the cytocompatibility was also improved which promoted faster healing. Yet, the incorporation of polyaniline into the hydrogel further improved the healing properties likely due to antioxidant and electroactivity facilitating the growth and tissue migration. The sum of all these properties provided by polyaniline resulted in a thicker granulation of tissue compared with the other tissue dressings (Figure 2.5c), which is the accumulation of extracellular components that enhance the wound-healing process.

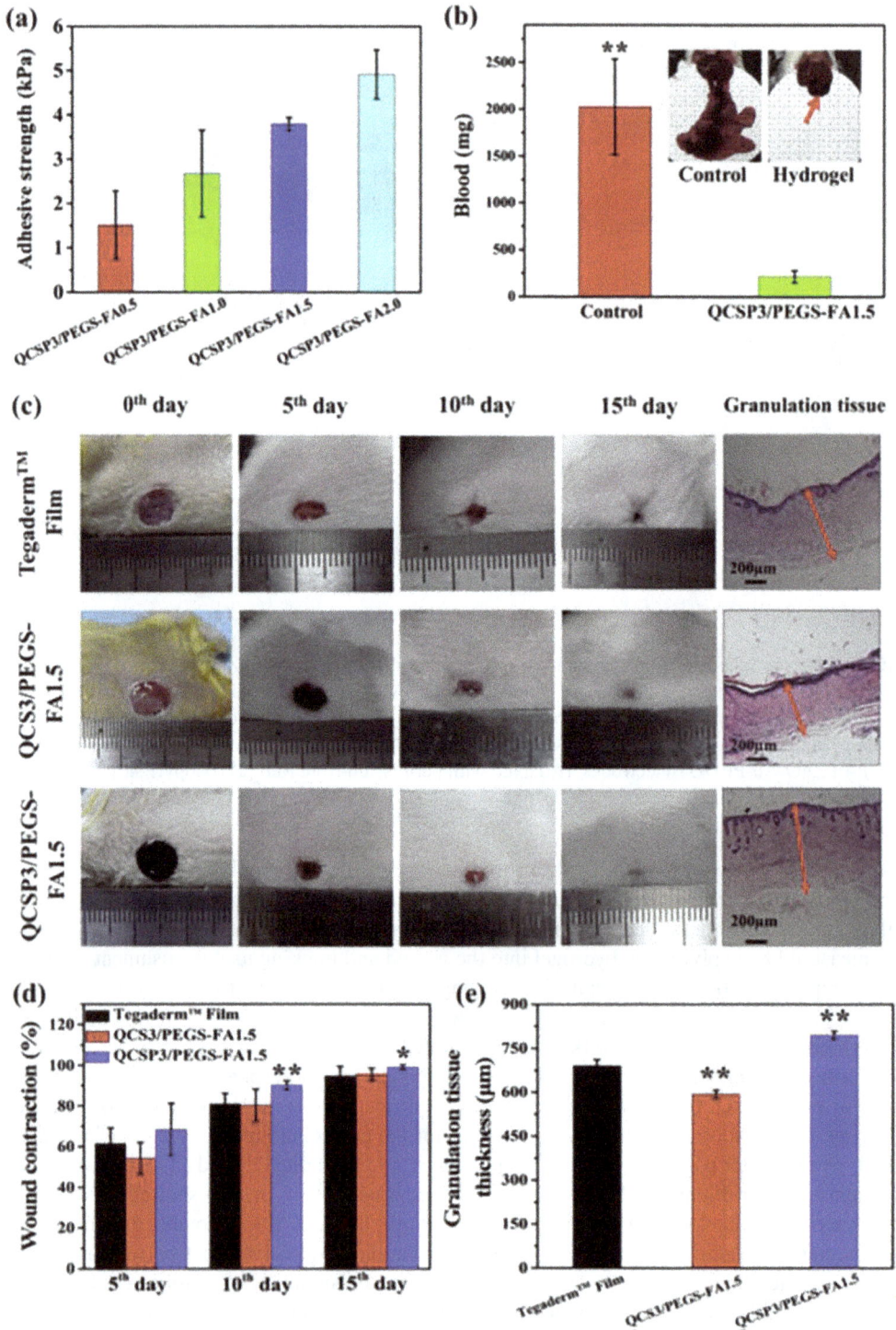

FIGURE 2.5 (a) Hydrogel's adhesion strength. (b) Performance in blood clot formation. (c) Photocopies for the wound-healing process and granulation tissue up to 15 days while comparing commercial counterpart (Tegaderm™), hydrogel without polyaniline, and hydrogel with polyaniline. (d) Histogram for would regeneration from 5 to 15 days for the film dressings. (e) Granulation tissue's thickness for the hydrogels. * $P < 0.05$, ** $P < 0.01$. Scale bar: 200 μm. (Adapted with permission from Ref. [31], Copyright 2017, Elsevier.)

An important field of application for CPs is their use as biosensors that are capable of detecting specific biocomponents. Biosensors require a component for detection and a transducer. CPs can conduct electrons to identify a signal related to the hybridization of a biomolecule or the attachment of a bioactive component into an active site of an enzyme for example antibodies, DNA fragments, enzymes, and cell receptors that can interact with the analyte. Based on that, composites used as substrates for biosensors have been developed. For example, Gao and his team used CNT functionalized with PPy to identify glucose [32]. The biosensor functioned based on the high surface area provided by the CNT and the conductibility and good interaction that PPy has with glucose oxidase. Pyrrole was electrochemically deposited over the surface of CNT in a solution that contained glucose oxidase. Through that, the enzyme was physically entrapped within the porous structure of CNT-PPy composite, and it was able to transduce the electric signal whenever hydrogen peroxide was released as a by-product. This example shows a viable strategy that can be applied on several CPs, making them suitable for biosensor applications. One of the reasons for that is the thickness control that can be made during an electrochemical polymerization to maintain good conductivity or deposit the CP over a substrate during CVD. These processes can promote entrapment or immobilization of biomolecules that can target specific bio analytes. Another example from the literature in that regard is a ternary composite of poly(ionic liquids), PPy, and GO nanolayered structure which was used for the identification of dopamine and ascorbic acid [33]. The proper interaction between GO and PPy due to π bonds alignment and biocompatibility of the poly(ionic liquids) provided high sensitivity along with a wide linear range were the main factors for the identification of the analytes with a low detection limit. The schematic for the fabrication of the ternary composite is presented in Figure 2.6. The incorporation of poly(ionic liquids) also improved the stability of the composite in water as it remained dispersed after one day meanwhile GO and GO/PPy precipitated after a few hours. On top of that,

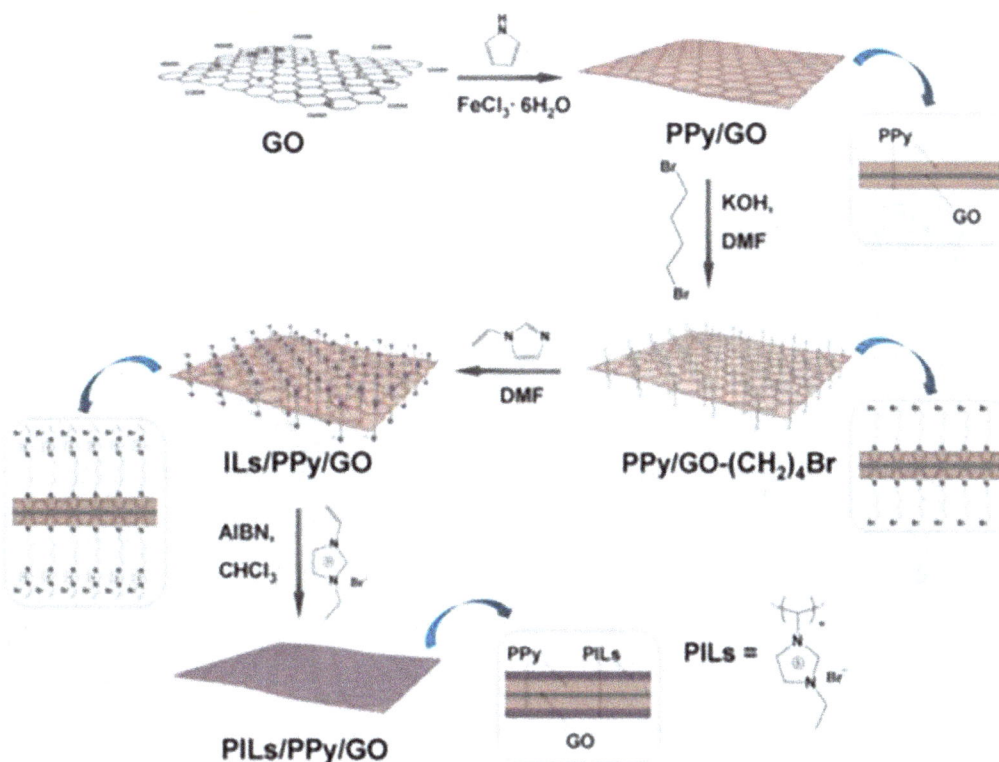

FIGURE 2.6   Scheme for the synthesis of a ternary composite of poly(ionic liquids) (PILs)/PPy/GO. (Adapted with permission from Ref. [33], Copyright 2015, Elsevier.)

poly(ionic liquids) increased the charged surface, which enhanced the electron transfer step during the oxidation of dopamine and ascorbic acid, providing a stronger signal for their identification.

Drug delivery system is a largely studied field within biomedical applications as it carries the important premise of decreasing the side effects of medications by slowly releasing them into the body as well as using strategies to target the delivery of the drug to a specific region to improve is efficacy. Through that, medical treatment that requires the use of strong drugs can be considerably improved. CPs can be used to trigger the release of a drug or bioactive material through electric stimuli. This effect was studied by Wadhwa and his team through the release of a drug from an electrode made of plastic with a thin layer of Au (40-nm thick) deposited over it [34]. After that, the electrode was coated with PPy. Dexamethasone was the drug as well as a dopant for PPy. A controlled release of the drug of around 0.5 μg/cm^2 was observed during the cyclic voltammetry study. It was further noted that the amount of drug released depends linearly on the number of cycles. The small amount of drug released by the system had positive effects on neuronal growth *in vitro* testing.

One of the challenges that scientists need to overcome to successfully perform treatments is to decrease the inflammatory reaction due to the insertion of a foreign agent into the patient's body. Hence, stable connections between neural cells and electrodes are needed. The controlled release of anti-inflammatory drugs is one strategy to prevent swelling and provide a stable insertion of devices into the tissue to further promote cell growth. Boehler and his colleagues fabricated a flexible microelectrode capable of controlled release of Dexamethasone, an anti-inflammatory drug [35]. The microelectrode was implanted into a rat's hippocampus to diminish neural swelling or rejection from the brain tissue through a less invasive process. The neural probes had the drug stored over its surface through electrical deposition of DEPOT performed in a solution containing the monomer with the drug. The drug could be released weekly through cyclic voltammetry in a three-electrode system in awakened animals. It demonstrated a low degree of inflammation and therefore provided a stable junction of flexible neural electrodes with polymeric conductive coatings that allowed the controlled and gradual release of drugs. Figure 2.7 demonstrates the electrode system and its implant into the hippocampus.

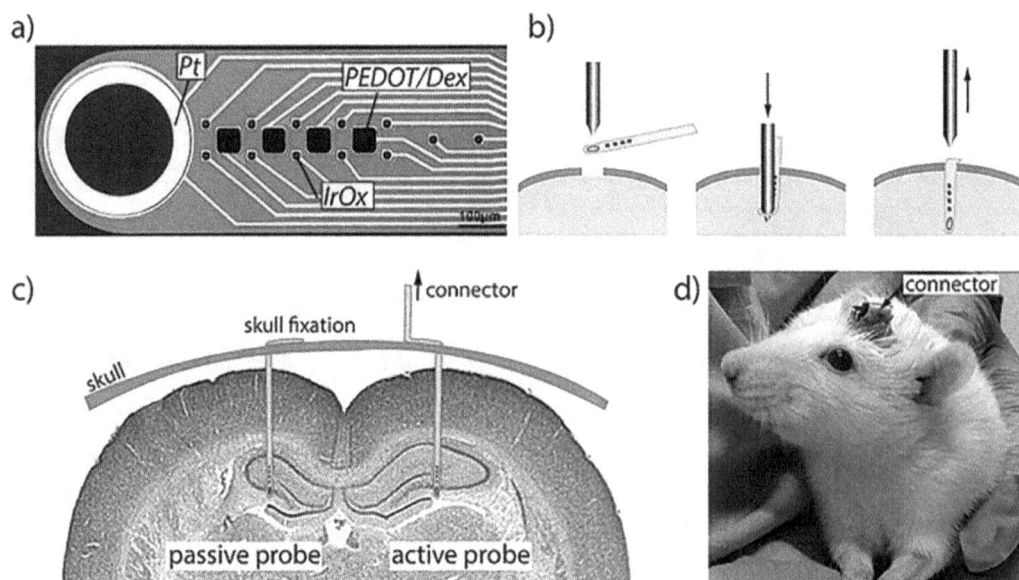

FIGURE 2.7 (a) Microscopic image for the three-electrode system for the neural probe containing four sites of PEDOT/Dex coated as the working electrode. (b) Probe insertion was performed with an optical fiber as a tool for place referring. (c) Passive (control) and active (functionalized) probes placed at the skull's surface. (d) A connector placed on top of the microelectrode was used to collect information and stimulate the tissue. (Adapted with permission from Ref. [35], Copyright 2017, Elsevier.)

2.4 CONCLUSION

CPs were introduced to the scientific community relatively early as the very first report on this type of material was published in 1834 describing the synthesis of polyaniline. After that, several CPs such as PPy, polyphenylene, polyacetylene, polyphenylene vinylene, and polythiophene were synthesized. Further research was performed leading to several other CPs that were mostly derivatives of these polymers. After that, their conducting properties found applications in organic photovoltaic cells, organic light-emitting diodes, field-effect transistors, electrochromic devices, sensors, supercapacitors, among others. Furthermore, the biomedical field received great benefits after the introduction of these materials as they were found to synergize well with cells due to their biocompatibility, promote adhesion, cell growth, and facilitate a stimulus through electrical current whereas showing fewer inflammatory reactions compared with metal implants. Such conveniences sparked the research on biological tissues such as cardiac cells, muscles, bones, nerves, etc. As studies progressed the incorporation of CPs for applications into the biomedical field demonstrated an effective and feasible path to improve the medical treatment in several ways such as promoting regeneration of myocadiac and/or neural tissue, as well as the development of synthetic skins which can drastically improve one's quality of life as the use of these conducting materials have potential to improve the quality treatment for several types of heart and brain degenerative diseases. These applications became possible primarily due to their tunable properties such as conductivity, flexibility, and smooth surface, which makes them adaptable to the desired conditions. On top of that, there are several ways to synthesize CPs such as chemical, electrochemical, vapor chemical deposition, electrospinning, along with other methods. The vast number of approaches loaded the scientific community with more possibilities and control over the reaction system. Because of that, the CPs could be synthesized in the form of nanoneedles which are desirable for better accommodation with the tissue and decreases inflammation. Also, when obtained in the form of thin films they can be used as devices to track biological activity. On the other hand, a morphology with a high surface area can be employed as a scaffold to promote the growth and migration of cells. Despite these advantages, CPs are not biodegradable, which can lead to infections over long-term exposure. To address that the scientific community is employing different strategies such as developing composites that combine conducting and biodegradable polymers or chemically attaching biodegradable sessions into a CP. Also, the fabrication of nanocomposites by incorporating CNTs, metal oxide nanoparticles, along other materials demonstrated to be a promising way to develop materials with a specific area of application within the biomedical field. These studies are leading to appreciable uses for these conducting polymeric composites such as (a) synthetic skin that work as an important component for prostheses, (b) biosensors for a more accurate and fast diagnosis of hormonal imbalances or identifiable diseases, (c) drug delivery systems that can greatly optimize the efficacy of strong drug treatments by decreasing the number of side effects due to the targeted action of the medicine into the desired area, (d) neural electrodes for the tracking of electric behavior of neurons during surgery or controlled release of anti-inflammatory drugs. Based on that, several possibilities and applications CPs are leading to, yet there is plenty of room for optimization and development of novel composites.

REFERENCES

1. Rasmussen SC (2011) Electrically conducting plastics: revising the history of conjugated organic polymers. In: ACS Symposium Series. American Chemical Society, pp. 147–163
2. Rasmussen SC (2020) Conjugated and conducting organic polymers: the first 150 Years. ChemPlusChem 85:1412–1429
3. Reynolds JR, Thompson BC, Skotheim TA (2019) Handbook of Conducting Polymers, Fourth Edition – 2 Volume Set., Boca Raton, FL, CRC Press
4. Rasmussen SC, Evenson SJ, McCausland CB (2015) Fluorescent thiophene-based materials and their outlook for emissive applications. Chem Commun 51:4528–4543

5. Boni R, Ali A, Shavandi A, Clarkson AN (2018) Current and novel polymeric biomaterials for neural tissue engineering. J Biomed Sci 25:90

6. Kenry, Liu B (2018) Recent advances in biodegradable conducting polymers and their biomedical applications. Biomacromolecules 19:1783–1803

7. Nezakati T, Seifalian A, Tan A, Seifalian AM (2018) Conductive polymers : opportunities and challenges in biomedical applications. Chem Rev 118:6766–6843

8. Upadhyay PK, Ahmad A (2010) Chemical synthesis, spectral characterization and stability of some electrically conducting polymers. Chin J Polym Sci 28:191–197

9. Toshima N, Hara S (1995) Direct synthesis of conducting polymers from simple monomers. Prog Polym Sci 20:155–183

10. He F, Omoto M, Yamamoto T, Kise H (1995) Preparation of polypyrrole–polyurethane composite foam by vapor phase oxidative polymerization. J Appl Polym Sci 55:283–287

11. Jang J, Oh JH (2002) Novel crystalline supramolecular assemblies of amorphous polypyrrole nanoparticles through surfactant templating. Chem Commun 2002: 2200–2201

12. Yoon H, Chang M, Jang J (2006) Sensing behaviors of polypyrrole nanotubes prepared in reverse microemulsions: effects of transducer size and transduction mechanism. J Phys Chem B 110:14074–14077

13. Alf ME, Asatekin A, Barr MC, Baxamusa SH, Chelawat H, Ozaydin-Ince G, Petruczok CD, Sreenivasan R, Tenhaeff WE, Trujillo NJ, Vaddiraju S, Xu J, Gleason KK (2010) Chemical vapor deposition of conformal, functional, and responsive polymer films. Adv Mater 22:1993–2027

14. Tran HD, Shin K, Hong WG, D'Arcy JM, Kojima RW, Weiller BH, Kaner RB (2007) A template-free route to polypyrrole nanofibers. Macromol Rapid Commun 28:2289–2293

15. Whang YE, Han JH, Nalwa HS, Watanabe T, Miyata S (1991) Chemical synthesis of highly electrically conductive polymers by control of oxidation potential. Synth Met 43:3043–3048

16. Fang Q, Chetwynd DG, Gardner JW (2002) Conducting polymer films by UV-photo processing. Sensors Actuat A Phys 99:74–77

17. Cruz GJ, Morales J, Castillo-Ortega MM, Olayo R (1997) Synthesis of polyaniline films by plasma polymerization. Synth Met 88:213–218

18. Groenewoud LMH, Engbers GHM, Terlingen JGA, Wormeester H, Feijen J (2000) Pulsed plasma polymerization of thiophene. Langmuir 16:6278–6286

19. Paosawatyanyong B, Tapaneeyakorn K, Bhanthumnavin W (2010) AC plasma polymerization of pyrrole. Surf Coat Technol 204:3069–3072

20. Subramanian A, Krishnan UM, Sethuraman S (2009) Development of biomaterial scaffold for nerve tissue engineering: Biomaterial mediated neural regeneration. J Biomed Sci 16:108

21. Chronakis IS, Grapenson S, Jakob A (2006) Conductive polypyrrole nanofibers via electrospinning: Electrical and morphological properties. Polymer (Guildf) 47:1597–1603

22. Aznar-Cervantes S, Roca MI, Martinez JG, Meseguer-Olmo L, Cenis JL, Moraleda JM, Otero TF (2012) Fabrication of conductive electrospun silk fibroin scaffolds by coating with polypyrrole for biomedical applications. Bioelectrochemistry 85:36–43

23. Hsiao C-W, Bai M-Y, Chang Y, Chung M-F, Lee T-Y, Wu C-T, Maiti B, Liao Z-X, Li R-K, Sung H-W (2013) Electrical coupling of isolated cardiomyocyte clusters grown on aligned conductive nanofibrous meshes for their synchronized beating. Biomaterials 34:1063–1072

24. Borriello A, Guarino V, Schiavo L, Alvarez-Perez MA, Ambrosio L (2011) Optimizing PANi doped electroactive substrates as patches for the regeneration of cardiac muscle. J Mater Sci Mater Med 22:1053–1062

25. Schmidt CE, Shastri VR, Vacanti JP, Langer R (1997) Stimulation of neurite outgrowth using an electrically conducting polymer. Proc Natl Acad Sci U S A 94:8948 LP–8953

26. Llorens E, Armelin E, Pérez-Madrigal M del M, del Valle LJ, Alemán C, Puiggalí J (2013) Nanomembranes and nanofibers from biodegradable conducting polymers. Polymers (Basel) 5:1115–1157

27. Ryu M, Yang JH, Ahn Y, Sim M, Lee KH, Kim K, Lee T, Yoo S-J, Kim SY, Moon C, Je M, Choi J-W, Lee Y, Jang JE (2017) Enhancement of interface characteristics of neural probe based on graphene, ZnO nanowires, and conducting polymer PEDOT. ACS Appl Mater Interfaces 9:10577–10586

28. Schwartz AB, Cui XT, Weber DJ, Moran DW (2006) Brain-controlled interfaces: movement restoration with neural prosthetics. Neuron 52:205–220

29. Schalk G, McFarland DJ, Hinterberger T, Birbaumer N, Wolpaw JR (2004) BCI2000: a general-purpose brain-computer interface (BCI) system. IEEE Trans Biomed Eng 51:1034–1043

30. Hochberg LR, Serruya MD, Friehs GM, Mukand JA, Saleh M, Caplan AH, Branner A, Chen D, Penn RD, Donoghue JP (2006) Neuronal ensemble control of prosthetic devices by a human with tetraplegia. Nature 442:164–171

31. Zhao X, Wu H, Guo B, Dong R, Qiu Y, Ma PX (2017) Antibacterial anti-oxidant electroactive injectable hydrogel as self-healing wound dressing with hemostasis and adhesiveness for cutaneous wound healing. Biomaterials 122:34–47

32. Gao M, Dai L, Wallace GG (2003) Biosensors based on aligned carbon nanotubes coated with inherently conducting polymers. Electroanalysis 15:1089–1094

33. Mao H, Liang J, Zhang H, Pei Q, Liu D, Wu S, Zhang Y, Song X-M (2015) Poly(ionic liquids) functionalized polypyrrole/graphene oxide nanosheets for electrochemical sensor to detect dopamine in the presence of ascorbic acid. Biosens Bioelectron 70:289–298

34. Wadhwa R, Lagenaur CF, Cui XT (2006) Electrochemically controlled release of dexamethasone from conducting polymer polypyrrole coated electrode. J Control Release 110:531–541

35. Boehler C, Kleber C, Martini N, Xie Y, Dryg I, Stieglitz T, Hofmann UG, Asplund M (2017) Actively controlled release of dexamethasone from neural microelectrodes in a chronic in vivo study. Biomaterials 129:176–187

3 Conducting Polymers
Fundamentals, Synthesis, Properties, and Applications

Suba Lakshmi Madaswamy, N. Veni Keertheeswari, and Ragupathy Dhanusuraman

Nano Electrochemistry Lab (NEL), Department of Chemistry, National Institute of Technology Puducherry, Karaikal, India

CONTENTS

DOI: 10.1201/9781003205418-3

3.1   CONDUCTING POLYMERS-HISTORICAL OVERVIEW

Polymers are long-chain macromolecules composed of smaller molecules (monomers) that can be linked together in a variety of ways to produce various microstructures such as linear chains, branched chains, densely interconnected networks, and so on. They are long been a vital research topic in academia and industry [1]. Polymers, by definition, are insulators since they do not conduct electricity. However, it has significant advantages in electronic applications. Copper wires are typically covered by polymers to prevent the metal wires from short circuits owing to the insulating feature [2]. However, the discovery of conductive polymers by Hideki Shirakawa, Alan MacDiarmid, and Alan Heeger has changed the perception of polymers (Figure 3.1). There are a lot of organic conductive compounds, discovered by K. Bechgaard (Copenhagen) and Jerome (Paris) are well-known for their superconductive nature relatively at elevated temperatures (Tc around 10 K). They are organic donors composed of large, delocalized electron systems that solidify into coin-pile stacks and salts of inorganic acceptors [3–6]. In 1958, Natta and collaborators dealt with polyacetylene (PA) by polymerizing acetylene in hexane using catalysts ($Et_3Al/Ti(OPr)_4$ (Et=ethyl, Pr=propyl)) [7]. Despite being having a regular structure and highly crystalline nature, the resulting material was an insoluble powder, air-sensitive, black, and infusible.

Shirakawa and colleagues adapted the method in the early 1970s to produce well-defined films of PA. Shirakawa made a significant discovery by introducing $Ti(OBu)_4$ and then Et_3Al to a little quantity of toluene in a passive environment. The solution was aged for 45 minutes at 20°C before being chilled to −78°C. The reaction vessel was evacuated, and acetylene gas was passed and permitted to react with a catalyst film that had already formed on the reaction vessel's walls. There was an instant formation of a PA film [5]. Unreacted acetylene gas was evacuated to control the reaction. This method produced a copper-colored all-*cis*-PA film with a 95% yield. By conducting the reaction in *n*-hexadecane at 150°C, Shirakawa's technique permitted the production of silvery all-*trans*-PA, as illustrated in Figure 3.2. PA was already known as a black powder until Shirakawa MacDiarmid

FIGURE 3.1   Hideki Shirakawa, Alan MacDiarmid, and Alan Heeger. (Reproduced with permission from Ref. [3]. Copyright (2003) Royal Society of Chemistry.)

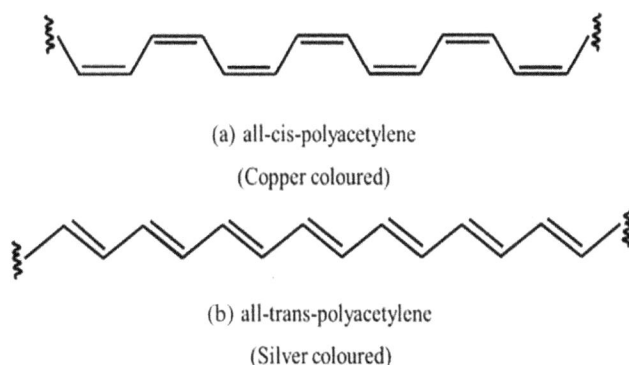

(a) all-cis-polyacetylene

(Copper coloured)

(b) all-trans-polyacetylene

(Silver coloured)

FIGURE 3.2   (a) All-*cis*-polyacetylene and (b) all-*trans*-polyacetylene.

and Heeger used a Ziegler-Natta catalyst to make a silvery film from acetylene in 1974. (K. Ziegler and G. Natta, Nobel Prize in Chemistry 1966). It is not a conductor, despite its metallic outlook [3].

3.2   MECHANISM OF CONDUCTION IN CPs

Conducting polymers (CPs), in general, have a delocalized electron system. The electronic structure of a substance is used to determine its electrical conductivity. The band theory may be used to compare and contrast insulators, conductors, and semiconductors. The bandgap is the energy difference between a material's valence and conduction bands when the valence band overlaps the conduction band and the valence electrons are free to migrate and propagate in the conduction band. When electrons in semiconductors are excited, they can cross the minimum energy gaps and reach the conduction band, creating a hole in the process. This allows current to flow by enabling hole and electron charge transfer. The bandgap is too high for electrons to pass in insulators; thus, they do not conduct electricity. The theory of bandgap of conductors, semiconductors, and insulators is depicted in Figure 3.3 [7, 9].

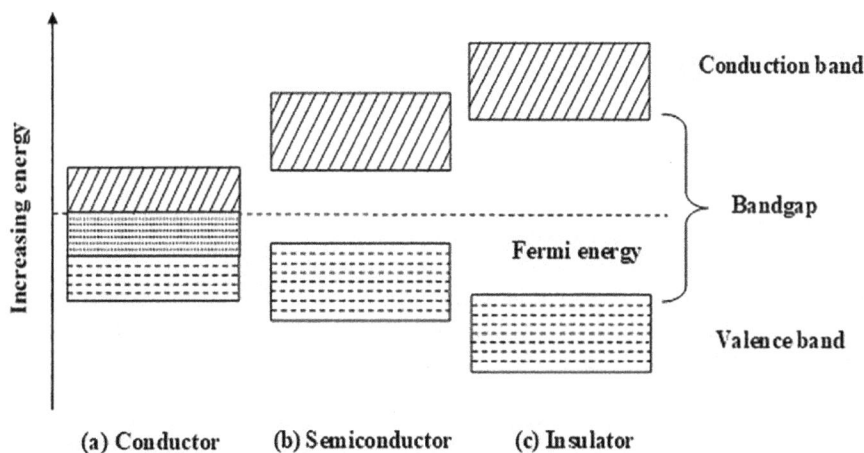

(a) Conductor   (b) Semiconductor   (c) Insulator

FIGURE 3.3   A bandgap diagram shows the difference between bandgap for (a) conductor, (b) semiconductor, and (c) insulator.

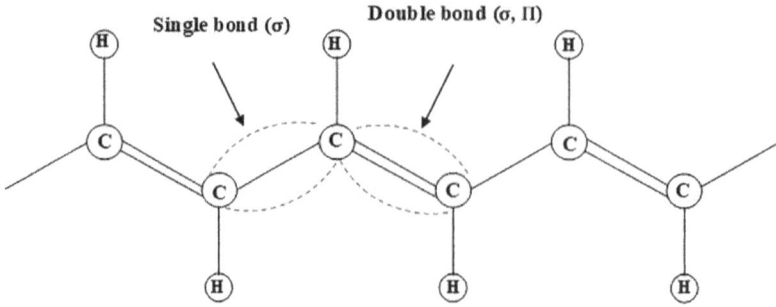

FIGURE 3.4 The structure of polyacetylene.

On the other hand, band theory does not explain the conductivity of organic materials. Many studies have been conducted at the molecular level to investigate the transport properties of CPs. Due to the simple chemical structure and remarkably high electrical conductivity of PA, it serves as an illustration of the concepts of conduction in CPs (Figure 3.4). According to chemists, conjugated single and double bonds along the polymer skeleton are a typical electrical characteristic of pure CPs. There is a transfer of π-bonds from the initial and second carbon atoms to the second and third carbon atoms. There is a further transfer of π-bonds from third and fourth to fourth and fifth carbon, and so on. Consequently, the electrons in the double bonds travel along the carbon chain allowing electric current to flow [6, 8].

The band theory can be supported by using molecular orbital theory. Two new molecular orbitals can be formed by combining one p orbital from one carbon atom and another p orbital from another carbon atom. The bonding molecular orbital has lower energy than the other orbitals. The orbital with the highest energy is known as the anti-bonding molecular orbital (ABMO). Two electrons from two carbon atoms fill the bonding molecular orbital because electrons in this orbital are more stable. A pi bond is formed by these two electrons sharing the same orbital. The bonding molecular orbital has the highest occupied molecular orbital (HOMO) in this case, while the ABMO has the lowest unoccupied molecular orbital (LUMO). The bandgap energy, or the energy required for electron activation, is the difference in energy between HOMO and LUMO. An unlimited number of p orbitals exist in CPs with an indefinitely long polymer chain, and the innumerable molecular orbitals eventually form a band between the conduction and valence bands. Electrons may move from one band to another with less activation energy as the bandgap comes closer, and the polymer finally becomes metallic [9].

The well-known free-electron molecular orbital model provides the necessary components for quantitatively explaining an insulator, semiconductor, or conductor made out of a linear chain of atoms, such as PA. Assume an N-atom row separated by d, so the total length of the chain is (N–1)d or, for large N, approximately Nd. In the quantum-mechanical model for a free particle in a one-dimensional box, the wave functions (Figure 3.5) correspond to a ladder of eigenvalues (potential zero inside the box and infinity outside).

$$E_n = n^2 h^2 / 8 \, m (Nd)^2, \text{ with } n = 1, 2, 3\ldots, \tag{3.1}$$

In Equation (3.1), h is Planck's constant, m is the electron mass and n is a quantum number.

The HOMO has energy as indicated in Equation (3.2) if the electrons from the N p-orbitals are filled into this ladder with two electrons per molecular orbital (according to the Pauli principle).

$$E(HOMO) = (N/2)^2 \, h^2 / 8 \, m (Nd)^2 \tag{3.2}$$

and the LUMO has the energy as given in Equation (3.3).

$$E(LUMO) = (N/2 + 1)^2 \, h^2 / 8 \, m (Nd)^2 \tag{3.3}$$

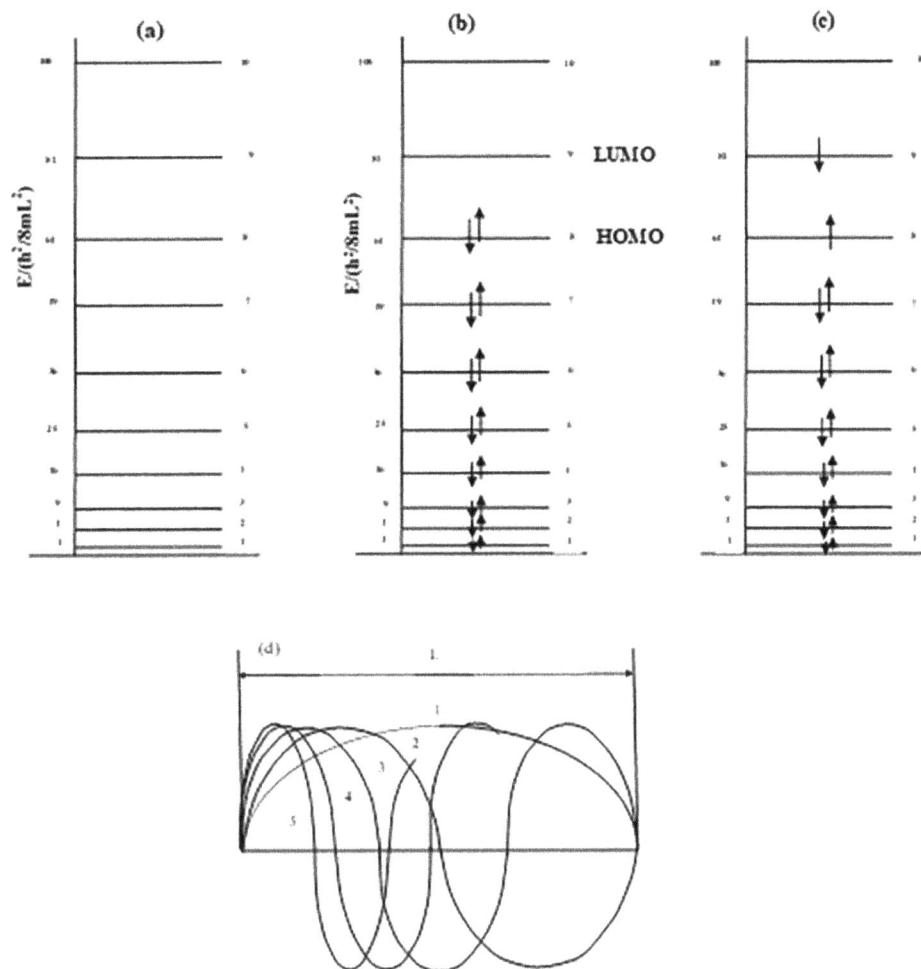

FIGURE 3.5 Free-electron model (one-dimensional box, length L): (a) energy levels, (b) 16 electrons occupy the first eight molecular orbitals in ground state level, (c) excited state of energy levels, and (d) wave functions.

Equation (3.4) gives the energy necessary to excite an electron from HOMO to LUMO.

$$\Delta E = E(LUMO) - E(HOMO) = (N+1)h^2/8 \, m(Nd)^2 \approx \left[h^2/8 \, md^2\right]/N \text{ for large } N (4) \quad (3.4)$$

With increasing polymer length, the bandgap is expected to diminish as $1/N$ and become insignificant in bulk dimensions.

Other models, such as the Hückel model for the electron system, or more advanced semi-empirical models and data programs, or ab initio models (e.g., the Gaussian programs of John Pople, Nobel Prize in Chemistry 1998), provide a qualitatively similar picture, with a more or less regular energy ladder for the filled orbitals followed by a ladder of empty orbitals. The overall wave function of a multi-electron system occupies a state and its energy is the entire system energy (in the simplest model it is a product of all the one-electron wave functions, that is to say, the filled orbitals). With an even number of electrons and two electrons occupying each of the lower energy ladder orbitals, we have the situation depicted in Figure 3.5b, which corresponds to the ground state because all electrons are paired, the total spin angular momentum S is zero, indicating that the substance is not magnetic. When an electron from one of the filled molecular orbitals moves up into one of the empty molecular orbitals (Figure 3.5c), there is an excited electron configuration and a corresponding excited state with higher energy than the ground state. The minimum energy difference between the ground and excited states – the bandgap – corresponds to the energy required to form a charge pair with one electron in the upper (empty)manifold of orbitals and one positive charge or "hole" in the lower (filled)manifold of orbitals. If the spins of the remaining single electron in the lower orbital and the excited electron are opposite each other, so that S remains zero (S=1/2–1/2=0), the excited state is a singlet state (non-magnetic), whereas if the two spins are parallel, S=1/2+1/2=1, the state is a triplet state (magnetic) (paramagnetic, i.e., that the molecule has a permanent magnetic dipole moment due to electron spin). As we have seen in the basic free-electron model (Eq. 3.4) that the bandgap vanishes for a sufficiently long chain, so PA is expected to behave as a conductor. The bandgap is experimentally related to the wavelength of the substance's first absorption band in its electronic spectrum. Thus, if the energy condition is met, a photon with wavelength can excite an electron from HOMO to LUMO level which is given in Equation (3.5).

$$\Delta E = E(LUMO) - E(HOMO) = h\nu = h\, c/\lambda, \tag{3.5}$$

where h is Planck's constant and ν the frequency of light (the third equality comes from c=$\nu\,\lambda$, with c the velocity of light). For polyenes, the wavelength of the absorption thus lengthens with the increasing length of the polyene: the bandgap ΔE decreases when double bonds are added to form molecules with lengthening conjugations in the progression from ethene to butadiene to hexatriene, etc., but not as predicted by Equation (3.4). Hence, there seems to be an upper limit beyond which no change will result from further conjugation into an infinite linear polyene. This convergence has also been confirmed by several theoretical predictions, the first by Lennard-Jones as early as 1937. PA is not expected to be a conductor due to its large bandgap and filled valence orbitals. It was discovered to be a semiconductor with intrinsic conductivity ranging from 10^{-5} to 10^{-7} S m^{-1}. Conjugated bonds, on the other hand, do not make polymeric materials highly conductive. Shirakawa, Heeger, and MacDiarmid made a breakthrough by utilizing the complementary ways in which chemists and physicists think about conductance [9, 10].

3.3 DOPING IN CPs

If the polymer structure has no charge carriers, the polymer is not conductive. The majority of conductive organic polymers deficiency intrinsic charge carriers. As a result, external charge carriers must be introduced into polymers to make them conductive. CPs can be partially reduced or partially oxidized by electron donors or electron acceptors. Researchers are focusing on a system in which a halogen dopant removes an electron from a delocalized bonding arrangement, resulting in the formation of a hole. Then, an adjacent electron leaps into the hole and fills it, forming a new hole and allowing charge to flow across the polymer chain. This creation of nonlinear local excitations (e.g., solitons, polarons, and bipolarons) as charge carriers has been attributed to the formation of several ideas about the conductivity of CPs [11].

The dopant's function is to either remove or add electrons to the polymer. Iodine (I_2), for example, will absorb an electron during the formation of an I^{3-} ion. When one electron is removed from the top of a semiconductive polymer's valence band, such as PA or polypyrrole (PPy), the vacancy (hole) that follows does not completely delocalize, as the classical band theory would predict. If one electron is removed locally from one carbon atom, then a radical cation is formed. The radical cation (also known as a "polaron") is localized, partly due to Coulomb attraction to its counterion (I^{3-}), which normally has very low mobility, and partly due to a local change in the radical cation's equilibrium geometry relative to the neutral molecule. A polaron's mobility along a PA chain is high, and the charge is carried along as shown in Figure 3.6. However, because the counterion to the positive charge is not very mobile, a high concentration of counterions is required for the polaron to move in the field of close counterions. This helps to explain why so much doping is required. If a second electron is removed from an already oxidized region of the polymer, a second independent

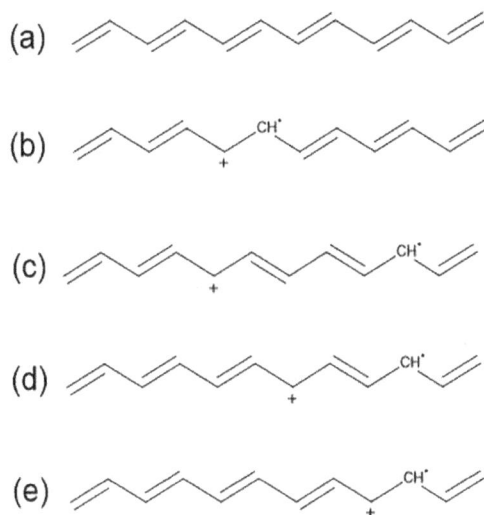

FIGURE 3.6 The mobility of a polaron through the polyacetylene chain. (a) Polyacetylene chain, (b) polaron formation of polyacetylene chain, and (c–e) successive polaron migration of polyacetylene chain.

polaron or a bipolaron is created, depending on whether the first polaron's unpaired electron is removed. In bipolarons, the two positive charges are not independent, but move as a pair, similar to the Cooperpair in theory of superconductivity [1, 11, 12]. The bulk conductivity of CPs should be composed of contributions from intra-chain, interchain, and inter-domain electron transports. The specifics of transporting process and their relative importance are still unknown. Some of the factors that influence conductivity have been identified. As shown in the preceding discussion, the doping process is the most obvious factor dominating the conductivity of conjugated polymers. Some considerations include conjugated polymers' orientation, crystallinity, and purity. The following are descriptions of some of the most commonly used doping methods in CPs.

3.3.1 CHEMICAL DOPING

By using electron acceptors or electron donors, almost all conjugated polymers can be partially oxidized (p-type redox doping) or partially reduced (n-type redox doping). For example, treating *trans*-PA with an oxidizing agent such as iodine causes the doping reaction as given in Equation (3.6) and an increase in conductivity ranging from 10^{-5} to 10^2 S cm^{-1}.

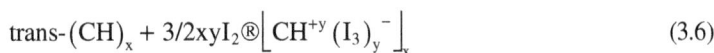

$$\text{trans-}(CH)_x + 3/2xyI_2 \circledR \left[CH^{+y} (I_3^-)_y^- \right]_x \tag{3.6}$$

Similarly, most conjugated polymers like *trans*-PA can be doped with electron donors (n-doping) as given in Equation (3.7) to gain high conductivities.

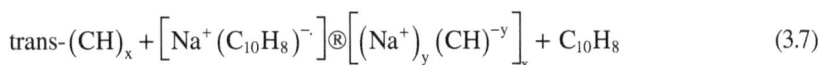

$$\text{trans-}(CH)_x + \left[Na^+ (C_{10}H_8)^- \right] \circledR \left[(Na^+)_y (CH)^{-y} \right]_x + C_{10}H_8 \tag{3.7}$$

Also, it was found that the n- and p-type dopants could compensate for each other [13].

3.3.2 ELECTROCHEMICAL DOPING

Conjugated polymers can also be readily reduced (n-doping) or oxidized (p-doping) electrochemically due to extensive conjugation of electrons, with the conjugated polymer acting as either an electron source or an electron sink. A DC power source, in particular, can be used to drive the doping reaction between a *trans*-PA-coated positive electrode and a negative electrode. They are both immersed in $LiClO_4$ solution in propylene carbonate, the reaction of which is given in Equation (3.8) [14].

$$\text{trans-}(CH)_x + (xy)(ClO_4)^- \rightarrow \left[CH^{+y}(ClO_4)_y^- \right]_x + (xy)e^- \ (y \leq 0.1) \tag{3.8}$$

Electrochemical doping has several unique advantages over chemical doping. To begin, the amount of current passed allows for precise control of the doping level. Second, doping and undoping are highly reversible, with no chemical products that must be removed. Finally, p- or n-type doping can be achieved even with dopant species that cannot be created chemically. Counter "dopant" ions are introduced in both cases to stabilize the charge along the polymer backbone. The presence of counter ions has its pros and cons, because counter ions may cause undesirable structural distortion and a decrease in conductivity, make conjugated CPs more suitable for actuation applications [14]. To avoid the inclusion of counterions, "photo-doping" and "charge-injection doping" methods were used to achieve the redox doping effects. Furthermore, some conjugated polymers, such as poly(heteroaromatic vinylenes) and polyaniline (PANI), can acquire high conductivities by protonating imine nitrogen atoms without any electron transfer between the polymer and the "dopants" – a process known as "non-redox doping."

3.3.3 PHOTODOPING

Radiated by an energy beam that is larger than the band gap of a conjugated polymer (e.g., *trans*-PA), macromolecule might be used to boost electrons from the valence band into the conduction band, as stated in Equation (3.9) [15].

$$\text{Trans} - (CH)_x \xrightarrow{h\upsilon} \qquad \cdots\cdots \tag{3.9}$$

Although the photogenerated charge carriers may vanish once the irradiation stops, applying an appropriate potential during irradiation could separate electrons from holes, resulting in photoconductivity.

3.3.4 CHARGE-INJECTION DOPING

Charge carriers can be injected into the conjugated polymers of bandgap [e.g., PA, poly(3-hexylthiophene) (P3HT)] using a field-effect transistor (FET) geometry applying a corresponding potential for the multilayer structure metal/insulator/polymer utilizing FET geometry [16]. Like photodoping, charge-injection doping does not create counter ions, allowing for a systematic investigation of electrical characteristics based on the charging density of the carriers with little structural distortion. A thin P3HT self-assembled film exhibits a metal-insulator transition with a metallic-like temperature dependence using the charge injection doping method, which is successfully proved. When the charge density exceeds $2.5 \times 10^{-14} \ cm^{-2}$ at temperatures below 2.35 K, superconductivity is observed. This finding of superconductivity seems strongly linked with the two-dimensional charge transfer of a self-assembled polymer film [17].

3.3.5 NON-REDOX DOPING

Non-redox doping results rather than changing the number of electrons, connected with the polymers of the backbone, unlike redox doping. The most widely used doping technique of this type is the protonic doping of the PANI emeraldine base (PANI-EB) using aqueous protonic acids such as HCl (Equation (3.10)), *d*, L-camphor sulfonic acid (HCSA), $(C_6H_5)SO_3H$, and p-CH_3-$(C_6H_4)SO_3H$ to produce conducting polysemiquinone radical cations is the most considered doping process of this type [18].

$$\cdot \left[\begin{array}{c} \text{structure with } 2n\ HCl \end{array} \right] \longrightarrow \cdot \left[\begin{array}{c} \text{product structure} \end{array} \right] \quad \ldots\ldots(3.10)$$

3.4　SYNTHESIS OF CONDUCTING POLYMERS BY DIFFERENT METHODS

3.4.1　Chemical Method

Chemical synthesis of polymerization was done by either reduction or oxidation of consistent monomers. Chemical polymerization occurs when conjugated monomers are mixed with an excess of oxidant in a suitable solvent, such as acid. The process of polymerization occurs quickly but requires constant stirring. A numerical analysis was used to improve the quality and yield of the synthesized product obtained through the oxidative polymerization technique. Most of the CPs like PANI and PPy are synthesized by chemically using oxidizing agents like ammonium persulfate ($(NH_4)_2S_2O_8$), hydrogen peroxide (H_2O_2), and potassium permanganate ($KMnO_4$). One of its benefits is the possibility of mass manufacturing at a reasonable cost [19]. For example, PANI is synthesized by the oxidative polymerization method. When aniline monomer is mixed with hydrochloric acid, then ammonium persulfate is used as an oxidizing agent, a dark green color PANI is obtained.

3.4.2　Electrochemical Method

Among the various reported methods for synthesis of conductive polymers, especially, electrochemical synthesis is very important because it is a simple, fast, and cost-effective process that can be performed by applying a voltage to a single section of a glass cell, and the fabricated films have the required uniformity and thickness. Mostly, an electrochemical technique used for the synthesis of CPs is anodic oxidation of suitable electroactive functional monomers; cathodic reduction is utilized considerably less often. In anodic oxidation, the potential for monomer oxidation leads to polymerization which is greater than that of charging oligomeric (intermediate polymer). For example, electrochemical polymerization was adapted for the synthesis of polythiophene (PTh). First, the monomer, thiophene is mixed with tetrabutylammonium tetrafluoroborate (Bu_4NBF_4) salt as the supporting electrolyte. The two solutions were constantly combined to produce a static organic/inorganic interface. Afterward, reference and counter electrodes were immersed in a solution mixture. Applying a potential voltage of −0.6 to 2 V at a moderate scan rate of 25 mV s^{-1} thin layers of PTh are formed [20, 21].

3.4.3　Photochemical Method

Some CPs are synthesized by photochemical technique. For example, pyrrole has been successfully polymerized to PPy using visible light irradiation either as a photosensitizer or a suitable electron acceptor. This technique provides simple and affordable synthesis, which is not harmful to the environment. Likewise, Suba and co-workers synthesized PANI via the solar irradiation approach [22, 23].

3.4.4　Concentrated Emulsion Method

The emulsion polymerization technique is a heterophase polymerization approach that may be divided into three segments: water segment, latex particle segment, and monomer droplet segment. Its basic mechanism is radical polymerization, but bulk and solution polymerization are methods in which only one segment is present in the arrangement, where the monomer in the solvent and initiator are in the same segment. The prepared polymer remains soluble either in the monomer or in the solvent until substantial modification. This technique typically includes a combination of micelle-forming surfactant, a water-soluble initiator, and a water-insoluble monomer [24].

3.4.5 INCLUSION METHOD

Inclusion polymerization is used to synthesis composite materials on an atomic or molecular scale. As a result, this type of polymerization can pave the way for novel low-dimensionality composite materials with great promise. For example, the inclusion of an electroconductive polymer may result in the formation of a molecular wire. Based on inclusion, composites of such polymers with organic hosts have been produced. According to Miyata et al., this polymerization may be regarded as a common space-dependent polymerization and should not be evaluated just from the standpoint of stereoregular polymerization [25].

3.4.6 PYROLYSIS METHOD

The chemical breakdown of organic compounds at high temperatures is defined as Pyrolysis. It has become a valuable technique for the research and detection of organic polymeric compounds in rubber and plastic manufacture, dentistry, environmental shelter, and failure testing. This approach allows for the direct examination of extremely tiny sample quantities without the need for time-consuming sample preparation. Spectroscopic techniques can identify the monomeric species present; nevertheless, pyrolytic degradation plays an important role in the final structure assignment. Pyrolysis gas chromatography is widely used to analyze synthetic and natural polymers [26]. For example, poly(o-methylaniline) was synthesized by a pyrolysis method. The monomer o-Methylaniline is dissolved in phosphoric acid. In this synthesis, hydrogen peroxide and ferric chloride hexahydrate were used as oxidizing agent. Then, the above mixture is taken in an autoclave and heated to 140°C for 6 hours in a temperature-controlled oven [27]. Finally, poly(o-methyl aniline) products were obtained.

3.4.7 PLASMA POLYMERIZATION

It is a new method for producing thin films from a variety of organometallic and organic precursor materials. Plasma polymerized films have no pinholes and are strongly cross-linked, thus, chemically inert, mechanically robust, insoluble, and thermally stable. Furthermore, such films are very coherent and sticky to a range of substrates including traditional polymer, metal, and glass surfaces. Due to their exceptional characteristics, they have been widely employed in recent years for a variety of applications such as protective shells, biological materials, selective membranes, optical, electrical devices, and adhesion supports [28].

3.4.8 SOLID-STATE METHOD

In this polymerization process, the polymer chain length increases due to the removal of oxygen and water either by vacuum or removal with an inert gas to remove away by-products. The reaction is regulated by pressure, temperature, and diffusion of by-products. It is frequently employed by following melt-polymerization to improve the rheological and mechanical characteristics of polymers before injection blow moulding. The solid-state technique is widely used in the commercial manufacture of polyethylene terephthalate (PET) bottles, films, and innovative industrial fibers. The main benefit of the solid-state method is the use of modest and inexpensive apparatus and eluding several difficulties associated with traditional polymerization processes [29].

3.5 CONDUCTIVE MECHANISM OF CONDUCTING POLYMERS

3.5.1 CONDUCTION MECHANISM OF POLYACETYLENE (PA)

The mechanism of PA conduction is revealed in Figure 3.7. PA is the most basic polymer with a high conductivity value. The radical cation (called polaron) is located partially due to Coulomb's attraction to its counterion, which generally has very limited mobility, partly because it changes the local geometry of the radical molecular equilibrium in comparison with the neutral molecule. Polaron mobility along the PA chain can be high, and the charge is transported along the chain. It also mentions the possibility of an electronic changeover [25].

3.5.2 CONDUCTION MECHANISM OF POLYPYRROLE (PPy)

The mechanism of PPy conduction is depicted in Figure 3.8. PPy is synthesized by a chemical or electrochemical method using pyrrole as a monomer. The oxidative polymerization process of pyrrole to PPy begins with the oxidation of pyrrole by a single electron into a radical cation which then combines to a different radical cation to produce a 2,2″ bipyrrole. The reaction chain continues with dication. Finally, PPy is formed [25].

3.5.3 CONDUCTION MECHANISM OF POLYANILINE (PANI)

Figure 3.9 represents the conduction mechanism of PANI. With the oxidation of aniline monomer, the aniline polymerization begins to create a radical cation.

Then, coupling of radical cation takes place with the anilinium ion of the amine group to form a dimer. The dimers are oxidized immediately after formation, followed by further oxidation and deprotonation to produce the trimers with a monomer of an aniline, via an electrophilic aromatic substitution. The process eventually leads to PANI formation [30].

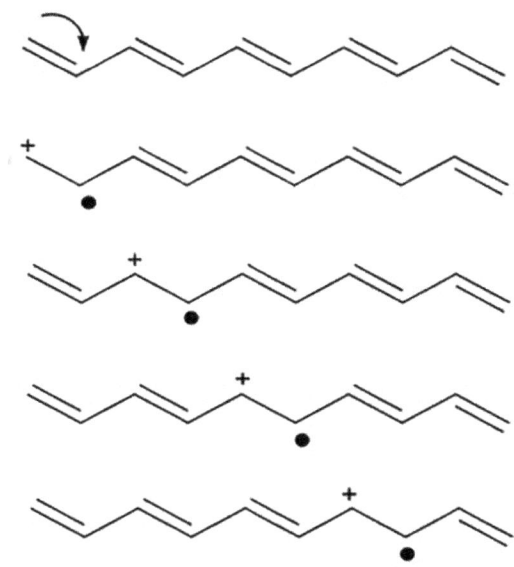

FIGURE 3.7 Conduction mechanism of polyacetylene.

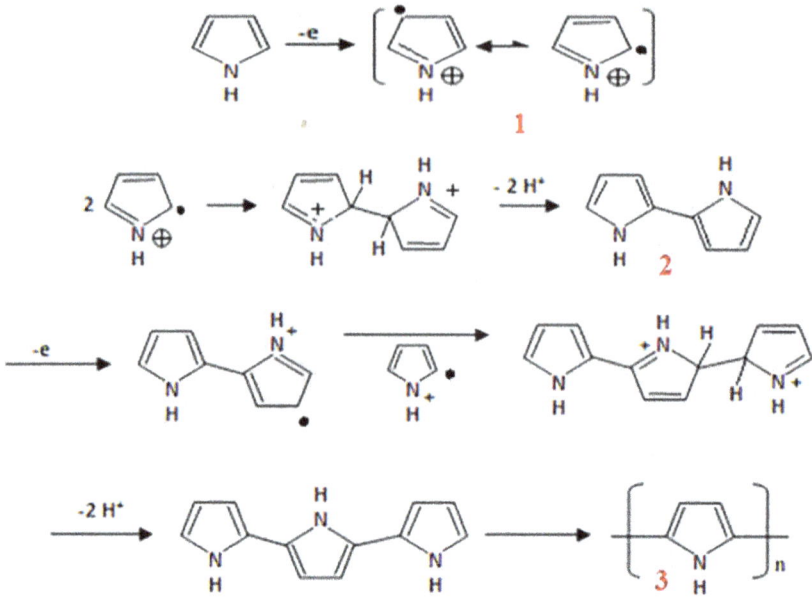

FIGURE 3.8 Conduction mechanism of polypyrrole. (The image is licensed for use from reference [25]. International Advanced Research Journal in Science, Engineering, and Technology.)

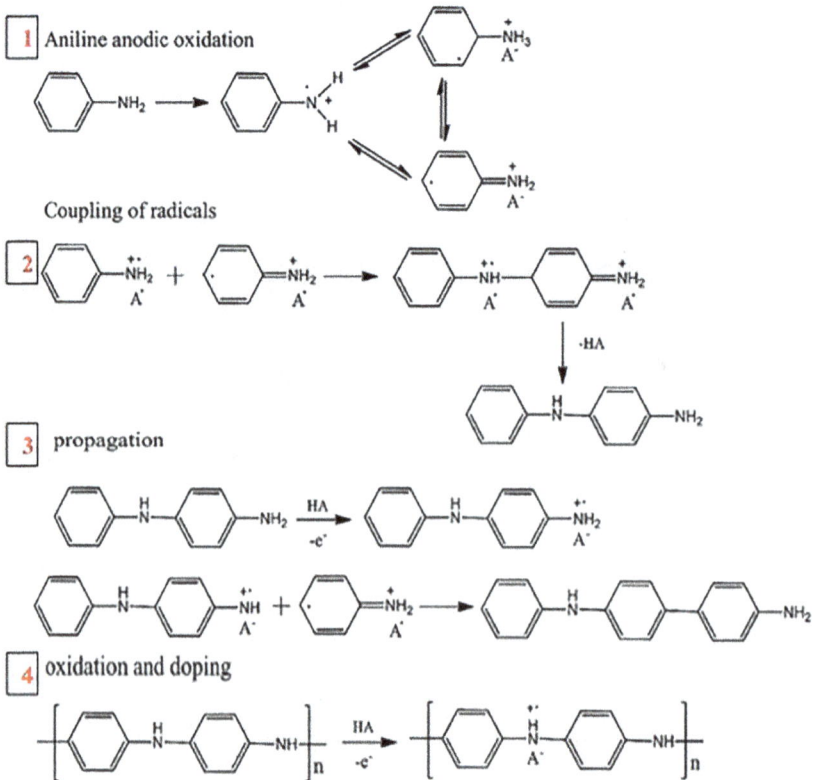

FIGURE 3.9 Conduction mechanism of polyaniline. (Reproduced with permission from Ref. [30]. Copyright (2017) Elsevier.)

1-Oxidation of monomer

2- Radical coupling

3-Chain propagation

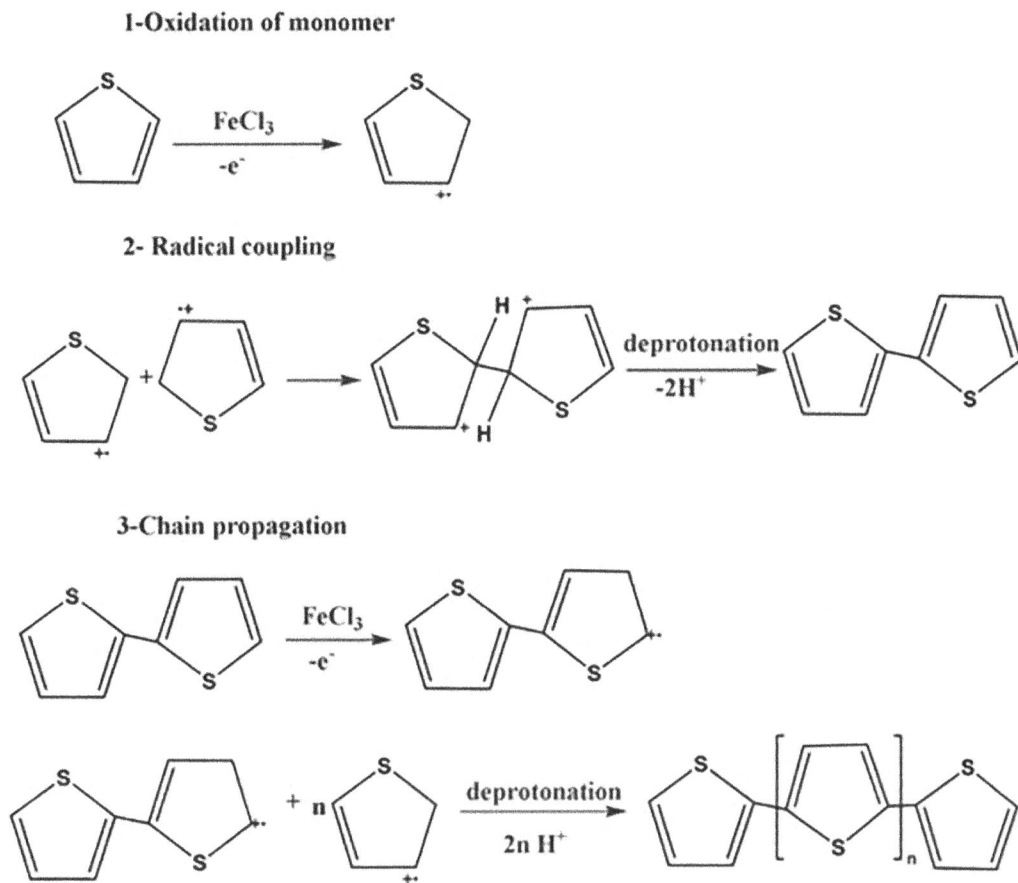

FIGURE 3.10 Conduction mechanism of polythiophene. (Reproduced with permission from Ref. [31]. Copyright (2021) Elsevier.)

3.5.4 CONDUCTION MECHANISM OF POLYTHIOPHENE (PTh)

Figure 3.10 depicts the PTh conduction mechanism. The oxidation of the thiophene monomer produces radical cations ($C_4SH_4^+$) in the first stage. Radical-radical coupling takes place between two radical cations, followed by a deprotonation reaction that results in the formation of a dimer species (bithiophene) is the second step. The dimer species is reoxidized and combines with additional radical cations to produce PTh during the chain propagation step [31].

3.5.5 CONDUCTION MECHANISM OF POLY(3,4-ETHYLENEDIOXYTHIOPHENE) (PEDOT)

Figure 3.11 illustrates poly(3,4-ethylenedioxythiophene) (PEDOT) conduction mechanism. An electron is removed from EDOT monomer, thus forming a radical cation intermediate. This radical cation combines with another radical cation to form a dimer and also two protons are eliminated. These dimers react with another cation and the reaction process is repeated many times. Finally, PEDOT is formed [32].

3.6 PROPERTIES OF CONDUCTING POLYMERS

The CPs have unique electrical, biological, and physicochemical properties like electronic, mechanical, wetting, magnetic, optical, and microwave-absorbing properties.

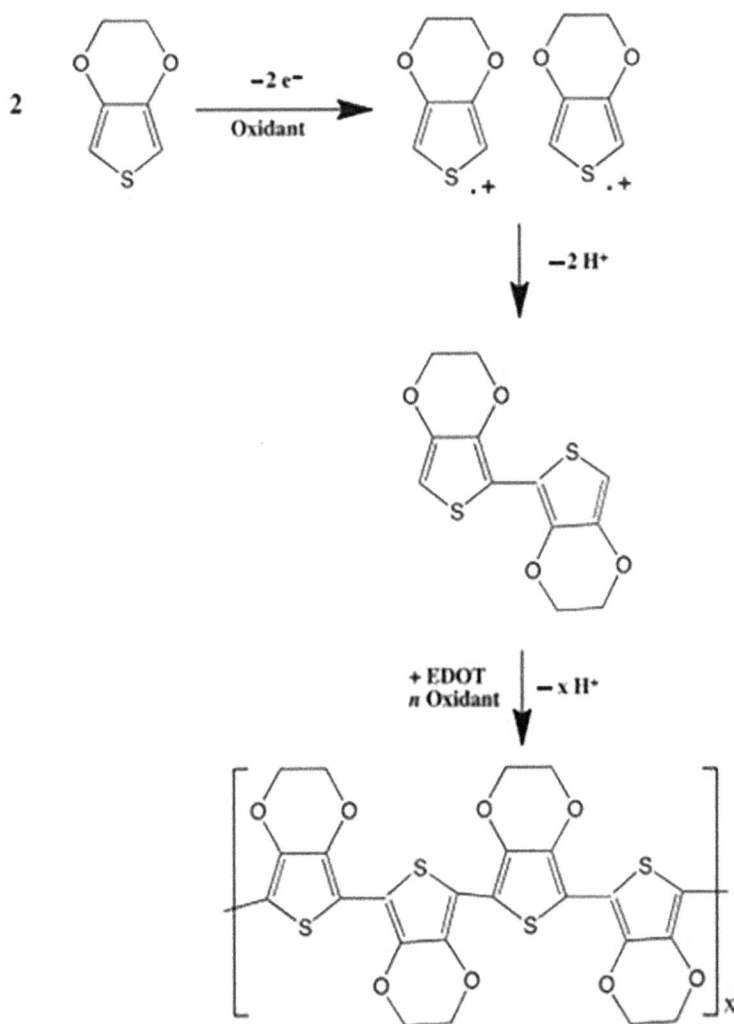

FIGURE 3.11 Conduction mechanism of poly(3,4-ethylenedioxythiophene). (Reproduced with permission from Ref. [32]. Copyright (2019) Elsevier.)

3.6.1 ELECTRICAL PROPERTIES

Polymer conductivity depends on the conjugation length, doping percentage, polymer chain arrangement, and sample purity. The molecular nature of the electrically conducting polymers (ECPs) is of long-range order. Polymer's molecular structure causes electrical mobility around individual macromolecules. The nature of the techniques used to create high conductivity differs for inorganic semiconductors and polymers. The preeminent conductivities rely on doping in polymers which are linked with the production of self-localized excitons such as polarons, bipolarons, and solitons. These particles result from a strong interaction between the charges on the chain as a result of doping. CPs having degenerate ground state for example *trans*-PA, charged solitons are the charge carriers, whereas in CPs with the non-degenerate ground state for example *cis*-PA, PPy, PTh primarily polarons are produced on doping. These polarons combine to produce spinless bipolarons, which serve as charge carriers. The ability of molecular engineering for properties, as well as the affordability of the polymers, have made them extremely appealing materials for electrically conductive applications. The electrical conductivity of doped PA was investigated by Hideki Shirakawa [6]. In polymeric material, doped PA has higher conductivity [7]. However, due to its low stability and processability issues, commercialization has not been possible. Electrical conductivity shows a certain impact on the following:

- Temperature
- Their mobility
- The direction
- The density of charge carriers
- Presence of doping materials

3.6.2 MAGNETIC PROPERTIES

CPs have significant value because of their extraordinary magnetic behavior in technical applications. Furthermore, the magnetic and structural properties of nanomaterials can be altered by the addition of transition metal oxide nanoparticles. In 2012, Nandapure et al. studied the magnetic and transport characteristics of conducting PANI/nickel oxide [33]. Yan et al. used the anodic-oxidation method to create a CP/ferromagnet film [34]. A ferromagnetic PPy composite was created by a chemical technique using p-dodecyl benzenesulfonic acid sodium salt as a surfactant and dopant. A ferromagnetic PPy composite was created using a chemical technique using p-dodecyl benzenesulfonic acid sodium salt acts as a dopant and surfactant. The magnetic characteristics of the final composite demonstrated ferromagnetic behavior, with high saturated magnetization.

3.6.3 OPTICAL PROPERTIES

The electronic structure of conjugated polymers is anisotropic and quasi-one-dimensional due to the presence of p bonds in the polymer backbone by utilizing electron-phonon interactions. Due to their advantages in nanophotonic devices, the unique optical characteristics of CPs have received a great deal of attention. Optical changes in CPs occur as a result of chemical reactions as well as extrinsic variables such as strain effect on the polymer matrix and a shift in planarity. Several physical and chemical factors influence optical property such as phonons present in the polymer, excitation and ionization impurities, excitons, recombination, strain factor, etc. Three particular chemical factors that influence the optical property are conjugation length, degree of anisotropy, and topochemical reaction. Chang discovered a precise formulation of the third-order nonlinear optical susceptibilities of CPs using the Genkin-Mednis method [35].

3.6.4 MECHANICAL PROPERTIES

The mechanical features of a polymer are determined by the arrangement and crystallinity of monomers. Also, the mechanical property of a polymer is significantly influenced by its molecular weight; the strength parameters and toughness are related to molecular weight and chain bonding, respectively. Crystalline polymers have greater mechanical characteristics than amorphous semi-crystalline polymers. The macroscopic mechanical characteristic of CPs depends on the molecular mobility of macromolecules. Molecular mobility relies on parameters such as the structure of branching polymer conformations and macroscopic properties like temperature, pressure, etc. Amorphous polymers have more molecular mobility, and when the temperature reaches Tg, the polymer transforms from a glassy to a rubbery state which leads to a change in mechanical properties [36].

3.6.5 MICROWAVE ABSORBING PROPERTIES

CPs have been explored as novel microwave fascinating materials, because of their easy processability, low density, and moderate price. Ting investigated characteristics of microwave absorption in the microwave frequency range using the free space technique, reflection loss, complex permeability, and determining complex permittivity [37]. The conductive polymer PANI addition proved useful for realizing high absorption across a wide frequency range. Phang et al. [38] studied the PANI-TiO$_2$ nanocomposites' microwave absorption properties.

3.6.6 WETTABILITY

Surface wettability is an important factor in material performance. Surface wettability control is vital for a wide number of applications, like waterproof textiles, Automotive self-cleaning windows and aerospace industries, cookware coatings, liquid transportation, optical equipment characteristics of anti-fingerprint or anti-reflexive, antibacterial adhesion, separation membrane, and cell. San et al. studied the surface wettability of polymer composite bipolar plates at different additives/binders and additives/fillers proportions for fluoropolymer-based additives [39].

3.6.7 BIOLOGICAL PROPERTIES

Biocompatibility and biodegradability are the two most important factors in determining the effect of a material on a living system. Biocompatibility is known as the interaction of a biological system or tissue with a substance used for medical treatment or completed medical equipment [40]. In other words, biocompatible materials have no negative effects on living tissues or organisms. Lots of CPs are presently being investigated for biocompatibility enhancement by combining with diverse biopolymers such as alginate, cellulose, chitosan, and so on. PPy was shown to be biocompatible with biological tissues and cells, but PANI shows preferential biocompatibility to some cell lines. Thus, the strategies used to enhance PANI's biocompatibility are appropriate for tissue engineering applications. The techniques used to improve PANI's biocompatibility include combining PANI with other biocompatible polymers such as poly-e-caprolactone (PCL) and gelatine to form a compound system, thereby mitigating any potential cytotoxic effects of PANI, and PPy/Chitosan composite reveals high Because of their high biocompatibility, CPs are also used for neural implants and tissue engineering applications. The property of biodegradability of CPs played a critical role in agriculture and biomedical applications. Although CPs (e.g., PPy and PTh) are not inherently biodegradable, they can be made biodegradable by combining CPs with degradable polymers such as polycaprolactone fumarate (PLF), poly(lactide-co-glycolide) (PLGA), PCL, and polylactide (PLA). A highly biodegradable conducting biocomposite consisting of 5% conductive PPy and 95% biodegradable poly(l-lactide) (PPy/PLLA) was created using a water in oil emulsion system and an aqueous dodecylbenzene sulfonic acid sodium salt solution in chloroform, followed by a specific amount of pyrrole. The co-solution casting method was used to create a biodegradable conductive membrane composed of conductive PPy nanoparticles (PPy, 2.5 wt%) and biodegradable chitosan (97.5 wt%). CPs are used as scaffolds for tissue engineering and drug delivery agents for targeted drug delivery due to their good biodegradability [40–44].

3.7 COMMON CONDUCTING POLYMERS AND THEIR APPLICATIONS

CP has received significant interest for its electrical property, environmental stability, inexpensive and potentially processable nature. CPs find potential applications in different fields, which is given in Figure 3.12. The various CPs and their applications are discussed below.

3.7.1 POLYACETYLENE (PA)

In the CP, PA was the first polymer that reveals a higher conductivity than metal, which was discovered by Shirakawa in the 1970s [4]. The repeating unit of this organic polymer is (C$_2$H$_2$)n. PA has conductivity at the range of 10^{-5} S cm^{-1}, but the doped PA conductivity is drastically changed

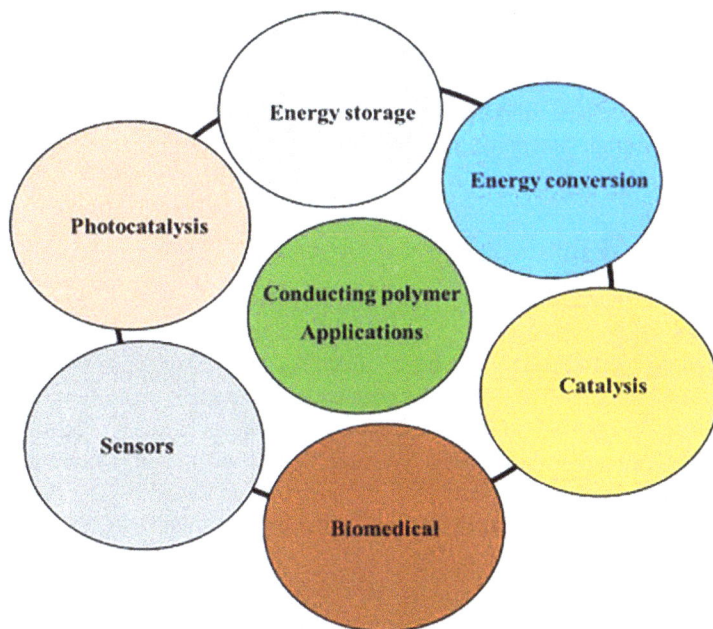

FIGURE 3.12   Illustration of conducting polymer applications.

to 10^2 to 10^3 S cm^{-1}. Also, its characteristics alter based on the dopant material, giving it tunable properties such as optical, mechanical electrochemical properties, etc. The discovery of highly conducting PA sparked a frenzy of study into the creation of new CPs. Because of the excellent electrical conductivity of these polymers, there has been a lot of interest in using organic compounds in microelectronics. A future objective would be to combine the outstanding processing characteristics of various doped and undoped conjugated polymers (e.g., PPy, PANI, and PTh) and their derivatives with the outstanding electric properties of the PA device [5–7, 25].

3.7.2   POLYANILINE (PANI)

PANI is a semi-flexible CP of the rod polymer family. This oldest known polymer also referred to as aniline black was first discovered by Henry Lethe in the 19th century. Amongst the CPs, PANI is widely used owing to its unique features such as low-cost monomers, high conduction, high solubility, flexibility, stability, and lightweight. Also, PANI has redox reversible behavior such as fully oxidized pernigraniline base, completely reduced leucoemeraldine base, half oxidized and reduced emeraldine base with alternating oxidized and reduced repeat units in its structure. In these forms, emeraldine structure is highly conductive than other structures. PANI has an electrical conductivity ranging from 10 to 100 S cm^{-1}. PANI has a lower nobleness than silver but a higher nobleness than copper, this makes it often used for manufacturing and corrosion prevention of printed platforms [8, 14, 19, 24, 43]. The primary obstacle for utilizing PANI and its derivatives like polydiphenylamine, poly-2,5-dimethoxy aniline has several potential applications such as supercapacitor, fuel cell, batteries, neural probes, controlled drug delivery, biosensors, and tissue engineering with auspicious results [45–48].

3.7.3   POLYPYRROLE (PPy)

PPy is the first organic polymer produced via pyrrole polymerization, which was discovered by Donald Weiss in the 1960s. PPy is widely used CP due to the simplicity with which it converts charges/signals generated by the interaction of different analytes and analyte-recognition sites, as well as excellent redox characteristics, simplicity of preparation, stabilized oxidized form, with soaring conductivity, commercially available, water-solubility, excellent environmental stability, and use electrical and optical characteristics. The electrical conductivity of PPy films can be attained up to ~10^3 S cm^{-1} depending on the type and amount of dopant. PPy has been explored for wide usage in energy storage and conversion as well several biomedical applications like bio actuator, biosensors, tissue engineering, and drug delivery [22, 25, 45].

3.7.4 POLYTHIOPHENE (PTh)

PTh is an electroluminescent polymer. It has received a lot of interest in research and industrial fields due to its significant properties like better thermal stability, high environmental stability, and lower bandgap energy. Furthermore, they have intriguing features such as semiconducting, electrical, and optical activities, as well as improved mechanical characteristics, simplicity of processing, and stability. PTh is likewise very conductive (it typically has a conductivity of more than 100 S cm^{-1}). PTh and its derivatives have been intensively researched experimentally and computationally for application in electrical devices like memory devices, hydrogen storage, water purification devices, light-emitting diodes, transistors, imaging material, and biosensors [20, 21, 31, 45].

3.7.5 POLY(3,4-ETHYLENEDIOXYTHIOPHENE) (PEDOT)

PEDOT is derived from 3,4-ethylene dioxylthiophene monomer. PEDOT offers many benefits, including excellent stability, an acceptable bandgap, low redox potential, and optical transparency in thin oxidized films. It has many applications such as nano-fiber electrodes for unit stimulation, organic light-emitting diode displays, printing wiring panels, high conductive shell, antistatic coating of polymers and glass transparent electrodes for thick-film electroluminescence, textile fibers with color varying properties, cathode material in electrolytic capacitors, solar cells, organic semiconductors, and biocompatible interfaces [32, 45].

3.7.6 POLYPARAPHENYLENE (PPP)

Polyparaphenylene (PPP) is a stiff and rod shape polymer formed by repetition of p-phenylene units and transformed to a conducting state with the help of a dopant or an oxidant. PPP was doped to achieve conductivity similar to PA. The primary example of the conducting properties of a nonacetylene hydrocarbon polymer doped with an electron acceptor or donor is PPP. PPP is a significant polymer used in a number of electronic applications, like photovoltaic devices and light-emitting diodes [45, 49].

3.8 CONCLUSION

In this chapter, we have summarized the history of CP, conductive theory, conductive mechanism, synthesis methods, doping process, properties, and their applications. Among the CPs, PA, PPy, PThs, PEDOT, and PANI show promising characteristics in all fields such as energy storage and conversion devices for futuristic applications as well as biomedical applications which includes artificial muscles, tissue engineering, biosensors, gene delivery, and drug delivery. CPs have been used as an essential component in these fields.

REFERENCES

1. Young, R. J., and P. A. Lovell. Introduction to polymers. CRC Press, Boca Raton, FL, 2011.
2. Namazi, H. "Polymers in our daily life." BioImpacts 7 (2017): 73.
3. Shirakawa, H., A. McDiarmid, A. Heeger. "Twenty-five years of conducting polymers." Chem. Commun. 1, (2003): 1–2.

4. Shirakawa, H., E. J. Louis, A. G. MacDiarmid, C. K. Chiang, and A. J. Heeger. "Synthesis of electrically conducting organic polymers: halogen derivatives of PA,(CH) x." J. Chem. Soc. Chem. Commun. 16 (1977): 578–580.

5. Ito. T., H. Shirakawa, and S. Ikeda. "Simultaneous polymerization and formation of PA film on the surface of concentrated soluble Ziegler-type catalyst solution." J. Polym. Sci. Polym. Chem. Ed. 12 (1974) 11–20.

6. Chiang, C. K., M. A. Druy, S. C. Gau, A. J. Heeger, E. J. Louis, A. G. MacDiarmid, Y. W. Park, and H. Shirakawa. "Synthesis of highly conducting films of derivatives of PA,(CH) x." J. Am. Chem. Soc. 3 (1978): 1013–1015.

7. Stafström, S., and K. A. Chao. "Polaron-bipolaron—soliton doping in PA." Phys. Rev. B 4 (1984): 2098.

8. Molapo, K. M., P. M. Ndangili, R. F. Ajayi, G. Mbambisa, S. M. Mailu, N. Njomo, M. Masikini, P. Baker, and E. I. Iwuoha. "Electronics of conjugated polymers (I): polyaniline." Int. J. Electrochem. Sci. 7 (2012) 11859–11875.

9. Heeger, A. J., MacDiarmid, A. G. and Shirakawa, H. "The Nobel Prize in chemistry, 2000: conductive polymers". Royal Swedish Academy of Sciences, Stockholm, 2000, 1–16.

10. Roth, S., and D. Carroll. "One-dimensional metals: conjugated polymers, organic crystals, carbon nanotubes and graphene". John Wiley & Sons, Weinheim, 2015.

11. Bredas, J.-L., B. Themans, J. M. Andre, R. R. Chance, and R. Silbey. "The role of mobile organic radicals and ions (solitons, polarons and bipolarons) in the transport properties of doped conjugated polymers." Synth. Met. 2 (1984): 265–274.

12. Chiang, C. K., S. C. Gau, C. R. Fincher Jr, Y. W. Park, A. G. MacDiarmid, and A. J. Heeger. "PA,(CH) x: n-type and p-type doping and compensation." Appl. Phys. Lett. 1 (1978): 18–20.

13. Shirakawa, H., and T. Kobayashi. "Chemical doping of PA by transition metal halides." J. Phys. Colloq. 3 (1983): 3–3.

14. MacDiarmid, A. G., and A. J. Epstein. "Secondary doping in polyaniline." Synth. Met. 3 (1995): 85–92.

15. Heeger, A. J., S. Kivelson, J. R. Schrieffer, and W-P. Su. "Solitons in conducting polymers." Rev. Mod. Phys. 3 (1988): 781.

16. Garnier, F., R. Hajlaoui, A. Yassar, and P. Srivastava. "All-polymer field-effect transistor realized by printing techniques." Science 265 (1994): 1684–1686.

17. Schön, J. H., A. Dodabalapur, Z. Bao, C. Kloc, O. Schenker, and B. Batlogg. "Gate-induced superconductivity in a solution-processed organic polymer film." Nature 410 (2001): 189–192.

18. Dai, L., J. Lu, B. Matthews, and A. W. H. Mau. "Doping of conducting polymers by sulfonated fullerene derivatives and dendrimers." J. Phys. Chem. B 21 (1998): 4049–4053.

19. Ragupathy, D., S. C. Lee, S. S. Al-Deyab, and A. Rajendran. "Simple and rapid synthesis of polyaniline microrods and its electrical properties." J. Ind. Eng. Chem. 4 (2013): 1082–1085.

20. Kaneto, K., K. Yoshino, and Y. Inuishi. "Electrical and optical properties of polythiophene prepared by electrochemical polymerization." Solid State Commun. 5 (1983): 389–391.

21. Ambade, R. B., S. B. Ambade, N. K. Shrestha, R. R. Salunkhe, W. Lee, S. S. Bagde, J. H. Kim, F. J. Stadler, Y. Yamauchi, and S.-H. Lee. "Controlled growth of polythiophene nanofibers in TiO_2 nanotube arrays for supercapacitor applications." J. Mater. Chem. 1 (2017): 172–180.

22. Deronzier, A., and J.-C. Moutet. "Polypyrrole films containing metal complexes: syntheses and applications." Coordin. Chem. Rev. 147 (1996): 339–371.

23. Lakshmi, M. S., S. M. Wabaidur, Z. A. Alothman, and D. Ragupathy. "Novel 1D polyaniline nanorods for efficient electrochemical supercapacitors: a facile and green approach." Synth. Met. 270 (2020): 116591.

24. Pan, C.-y., Z.-h. Chen, Y.-l. Huang, and K.-l. Huang. "Effects of water phase concentration on the emulsion polymerization of polyaniline." J. Cent. South Univ. 8 (2001): 140–142.

25. Kumar, R., S. Singh, and B. C. Yadav. Conducting polymers: synthesis, properties and applications. Int. Adv. Res. J. Sci. Eng. Technol. 2 (2015) 110–124.

26. Tarmidzi, F. M., and S. B. Sasongko. "Highly conductive PEDOT: PSS flexible film with secondary doping and spray pyrolysis method." IJAER 13 (2018): 10234–10239.

27. He, Y., X. Han, Y. Du, B. Song, P. Xu, and B. Zhang. "Bifunctional nitrogen-doped microporous carbon microspheres derived from poly (o-methylaniline) for oxygen reduction and supercapacitors." ACS Appl. Mater. Interfaces 6 (2016): 3601–3608.

28. Arefi, F., V. Andre, P. Montazer-Rahmati, and J. Amouroux. "Plasma polymerization and surface treatment of polymers." Pure Appl. Chem. 5 (1992): 715–723.

29. Wu, F., Z. Xu, Y. Wang, and M. Wang. "Electromagnetic interference shielding properties of solid-state polymerization conducting polymer." Rsc. Adv. 73 (2014): 38797–38803.

30. Naveen, M. H., N. Ganesh Gurudatt, and Y.-B. Shim. "Applications of conducting polymer composites to electrochemical sensors: a review." Appl. Mater. Today 9 (2017): 419–433.
31. AL-Refai, H. H., A. A. Ganash, and M. A. Hussein. "Polythiophene and its derivatives–based nanocomposites in electrochemical sensing: a mini review." Mater. Today Commun. 26 (2021): 101935.
32. Sanchez-Sanchez, A., I. del Agua, G. G. Malliaras, and D. Mecerreyes. "Conductive poly (3,4-ethylenedioxythiophene) (PEDOT)-based polymers and their applications in bioelectronics." Smart Polymers and Their Applications, Woodhead Publishing, Oxford (2019): 191–218.
33. Nandapure, B. I., S. B. Kondawar, M. Y. Salunkhe, and A. I. Nandapure. "Magnetic and transport properties of conducting polyaniline/nickel oxide nanocomposites." Adv. Mater. Lett. 2 (2013): 134–140.
34. Yan, F., G. Xue, J. Chen, and Y. Lu. "Preparation of a conducting polymer/ferromagnet composite film by anodic-oxidation method." Synth. Met. 123 (2001): 17–20.
35. Wu, C.-Q., and X. Sun. "Nonlinear optical properties of conducting polymers." Synth. Met. 2 (1991): 3213–3216.
36. Machado, J. M., M. A. Masse, and F. E. Karasz. "Anisotropic mechanical properties of uniaxially oriented electrically conducting poly (p-phenylene vinylene)." Polymer 11 (1989): 1992–1996.
37. Ting, T. H., Y. N. Jau, and R. P. Yu. "Microwave absorbing properties of polyaniline/multi-walled carbon nanotube composites with various polyaniline contents." Appl. Surf. Sci. 7 (2012): 3184–3190.
38. Phang, S. W., M. Tadokoro, J. Watanabe, and N. Kuramoto. "Microwave absorption behaviors of polyaniline nanocomposites containing TiO_2 nanoparticles." Curr. Appl. Phys. 3 (2008): 391–394.
39. San, F. G. B., and I. Isik-Gulsac. "Effect of surface wettability of polymer composite bipolar plates on polymer electrolyte membrane fuel cell performances." Int. J. Hydrog. Energy 10 (2013): 4089–4098.
40. Dubey, N., C. S. Kushwaha, and S. K. Shukla. "A review on electrically conducting polymer bionanocomposites for biomedical and other applications." Int. J. Polym. Mater. 11 (2020): 709–727.
41. Rivers, T. J., T. W. Hudson, and C. E. Schmidt. "Synthesis of a novel, biodegradable electrically conducting polymer for biomedical applications." Adv. Funct. Mater. 1 (2002): 33–37.
42. da Silva, A. C., and S. I. Córdoba de Torresi. "Advances in conducting, biodegradable and biocompatible copolymers for biomedical applications." Front. Mater. Sci. 6 (2019): 98.
43. Zare, E. N., P. Makvandi, B. Ashtari, F. Rossi, A. Motahari, and G. Perale. "Progress in conductive polyaniline-based nanocomposites for biomedical applications: a review." J. Med. Chem. 1 (2019): 1–22.
44. Kliem, S., M. Kreutzbruck, and C. Bonten. "Review on the biological degradation of polymers in various environments." Materials 20 (2020): 4586.
45. Namsheer, K., and C. S. Rout. "Conducting polymers: a comprehensive review on recent advances in synthesis, properties and applications." RSC. Adv. 10 (2021): 5659–5697.
46. Madaswamy, S. L., S. M. Wabaidur, M. R. Khan, S. C. Lee, and R. Dhanusuraman. "Polyaniline-graphitic carbon nitride based nano-electrocatalyst for fuel cell application: a green approach with synergistic enhanced behaviour." Macromol. Res. 6 (2021): 411–417.
47. Ragupathy, D., A. I. Gopalan, K.-P. Lee, and K. M. Manesh. "Electro-assisted fabrication of layer-by-layer assembled poly(2,5-dimethoxyaniline)/phosphotungstic acid modified electrode and electrocatalytic oxidation of ascorbic acid." Electrochem. Commun. 4 (2008): 527–530.
48. Lakshmi, M. S., E. Muthusankar, S. Mohammad Wabaidur, Z. Abdullah Alothman, V. Kumar Ponnusamy, and D. Ragupathy. "Development and characterization of polydiphenylamine/CuO nanohybrid electrode and its improved electrochemical properties." Sens. Lett. 1 (2020): 5–11.
49. Froyer, G., J. Y. Goblot, J. L. Guilbert, F. Maurice, and Y. Pelous. "Poly (para phenylene): some properties related to the synthesis method." J. Phys. Colloq. 44 (1983): 745–748.

4 Chemistries and Biodegradability of Conducting Polymers

Shuo Chen and Zhengwei You
State Key Laboratory for Modification of Chemical Fibers and Polymer Materials, Shanghai Belt and Road Joint Laboratory of Advanced Fiber and Low Dimension Materials, College of Materials Science and Engineering, Donghua University, Shanghai, PR China

CONTENTS

4.1   INTRODUCTION

A conceptual change has been seen from bio-stable materials to biodegradable materials for electronics, biomedicine other related applications in recent years [1–4]. There are two main reasons biodegradable materials receive more attention than bio-stable materials [5]. First is waste management, especially electronics waste (e-waste). Over 50 million metric tons of e-waste were thrown away globally every year [3]. Few amounts of the e-waste are recycled, while most of them go to landfills and incineration. It would be great damage to the environment if the toxic and harmful ingredients were released, e.g., lead, cadmium, chromium, or polychlorinated biphenyls. Another driving force is the biocompatibility concerns of the long-time implanted materials as well as the ethical and technical concerns induced by the revision surgeries [4]. The biodegradable implants that could degraded and be metabolized by the body after completing their functions would avoid the secondary removal surgery and reduce the probability of the occurrence of infection.

Compared with inorganic materials, the mechanical properties of polymers could better mimic the properties of biological soft tissues, enabling conformal touch with tissue surface [6]. Besides, the electronic, mechanical, and biodegradable properties of polymers also could be readily tuned by modifying their chemical structures. Thus, this chapter would focus on polymers as electronics materials for biodegradable applications. The enzymatically degradable natural polymers (e.g., cellulose, collagen) have been brought into use thousands of years ago. However, the use of synthetic biodegradable polymers is later than natural polymers, which began in the 1960s. In the

DOI: 10.1201/9781003205418-4

21st century, the significant development of the technologies of the internet consequentially increases the relationship between people and electronics. Meanwhile, the rapid upgrading and renewing of electronics (months generation) lead to serious e-waste, which poses an urgent need for biodegradable electronics. Biodegradable electronics is an important class of "transient electronics" that could be totally or partially absorbed after use under mild environmental or physiological conditions, which have been proven great potential for e-waste management and biomedical applications [6, 7].

However, progress in biodegradable electronics has largely been hindered by the sluggish development of electronics materials, partly due to the unique challenges in the balance of degradability and other chemicals, physical, and biological properties. Important properties of biodegradable polymers are summarized as follows [4]: (a) The degradation component of materials should have no side effect on the environment. For the implanted applications, the degradation component should not raise a persistent inflammatory or toxic reaction and be able to be metabolized and cleared from the body; (b) The degradation time of the materials should match the functional duration of the devices; (c) The materials should have acceptable shelf time and a controllable degradation rate; (d) The material should have suitable mechanical properties and the processibility for the indicated application. In general, the construction and regulation of biodegradability properties is the key point for the design and preparation of biodegradable polymers.

The degradability of the polymers is affected by some inherent properties, such as the chemical structure, molecular weight, hydrophilicity, degradations and erosion mechanism. Considered the complexity of constructing biodegradable materials and the narrow range of biodegradable materials are currently used, it is necessary to specifically design and synthesize polymers that can be appropriately matched to the specific requirements of each application. Investigating the relationship between degradation mechanisms and the chemical structures involved in biodegradable materials is important for materials' structure design. In this chapter, we discuss the mechanism and influence factors of materials degradation; the representative degradable conductive and semiconductive polymers as well as their preliminary applications.

4.2 THE BIODEGRADATION OF POLYMERIC MATERIALS

Based on the specific degradation initiation methods and mechanism, the polymer degradation could broadly be divided into chemical, mechanical, thermal, and biodegradation [8]. For chemical degradation, the active reagents in the environment tend to react with polymer materials to change their chemical structures. Mechanical degradation is regarded as the molecular fracture induced by mechanical stress. These stresses contain various mechanical forces especially during processing, e.g., compression, shear, or tension forces. Thermal degradation refers to the decomposition of the polymers in a high-temperature environment. Unlike the former three kinds of degradation, biodegradation features an environmentally friendly method of degrading polymers, which shows the great potential of green and sustainable development.

Mikos and Temenoff defined the "biodegradation" of materials as the "chemical breakdown of a material mediated by any component of the physiological environment (e.g., water, ions, proteins, cells, bacteria), into smaller constituents of low molecular weight products which are then processed, resorbed or cleared by the body" [3]. It means that the biodegradation process of polymers involves chemical or biological breakdown into smaller fragments that could be dissolved or metabolized. Polymeric biomaterials used *in vivo* would contact the body fluid in the biological environment and be degraded by the enzymes with the mechanism of either hydrolysis or oxidation. The polymers in the environment are degraded mainly by the metabolic activity of microbial communities. Regardless of the surrounding environment, the biodegradation process is always first fragmented into small fractions with the help of macrophage or biodegrading microorganisms. Subsequently, the backbones of polymer are cleaved by the hydrolytic or the oxidation effect, resulting in the gradual reduction of the molecular weight.

Biodegradable polymeric materials contain two major categories: natural polymeric materials and synthetic polymeric materials [4]. The natural-derived materials, such as polysaccharides derived from plants (e.g., cellulose, alginate) and polymers derived from animals (e.g., silk, collagen, chitosan), have been extensively used as biodegradable polymeric materials due to their enzymatic degradability as well as biocompatibility. However, these natural polymers have their downsides, including the complexities of purification, batch-to-batch variation, the risk of disease transmission, and the immunogenic response. In contrast, synthetic polymers are usually biologically inert, with predictable and tailored physical properties, which are widely used as drug delivery vehicles and tissue engineering scaffolds. The biomaterials discussed in this chapter can be classified into hydrolytically degradable polymers and oxidized degradable polymers [28]. This chapter focuses on the degradation modes of the corresponding polymers.

4.2.1 Hydrolytic Degradation

Polymers that have hydrolytically unstable chemical bonds in their molecular backbone are often referred to as hydrolytically degradable polymers. Hetero-chain polymers, especially those containing oxygen and/or nitrogen atoms in the backbone, are preferred to be hydrolyzed [2]. The cleavage of these chemical bonds is induced by water, which could be catalyzed by bases, acids, or enzymes. The determinant factors of polymer's hydrolytic degradation are the chemical structures and the surrounding environment. As shown in Figure 4.1, the common groups prefer to hydrolysis contain

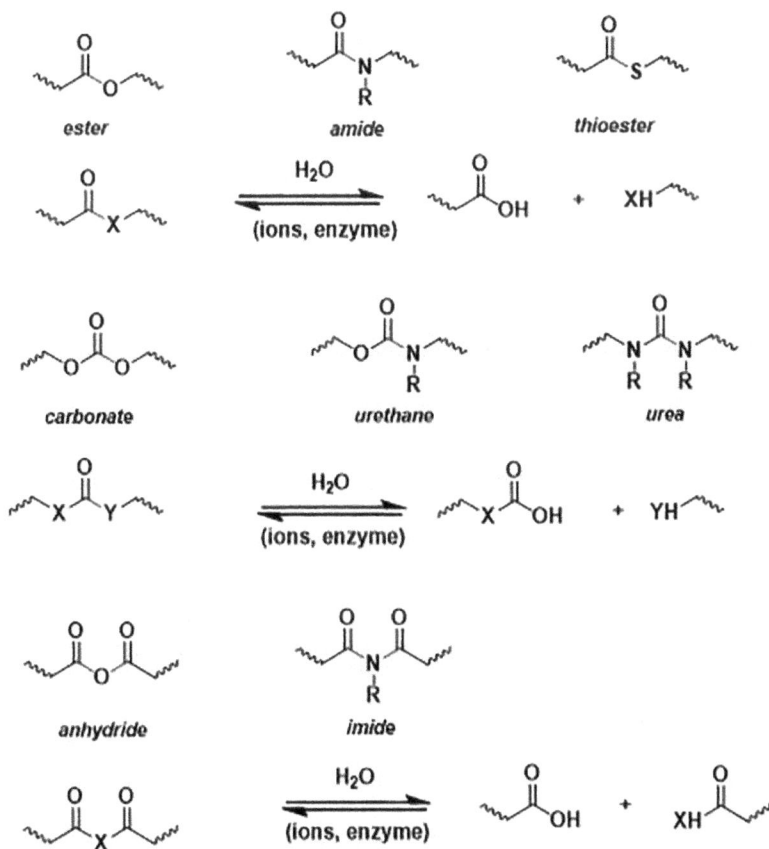

FIGURE 4.1 The hydrolyzable groups containing C=O bonds and the possible chemistry reactions. The hydrolyzable bonds are marked with red color. (Adapted with permission from Ref. [2]. Copyright (2008) Elsevier.)

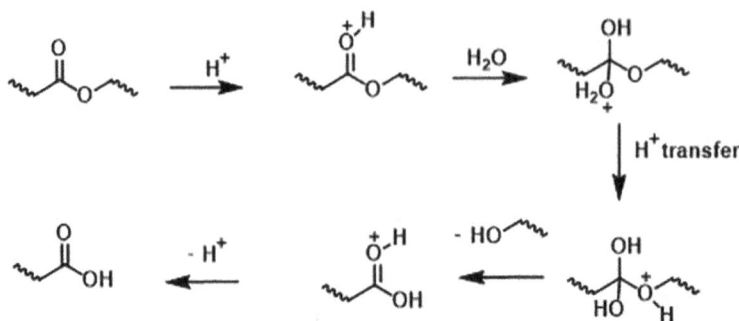

FIGURE 4.2   The mechanism for acid-catalyzed ester bonds hydrolysis, which is cleaved to yield carboxylic acid and alcohol. (Adapted with permission from Ref. [2]. Copyright (2008) Elsevier.)

esters, thioesters, carbonates, urethanes, anhydrides, amides, etc. [2]. These functional groups generally contain C=O bonds linked to other hetero-elements (oxygen, nitrogen, sulfur) and provide cleavage sites of the molecular backbone in biologically benign environments. Another category of functional groups, such as ethers, sulfonates, polyphosphonates, and cyanoacrylates are also readily hydrolyzed by a catalyzed acid or base.

The rate of the hydrolysis degradation process is affected by multiple factors. The primary determinant of the hydrolysis rate is the chemical structures of hydrolyzable groups in the materials. The properties including hydrophilicity and crystallinity would influence the hydrolysis rate by determining the rate of water uptake, while the surface area would determine the real contact area between materials and the surrounding environment. Generally, high hydrophilicity, low crystallinity, less-crosslinked structure, and high surface area (porous materials or materials with a rough surface) would effectively increase the hydrolysis rate of materials.

Except for the intrinsic factors of materials, the surrounding environment also greatly influences the rate of hydrolysis degradation. Previous reports investigate the catalysis of anions and cations in the surrounding environment on hydrolysis [9]. The biological and natural fluids typically contain ions, e.g., H^+, Na^+, K^+, Mg^{2+}, Ca^{2+}, OH^-, HCO_3^-, Cl^-, $H_2PO_4^{2-}$, and SO_4^{2-}. Salt diffusion into polymers is related to the hydrophilicity of the materials. For the hydrophobic materials, ion-mediated hydrolysis is limited on the polymer surface. For the hydrophilic materials, the salts can penetrate with water into the deeper layers of materials. The hydrolysis of ester and amide bonds may be catalyzed considerably by the pH of salt and the environment (Figure 4.2). Previous studies demonstrated the catalytic effect of phosphate buffer saline (PBS) on hydrolytic biodegradation of polycaproamide and polyglycolide [10]. Phosphate ions are particularly effective hydrolytic degradation catalysts for polymers containing carbonyl groups. Enzymes can greatly promote hydrolytic degradation both *in vivo* and *in vitro* [10]. Because the large size of enzyme molecules hinders them penetrate deeply into the material, they usually participate in degradation from the surface. In general, utilizing hydrolyzable groups as linkages to construct electronic biodegradable polymers degrading in physiological, aqueous conditions is thus far the most common method. Comprehensively considering the influence factors, the degradation times of the materials can be readily adjusted within a range of a few days to several years for specific applications.

4.2.2   Oxidative Degradation

Oxidation is a biologically relevant method of the degradation of polymers, by which the free radicals generated chemically and enzymatically directly induce the degradation of the polymer. As shown in Figure 4.3, the sites that contain the carbon substituted by allylic, ethers, phenols, alcohols,

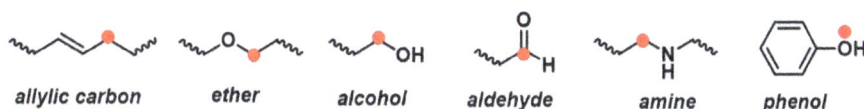

FIGURE 4.3 Chemical structures of moieties susceptible to oxidative. The red circle represents a susceptible point. (Adapted with permission from Ref. [2]. Copyright (2008) Elsevier.)

aldehydes, and amines group are the most likely to degrade by oxidation [2]. As is well-known that the process of oxidative degradation includes three stages: initiation, propagation, and termination [11]. The absorbed energy from an external source would trigger the initiation reaction, by causing the cleavage of the covalent bonds. In the homo-chain polymers, the C–C bonds are the cleaved sites. In the polyethers and polyesters, the C–O bonds tend to start the initiation route. After the initiation reaction started, the propagation may take place by unzipping. The cleavage of covalent bonds would generate radicals, which could be transferred to another chain or abstract the neighboring hydrogen atoms to generate new radicals in the same chain.

The triggered energy for oxidative degradation of the hydrolytic stable polymers ranges from 30 to 90 kcal/mol [11]. Thus, the reaction usually need an external energy source, such as heat, light, or radiation. It seems that the oxidative degradation of polymers would not happen in the human body. On the contrary, plenty of the implanted polymers would suffer oxidative degradation *in vivo*. The implantation of materials would trigger an immune response in the body. The activated phagocytes, such as polymorphonuclear neutrophils and macrophages, swallow and try to digest the foreign materials. Studies have found that the macrophages and neutrophils can release reactive oxygen species (ROS) around the implanted materials during the phagocytosis process [12, 13], which is caused by the one-electron reduction of oxygen to superoxide anion (O^{2-}) catalyzed by the nicotinamide adenine dinucleotide phosphate (NADPH) oxidase and using NADPH as substrate:

$$2O_2 + NADPH \rightarrow 2O_2^- + NADP^- + H^+$$

Specific enzymes (e.g., peroxysome) are capable to transfer the O_2^- to hydrogen peroxide (H_2O_2) by the superoxide dismutase:

$$2O_2^- + 2H^+ \rightarrow H_2O_2 + O_2$$

These substances are relatively harmless on their own but can be converted to hydroxyl radicals (OH•) in the presence of iron or other transition metal catalysts and trigger oxidation on the polymer surface [11]. Besides, the potent oxidant hydroxyl radicals also play an important role as protectors to eliminate the invading microorganisms by the phagocytes. Hydrogen peroxide is converted to hypochlorous acid (HOCl) by myeloperoxidase (MPO) produced by lysosomes in neutrophils. The HOCl could oxidize the nitrogen functional groups (amide, urethane) and possibly break these bonds [14].

4.3 THE BIODEGRADABLE CONDUCTING POLYMERS

Depending on the electric conductivity in electronic devices, polymers are classified into two groups: insulators and conductors. The insulating materials usually serve as the substrate or dielectrics components in organic electronics, while the conjugated polymers function as the semiconductors or conductors. Conjugated polymers can be either semiconducting or conducting, depending on their Fermi level. Owing to the decent electronic conductivity after doping, the utility of possessing both electronic conduction, and the higher mechanical flexibility, the organic conductive materials gain prominence for the fabrication of flexible and conformal electronic devices.

4.3.1 Biodegradable Conjugated Conducting Polymers

The conducting polymers generally act as electrodes, which are responsible for device interconnects or contacts with the external circuit. The conjugated polymers with a doped conducting state are the representative conducting materials. Figure 4.4 depicts the common conjugated polymers includes polypyrrole (Ppy), poly(3,4-ethylenedioxythiophene) (PEDOT), polyaniline (PANi) and so on [1]. These polymers are not only widely used in an electronic circuit to connect different components but also collect electrical signals of body tissues, such as neurons and cardiac tissues due to their stable doped and conducting states.

However, the ordered and rigid conjugated regions are the basic demand for electronic conduction in conducting polymers. As the consequence, the conductive polymers often exhibited brittle and stiff mechanical performance and are seldom biodegradable. Two strategies are developed to fabricate the biodegradable conducting polymers [5]. One is to mix conjugated conducting polymers with flexible, biodegradable, and insulating polymers, by which the excellent electrical properties and mechanical properties can be accomplished at the same time. Conductive composites fabricated in this way can disintegrate in mild conditions based on the biodegradation of the host polymers, even though the conjugated polymers cannot be completely broken down into monomers. Due to the nondegradation of conjugated polymers, it is necessary to maximize the conductivity while keeping the conjugated polymers at a relatively lower concentration. However, the concentration of blending conjugated polymers should be higher than the percolation threshold to form the conductive networks in the composites. Another strategy is to achieve the degradation by breaking the conjugated region in the polymer backbone, where flexible and biodegradable linkages are introduced in the molecular backbones. This strategy has been widely used to improve the physical properties (e.g., processability and mechanical properties) and endow novel properties (e.g., self-healing) of conducting polymers without compromising the electronic performance. Analogous to the first strategy, the short conjugated regions in copolymers also show decent charge transport if the conjugated segments could form stable and continuous networks. However, the conducting copolymers prepared with this method exhibit inferior conductivities but better biodegradability compared with the blending composites counterparts. The conjugated segments in the conducting copolymers could degrade into oligomers during the degradation process, which could be considered biocompatible degradation products. Because the sizes of these oligomer fragments are small, which could be phagocytized by macrophages and be eliminated from the body relied on the body's innate immune response, the copolymer conductors have shown great promise for biomedical applications, such as implanted sensors and electrodes.

4.3.1.1 Polythiophene

Polythiophene is widely used due to its stability and high conductivity (10^3 S/cm), which changes with the dopant and polymerization [15]. Previously reported studies have confirmed that the oligomers with 11 thiophene units exhibit analogous conductivity with those of higher molecular weight polythiophene [16]. Very early, Rivers and coworkers reported a kind of biodegradable polymer

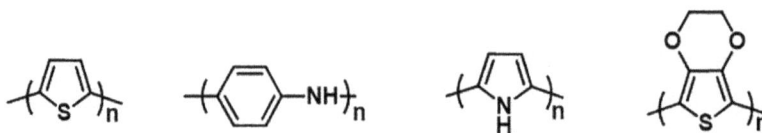

FIGURE 4.4 The chemical structure of common conductive polymers.

made from conducting oligomers of pyrrole and thiophene connected via aliphatic chains connected by ester bonds, as shown in Figure 4.5a [17]. Iodine was used as the doping agent. The conducting copolymer showed conductivities of 10^{-4} S/cm and exhibited a surface erosion degradation process in PBS solution with the esterase at 37°C. *In vitro* and *in vivo* biocompatibility tests demonstrated that no toxic effect of the conducting polymers or their degraded byproducts. Guimard and coworkers synthesized conducting copolymers with alternating biodegradable ester units and electroactive quaterthiophene units [18]. The nontoxic ions $FeCl_3$ and $Fe(ClO_4)_3$ were used as doping agents. The

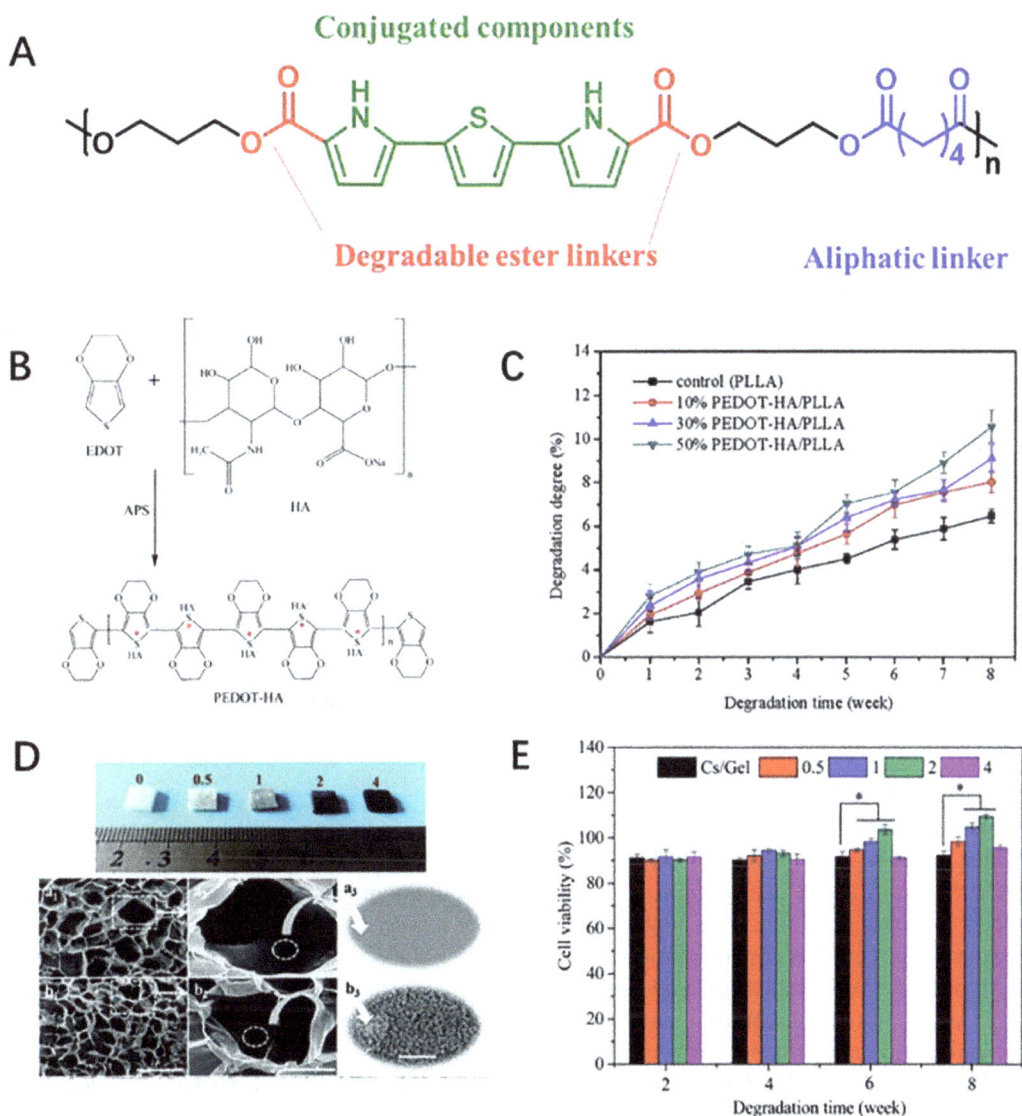

FIGURE 4.5 (a) Schematic diagram of the degradable conductive polymer. (b) Schematic diagram for chemical oxidative polymerization of EDOT with hyaluronic acid (HA). (c) *In vitro* biodegradation of PEDOT-HA/PLLA films. (d) Image of PEDOT/Cs/Gel scaffolds polymerized with different molar ratios of ammonium persulfate (APS) to EDOT (top). SEM images (bottom) of the Cs/Gel scaffold and PEDOT/Cs/Gel scaffold. a: Cs/Gel, b: 2PEDOT/Cs/Gel. (e) Cytotoxicity of conductive scaffolds during biodegradation. The degradation solution was obtained to culture PC12 cells (4×10^5 cells/mL) for 48 h. *$P < 0.05$. (Adapted with permission from Refs. [17, 22, 23]. Copyright (2002) Wiley, (2016) Elsevier, (2013) Royal Society of Chemistry.)

redox activity using cyclic voltammetry and UV–vis spectroscopy were performed to confirm the doping and electroactivity of the copolymers. The copolymers also processed a surface erosion degradation within 1–2 weeks with the presence of cholesterol esterase. The *in vitro* biocompatibility results have confirmed the compatibility of the conducting copolymers, leading to the potential as a biodegradable conducting polymer for biomedical applications.

Poly 3,4-ethylenedioxythiophene (PEDOT), the most common polythiophene derivative, has been proved as a biocompatible conductive polymer for biomedical applications [19]. PEDOT can be prepared by electrochemical polymerization or chemical-oxidative polymerization. In comparison to other conductive polymers, PEDOT shows excellent chemical and thermal stability. Thus, PEDOT has been proven great potential as an interfacing agent for collecting biological signals. The electrical properties and morphologies of PEDOT polymers are influenced by the counterion incorporation [20]. The most used counterions include inorganic salts (e.g., sodium chloride [NaCl], lithium perchlorate [LiClO$_4$], and ions in PBS) and organic salts (e.g., poly(sodium 4-styrene sulfonate)). In addition, the amino acids, polysaccharides, and proteins, are also used as dopants of the PEDOT for specific biomedical applications. The polyanion PSS incorporated into PEDOT could form a uniform and stable aqueous solution, which have been commercialized due to their excellent processability and biocompatibility [21].

Wang and coworkers have reported a conductive composite consisted of poly(L-lactic acid) (PLLA) and PEDOT nanoparticles doped by hyaluronic acid (PEDOT-HA) with a conductivity of 4.7×10^{-3} S/cm, as shown in Figure 4.5b and c [22]. Degradation rate and cytotoxicity of intermediate products of conductive composite films were studied. Due to the increased water penetration resulting from the hydrophilicity of the HA component, the conductive composites showed a relatively rapid degradation rate than pure PLLA films. The extraction solutions had no toxic influence on the proliferation and vitality of cells. *In vitro* cytocompatibility studies confirmed that the conductive composites could better support the adhesion and spread of cells. Additionally, PC12 cells cultured on the conductive composite films exhibited a better neurite outgrowth under the current simulation. Wang and coworkers reported conductive scaffolds that assembled PEDOT on the surface of a porous chitosan/gelatin (Cs/Gel) scaffold via *in situ* interfacial polymerization (Figure 4.5d and e) [23]. The addition of PEDOT improves the conductivity, hydrophilicity, biodegradability, mechanical properties, and thermal stability of the scaffolds. PEDOT/Cs/Gel scaffolds can not only support the adhesion of PC12 cells but also maintain more active cell proliferation and neurite growth.

4.3.1.2 Polyaniline

Polyaniline (PANi) is one of the widely used conjugated conducting polymers on account of its simple preparation, high conductivity and good environmental stability [24]. The conductivity of PANi derives from the π-electron conjugated structure in the molecule chains. Such nonlocalized π-electron conjugated structures could enable p-type or n-type states after doping. Unlike the doping mechanism of other conductive polymers, the number of electrons would not change in the doping process of PANi, but the decomposition of doped protonic acid produces H$^+$ and pairs of anions (such as Cl$^-$, sulfate, phosphate, etc.) into the main chain [25]. The electrons bond with N atoms in the amine and imine groups to form poles or dipoles, delocalizing into the π-bond of the molecular chain, thus endowing the PANi high electrical conductivity. This special mechanism allows the doping and de-doping process of PANi reversible. The doping process depends on many factors, such as pH value, potential, and shows different colors in a different state. Thus, PANi shows electrochemical activity and electrochromic characteristics. These characteristics make PANi appropriate to be used for biomedical applications. Tremendous development has been done in PANi-based biopolymer preparation and application [26].

Ma and coworkers reported biodegradable electron-conducting hydrogels fabricated by grafting polyaniline to gelatin and then *in situ* crosslinking by genipin at body temperature (Figure 4.6a) [27]. The camphor sulfonic acid (CSA) used as a dopant leads to the conductivity of 10^{-4} S/cm. The *in vitro* cytocompatibility tests with mesenchymal stem cells and C2C12 myoblast cells demonstrated supporting of the hydrogel to the adhesion and proliferation of cells. As shown in Figure 4.6b, the hydrogel also exhibited excellent biodegradability in PBS solutions. Thus, the application of drug delivery was demonstrated by using the diclofenac sodium (DS) as a model molecule. Though the PANi-based hydrogels are good candidates as tissue engineering scaffolds, preparing the films is extremely vital for the standard manufacturing process of devices. As shown in Figure 4.6c and d, polycondensation of hydroxyl-capped poly(L-lactide) (PLA) and carboxyl-capped aniline pentamer (AP) has been reported to prepare the electroactive and biodegradable copolymer PLAAP [28]. *In vitro* cytocompatibility tests have shown that the electrical signal stimulation applied to the polymers could make the PC12 cells differentiated faster. Guo and coworkers reported conductive copolymers consisted of AP and polycaprolactone (PCL) with different branched structure [29].

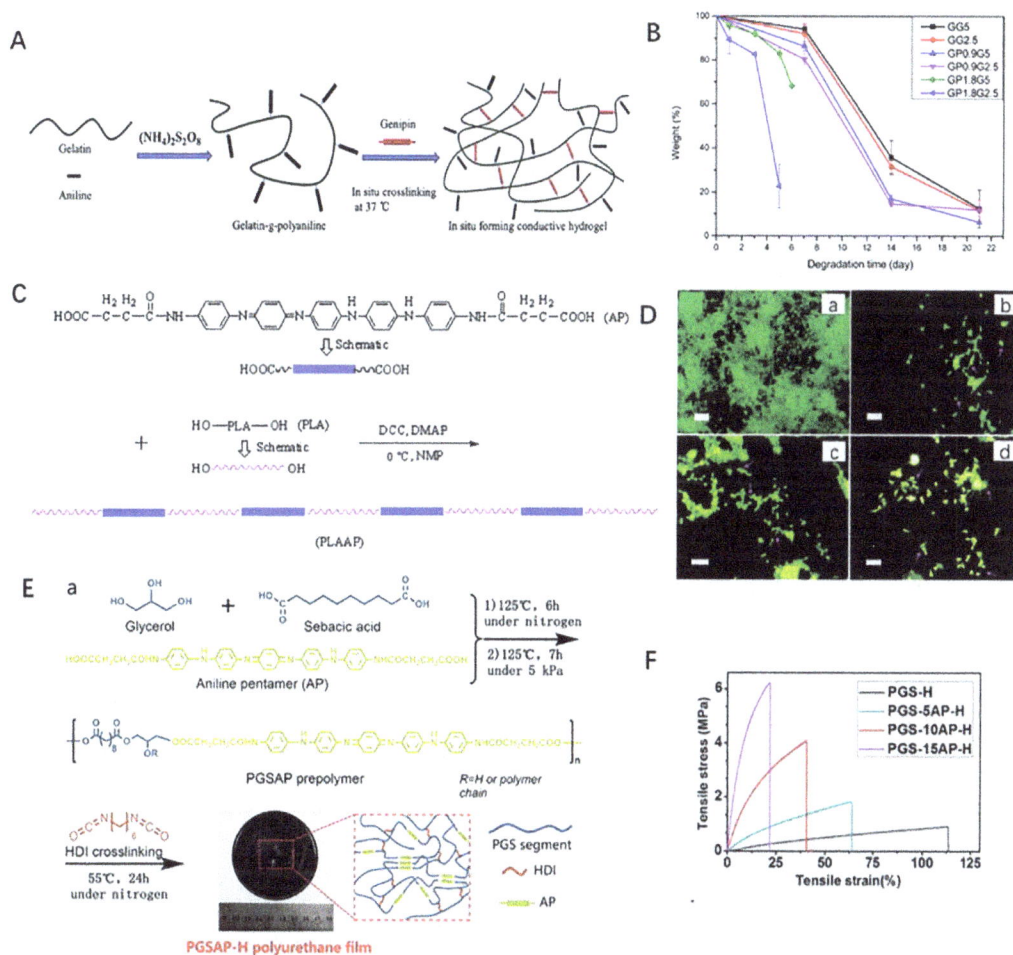

FIGURE 4.6 (a) *In situ* synthesis of preparing gelatin-graft-polyaniline-genipin hydrogels. (b) Degradation profiles of the hydrogels. (c) Schematic diagram of synthesis route and structure of PLAAP copolymer. (d) Representational fluorescence micrographs of PC-12 cell for the substrates (A) TCPS (−) (TCPS without electrical stimulation), (B) TCPS (+) (TCPS exposed to electrical stimulation), (C) PLAAP (−) (PLAAP doped with CSA without electrical stimulation), (D) PLAAP (+) (PLAAP doped with CSA exposed to electrical stimulation) on day 4. Scale bar = 50 μm. (e) Schematic synthesis of PGSAP pre-polymer. (f) Typical strain and stress curves of the PGSAP-H polyurethane. (Adapted with permission from Refs. [27, 28, 30]. Copyright (2010) Royal Society of Chemistry, (2008) American Chemical Society, (2016) Elsevier.)

Authors have demonstrated that compared with the linear structures, the hyperbranched molecular structures are favorable for improving electrical conductivity. However, the conductivity of the aniline-based copolymers still decreased by nearly three orders of magnitude compared with pure AP thin films (10^{-2} S/cm vs. 10^{-5} S/cm). Combining AP with flexible and biodegradable segments would remove the limitation of the mechanical rigidity and nonbiodegradability, which increases their application potential in biomedical applications [5]. For example, Ma and coworkers reported polyurethane with high conductivity and excellent biodegradability through the poly-condensation of poly(glycerol sebacate) (PGS) and AP (Figure 4.6e) [30]. The conductivities of prepared polyurethanes doped with CSA range from 1.4×10^{-6} to 8.5×10^{-5} S/cm. As shown in Figure 4.6f, the flexible segments of PGS provided degradability and elasticity to the copolymers with strength range from 1.7 to 5.3 MPa and strain range from 17% to 55%. The elastic nature involved with conductivity could effectively improve Schwann cells' myelin gene expressions and the corresponding neurotrophin secretion.

4.3.1.3 Polypyrrole

Polypyrrole (Ppy) is a common conductive polymer with heterocyclic conjugated structures [31]. The conductive film could be made by electrochemical oxidation polymerization in an acidic aqueous solution and a variety of organic electrolytes. The oxidants are usually $FeCl_3$, ammonium persulfate, etc. [32]. The overlapping of clouds of π electrons of the double bonds within the molecule are not fixed and can be translocated among the carbon atoms and tend to extend throughout the molecular backbones. Under the electric field, the π-electrons can move along the molecular backbones, by which the Ppy shows excellent conductivity. The conductivity and mechanical strength depend largely on the conditions of polymerization, e.g., anion, solvent, pH value, and temperature. Due to the structure of conjugated chain oxidation and corresponding anion doping, Ppy obtains a conductivity at the order of 10^{-4} to 10^2 S/cm and exhibits electrochemical oxidation–reduction reversibility [33].

Shi and coworkers reported conductive and biodegradable composites consisting of oxidized polypyrrole nanoparticles and poly(D,L-lactide) (PDLLA) using emulsion polymerization in the presence of $FeCl_3$ (Figure 4.7a) [34]. The Ppy nanoparticles aggregated in a matrix of PDLLA to form the conductive network. The resultant Ppy/PDLLA membrane obtains a surface resistivity of 1×10^3 Ω/square with 3% Ppy loading. Authors have demonstrated the stability by performing a biologically meaningful electrical conductivity (100 mV) on the membrane in the environment of cell culture for 1000 h. Additionally, the fibroblasts seeded on the copolymers were upregulated under the stimulation DC, indicating the potential for tissue engineering applications (Figure 4.7b). Runge and coworkers reported a kind of biodegradable composites made from the polymerization of pyrrole in preformed PCLF scaffolds (PCLF-Ppy) [35]. As shown in Figure 4.7c and d, PCLF-Ppy was synthesized under different conditions with various anions, e.g., naphthalene-2-sulfonic acid sodium salt (NSA), dodecylbenzene sulfonic acid sodium salt (DBSA), dioctyl sulfosuccinate sodium salt (DOSS), and lysine to study the impact of composition on conductivity and cellular compatibility. The PCLF-Ppy showed a conductivity of 6 mS/cm with 13.5% polypyrrole. *In vitro* cell test results indicated that PCLF-Ppy prepared with DBSA or NSA could support the attachment and proliferation of the DRG and PC12 cells. Due to the mild polymerization conditions, the Ppy was widely used to endow the biopolymers' conductivity by surface modification technology without damaging their intrinsic morphologies and structures. For instance, an *in situ* oxidative polymerization of Ppy was performed to modified silk fabrics using $FeCl_3$ as a catalyst in the aqueous solutions [36]. The polymerization did not affect the molecular conformation and the intrinsic crystal structure of the silks, but with the improved thermal stability and electrical characteristics. As shown in Figure 4.7e–g, Lee and coworkers reported biodegradable conducting scaffolds prepared by electrospinning poly(lactic-*co*-glycolic acid) (PLGA) and subsequently polymerizing conducting monomers *in situ* [37]. Compared with the uncoated PLGA control samples, Ppy in the scaffolds promoted the proliferation and differentiation of PC12 cells.

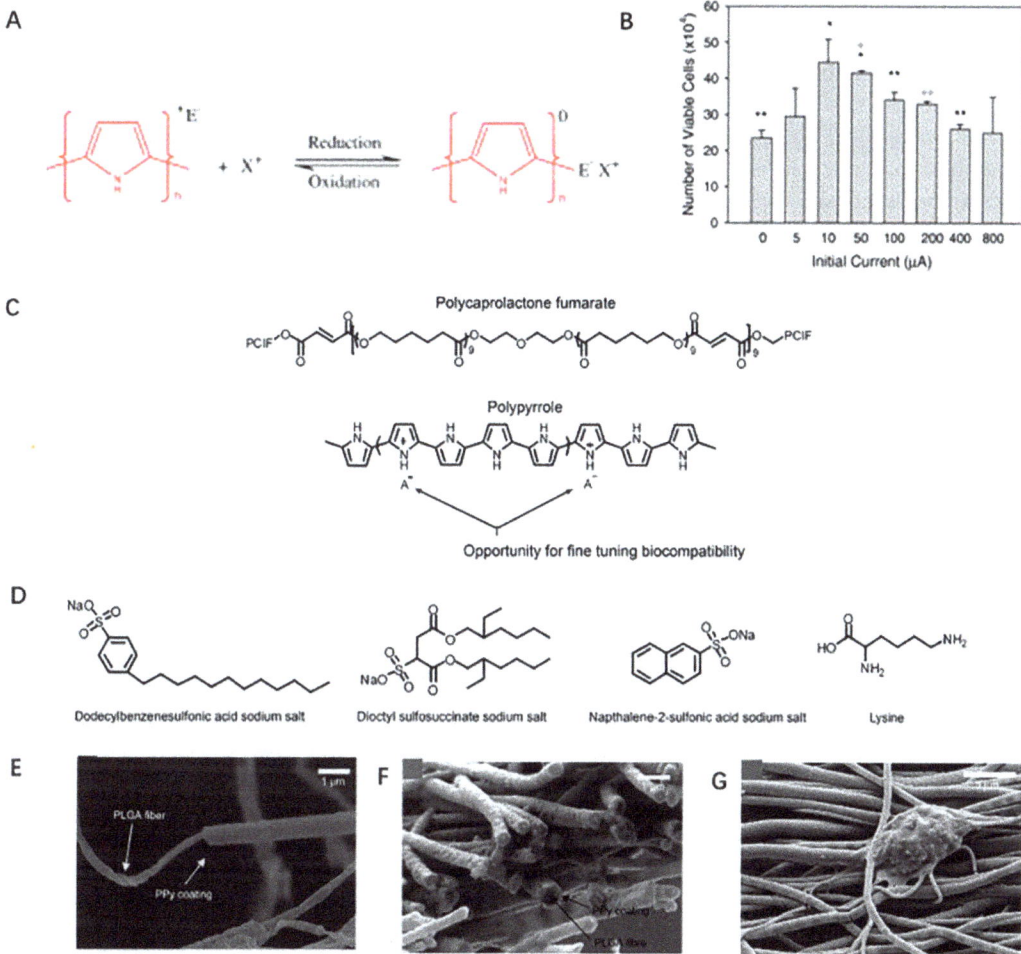

FIGURE 4.7 (a) Schematic diagram of reversible conversion between reduced state and oxidation state of Ppy. (b) The effect of DC stimulation on cell growth showed that medium-range current intensity could upregulate cell proliferation. Significant difference between single and double labeled data pairs (*t*-test, $P < 0.05$; DF = 2–4). (c) Chemical structures of polycaprolactone fumarate and polypyrrole. (d) Anions used in the synthesis of polypyrrole to modify the chemical composition of the resulting PCLF-Ppy scaffolds. (e) SEM micrograph of single strands of Ppy–PLGA fibers. (f) SEM image of section of the Ppy–PLGA meshes. (g) SEM images of PC12 cells cultured on aligned Ppy–PLGA fibers for 2 days. (Adapted with permission from Refs. [34, 35, 37]. Copyright (2004) Elsevier, (2010) Elsevier, (2009) Elsevier.)

4.3.2 BIODEGRADABLE SEMICONDUCTORS

Semiconductors are known as the active materials in electronic devices and enable the switching mechanism of organic transistors by the manageable conductivity. The most important parameter of the semiconducting materials is the charge carrier mobility (μ), which is characterized by the velocity of the free charge moving through the material under the applied electric field [5]. Mobility represents the conductivity of semiconductor and depends on the carrier (electron or hole) concentration in the materials. Mobility is usually presented as $cm^2/V\cdot s$ and could be obtained by testing the performance of the electronics, such as organic field-effect transistors. Most used semiconductive polymers include polythiophenes, such as poly(3-hexylthiophene) (P3HT) [38], and donor–acceptor copolymers, such as diketopyrrolopyrroles-derived polymers [39].

The previous purpose of the development of electronic active polymers was the detection of tissues' electrical signals, which were confined to the *in vitro* cell experiments or simple verification of biocompatibility *in vivo*. The integration and signal processing of electronic devices were inevitably ignored. Thus, there are fewer previous works focus on biodegradable semiconductors. However, with the development of wearable devices, the studies of semiconductors used in wearable electronics developed rapidly. Compared with inorganic semiconductors, the semiconducting polymers show great advantage in wearable electronics due to their flexible mechanical properties, solution processing, and preparation at room temperature. Analogous to the conducting polymers, the challenge of making semiconducting polymers biodegradable also arises from their conjugated molecular structures. To balance the relationship of their rigid, crystallized aggregates and the biodegradable, flexible functional groups is the key to utilize traditional semiconducting polymers in flexible and biodegradable electronic devices. Some works have deepened the understanding of the charge transporting of semiconducting polymers, which may offer inspiration for designing biodegradable semiconducting polymers. Rivnay and coworkers demonstrated that short-range intermolecular aggregation is sufficient for efficient long-range charge transport, which means that the high mobility and flexible nature could be achieved simultaneously in macro-disordered semiconductor polymers with interconnected conjugated aggregates (Figure 4.8a) [40]. After a short while, some works have demonstrated that the composites consisted of conjugated semiconducting polymers and an insulating polymer matrix could retain their high charge carrier mobility as well as the physical properties of the polymers matrix. In principle, tuning the concentration and compatibility of semiconductors in the polymer matrix to form interconnected conjugated areas with efficient charge transportation is crucial for the successful construction of these results. Thus, the strategies used in the preparation of biodegradable conducting polymers are also compatible with the semiconducting polymer.

The biodegradable polymers (e.g., PCL and PLGA) were reported as the biodegradable matrix to semiconducting composites with P3HT [41, 42]. The semiconducting composites were electrospun as the conducting nanofibers for further measurement (Figure 4.8b). The mobility of P3HT/PCL electrospun nanofibers is 1.7×10^{-2} cm^2/V·s, which decreased with the improved concentration of the insulating component. These results have been demonstrated that the success of the construction strategy for biodegradable semiconducting polymers, though the lower mobilities resulting from grain boundaries of the aggregates formed by phase segregation, analogous to the first strategy of conducting polymers. However, the aforementioned examples generally achieved their biodegradability via hydrolyzable linkages like ester bonds of the insulting matrix, by which the semiconductor cannot be completely degraded.

Recent breakthroughs have been made by using hydrolyzable bonds as linkages in the semiconducting polymers conjugated molecular chains to enable complete degradation similar to the second strategy of above conducting biodegradable polymers. Lei and coworkers recently reported completely disintegrable semiconducting polymers for thin-film transistors using hydrolyzable and reversible imine bonds as conjugated linkages between DPP and p-phenylenediamine (Figure 4.8c) [43]. The DPP-based semiconducting polymer showed a high charge carrier mobility (0.34 cm^2/V·s) and solution processing capabilities comparable to traditional conjugated polymers. The resultant semiconducting polymers could degrade in acidic conditions (pH = 4.6) due to the hydrolyzable imine bonds, with the byproducts of aldehyde and amine precursors. As shown in Figure 4.8d and e, completely disintegrable transistors based on the semiconducting polymers consisting of an ultrathin biodegradable substrate (the cellulose with a thickness of 800 nm), Al$_2$O$_3$ as dielectric, and iron as electrodes, were successfully fabricated. The devices could completely degrade within 30 days. Furthermore, the pseudo-complementary metal-oxide semiconductor (CMOS) flexible circuits were demonstrated, proposing the potential of the biodegradable semiconductor polymers transient electronics. It is worth mentioning that the condition of pH 4.6 or lower exists in many biological processes. Therefore, this work developed an innovative strategy to generate recyclable semiconductors, which tend to decompose under mild conditions with almost no impact on the environment.

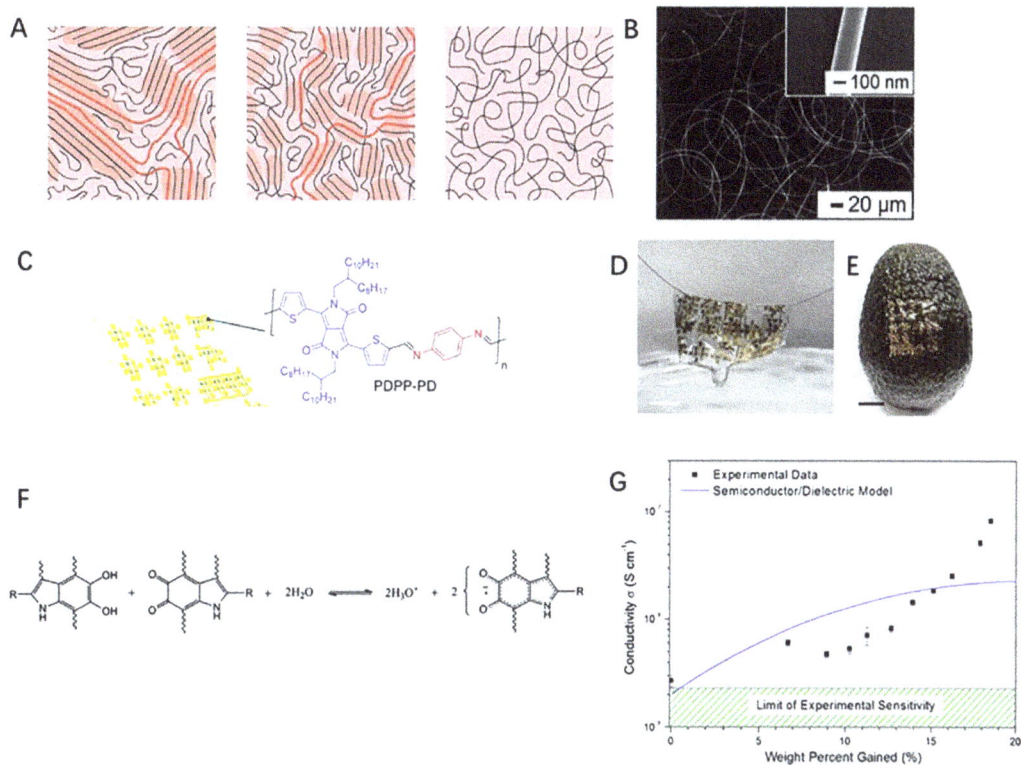

FIGURE 4.8 (a) Schematics of the microstructure of a semi-crystalline polymer film, P3HT (left), disordered aggregates (middle), and a completely amorphous film (right). The darker shadowed areas represent ordered regions. If the molecular weight is high enough and the density of the ordered material is large enough, long polymer chains (red areas) can join the ordered regions without loss of conjugation, greatly improving charge transport. (b) SEM images of blend fibers of P3HT and poly(3-caprolactone) (PCL) (50:50, w/w) using a single nozzle. (c) The chemical structure of PDPP-DP containing hydrolyzable imine bonds under acidic conditions. (d) Disintegrable pseudo-CMOS circuits based on PDPP-PD picked up by a human hair. (Scale bar: 5 mm.) (e) Device was transferred onto the rough surface of an avocado. (Scale bar: 10 mm.) (f) The reversible reaction between melanin and water has been proposed to explain the doping effect of water on melanin. (g) The conductivity of melanin as a function of water content. (Adapted with permission from Refs. [40, 41, 43, 45]. Copyright (2013) Springer, (2009) Royal Society of Chemistry, (2012) National Academy of Sciences, (2012) National Academy of Sciences.)

Except in synthetic materials, the conjugated molecules found in nature also exhibited semiconductivity and were used as active components for biocompatible and biodegradable electronics. The natural pigments with conjugated structure always exhibited electronic behavior. A common example is the bio-based p-conjugated molecule, e.g., carotenoids, which could serve as functional electron donor molecules [44]. These naturally conjugated molecules have low toxicity and exhibit natural degradation behaviors, ensuring the biocompatibility and biodegradability of devices over a wider life cycle. Melanin is another natural pigment with plate-like structures including randomly aggregated oligomers. The charge transport characteristic of melanin was investigated by Mostert and coworkers in 2012 (Figure 4.8f and g) [45]. Melanin is intrinsically insulting but exhibits conductivity through doping via water absorption, which causes a proportional reaction that produces free electrons and protons, making the double ions and electrons conductivity. This special character would extend melanin's usage of promising materials for biomedical fields. Bettinger and coworkers demonstrated the application of hydrated melanin as scaffold materials [46]. The completely hydrated melanin films showed a conductivity of 7×10^{-5} S/cm and could enhance the proliferation of Schwann cells and PC12 neurites *in vitro*. *In vivo* implantation also demonstrated that the melanin could be completely resorbed after 8 weeks. Other notable dye molecules are another category of natural small molecules. Their conjugated structure with massive fused benzenes enables massive π–π stacking. Indigo, produced from plants, is a semiconductor that exhibits superior anisotropic charge transfer property due to intermolecular hydrogen bonds enhanced π-conjugation along the b-axis of the crystal [47]. The indigo-based OFET devices consisting of natural resin shellac as substrate, Al as a gate, and Al_2O_3 as the dielectric were successfully prepared and demonstrated. Though possessing exciting potential, these natural semiconducting materials require in-depth research and characterization (such as the mechanism of charge transport, film formation, and electrochemical properties) for devices with superior performance.

4.4 CONCLUSION

Biodegradable conducting polymers have not only proven as a great solution for e-waste management but also provide the basis for the applications of transient electronic devices in biomedical fields. With the development of material and biological science, biodegradable conducting polymers have become a new type of biomaterials featuring their unique electrical properties. While innovative chemical and processing technologies have led to many applications of natural and synthetic polymers in different biomedical fields, such as tissue engineering, drug delivery, artificial skin, and diagnostic applications, there still is an urgent need for more types of conductive polymers with further improved electrical and mechanical properties. Designing conductive polymers with desired properties remains a challenge, which has demonstrated the basis of biodegradability and biocompatibility to reduce the inflammatory response to host tissue, and has sufficient electrical conductivity and appropriate mechanical properties.

Synthetic materials offer more tunability in physical, chemical, mechanical, and biological properties and, therefore, show a great advantage to fabricate degradable devices for biomedical applications. The design strategies like compositing and copolymerization have been demonstrated availability to fabricate conductors and semiconductors. Despite the great progress of biodegradable electroactive polymers, the trade-off between degradability and electrical performance still is an important and urgent problem. The construction of fully conjugated and degradable semiconductor has demonstrated the possibility of using unconventional degradable bonds as linkage to connect the conjugated monomers. Thus, it is intriguing to choose the suitable degradable and conductive linkage depending on the actual physiological applied environments in future research. The processability of conductive polymers is also challenging due to the difficulty to combine them with the standard electronic device fabrication process. The potential toxicity of the dopants and involved organic solvents would be harmful to the biocompatibility of these materials. Further exploration of biodegradable chemistry would provide the foundation to explore a new class of conductive polymer systems, adding further contributions and great opportunities in the era of biomedical science.

REFERENCES

1. Jadoun, S., Riaz, U. & Budhiraja, V. Biodegradable conducting polymeric materials for biomedical applications: A review. *Medical Devices & Sensors* **4**, e10141 (2021).
2. Mabilleau, G. & Sabokbar, A. In vitro biological test methods to evaluate bioresorbability. in *Degradation Rate of Bioresorbable Materials* (ed. Buchanan, F.) 145–160 (Woodhead Publishing, 2008, Lodon).
3. Cao, Y. & Uhrich, K. E. Biodegradable and biocompatible polymers for electronic applications: A review. *Journal of Bioactive and Compatible Polymers* **34**, 3–15 (2018).
4. Nair, L. S. & Laurencin, C. T. Biodegradable polymers as biomaterials. *Progress in Polymer Science* **32**, 762–798 (2007).
5. Feig, V. R., Tran, H. & Bao, Z. Biodegradable polymeric materials in degradable electronic devices. *ACS Central Science* **4**, 337–348 (2018).

6. Tan, M. J., Owh, C., Chee, P. L., Kyaw, A. K. K., Kai, D. & Loh, X. J. Biodegradable electronics: Cornerstone for sustainable electronics and transient applications. *Journal of Materials Chemistry C* **4**, 5531–5558 (2016).

7. Irimia-Vladu, M. "Green" electronics: Biodegradable and biocompatible materials and devices for sustainable future. *Chemical Society Reviews* **43**, 588–610 (2014).

8. Banerjee, A., Chatterjee, K. & Madras, G. Enzymatic degradation of polymers: A brief review. *Materials Science and Technology* **30**, 567–573 (2014).

9. Zaikov, G. E. Quantitive aspects of polymer degradation in the living body. *Journal of Macromolecular Science, Part C* **25**, 551–597 (1985).

10. Smith, R., Oliver, C. & Williams, D. F. The enzymatic degradation of polymers in vitro. *Journal of Biomedical Materials Research* **21**, 991–1003 (1987).

11. Ali, S. A. M., Doherty, P. J. & Williams, D. F. The mechanisms of oxidative degradation of biomedical polymers by free radicals. *Journal of Applied Polymer Science* **51**, 1389–1398 (1994).

12. Rajagopalan, S., Meng, X. P., Ramasamy, S., Harrison, D. G. & Galis, Z. S. Reactive oxygen species produced by macrophage-derived foam cells regulate the activity of vascular matrix metalloproteinases in vitro. Implications for atherosclerotic plaque stability. *The Journal of Clinical Investigation* **98**, 2572–2579 (1996).

13. Rosen, G. M., Pou, S., Ramos, C. L., Cohen, M. S. & Britigan, B. E. Free radicals and phagocytic cells. *The Federation of American Societies for Experimental Biology Journal* **9**, 200–209 (1995).

14. Locksley, R. M., Wilson, C. B. & Klebanoff, S. J. Role for endogenous and acquired peroxidase in the toxoplasmacidal activity of murine and human mononuclear phagocytes. *The Journal of Clinical Investigation* **69**, 1099–1111 (1982).

15. Kaloni, T. P., Giesbrecht, P. K., Schreckenbach, G. & Freund, M. S. Polythiophene: From fundamental perspectives to applications. *Chemistry of Materials* **29**, 10248–10283 (2017).

16. Hempenius, M. A., Langeveld-Voss, B. M. W., Haare, J. A. E. H., van Janssen, R. A. J., Sheiko, S. S., Spatz, J. P., Möller, M. & Meijer, E. W. A polystyrene–oligothiophene–polystyrene triblock copolymer. *Journal of the American Chemical Society* **120**, 2798–2804 (1998).

17. Rivers, T. J., Hudson, T. W. & Schmidt, C. E. Synthesis of a novel, biodegradable electrically conducting polymer for biomedical applications. *Advanced Functional Materials* **12**, 33–37 (2002).

18. Guimard, N. K. E., Sessler, J. L. & Schmidt, C. E. Toward a biocompatible and biodegradable copolymer incorporating electroactive oligothiophene units. *Macromolecules* **42**, 502–511 (2009).

19. Nezakati, T., Seifalian, A., Tan, A. & Seifalian, A. M. Conductive polymers: Opportunities and challenges in biomedical applications. *Chemical Reviews* **118**, 6766–6843 (2018).

20. Spanninga, S. A., Martin, D. C. & Chen, Z. X-ray photoelectron spectroscopy study of counterion incorporation in poly(3,4-ethylenedioxythiophene). *The Journal of Physical Chemistry C* **113**, 5585–5592 (2009).

21. He, H., Zhang, L., Guan, X., Cheng, H., Liu, X., Yu, S., Wei, J. & Ouyang, J. Biocompatible conductive polymers with high conductivity and high stretchability. *ACS Applied Materials & Interfaces* **11**, 26185–26193 (2019).

22. Wang, S., Guan, S., Wang, J., Liu, H., Liu, T., Ma, X. & Cui, Z. Fabrication and characterization of conductive poly (3,4-ethylenedioxythiophene) doped with hyaluronic acid/poly (l-lactic acid) composite film for biomedical application. *Journal of Bioscience and Bioengineering* **123**, 116–125 (2017).

23. Wang, S., Sun, C., Guan, S., Li, W., Xu, J., Ge, D., Zhuang, M., Liu, T. & Ma, X., Chitosan/gelatin porous scaffolds assembled with conductive poly(3,4-ethylenedioxythiophene) nanoparticles for neural tissue engineering. *Journal of Materials Chemistry B* **5**, 4774–4788 (2017).

24. Bhadra, S., Khastgir, D., Singha, N. K. & Lee, J. H. Progress in preparation, processing and applications of polyaniline. *Progress in Polymer Science* **34**, 783–810 (2009).

25. Chaudhari, H. K. & Kelkar, D. S. Investigation of structure and electrical conductivity in doped polyaniline. *Polymer International* **42**, 380–384 (1997).

26. Zare, E. N., Makvandi, P., Ashtari, B., Rossi, F., Motahari, A. & Perale, G. Progress in conductive polyaniline-based nanocomposites for biomedical applications: A review. *Journal of Medicinal Chemistry* **63**, 1–22 (2020).

27. Li, L., Ge, J., Guo, B. & Ma, P. X. In situ forming biodegradable electroactive hydrogels. *Polymer Chemistry* **5**, 2880–2890 (2014).

28. Huang, L., Huang, L., Zhuang, X., Hu, J., Lang, L., Zhang, P., Wang, Y., Chen, X., Wei, Yen. & Jing, X. Synthesis of biodegradable and electroactive multiblock polylactide and aniline pentamer copolymer for tissue engineering applications. *Biomacromolecules* **9**, 850–858 (2008).

29. Guo, B., Finne-Wistrand, A. & Albertsson, A.-C. Enhanced electrical conductivity by macromolecular architecture: Hyperbranched electroactive and degradable block copolymers based on poly(ε-caprolactone) and aniline pentamer. *Macromolecules* **43**, 4472–4480 (2010).
30. Wu, Y., Wang, L., Guo, B., Shao, Y. & Ma, P. X. Electroactive biodegradable polyurethane significantly enhanced Schwann cells myelin gene expression and neurotrophin secretion for peripheral nerve tissue engineering. *Biomaterials* **87**, 18–31 (2016).
31. Vernitskaya, T. V. & Efimov, O. N. Polypyrrole: A conducting polymer; its synthesis, properties and applications. *Russian Chemical Reviews* **66**, 443–457 (1997).
32. Yussuf, A., Al-Saleh, M., Al-Enezi, S. & Abraham, G. Synthesis and characterization of conductive polypyrrole: The influence of the oxidants and monomer on the electrical, thermal, and morphological properties. *International Journal of Polymer Science* 2018, 4191747 (2018).
33. Kaynak, A., Rintoul, L. & George, G. A. Change of mechanical and electrical properties of polypyrrole films with dopant concentration and oxidative aging. *Materials Research Bulletin* **35**, 813–824 (2000).
34. Shi, G., Rouabhia, M., Wang, Z., Dao, L. H. & Zhang, Z. A novel electrically conductive and biodegradable composite made of polypyrrole nanoparticles and polylactide. *Biomaterials* **25**, 2477–2488 (2004).
35. Runge, B. M., Dadsetan, M., Baltrusaitis, J., Knight, A. M., Ruesink, T., Lazcano, E. A., Lu, L., Windebank, A. J. & Yaszemski, M. J. The development of electrically conductive polycaprolactone fumarate–polypyrrole composite materials for nerve regeneration. *Biomaterials* **31**, 5916–5926 (2010).
36. Cucchi, I., Boschi, A., Arosio, C., Bertini, F., Freddi, G. & Catellani, M. Bio-based conductive composites: Preparation and properties of polypyrrole (PPy)-coated silk fabrics. *Synthetic Metals* **159**, 246–253 (2009).
37. Lee, J. Y., Bashur, C. A., Goldstein, A. S. & Schmidt, C. E. Polypyrrole-coated electrospun PLGA nanofibers for neural tissue applications. *Biomaterials* **30**, 4325–4335 (2009).
38. Bao, Z., Dodabalapur, A. & Lovinger, A. J. Soluble and processable regioregular poly(3-hexylthiophene) for thin film field-effect transistor applications with high mobility. *Applied Physics Letters* **69**, 4108–4110 (1996).
39. Chen, H., Guo, Y., Yu, G., Zhao, Y., Zhang, J., Gao, D., Liu, H. & Liu, Y. Highly π-extended copolymers with diketopyrrolopyrrole moieties for high-performance field-effect transistors. *Advanced Materials* **24**, 4618–4622 (2012).
40. Noriega, R., Rivnay, J., Vandewal, K., Koch, F. P. V., Stingelin, N., Smith, P., Toney, M. F. & Salleo, A. A general relationship between disorder, aggregation and charge transport in conjugated polymers. *Nature Materials* **12**, 1038–1044 (2013).
41. Lee, S., Moon, G. D. & Jeong, U. Continuous production of uniform poly(3-hexylthiophene) (P3HT) nanofibers by electrospinning and their electrical properties. *Journal of Materials Chemistry* **19**, 743–748 (2009).
42. Subramanian, A., Krishnan, U. M. & Sethuraman, S. Axially aligned electrically conducting biodegradable nanofibers for neural regeneration. *Journal of materials science. Materials in medicine* **23**, 1797–1809 (2012).
43. Lei, T., Guan, M., Liu, J., Lin, H., Pfattner, R., Shaw, L., McGuire, F. A., Huang, T., Shao, L., Cheng, K., Tok, J. & Bao, Z., Biocompatible and totally disintegrable semiconducting polymer for ultrathin and ultralightweight transient electronics. *Proceedings of the National Academy of Sciences* **114**, 5107 (2017).
44. Wang, X. F., Wang, L., Wang, Z., Wang, Y., Tamai, N., Hong, Z. & Kido, J. Natural photosynthetic carotenoids for solution-processed organic bulk-heterojunction solar cells. *The Journal of Physical Chemistry C* **117**, 804–811 (2013).
45. Mostert, A., Mostert, B., Benjamin, J. P., Francis, L., Pratt, G. R., Hanson, T. S., Ian, R. G. & Meredith, P. Role of semiconductivity and ion transport in the electrical conduction of melanin. *Proceedings of the National Academy of Sciences* **109**, 8943 (2012).
46. Bettinger, C. J., Bruggeman, J. P., Misra, A., Borenstein, J. T. & Langer, R. Biocompatibility of biodegradable semiconducting melanin films for nerve tissue engineering. *Biomaterials* **30**, 3050–3057 (2009).
47. Irimia-Vladu, M., Głowacki, E. D., Troshin, P. A., Schwabegger, G., Leonat, L., Susarova, D. K., Krystal, O., Ullah, M., Kanbur, Y., Bodea, M. A., Razumov, V. F., Sitter, H., Bauer, S. & Sariciftci, N. S. Indigo – A natural pigment for high performance ambipolar organic field effect transistors and circuits. *Advanced Materials* **24**, 375–380 (2012).

5 Strategies to Synthesize Biodegradable Conducting Polymers

Shagun Kainth,[1] *Piyush Sharma,*[1] *and Pawan Kumar Diwan*[2]

[1]Virginia Tech Center of Excellence in Emerging Materials,
Thapar Institute of Engineering and Technology, Patiala, India

[2]Department of Applied Science, UIET, Kurukshetra University,
Kurukshetra, India

CONTENTS

5.1 INTRODUCTION

Boosted by innovations in material science, the scientific society turned its focus to develop functional materials for futuristic technologies [1]. These materials possess innate properties and demonstrate unique functions of their own. The functional materials emerged as promising in various fields including medical science, biotechnology, energy sector, water treatment, and defense [2]. Rapid expansion in the functional materials based on metal, ceramics, polymers, and organic materials has open new gateways to tune materials for desired applications. Among these materials, conducting polymers (CPs) have gained significant attention due to their exceptional electrical and optical properties [3]. In general, CPs are referred to as any π-conjugated polymer. These polymers possess a backbone of single and double or triple covalent bonds and are capable to transport charges irrespective of their inherent conductivity and charge transport properties. Figure 5.1 demonstrates the backbone structure of CPs, where the π bond favors the electron's delocalization and the σ bond upholds the chain strength. The journey of CPs begins with the breakthrough contribution of Shirakawa, MacDiarmid, and Heeger that renewed the opinion about organic polymers [4]. CPs shares many properties with metals and inorganic semiconductors. The unique coupling of these features within vitro and in-vivo biocompatibility brought new opportunities to develop biodegradable CPs. Therefore, CPs emerged as a potential candidate for biomedical applications and clinical translations [5]. However, several limitations restrict their utilization such as poor solubility and

DOI: 10.1201/9781003205418-5

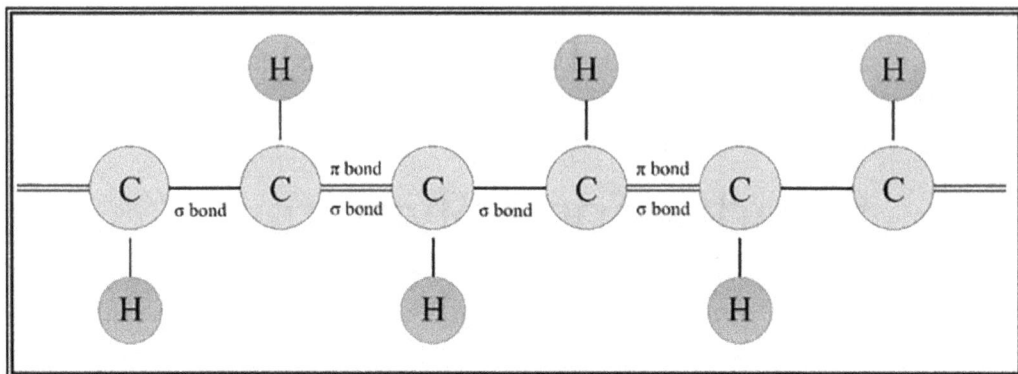

FIGURE 5.1 The backbone structure of CPs.

biocompatibility of conventional CPs [6]. The nonbiodegradable CPs may affect adversely when used for implantation in tissue engineering applications [7]. Moreover, CPs may demonstrate an unwanted inflammatory response due to their incapability to degrade.

Therefore, the scientific community aims to improve the biodegradability of CPs. Numerous efforts have been made to find suitable design and synthesis protocols to develop CPs with superior biodegradability [8]. Currently, there are three synthesis strategies have been opted to obtain biodegradable CPs. Initially, early attempts were made to achieve partial biodegradability in CPs [9]. It involves blending biodegradable polymers (BPs) and CPs to obtain partially biodegradable CP composites. In this strategy, the CPs were blended with BPs, i.e., polylactide (PLA), polycaprolactone (PCL), polyglycolide (PGA), poly(lactic-co-glycolic acid) (PLGA), and polyurethane (PU). Nevertheless, the prepared composites were still far behind the standard material used for real-world biomedical applications. The second strategy includes conducting oligomer-based synthesis of biodegradable CPs with different architectures (linear, star-shaped, hyperbranched, and other complexes). The other synthesis strategy includes modification through monomers or biodegradable monomers and conducting linkers. Such growth in synthesis protocols unveils new possibilities to commercialize CPs for biomedical applications. This chapter provides detailed insight related to the synthesis of BP, CPs and strategies opted to obtain biodegradable CPs. The chapter aims to provide an understanding of various aspects involved during the preparation of biodegradable CPs.

5.2 BLEND OF CONDUCTING POLYMERS AND BIODEGRADABLE POLYMERS

The synthesis of biodegradable CPs based on the direct blending of CPs and BPs is one of the oldest strategies. This strategy is simple to fabricate partially biodegradable CP composites. The foremost benefits of using direct blending are the ease of selecting suitable polymers and varying ratios of polymers to attain desirable conductivity and degradation rate of biodegradable CP composite [9]. This has open new gateways to utilize biodegradable CPs for a wide range of biomedical applications especially tissue engineering and regenerative medicine applications. The direct blending of polymers improves the strength and offers to tune properties of the final product for target application [10]. The extensively studied CPs for the fabrication of biodegradable CPs are polypyrrole (PPy), poly(3,4-ethylene dioxythiophene) (PEDOT), and polyaniline (PANi) (Figure 5.2). PPy is commonly blended with several BPs such as polyglycolide (PGA), poly(l-lactide) (PLA), poly(lactic-co-glycolic acid) (PLGA), poly(d,l-lactide) (PDLA), polyurethane (PU), and polycaprolactone (PCL), to obtain nanosized partial biodegradable CPs [11]. The chemical structure of these CPs and BPs is presented in Figure 5.2.

The partial biodegradable CPs are prepared by using electrospinning, electrochemical, and emulsification polymerization [12, 13]. In the electrospinning polymerization (Figure 5.3a), the liquid

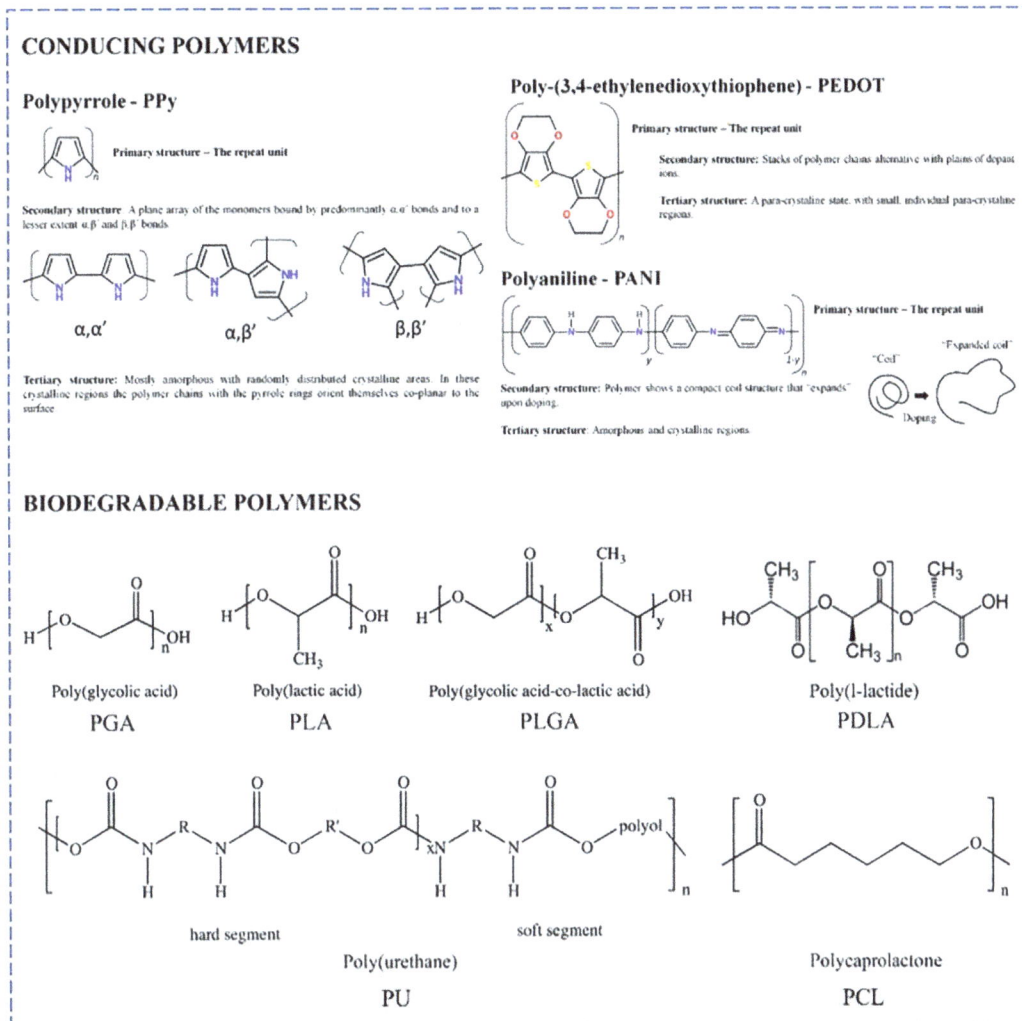

FIGURE 5.2 The chemical structure of conducting polymers: polypyrrole (PPy), polyaniline (PANi), and poly(3,4-ethylenedioxythiophene) (PEDOT). (Adapted with permission from Ref. [11]. Copyright (2021) Copyright the Authors, some rights reserved; exclusive licensee [Elsevier]. Distributed under a Creative Commons Attribution License 3.0 (CC BY) https://creativecommons.org/licenses/by/3.0/) and biodegradable polymers: poly(*d*,*l*-lactide) (PDLA). (Adapted with permission from Ref. [16]. Copyright (2021) Copyright The Authors, some rights reserved; exclusive licensee [MPDI]. Distributed under a Creative Commons Attribution License 4.0 (CC BY) https://creativecommons.org/licenses/by/4.0/), poly(*l*-lactide) (PLA), polyglycolide (PGA), polycaprolactone (PCL), poly(lactic-*co*-glycolic acid) (PLGA), and polyurethane (PU) (Adapted with permission from Ref. [17]. Copyright (2021) Publisher Elsevier).

droplet of a polymeric solution is drawn from the nozzle with the aid of high voltage. The optimum applied voltage charges the liquid droplet and electrostatic repulsion counters the surface tension resulting in stretching of the liquid droplet. The nanosized fiber of the polymeric blend can be obtained when the stretching of the droplet reached a critical point of liquid stream eruption known as the Taylor cone [14]. In addition, electrochemical polymerization (Figure 5.3b) is also used to developed polymeric films. In this method, electrical current is applied through an electrode which is dipped in a mixture of monomers, solvent, and a dopant. Thin-film of polymers of thickness ~20 nm can be deposited on a substrate [11]. While the emulsification polymerization (Figure 5.3c) involves

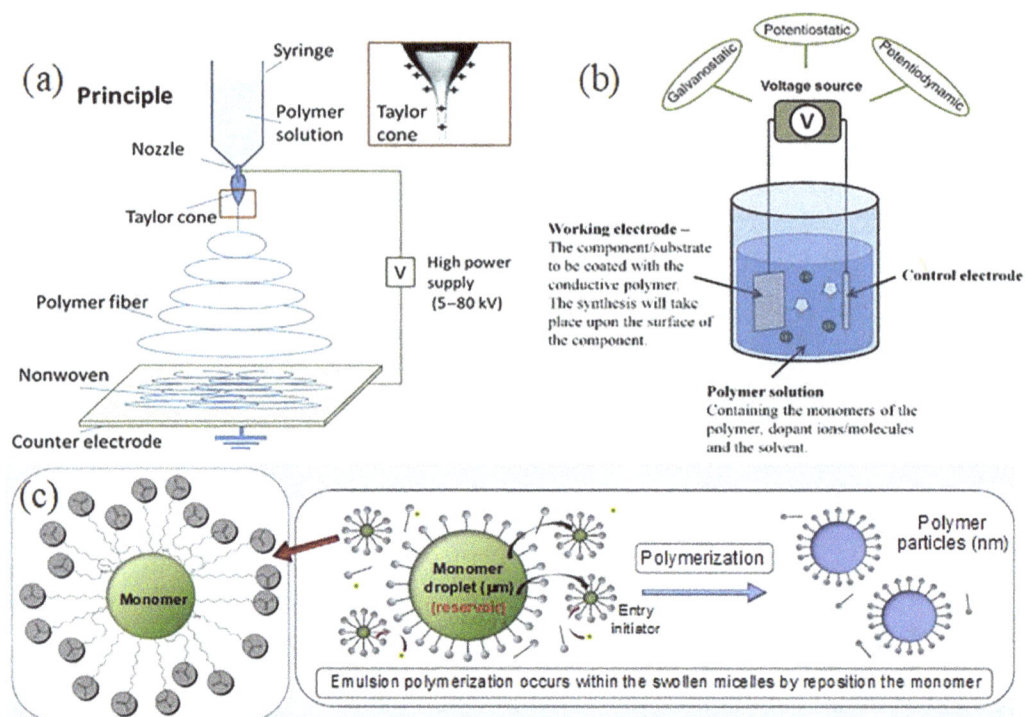

FIGURE 5.3 Schematic representation of (a) electrospinning, (b) electrochemical, and (c) emulsification polymerization. (a) Adapted with permission from Ref. [18]. Copyright (2021) Publisher Elsevier. (b) Adapted with permission from Ref. [11]. Copyright (2021) Copyright the Authors, some rights reserved; exclusive licensee [Elsevier]. Distributed under a Creative Commons Attribution License 3.0 (CC BY) https://creativecommons.org/licenses/by/3.0/. (c) Adapted with permission from Ref. [19]. Copyright (2021) Publisher Elsevier.

a free-radical polymerization of a water-insoluble monomer or mixture of monomers in a surfactant aqueous solution, the nucleation of monomer-swollen micelle occurs when the amount of surfactant is above the critical micelle amount. The formation of initiator radical or organophilic oligomers occur due to the reaction between the free radical of the initiator and monomer. This initiator radical enters the monomer-swollen micelles and result in the nucleation of a new polymer [15].

The unique blend of CPs and BPs has attracted many researchers to develop partial biodegradable CPs and studied their biomedical applications. For instance, Liu and coworkers [20] electrochemically prepared chondroitin sulfate-doped PPy substrates. These prepared substrates were coupled with collagen type I. This resulted in the formation of a 3D fibrillar matrix at the interface of PPy. The results indicated that the prepared biodegradable CP is highly suitable to improve the interface of neural and implant electrodes by promoting nerve cell differentiation and attachment. The nanosized PPy can be obtained by oxidizing PPy via a microemulsion system [21]. The nanoparticles of PPy possess higher electrical conductivity. However, the nanoparticles of PPy are not biocompatible. Therefore, it is critically important to keep the content of PPy as low as possible in the composite. The composite of PPy nanoparticles and poly (d,l-lactide) (PDLLA) was prepared through emulsion polymerization. The membrane of the PPy/PDLLA composite was prepared through the casting method on a poly(tetrafluoro-ethylene) plate. The performance of this membrane was also testified for the human cutaneous fibroblasts. A composite of 5% PPy and 95% biodegradable poly(l-lactide) (PLLA) was also prepared using the same method as that of PPy/PDLLA composite [22]. The composite (PPy/PLLA) also supports the adhesion, dispersion, and growth of human cutaneous fibroblasts with and without electrical stimulation (ES). To further improve the biocompatibility, a composite of PPy and chitosan was prepared through the co-solution casting method. Chitosan is a bioactive polymer and is extensively used for drug delivery and tissue engineering. In the composite, the content of PPy and chitosan was 2.5% and 97.5%, respectively [23]. The outcome of these studies brought new opportunities to improve nerve regeneration and neurotrophin secretion in ES-induced biodegradable CP membranes.

Polyurethane (PU) is also blended with PPy to prepare biodegradable CPs. PU possesses good mechanical properties as well as superior biocompatibility and is widely used in the sector of biomedical [24]. The composite of PPy/PU was prepared in different ratios through chemical polymerization of PPy in PU emulsion [25]. The composite has shown a combination of elastomeric properties and electrical conductivity. This type of composite was found to be cytocompatible with C2C12 myoblast cells. Furthermore, nanomembranes with thickness ~20–80 nm are also considered for drug delivery and tissue engineering. The biodegradable nanomembranes comprising the blend of conducting poly(3-thiophene methyl acetate) and biodegradable poly(tetramethylene succinate) are prepared through spin coating [26]. This suggested that nanomembranes are suitable to develop bioactive substrates for tissue regeneration.

In tissue engineering, porous scaffolds are highly recommended and studied extensively. A porous scaffold based on partial biodegradable CP is prepared by blending PPy in a chitosan matrix [27]. In this scaffold, the pores are created by the phase separation technique. The structure of pores in the scaffolds is controllable by varying the content of NaOH in an aqueous solution. In addition, the mechanical properties of scaffolds are tunable by changing pores parameters. The conductivity ~10^{-3} S/cm was achieved in the scaffold when the content of PPy is set to 2%. The integration of PPy in biodegradable scaffolds through electrospinning polymerization has been widely explored to prepare biodegradable CPs [28]. The biodegradable CPs based on PPy prepared through electrospinning polymerization possess improved mechanical properties.

Alongside, PANi and PEDOT CPs were also blended with BPs to develop partial biodegradable CPs. PANi CP offers superior electrical conductivity, ease of fabrication, environmental stability, and tunable electrical conductivity [29]. These unique features made PANi suitable for conductive substrates needed in tissue engineering applications. The co-electrospun of PANi/gelatin in nanofibers resulted in the formation of partial biodegradable PC [30]. The lower content of PANi (<3 wt%) in the blend leads to uniform fibers and no phase segregation. Interestingly, the content PANi (0–5 wt%) significantly affects the average fiber length and tensile modulus. As the content of PANi decreased the length of the fiber and increased the tensile modulus, these fibers favored the attachment and propagation of H9c2 cells up to a similar extent in comparison to control tissue culture-treated polymers. The choice of various degradable polymers provides more opportunities to tune the properties of scaffold for a particular application. It is advantageous to develop a biodegradable CP scaffold that offers a good combination of electrical and mechanical properties similar to the natural extracellular matrix (ECM). This type of scaffolds could be highly suitable for soft tissue regeneration including skin, blood vessels, and skeletal muscle. A blend of PANi with PLA resulting into formation of nanofiber through electrospinning demonstrated improved growth rate of mammalian cell. Camphorsulfonic acid (CSPA)-doped PANi blended through electrospinning with poly(*l*-lactide-*co*-caprolactone) (PLCL) showed even fiber structure and diameter ~100–700 nm, by varying the content of PANi (0–30 wt%) [31]. The length of these fibers could be extended to 391.54 ± 9.20–$207.85 \pm 6.74\%$ owing to good elastic properties. In addition, the electrical conductivity was found to be 0.0015–0.0138 S/cm depending on the content of PANi [32].

Furthermore, hydrogels of PANi grafted with gelatin and cross-linked with genipin were developed with conductivity 4.54×10^{-4}–2.41×10^{-4} S/cm [33]. PEDOT was blended with PLGA to develop biodegradable conducting microfibers. The conductivity of this blend was found to be in the range of 7×10^{-2}–2.8×10^{-1} S/cm depending on the content of PEDOT [34]. However, it is worthwhile to mention that these blends may not retain the original electrical conductivity and biocompatibility of their parents. For example, nonbiodegradable CP content is always kept lower in the blend to get good biocompatibility. As a result, the blend possesses inferior electrical conductivity for commercialization. Moreover, the small content of CP in the blend may minutely decrease the overall biocompatibility. It is expected that CP in the blend must be chemically inert and could sustain in the physiological environment when introduced in the body. Consequently, direct blending is not a suitable strategy to develop biodegradable CPs in such a way that complete removal of nondegradable CPs could be achieved from the body after the decomposition of biodegradable CPs. Therefore, increasing efforts in recent years have been devoted to develop alternative strategies that can enable the complete disintegration of biodegradable CP composites.

5.3 BIODEGRADABLE CPs BASED ON THE CONDUCTING OLIGOMERS

The conducting oligomers are short chains of conducting monomers. These oligomers offer flexibility in fabrication, better biocompatibility, solubility and decays rapidly in the body. In addition, oligomers own well-confined structures with good electrical conductivity and share redox reaction responses with CPs. These unique features of conducting oligomers draw the attention of several research groups to develop biodegradable CPs [35, 36]. Literature evidenced that the oligomers of pyrrole, aniline, and thiophene were easy to copolymerized with BPs to attain better biodegradability in comparison to their counterpart CPs [37]. The strategies involved for the synthesis of biodegradable CPs based on the conducting oligomers have been categorized as linear degradable CPs, grafted degradable CPs, star-shaped and hyperbranched degradable CPs, degradable conducting hydrogels, and self-assembled degradable CPs.

5.3.1 LINEAR BIODEGRADABLE CPs

In the 1970s, Heeger et al. [38] doped polyacetylene $(CH)_x$ with iodine and observed high electrical conductivity. After this remarkable discovery, several polymers were developed owing combination of metallic and semiconducting properties. It was found that the electronic properties of CP significantly depend on the linear π-conjugated systems. Oligomers based on pyrrole were prepared that controls the fabrication and π-conjugated system. Besides pyrrole oligomer, several other oligomers were also developed based on thiophene, aniline oligomers, primarily aniline trimer, aniline tetramer, and aniline pentamer (AP) [8]. These oligomers possess convenient synthesis procedures and processing routes. One of the major advantages of these oligomers is the ease of consumption and adoption in comparison to the macrophages and subsequent removal by the kidneys [39]. Consequently, these oligomers are highly suitable to achieve the goal of fabricating fully biodegradable as well as conducting polymeric composites. A schematic synthesis representation of electroactive oligomers and block copolymers, electroactive macromonomers and graft copolymers and chemical structure of electroactive and biodegradable oligomers and macromonomers is presented in Figure 5.4.

Schmidt et al. [39] prepared polymeric composite with oligomers of pyrrole and thiophene and degradable ester linkers to obtain multi-blocks biodegradable CP. It was the first report on the fabrication of pyrrole and thiophene ending with two hydroxyl groups and later adjoint with adipoyl chloride to attain biodegradability. This polymeric composite was found to be degradable with the enzymes naturally occurring in the human body. The in-vivo biocompatibility tests of these polymers revealed no toxicity. However, the electrical conductivity of the multi-block copolymer was in the order of 10^{-4} S/cm. To enhance the conductivity of the copolymer, a copolymer comprising of quarter thiophene oligomer and adipoyl chloride was designed [40]. The prepared copolymer could be doped with anions ferric chloride $(FeCl_3)$ or ferric perchlorate $(Fe(ClO_4)_3)$, Cl^- or ClO^{4-}. These anions counters biocompatibility and the remaining iron are the least toxic. As long as the prepared copolymer comes in contact with cholesterol esterase, erosion begins.

Furthermore, aniline oligomers (aniline trimer, tetramer, and pentamer) synthesis was simple in comparison to the complex synthesis of pyrrole and thiophene oligomers. The oligomers based

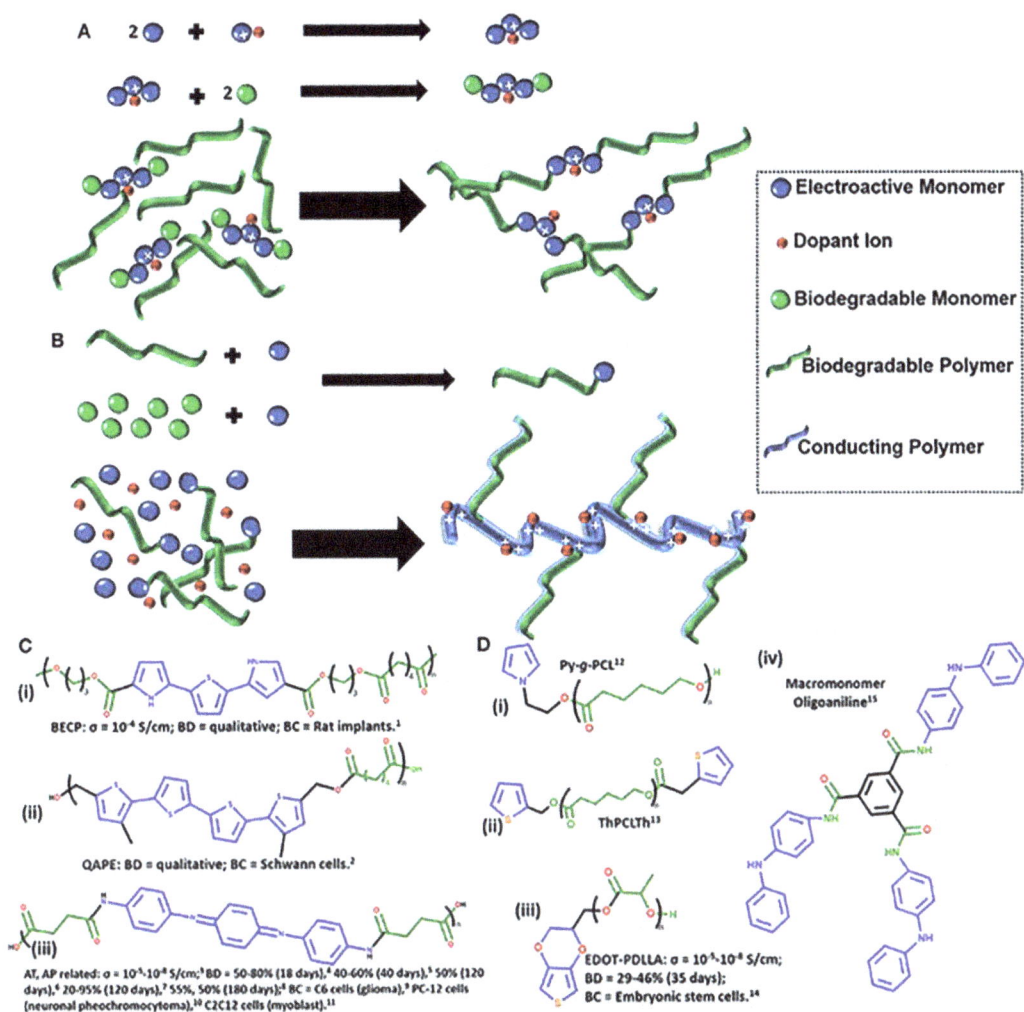

FIGURE 5.4 Schematic representation of synthesis protocol to obtain (a) electroactive oligomers and block copolymers, (b) electroactive macromonomers and graft copolymers. The chemical structure of (c) electroactive and biodegradable oligomers, (d) electroactive and biodegradable macromonomers. Here, σ is conductivity, BD and BC are biodegradability and biocompatibility, respectively. The color blue and green are assigned to conductive/electroactive molecules and biodegradable molecules, respectively. (Adapted with permission from Ref. [14]. Copyright (2021) Copyright The Authors, some rights reserved; exclusive licensee [Frontiers]. Distributed under a Creative Commons Attribution License 4.0 (CC BY) https://creativecommons.org/licenses/by/4.0/.)

on aniline open new pathways to fabricate and design well-defined functional copolymers [41]. A highly soluble triblock polymeric composite comprising of hydroxyl-capped PLA and carboxyl-capped AP was prepared through a coupling reaction [42]. The solubility of the copolymer was due to the coupling of PLA segments on both ends of the AP. This copolymer was found to be soluble in various compounds such as dimethylformamide (DMF), dimethyl sulfoxide (DMSO), N-methyl-2-pyrrolidone (NMP), chloroform (CHCl$_3$), tetrahydrofuran (THF), and toluene. When the copolymer was doped with camphor sulfonic acid then the conductivity was tested to be ~5 × 10^{-6} S/cm. Besides all these advantages, these low molecular weighted (2700–10,000 g/mol) copolymers exhibit poor mechanical properties. To improve the mechanical properties, the copolymer of hydroxyl-capped polylactide and carboxyl-capped AP was prepared to have a molecular weight of 66.8 kDa [43]. The tensile strength of copolymer was found to be 3.0 MPa along with 95% breaking elongation and the young modulus was 33 MPa. The copolymer doped with camphor-sulfonic acid (CSA) became hydrophilic and the water contact angle decreased from 90° to 50°.

In some studies, the linear biodegradable CPs were developed by preparing to conducting and biodegradable portions separately and later combined through a condensation reaction. This strategy is unfavorable due to complexity in the reaction, tedious purification, poor yield, and in some cases toxic compounds are used. Further, ring-opening polymerization was employed to prepare ABA block polymer [44]. These polymers possess good degradability as well as electrical conductivity. This strategy was easier than the previously discussed strategy to develop the linear biodegradable CPs. In this strategy, aniline trimer double amino-capped was used to trigger the e-caprolactone with Sn(Oct)$_2$ catalyst in anhydrous toluene. The prepared tri-block polymer demonstrated improved stability and good sensitivity. This copolymer has also shown potential as a sensing material.

Moreover, an approach to combine the ring-opening polymerization and oxidative coupling to prepare di or tri-block biodegradable CPs by following a two-step synthesis protocol was proposed [45]. To perform the ring-opening polymerization of CL, aniline dimer (AD) was used as initiator and Sn(Oct)$_2$ as a catalyst. The AD with lower molecular weight dissolve in the CL and thus, organic solvents were avoided for polymerization. The AD group was employed for post-polymerization of AD-PCL through oxidative coupling reaction. This resulted in the formation of AT at the end of the macromolecular chains. The electrical conductivity of AD-PCL copolymer was found to be 6.30×10^{-7}–1.03×10^{-5} S/cm. It was observed the conductivity of AD-PCL could vary with the content of AD. Also, the addition of crosslinkers such as 2,2-bis-(e-caprolactone-4-yl) propane, in the beginning, lead to the formation of a biodegradable and conductive network. A two-step synthesis protocol was used to develop a coil-rod-coil triblock copolymer in which AP block was in the middle of two polycaprolactones (PCL) bilateral blocks [46]. This protocol leads to the development of triblock polymer with well-defined structure and controlled properties. The tri-block copolymers exhibit better conductivity in comparison to a di-block polymer.

5.3.2 GRAFTED BIODEGRADABLE CPS

The grafted polymerization is the classic approach to synthesize polymers [14]. However, this approach is still new for CP because it involves the dilution of a nonconductive component. Grafting enables better solubility, nano-dimension morphology, and biocompatibility. In the previous section, the conducting oligomers were in the main chain and degradable oligomers were polyester (PLA, PGA, and PCL). In grafted biodegradable CPs, polyphosphazene inorganic polymer was used for the degradable segment [47]. A large number of functional substituent groups were attached to polyphosphazene polymer. The functionalization of polyphosphazene was performed with the help of parent AP and glycine ethyl ester through nucleophilic substitution reaction [48]. The modification of side chains offers ease to regulate the content of aniline oligomer and upholding the mechanical properties. AP promotes the conductivity, while the glycine ethyl ester group maintains the degradability of the polymer. The conductivity of the prepared protonic doped polymeric film was found to be 2×10^{-5} S/cm. The PLA was modified by following the strategy that involves two-step, i.e., functionalization and surface modification. It was observed the PLA prepared through this strategy exhibit improved hydrophilicity and incorporation of aniline oligomer on the surface of PLA maintains the cell response [49].

5.3.3 STAR-SHAPED AND HYPERBRANCHED BIODEGRADABLE CPS

The architecture of polymers significantly influences the overall properties [50]. There are linear, branched, and cross-linked networks of BPs that have been synthesized [51]. It is critically important to investigate the relation between polymer architecture and its properties. This will lead to the development of tunable biodegradable CPs with properties desired for a target application. The star-shaped polymers possess linear star arms such as polyesters. These types of polymers are generally prepared from 3-(4-hydroxyphenyl) propionic acid (HPPA) because it is commercially available, biodegradable, and liquid crystalline copolyester [52]. Albertsson and coworkers [53] developed a star-shaped BP by using star-shaped PLA and carboxyl-capped AT through coupling reaction. It was observed that the star-shaped biodegradable CP prepared through this strategy possesses better conductivity in comparison to PANi. Moreover, doping of HCl resulted in improved hydrophilicity than PLA films [54].

Furthermore, the hyperbranched polymers were first discovered by evaluating amylopectin and glycogen [55]. Flory [56] was the first to discuss theoretically branching and polycondensations that led to the formation of hyperbranched polymers. These polymers were prepared via copolycondensation of 3-trimethylsiloxy benzoyl chloride (A-B) and 3,5-bistrimethylsiloxyben-zoyl chloride (A_2B). The density of branching varied with a change in the molar ratio. Similarly, copolyesters with a variety of end groups were synthesized through copolycondensation of 3-acetoxybenzoic acid and 3,5-bisacetoxybenzoic acid [54]. This strategy of synthesizing linear and hyperbranched copolymers is termed as "$A_2 + B_n$ ($n = 2, 3, 4$)". The copolycondensation reaction occurred in between the hydroxyl and carboxyl groups of PCL and carboxyl-capped AP, respectively. In this reaction, dicyclohexylcarbodiimide (DCC) was used as a condensation agent and 4-dimethylaminopyridine (DMAP) as a catalyst. The results revealed the formation of copolymers with different architectures. It was observed that the conductivity of these polymers relied on 5.01×10^{-6} and 2.42×10^{-5} S/cm. The conductivity of hyperbranched CP was found to be 1.6–4.8 times more than its linear counterparts. Additionally, a variety of biodegradable conducting films were prepared by blending hyperbranched biodegradable CPs with linear polycaprolactone [57]. These blends were used to develop tubular porous nerve conduits through the casting or leaching method. The content of AP in the blend varied the conductivity. The prepared scaffolds were also confirmed to be nontoxic through cytotoxicity assay by using HaCaT cells. The biodegradable CP prepared through this strategy has shown great potential for neural tissue engineering applications.

5.3.4 BIODEGRADABLE CONDUCTING HYDROGELS

Hydrogels immerged as an important class of biomaterials due to their 3D crosslinked hydrophilic polymer networks, exceptional biocompatibility, rubbery nature-like tissue, offer ease to migrate O_2, nutrients, and other biomolecules [58]. These unique features of hydrogels placed them as a potential candidate for tissue engineering applications. In general, biodegradable conducting hydrogels are synthesized by blending the polymer with biomaterials. This lead development of biodegradable conducting hydrogels with a unique combination of properties of CPs and hydrogels. However, poor biodegradability restricts its utilization in tissue engineering. The synthesis protocol of these hydrogels involves the formation of the degradable network through a coupling of acrylate poly(d,l-lactide)-poly(ethylene glycol)-poly(d,l-lactide), glycidyl methacrylate (GMA), and ethylene glycol dimethacrylate [59]. To induce conduction in AT, GMA was incorporated [43]. The swelling of hydrogels could be adjusted by varying amounts of AT, degree of crosslinking, and pH value in the solution. This strategy has brought new opportunities to tune conductivity and hydration levels to achieve the goal of commercialization. Although the product formed after decomposition of polylactide was found to create an acidic environment in vivo, the amount of AT was kept lower to get high conductivity in biodegradable hydrogels. In addition, toxicity increased with higher content of aniline in a polymeric material. To overcome the drawback of prepared hydrogels, less acidic PCL-based hydrogels were functionalized with an AP. In this case, the reaction was performed by using hydroxyethylmethacrylate (HEMA) in GMA functionalized with polycaprolactone-poly(ethylene glycol)-polycaprolactone network and AP [60]. The water condensing agent and catalyst used were ethyl-3-(3-dimethylaminopropyl) carbodiimide hydrochloride and dimethyl aminopyridine

(DMAP), respectively. These hydrogels possess conductivity of about 2.02×10^{-4} S/cm when the amount of AP was 17 wt%.

The biodegradable conducting hydrogels were also synthesized by using natural polymers including gelatin and chitosan. Liu and coworkers [61] prepared hydrogels by blending aniline oligomer with gelatin. They were first to report grafting of N-hydroxysuccinimide-capped AP to gelatin. The grafted polymer was then frozen to develop scaffolds. The prepared scaffolds were crosslinked in ethanol with 1-ethyl-3-(3-dimethylaminopropyl) carbodiimide. The increased amount of AP in the hydrogel resulted in an alteration of its structure from honeycomb to bamboo raft. These hydrogels were found to be nontoxic and biocompatible when Rat Schwann cells 96 (RSC96) were cultured on them. The reason for biocompatibility was associated with the presence of gelatin. Another strategy involves a reaction between chitosan and AT [62]. In this strategy, glutaraldehyde was used as a coupling agent for AT and crosslinking agent for chitosan. The hydrogels of this combination were prepared in a one-pot reaction at ambient temperature. The prepared hydrogels films were found to be flexible, easy to fabricate, and free-standing. The electrical conductivity of these hydrogels could be varied by changing the content of AT. These hydrogels were also found to be sensitive to pH levels.

5.3.5 SELF-ASSEMBLY OF BIODEGRADABLE CPs

Self-assembly of biodegradable CPs has emerged as a resourceful tool for the development of novel functional materials with a unique combination of properties and responsive to external conditions. Rod-coil block copolymers with the self-assembling feature have attained significant attention from the research community [63]. The reason behind growing interest is due to the rigid-rod block functionality and the response of assembling is found to be distinct as compared to coil-coil amphiphilic diblock or triblock copolymers. The rigid conformation of aniline oligomer can be used for assembling rod-coil and coil-rod-coil block copolymers [64]. Self-assembly biodegradable CPs were synthesized by using poly(ethylene glycol) (PEG) and aniline oligomers [65]. This polymer was used to develop an electrically switchable vesicular system and brought new possibilities for utilizing the self-assembly of biodegradable CPs for drug delivery in biomedical science.

Furthermore, the micelles from CPs in the main chain are not advantageous due to the nondegradability of micelles [66]. The use of ester bonds in the main chain of CPs resulted in the formation of biodegradable CPs by self-assembly. Wang and coworkers [67] prepared biodegradable conducting deblock oligomers of tetraaniline-block-poly(l-lactide). They employed ring-opening polymerization of LLA by using an initiator tetraaniline. Self-assembly response was also examined in the presence of chloroform. Interestingly, the oligomer changes morphology from spherical micelles to ring-like aggregates when cast from chloroform. The spherical micelles were in the leucoemeraldine state, while ring-like aggregates were associated with emeraldine state. Triblock copolymers were also synthesized by using EMAP or LMAP with PCL [46]. These copolymers became self-assembled in the presence of chloroform. The self-assembly of these polymers in core-shell nanoparticles was also confirmed in CDCl$_3$ solution. These nanoparticles possess a diameter ~50–393 nm, and the size increased with a higher molecular weight polymer. In these triblock polymers, the size of the aggregates depends on the molecular weight and AP oxidation state. The prepared self-assemble of BPs with the tunable size is highly suited for drug delivery, sensing, and detection of biomolecules.

It was observed that the behavior of self-assembly was distinct in dendritic polymers in comparison to block polymers [68]. The dendritic polymers were made of ester dendron and AT and self-assembled in the presence of THF. The self-assembly of these polymers significantly depends on various factors such as amphiphilic interactions, intermolecular interactions, and the intermolecular π-π stacking among oligomers. The amphiphilic interactions and intermolecular interactions occurred due to building blocks and easter dendrons, respectively. Later, a block oligomer of aniline having a dendron-rod-dendron dumbbell shape was synthesized [69]. In this case, the morphology was changed from fibrils to flat single-layer films. In addition, there was the formation of a porous network due to the conformational transition of the conductive oligomer during oxidation. There are some grafted biodegradable CPs that can undergo self-assembly [70]. For example, AP grafted chitosan self-assemble into micelles of size 200–300 nm. The reason behind self-assembly was associated with variation in pH and the presence of salt. It was also found that multialdehyde sodium alginate-graft-tetraaniline copolymer possesses a tendency to self-assemble in the nanosphere having a hydrophilic part as shell and hydrophobic part as a core. The aggregation of micelle was prevented by a negatively charged surface of micelle due to the presence of large-sized carboxylic ions in the multialdehyde sodium alginate-graft-tetraaniline copolymer [71]. These fascinating results open new doors to couple drugs or biomolecules in the presence of salt.

5.4 BIODEGRADABLE CPs BASED ON THE INTEGRATION OF BIO-ERODIBLE AND CPs

In this strategy, completely biodegradable CPs are synthesized by designing erodible CPs. These polymers degrade gradually through the dissolution process instead of classic degradation that involves breaking chemical bonds. Zelikin and coworkers [72] were the first to introduce this strategy. They modified the PPy monomer and polymerized it to obtain erodible PPy. The modification process involves a synthesis of β-substituted PPy monomer by incorporating ionizable and hydrolyzable to the backbone of the monomer. Thereafter, electrochemical oxidation of the modified PPy monomer was performed followed by ferric-chloride-mediated chemical polymerization to fabricate erodible PPy. The average resistance of 300 Ω was testified when a thin film of erodible PPy was prepared via acid functionalization. The prepared erodible polymer was found to degrade gradually in physiological conditions and supported the propagation and differentiation of cells.

Furthermore, thiophene-based erodible CPs were also investigated for medical applications [73]. The thin films of thiophene-based biodegradable CPs were developed by a layer-by-layer method. These films showed higher resistance 7.82×10^{-3}–2.76×10^{-2} S/cm and supported the attachment and propagation of muscle cells. It was also observed that these films undergo complete degradation in 83–130 days in physiological conditions [74]. Besides the modification of the monomer, degradable conjugated linkers were used to fabricate completely biodegradable CPs. Based on imine chemistry, polymer diketopyrrolopyrrole-phenylenediamine (PDPP-PD) was fabricated which was fully degradable [75]. Diketopyrrolopyrrole (DPP) dye was used to prepare biodegradable CP composite [76]. DPP dye possesses high mobility of charge carriers, superior biocompatibility, and offers ease to chemical functionalization. Imine bonds ($-C=N-$) were used linkers owing to their high stability at pH 7. These linkers facile hydrolysis in the presence of a mild acidic environment [77]. When two aldehyde groups were added to the DPP monomer, the formation of DPP-CHO was observed. Lastly, PDPP-PD biodegradable CP was fabricated by reacting DPP-CHO and p-phenylenediamine (PPD) by using p-toluenesulfonic acid (PTSA) as a catalyst. The average hole mobility of PDPP-PD was found to be 4.2×10^{-2}–3.4×10^{-1} cm^2/Vs. The stability of PDPP-PD was good in neutral as well as basic pH solution. However, degradation occurred in an acidic medium. Based on the absorption spectrum and color of PDPP-PD, it was observed that PDPP-PD degraded after 10 days in DPP-CHO monomer and completely disintegrated in 40 days. Moreover, a thin film of PDPP-PD was prepared through spin coating. The results revealed that PDPP-PD degraded gradually when introduced in a solution of pH 4.6. It was concluded that the total degradation of a polymer depends on the hydrolyzation via acid-catalyzed imine bond followed decomposition of DPP monomer by lactam ring hydrolyzation.

5.5 SUMMARY

Biodegradable CPs emerged as a new class of biomaterials owing to superior physicochemical properties including outstanding electrical conductivity and biocompatibility. Yet, this new class needs to be explored to meet the current demands of biodegradable conducting materials. Optimization of conductivity is crucial for the development of biodegradable CPs. Several strategies have been developed to synthesize a biodegradable CP with the least content of conducting species while maintaining sufficient conductivity for the target application. Architecture is also one of the ways to design and fabricate novel biodegradable CPs with improved biodegradability and conductivity. The major downfall of the CPs is processability that can be avoided by coupling CPs with degradable polymers. Also, the biodegradable CPs are found to be soluble only in organic solvents. However, these solvents are toxic. The possible solution is to design hydrophilic and water-soluble biodegradable CPs. The research efforts made on the development of biodegradable CPs based on oligomers are highly suited. It offers ease of fabrication and processing. It is still challenging to hunt novel routes fabricate pyrrole and thiophene combined with aniline oligomers to develop biodegradable CPs with tunable properties. In addition, coupling chemistry plays a vital role in the development of CPs based on aniline oligomer coupled with degradable polymer. However, coupling reactions are inefficient. The biodegradable CPs demonstrated excellent cell adhesion and propagation. This makes them appropriate for scaffold materials that can be used for tissue regeneration applications. It would be interesting to investigate the interaction between cells and biodegradable CPs, especially star-shaped and hyperbranched for commercialization in biomedical sciences, although there are many questions related to the development of biodegradable CPs that are yet to be explored by designing new synthesis strategies. Consequently, there is a need to find novel routes to make these reactions efficient for the synthesis of biodegradable CPs.

REFERENCES

1. H. Goesmann, C. Feldmann, Nanoparticulate functional materials, Angew. Chem. Int. Ed. 49 (2010) 1362–1395.
2. S.F. Anis, R. Hashaikeh, N. Hilal, Functional materials in desalination: a review, Desalination. 468 (2019) 114077.
3. J. Stejskal, Conducting polymers are not just conducting a perspective for emerging technology, Polym. Int. 69 (2020) 662–664.
4. N. Hall, Twenty-five years of conducting polymers, Chem. Commun. 1 (2003) 1–4.
5. A. Maziz, E. Özgür, C. Bergaud, L. Uzun, Progress in conducting polymers for biointerfacing and biorecognition applications, Sens. Actuators Rep. 3 (2021) 100035.
6. Y. Liu, V.R. Feig, Z. Bao, Conjugated polymer for implantable electronics toward clinical application, Adv. Healthc. Mater. 10 (2021) 2001916.
7. E. Cheah, Z. Wu, S.S. Thakur, S.J. O'Carroll, D. Svirskis, Externally triggered release of growth factors – a tissue regeneration approach, J. Control. Release. 332 (2021) 74–95.
8. M. Fan, B. Zhang, L. Fan, F. Chen, Q. Fu, Adsorbability of modified PBS nanofiber membrane to heavy metal ions and dyes, J. Polym. Environ. 29 (2021) 3029–3039.
9. A. Puiggalí-Jou, J. Ordoño, L.J. del Valle, S. Pérez-Amodio, E. Engel, C. Alemán, Tuning multilayered polymeric self-standing films for controlled release of L-lactate by electrical stimulation, J. Control. Release. 330 (2021) 669–683.
10. S. Jadoun, U. Riaz, V. Budhiraja, Biodegradable conducting polymeric materials for biomedical applications: a review, Med. Devices Sens. 4 (2021) e10141.
11. R. Balint, N.J. Cassidy, S.H. Cartmell, Conductive polymers: towards a smart biomaterial for tissue engineering, Acta Biomater. 10 (2014) 2341–2353.
12. R. Singh, M.J. Bathaei, E. Istif, L. Beker, A review of bioresorbable implantable medical devices: materials, fabrication, and implementation, Adv. Healthc. Mater. 9 (2020) 2000790.
13. C.Y. Chen, S.Y. Huang, H.Y. Wan, Y.T. Chen, S.K. Yu, H.C. Wu, T.I. Yang, Electrospun hydrophobic polyaniline/silk fibroin electrochromic nanofibers with low electrical resistance, Polymers (Basel). 12 (2020) 2102.
14. A.C. da Silva, S.I. Córdoba de Torresi, Advances in conducting, biodegradable and biocompatible copolymers for biomedical applications, Front. Mater. 6 (2019) 98.
15. S. Ebnesajjad, Handbook of Biopolymers and Biodegradable Plastics: Properties, Processing and Applications, 1st edition, Elsevier, 2012.
16. H.S. Park, C.K. Hong, Relationship between the stereocomplex crystallization behavior and mechanical properties of plla/pdla blends, Polymers (Basel). 13 (2021) 1851.

17. Z. Zhang, O. Ortiz, R. Goyal, J. Kohn, Biodegradable polymers, in Handbook of Polymer Applications in Medicine and Medical Devices (2014) 303–335.
18. Y. Zheng, Fabrication on bioinspired surfaces, in Bioinspired Design of Materials Surfaces (2019) 99–146.
19. A.T. Jensen, W.S. Neto, G.R. Ferreira, A.F. Glenn, R. Gambetta, S.B. Gonçalves, L.F. Valadares, F. Machado, Synthesis of polymer/inorganic hybrids through heterophase polymerizations, in Recent Developments in Polymer Macro, Micro and Nano Blends: Preparation and Characterisation (2017) 207–235.
20. X. Liu, Z. Yue, M.J. Higgins, G.G. Wallace, Conducting polymers with immobilised fibrillar collagen for enhanced neural interfacing, Biomaterials. 32 (2011) 7309–7317.
21. G. Shi, M. Rouabhia, Z. Wang, L.H. Dao, Z. Zhang, A novel electrically conductive and biodegradable composite made of polypyrrole nanoparticles and polylactide, Biomaterials. 25 (2004) 2477–2488.
22. G. Shi, Z. Zhang, M. Rouabhia, The regulation of cell functions electrically using biodegradable polypyrrole–polylactide conductors, Biomaterials. 29 (2008) 3792–3798.
23. J. Huang, X. Hu, L. Lu, Z. Ye, Q. Zhang, Z. Luo, Electrical regulation of Schwann cells using conductive polypyrrole/chitosan polymers, J. Biomed. Mater. Res. A. 93 (2010) 164–174.
24. I.H.L. Pereira, E. Ayres, P.S. Patrício, A.M. Góes, V.S. Gomide, E.P. Junior, R.L. Oréfice, Photopolymerizable and injectable polyurethanes for biomedical applications: synthesis and biocompatibility, Acta Biomater. 6 (2010) 3056–3066.
25. C.R. Broda, J.Y. Lee, S. Sirivisoot, C.E. Schmidt, B.S. Harrison, A chemically polymerized electrically conducting composite of polypyrrole nanoparticles and polyurethane for tissue engineering, J. Biomed. Mater. Res. Part A. 98A (2011) 509–516.
26. E. Armelin, A.L. Gomes, M.M. Pérez-Madrigal, J. Puiggalí, L. Franco, L.J. Del Valle, A. Rodríguez-Galán, J.S.D.C. Campos, N. Ferrer-Anglada, C. Alemán, Biodegradable free-standing nanomembranes of conducting polymer:polyester blends as bioactive platforms for tissue engineering, J. Mater. Chem. 22 (2012) 585–594.
27. Y. Wan, H. Wu, D. Wen, Porous-conductive chitosan scaffolds for tissue engineering, 1: Preparation and characterization, Macromol. Biosci. 4 (2004) 882–890.
28. J.Y. Lee, C.A. Bashur, A.S. Goldstein, C.E. Schmidt, Polypyrrole-coated electrospun PLGA nanofibers for neural tissue applications, Biomaterials. 30 (2009) 4325–4335.
29. S. Bhadra, D. Khastgir, N.K. Singha, J.H. Lee, Progress in preparation, processing and applications of polyaniline, Prog. Polym. Sci. 34 (2009) 783–810.
30. M. Li, Y. Guo, Y. Wei, A.G. MacDiarmid, P.I. Lelkes, Electrospinning polyaniline-contained gelatin nanofibers for tissue engineering applications, Biomaterials. 27 (2006) 2705–2715.
31. S.I. Jeong, I.D. Jun, M.J. Choi, Y.C. Nho, Y.M. Lee, H. Shin, Development of electroactive and elastic nanofibers that contain polyaniline and poly(L-lactide-co-ε-caprolactone) for the control of cell adhesion, Macromol. Biosci. 8 (2008) 627–637.
32. I. Jun, S. Jeong, H. Shin, The stimulation of myoblast differentiation by electrically conductive sub-micron fibers, Biomaterials. 30 (2009) 2038–2047.
33. S.N. Karri, S.P. Ega, P. Srinivasan, Synthesis of novel fluorescent molecule and its polymeric form with aniline as fluorescent and supercapacitor electrode materials, Polym. Adv. Technol. 31 (2020) 1532–1543.
34. S. Kee, P. Zhang, J. Travas-Sejdic, Direct writing of 3D conjugated polymer micro/nanostructures for organic electronics and bioelectronics, Polym. Chem. 11 (2020) 4530–4541.
35. J.P. Sadighi, R.A. Singer, S.L. Buchwald, Palladium-catalysed synthesis of monodisperse, controlled-length, and functionalized oligoanilines, J. Am. Chem. Soc. 120 (1998) 4960–4976.
36. Z. Wei, C.F.J. Faul, Aniline oligomers – architecture, function and new opportunities for nanostructured materials, Macromol. Rapid Commun. 29 (2008) 280–292.
37. N. Nasongkia, B. Chen, N. Macaraeg, M.E. Fox, J.M.J. Fréchet, F.C. Szoka, Dependence of pharmacokinetics and biodistribution on polymer architecture: effect of cyclic versus linear polymers, J. Am. Chem. Soc. 131 (2009) 3842–3843.
38. C.K. Chiang, C.R. Fincher, Y.W. Park, A.J. Heeger, H. Shirakawa, E.J. Louis, S.C. Gau, A.G. MacDiarmid, Electrical conductivity in doped polyacetylene, Phys. Rev. Lett. 39 (1977) 1098–1101.
39. T.J. Rivers, T.W. Hudson, C.E. Schmidt, Synthesis of a novel, biodegradable electrically conducting polymer for biomedical applications, Adv. Funct. Mater. 12 (2002) 33–37.
40. N.K.E. Guimard, J.L. Sessler, C.E. Schmidt, Toward a biocompatible and biodegradable copolymer incorporating electroactive oligothiophene units, Macromolecules. 42 (2009) 502–511.
41. C.U. Udeh, N. Fey, C.F.J. Faul, Functional block-like structures from electroactive tetra(aniline) oligomers, J. Mater. Chem. 21 (2011) 18137–18153.

42. L. Huang, J. Hu, L. Lang, X. Wang, P. Zhang, X. Jing, X. Wang, X. Chen, P.I. Lelkes, A.G. MacDiarmid, Y. Wei, Synthesis and characterization of electroactive and biodegradable ABA block copolymer of polylactide and aniline pentamer, Biomaterials. 28 (2007) 1741–1751.

43. L. Huang, X. Zhuang, J. Hu, L. Lang, P. Zhang, Y. Wang, X. Chen, Y. Wei, X. Jing, Synthesis of biodegradable and electroactive multiblock polylactide and aniline pentamer copolymer for tissue engineering applications, Biomacromolecules. 9 (2008) 850–858.

44. S. Liu, Y. Wu, Y. Zhang, Z. Chi, Y. Wei, J. Xu, Synthesis and characterization of functional ABA block polymer containing aniline trimer, Chem. Lett. 38 (2009) 840–841.

45. B. Guo, A. Finne-Wistrand, A.C. Albertsson, Universal two-step approach to degradable and electroactive block copolymers and networks from combined ring-opening polymerization and post-functionalization via oxidative coupling reactions, Macromolecules. 44 (2011) 5227–5236.

46. B. Guo, A. Finne-Wistrand, A.C. Albertsson, Simple route to size-tunable degradable and electroactive nanoparticles from the self-assembly of conducting coil-rod-coil triblock copolymers, Chem. Mater. 23 (2011) 4045–4055.

47. S. Lakshmi, D.S. Katti, C.T. Laurencin, Biodegradable polyphosphazenes for drug delivery applications, Adv. Drug Deliv. Rev. 55 (2003) 467–482.

48. Q.S. Zhang, Y.H. Yan, S.P. Li, T. Feng, Synthesis of a novel biodegradable and electroactive polyphosphazene for biomedical application, Biomed. Mater. 4 (2009) 035008.

49. B. Guo, A. Finne-Wistrand, A.C. Albertsson, Electroactive hydrophilic polylactide surface by covalent modification with tetraaniline, Macromolecules. 45 (2012) 652–659.

50. X. Zhu, Y. Zhou, D. Yan, Influence of branching architecture on polymer properties, J. Polym. Sci., B: Polym. Phys. 49 (2011) 1277–1286.

51. M. Hakkarainen, A. Höglund, K. Odelius, A.C. Albertsson, Tuning the release rate of acidic degradation products through macromolecular design of caprolactone-based copolymers, J. Am. Chem. Soc. 129 (2007) 6308–6312.

52. H.R. Kricheldorf, Star shaped and hyperbranched aromatic polyesters, Pure Appl. Chem. 70 (1998) 1235–1238.

53. B. Guo, A. Finne-Wistrand, A.C. Albertsson, Molecular architecture of electroactive and biodegradable copolymers composed of polylactide and carboxyl-capped aniline trimer, Biomacromolecules. 11 (2010) 855–863.

54. B. Guo, A. Finne-Wistrand, A.C. Albertsson, Enhanced electrical conductivity by macromolecular architecture: hyperbranched electroactive and degradable block copolymers based on poly(ε-caprolactone) and aniline pentamer, Macromolecules. 43 (2010) 4472–4480.

55. H.R. Kricheldorf, Q.Z. Zang, G. Schwarz, New polymer syntheses: 6. Linear and branched poly(3-hydroxybenzoates), Polymer (Guildf). 23 (1982) 1821–1829.

56. P.J. Flory, Molecular size distribution in three dimensional polymers. VI. Branched polymers containing A-R-Bf-1 type units, J. Am. Chem. Soc. 74 (1952) 2718–2723.

57. B. Guo, Y. Sun, A. Finne-Wistrand, K. Mustafa, A.C. Albertsson, Electroactive porous tubular scaffolds with degradability and non-cytotoxicity for neural tissue regeneration, Acta Biomater. 8 (2012) 144–153.

58. K.Y. Lee, D.J. Mooney, Hydrogels for tissue engineering, Chem. Rev. 101 (2001) 1869–1879.

59. B. Guo, A. Finne-Wistrand, A.C. Albertsson, Versatile functionalization of polyester hydrogels with electroactive aniline oligomers, J. Polym. Sci., A: Polym. Chem. 49 (2011) 2097–2105.

60. B. Guo, A. Finne-Wistrand, A.C. Albertsson, Degradable and electroactive hydrogels with tunable electrical conductivity and swelling behavior, Chem. Mater. 23 (2011) 1254–1262.

61. Y. Liu, J. Hu, X. Zhuang, P. Zhang, Y. Wei, X. Wang, X. Chen, Synthesis and characterization of novel biodegradable and electroactive hydrogel based on aniline oligomer and gelatin, Macromol. Biosci. 12 (2012) 241–250.

62. B. Guo, A. Finne-Wistrand, A.C. Albertsson, Facile synthesis of degradable and electrically conductive polysaccharide hydrogels, Biomacromolecules. 12 (2011) 2601–2609.

63. M. Moussa, M.F. El-Kady, D. Dubal, T.T. Tung, M.J. Nine, N. Mohamed, R.B. Kaner, D. Losic, Self-assembly and cross-linking of conducting polymers into 3D hydrogel electrodes for supercapacitor applications, ACS Appl. Energy Mater. 3 (2020) 923–932.

64. P. He, X. Li, M. Deng, T. Chen, H. Liang, Complex micelles from the self-assembly of coil-rod-coil amphiphilic triblock copolymers in selective solvents, Soft Matter. 6 (2010) 1539–1546.

65. H. Kim, S.M. Jeong, J.W. Park, Electrical switching between vesicles and micelles via redox-responsive self-assembly of amphiphilic rod-coils, J. Am. Chem. Soc. 133 (2011) 5206–5209.

66. K. Knop, R. Hoogenboom, D. Fischer, U.S. Schubert, Poly(ethylene glycol) in drug delivery: pros and cons as well as potential alternatives, Angew. Chem. Int. Ed. 49 (2010) 6288–6308.

67. H. Wang, P. Guo, Y. Han, Synthesis and surface morphology of tetraaniline-block-poly(L-lactate) diblock oligomers, Macromol. Rapid Commun. 27 (2006) 63–68.

68. W. Xiong, H. Wang, Y. Han, Fibrils formed by dendron-b-oligoaniline-b-dendron block co-oligomer, Macromol. Rapid Commun. 31 (2010) 1886–1891.

69. W. Xiong, H. Wang, Y. Han, Oxidation induced self-assembly transformation of dendron-b-oligoaniline-b-dendron dumbbell shape triblock oligomer, Soft Matter. 7 (2011) 8516–8524.

70. J. Hu, L. Huang, X. Zhuang, P. Zhang, L. Lang, X. Chen, Y. Wei, X. Jing, Electroactive aniline pentamer cross-linking chitosan for stimulation growth of electrically sensitive cells, Biomacromolecules. 9 (2008) 2637–2644.

71. Q. Wang, W. He, J. Huang, S. Liu, G. Wu, W. Teng, Q. Wang, Y. Dong, Synthesis of water soluble, biodegradable, and electroactive polysaccharide crosslinker with aldehyde and carboxylic groups for biomedical applications, Macromol. Biosci. 11 (2011) 362–372.

72. A.N. Zelikin, D.M. Lynn, J. Farhadi, I. Martin, V. Shastri, R. Langer, Erodible conducting polymers for potential biomedical applications, Angew. Chem. 114 (2002) 149–152.

73. H. Xiang, Y. Chen, Energy-converting nanomedicine, Small. 15 (2019) 1805339.

74. L.V. Kayser, D.J. Lipomi, Stretchable conductive polymers and composites based on PEDOT and PEDOT:PSS, Adv. Mater. 31 (2019) 1806133.

75. R. Rasouli, A. Barhoum, M. Bechelany, A. Dufresne, Nanofibers for biomedical and healthcare applications, Macromol. Biosci. 19 (2019) 1800256.

76. F.D. Bobbink, A.P. Van Muyden, P.J. Dyson, En route to CO_2-containing renewable materials: catalytic synthesis of polycarbonates and non-isocyanate polyhydroxyurethanes derived from cyclic carbonates, Chem. Commun. 55 (2019) 1360–1373.

77. B. Shu, X. Sun, R. Liu, F. Jiang, H. Yu, N. Xu, Y. An, Restoring electrical connection using a conductive biomaterial provides a new therapeutic strategy for rats with spinal cord injury, Neurosci. Lett. 692 (2019) 33–40.

6 Biodegradable Polymers
Synthesis to Advanced Biomedical Applications

Eman Abdallah Ismail,[1] *Mbuso Faya,*[1] *Edith Amuhaya,*[2]
Calvin A. Omolo,[1,2] *and Thirumala Govender*[1]

[1]Discipline of Pharmaceutical Sciences, School of Health
Sciences, University of KwaZulu-Natal, Durban, South Africa

[2]Department of Pharmaceutics and Pharmacy Practice,
School of Pharmacy and Health Sciences, United States
International University-Africa, Nairobi, Kenya

CONTENTS

6.1 INTRODUCTION

Biodegradable polymers (BPs) are being employed in various biomedical applications. They gained prominence in the 1980s, due to the environmental concerns and realization of the fact that petroleum resources are confined. They can be classified based on their origin, chemical composition, synthesis technique, applications, economic importance, etc. [1]. Classification based on their origin is the most common employed [2, 3] based on which they have been classified into natural polymers derived from natural sources, semisynthetic BPs, sourced from natural raw materials, but polymerized after chemical modification and synthetic BPs which are obtained through chemical synthesis [4, 5]. Despite their availability and biocompatibility, the application of natural BPs has been limited due to batch-to-batch differences in characteristics and microbial contaminations [6]. This was the major motive for the emergence of synthetic BPs that have gained prominence and become an active area of research due to the reproducibility of the polymers, high synthetic yields, and controlled biodegradation that can be modified to have a wide range of physical, chemical and mechanical characteristics based on the intended application [7, 8]. Moreover, the biodegradable materials can be morphed into nanoscale assemblies with attractive traits that can be employed in a variety of medical applications, from the delivery of drugs to surgical implants, and tissue regeneration [9].

BPs have been widely employed in biomedical fields such as drug carriers biomaterials, medical devices, and tissue engineering. With organ transplantation being greatly controlled by the dearth of organ donors, tissue engineering is being explored as an approach to address this problem as it offers flexibility in processing; biocompatibility, degradation can be controlled to be nontoxic by-products; three-dimensional (3D), porous, and well-organized pore network that can facilitate nutrient and waste transport. Moreover, the engineering of the BPs can be made to have appropriate mechanical integrity to support regenerations and have the suitable design and topography to adequately interact with cells [10–12].

The applications of BPs in the biomedical disciplines are continually growing and evolving. This is owed to their outstanding properties in terms of biocompatibility, low toxicity, and chemically tunable characteristics, most interestingly, the ability of such polymers to be degraded into by-products that are nontoxic [10–12]. Furthermore, these polymeric biomaterials have been utilized in orthopedic devices to replace blood vessels, bones, and surgical sutures [13], tissue engineering and scaffolds [14, 15], urology [16], and cardiology [17]. Extensive research has been done and is still ongoing in an attempt to engineer new BPs or modify existing ones to acquire the desired features.

This chapter is aimed to provide a thorough overview of the recent advancement in the strategies of engineering and fabricating BPs, their various application in the biomedical field. Additionally, the chapter highlights the unexplored avenues regarding types of linkages and bonds required to achieve degradability and chemical processes involved in the breakdown of such linkage. Finally, general biomedical applications of biodegradable polymeric materials in drug delivery, engineering of medical devices, tissue engineering, and future perspectives of the chemistry of BPs are thoroughly discussed.

6.2   SYNTHETIC ROUTES TO BIODEGRADABLE POLYMERS

Synthetic protocols have been developed to prepare different BPs with reproducible quality and purity, leading to their wide range of applications [18]. This section covers different synthetic approaches to BPs. These have been broadly categorized into (a) chemical synthesis of BPs, (b) enzyme-mediated synthesis of BPs, (c) synthesis of BPs by use of microorganisms, and (d) chemo-enzymatic synthesis of BPs [18, 19].

6.2.1   CHEMICAL SYNTHESIS OF BIODEGRADABLE POLYMERS

Advances in polymer science and technology have led to the development and preparation of a variety of BPs with a wide range of applications. Furthermore, the fact that these synthetic protocols can be scaled up for mass production with relative ease has made these types of preparations more desirable [20]. In this section, we review chemical synthetic approaches used to synthesize BPs.

6.2.1.1   Condensation Polymerization

Condensation polymerization, alternatively known as step-growth polymerization, is a reaction whereby two monomer units react and, in the process, release a small molecule such as water, ammonia, ethanol, or carbon dioxide. These reactions can be carried out in the presence or absence of a catalyst. The reaction requires that the monomers involved must not only bear functional groups such as $-NH_2$, $-OH$, or $-COOH$ but should also have at least two reactive sites (or growth points). For monomers with two reactive sites, linear polymers are obtained, while monomers with more than two reactive sites generate cross-linked 3D polymers [21, 22]. Scheme 6.1 shows the synthesis of some common polymers using condensation polymerization.

Polylactic acid (PLA) is a popular aliphatic polyester that can be prepared through a condensation reaction. It is a BP that degrades into lactic acid. Because of this, it is used in medical implants such as screws, rods, mesh, and pins. Two methods can be utilized to prepare PLA: direct polycondensation of the lactic acid and ring-opening polymerization. The latter is discussed in Section 6.2.1.3. In condensation polymerization, the monomer, lactic acid is dissolved in an appropriate solvent, and the reaction is allowed to proceed for a relatively long duration at moderate heat, as shown in Scheme 6.2. This leads to a polymeric material that is characterized by a low to intermediate molecular weight [22].

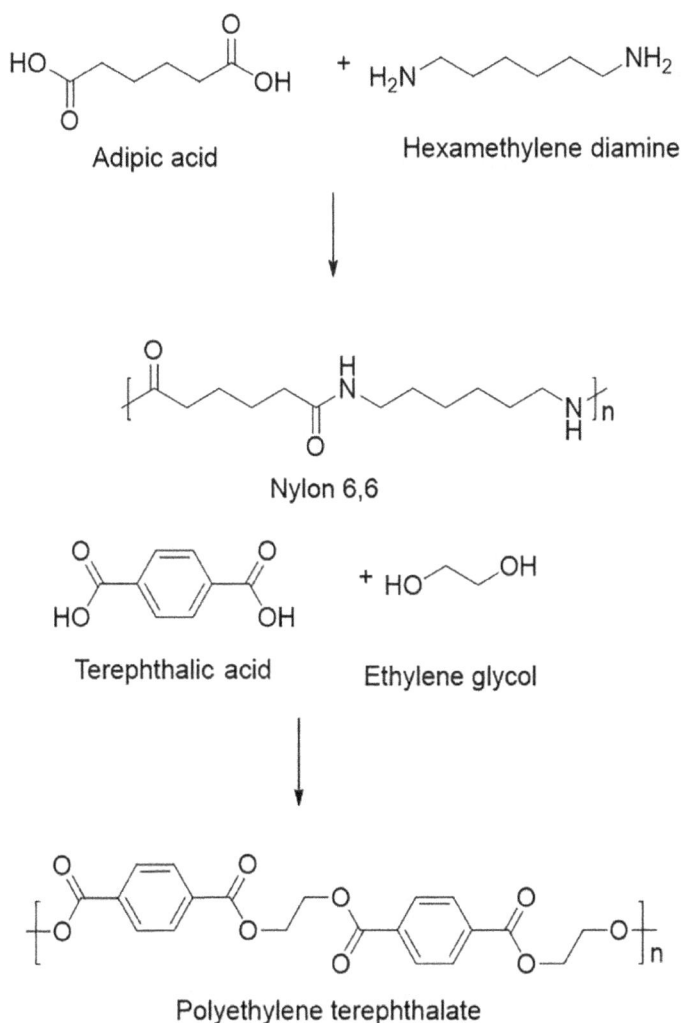

Adipic acid

Hexamethylene diamine

Nylon 6,6

Terephthalic acid

Ethylene glycol

Polyethylene terephthalate

SCHEME 6.1   (a) Synthesis of nylon 6,6 and (b) synthesis of polyethylene terephthalate.

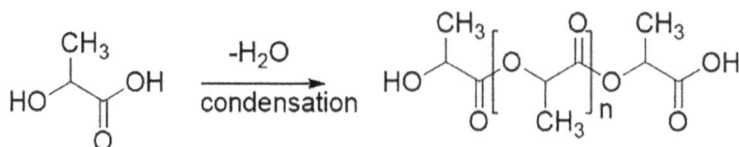

SCHEME 6.2   Synthesis of PLA using condensation polymerization.

6.2.1.2   Addition Polymerization

Addition polymerization is also defined as chain-growth polymerization. Unlike condensation polymerization, addition polymerization reactions do not involve the loss of a small molecule. The monomers that usually undergo addition polymerization are those that are unsaturated, such as aldehydes, olefins, and acetylenes. This reaction proceeds in a stepwise fashion by forming reactive intermediates. This polymerization is usually exothermic because it involves the conversion of a p bond in the monomer into a sigma bond in the polymer. Polymers that are synthesized through this process are characterized by high molecular weights. While this process is mainly used to prepare straight-chain polymers, cross-linking can be achieved by using monomers containing two double bonds [22]. There are three possible mechanisms through which polymers can be prepared: anionic polymerization, cationic polymerization, and radical polymerization. The exact mechanism will depend on the reaction conditions used as well as the presence of electron-withdrawing or electron-donating groups on the monomers.

Polycyanoacrylate is an example of an addition polymer that is made from the monomer a-cyanoacrylate. a-Cyanoacrylate rapidly undergoes polymerization in the presence of a hydroxide ion to form a high molecular weight, straight-chain polymer. Figure 6.1 shows the general formula for polycyanoacrylate. Generation of this polymer proceeds through an anionic mechanism since the monomer contains a −CN group which is electron-withdrawing. Their adhesive properties make them suitable for use in the medical field. It is however worth noting that cyanoacrylate monomers with large alkyl ester groups are preferred since smaller alkyl groups have been found to irritate tissues in the body. The use of cyanoacrylate medical adhesives has been adopted in some cases to replace sutures because of their cosmetic effects, lowered pain and reduced recovery periods in patients, and hence reduced associated costs. Cyanoacrylates have also been found useful in drug delivery and targeting systems, as well as in skin burns, bone and cartilage grafting. Finally, these materials have been used by dentists in dental fillings [23]. Other applications of addition polymers are shown in Table 6.1.

6.2.1.3   Ring-Opening Polymerization

Ring-opening polymerization (ROP) is a class of polymerization whereby a cyclic monomer is utilized to produce a polymer that lacks cyclic structures in its backbone [18, 22]. It is associated with higher molecular weight polymers compared to polycondensation. However, a major drawback is

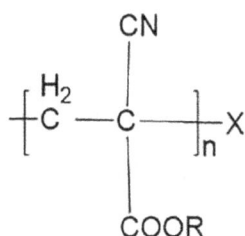

FIGURE 6.1   General structure of polycyanoacrylates.

TABLE 6.1

Some Common Addition Polymers' Properties and Their Biomedical Applications

Properties and Biomedical Uses of Some Common Addition Polymers

Polymer Name(s)	Properties	Biomedical Uses
Polyethylene low density (LDPE)	Soft, waxy solid	Films, blood bags
Polyethylene high density (HDPE)	Rigid, translucent solid	Hip joints
Polyvinyl chloride (PVC)	Strong rigid solid	Reinforcement of artery
Polytetrafluoroethylene (PTFE, Teflon)	Resistant, smooth solid	Heart pumps, reinforcement of artery and blood vessels
Polymethyl methacrylate (PMMA, Lucite and Plexiglas)	Hard, transparent solid	Contact lenses, heart pumps

Source: Adapted with permission from Ref. [22]. Copyright (2014) Elsevier.

the costly manufacturing of the cyclic oligomers at the industrial level [24]. It is a type of chain-growth polymerization in which the end of the polymer chain acts as a reactive site and is used to form a longer polymer chain by ring-opening of the cyclic monomers. Depending on the types of groups on the monomer, three types of reactive sites can be generated in ROP: anionic, cationic, or radical. This makes ROP a highly versatile method for synthesizing polymers. ROP proceeds with the use of metal or metal-ligand catalysts, using metals such as Li [25], Fe [26], Mg [27], and Zn [28], among others being used to successfully prepare a variety of polymers. ε-Caprolactone, D-valerolactone, and lactic acid are among the commonly used monomers [29].

As mentioned in a previous section, PLA can be prepared by polycondensation as well as ROP [22]. ROP of the cyclic dimer, known as lactide, proceeds in presence of a catalyst and it produces polymers with higher molecular weight compared to those obtained using the polycondensation process. Scheme 6.3 illustrates the synthesis of PLA via ROP. In addition to the production of higher molecular weight PLA using ROP, it is also possible to modify the sequence and ratio of D-and L-lactic acid units in the final polymer. This gives rise to PLA polymers with varying properties and therefore different applications [22].

6.2.2 SYNTHESIS OF BIODEGRADABLE POLYMERS BY USE OF MICROORGANISMS

This refers to the use of microorganisms to produce polymeric substances utilizing organic materials like glucose and starch as food sources. An example of polymers prepared through this method is polyhydroxyalkanoates (PHAs) (Figure 6.2). PHAs are synthesized by the bacterial fermentation of sugar or lipids. They are mainly produced from saturated and unsaturated hydroxy alkanoic acids (HAA) which can be branched or unbranched [19, 30]. PHAs are deemed as biocompatible and therefore can be utilized in biomedical applications such as drug encapsulation and tissue engineering [31].

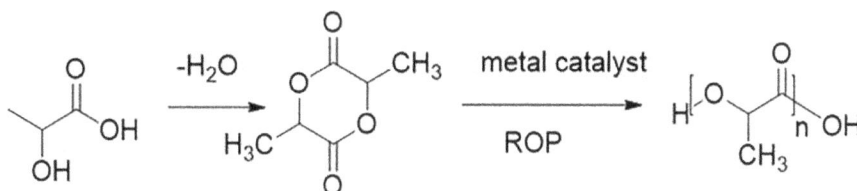

SCHEME 6.3   Synthesis of PLA via ROP.

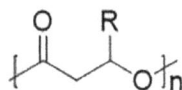

FIGURE 6.2 A general structure of PHAs.

One widely studied PHA is poly(hydroxybutyrate) (PHB), which is a BP produced by micro-organisms such as *Ralstonia eutropha* and *Bacillus megaterium* in response to physiological strain. It has attracted attention because it can be generated from low-cost renewable feedstocks. Furthermore, the polymerization process can be achieved under moderate conditions without causing too much effect on the environment [31]. Studies have shown that bacterial synthesized PHAs have much higher molecular weights, 500,000 to > 1,000,000, than those attained by typical poly-condensation polymerization reactions which utilize hydroxy acid monomers [18]. One example of the synthesis of PHB can be seen in Scheme 6.4 which shows the synthesis of PHB in the presence of PHB synthase [18, 32]. The synthesis starts with the preparation of 3-hydroxybutyryl-CoA, which is then polymerized by reacting it with mutant enzymes cultivated from *Aeromonas caviae*. The average molecular weight of the polymers synthesized range from 280,000 to 1,070,000 Da. PHB is marketed under the trade name Biopol and because of its nontoxicity and biodegradability, it is widely utilized in the medical industry for internal sutures which do not have to be removed after recovery [33]. Based on their varied properties, other PHAs have found applications in the medical field as summarized in Table 6.2.

6.2.3 Enzyme-mediated Synthesis of Biodegradable Polymers

This is a relatively novel method for the production of BPs. By taking advantage of an enzyme's unique properties, specialized polymerization reactions can be catalyzed to make a variety of polymers. A major advantage of this method is that the synthesis does not involve any by-products owing to the outstanding specificity of enzymes. This makes extraction and purification of the final polymeric product easy to carry out. Compared with conventional chemical polymerization reactions, these reactions have added advantages that include (a) mild reaction conditions, usually room temperature and

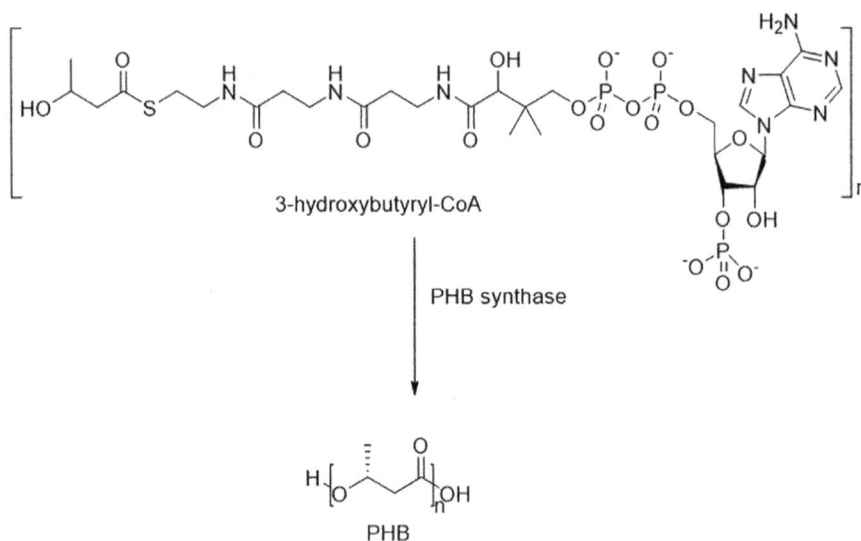

3-hydroxybutyryl-CoA

PHB synthase

PHB

SCHEME 6.4 Synthesis of PHB in presence of PHB synthase.

TABLE 6.2
Bacterial PHAs and Their Applications [34]

Polymer	–R Group	Applications
Poly(hydroxylbutyrate) PHB	$-CH_3$	Seam threads for the healing of wounds and blood vessels, creating bioplastics
Poly(hydroxyvalerate) PHV	$-CH_2-CH_3$	Controlled release of drugs, clinical repairs, orthopedic devices
Poly(hydroxyhexanoate) PHHex	$-(CH_2)_2-CH_3$	Medical implants, adhesion barriers, stents, vein valves
Poly(hydroxyoctanoate) PHO	$-(CH_2)_4-CH_3$	Biomedical grafting, bone marrow scaffolds
Poly(hydroxydecanoate) PHD	$-(CH_2)_6-CH_3$	Drug delivery
Poly(hydroxy phenylvalerate) PHPV	$-CH_2-C_6H_5$	Surgical implants, biofuels

atmospheric pressure; (b) enhanced control of enantioselectivity, regioselectivity, and chemoselectivity; (c) ability to recycle catalysts; and (d) ability to catalyze ROP of macrocyclic lactones.

An example of an enzyme-mediated synthesis of biopolymers is lipase-catalyzed polymerization [35]. Polyesters and polyamides have been prepared using this method to produce good-to-high yields of polymers [36]. The example in Scheme 6.5 shows the synthesis of poly(2-dimethylaminotrimethylene carbonate) (PDMATC), through the lipase-catalyzed ring-opening polymerization of 2-dimethylaminotrimethylene carbonate (DMATC). The polymerization reaction was done at 60°C for 24 hours and the resultant product was purified by precipitation to give the final polymer which had a molecular weight of 4200 Da as determined by GPC.

Copolymers have also been prepared through lipase-catalyzed polymerization. For example, Scheme 6.6 shows the synthesis of poly(hexamethylene *g*-ketopimelate-*co*-hexamethylene adipate) (poly(HK-*co*-HA)) by a *Candida antarctica* lipase B (CALB)-catalyzed polycondensation of diethyl *g*-ketopimelate (DEK), diethyl adipate (DEA), and 1,6-hexanediol (HDO) using different ratios. The polymerization reactions were carried out at 90°C at reduced pressure for 48 hours. The resulting crude polymers were purified by filtration. In all cases, the final polyesters were obtained at yields above 75%, with molecular weights ranging between 24,200 and 28,000 Da, with PDI values between 1.47 and 1.60 [38].

6.2.4 CHEMO-ENZYMATIC SYNTHESIS OF BIODEGRADABLE POLYMERS

While enzyme-catalyzed polymerization reactions have advantages over the conventional chemical polymerization process, it still suffers from major drawbacks, such as the absence of control over the structure of the polymer, which prohibits engineering of complex structures such as block and grafted copolymers [39]. This means that it is not possible to completely replace chemical polymerizations. This has led to the development of chemo-enzymatic polymerizations that utilize both biosynthetic and organic catalytic methods such that they occur concurrently in the same system

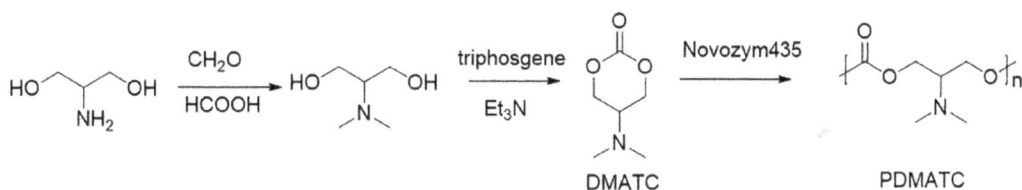

SCHEME 6.5   Synthesis of water-soluble aliphatic polycarbonate (PDMATC) [37].

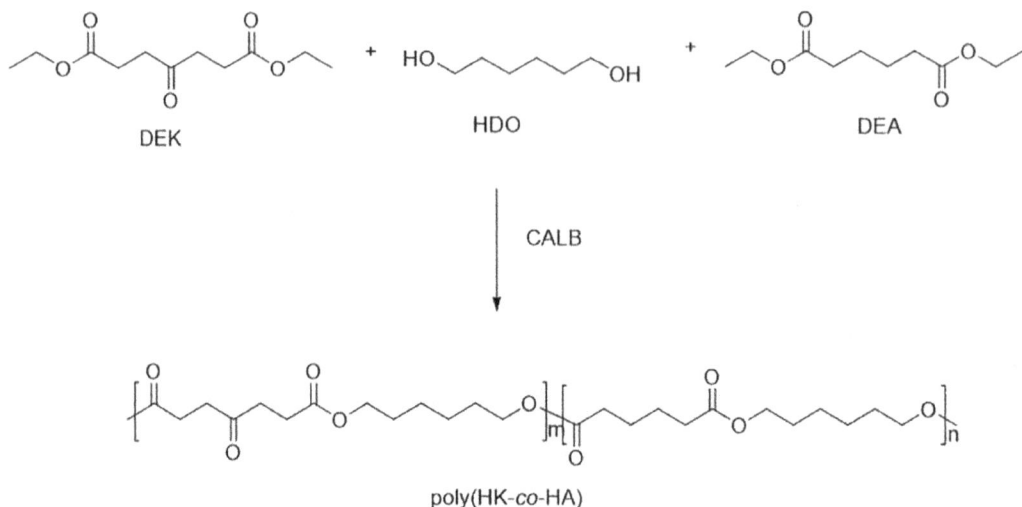

SCHEME 6.6 CALB-catalyzed synthesis of poly(HK-*co*-HA) copolymer [38].

[40]. This has led to the development of polymers that are otherwise difficult to prepare using the two methods independently.

Synthetic protocols that combine lipase-catalyzed polymerization with different chemical processes, such as atom transfer radical polymerization (ATRP) and kinetic resolution, have been developed. The most commonly used chemical route for combination with lipase-catalyzed polymerization is ATRP. This technique is highly successful due to the enzymatically catalyzed synthesis of end-functionalized polyesters, which can be achieved in two ways: the initiator method and terminator method [41]. In the initiator pathway, a nucleophile such as an amine or alcohol initiates the enzymatic polymerization to make up the end-functionalized polyesters. On the other hand, the terminator route utilizes vinyl esters, commonly divinyl sebacate, and methacrylic acid, which function as terminators in a one-step acylation of a hydroxyl group of polyester as shown in Scheme 6.7. The terminator technique is, however, of less efficiency since polymer molecular weight decreases with an increase in terminator concentration [41].

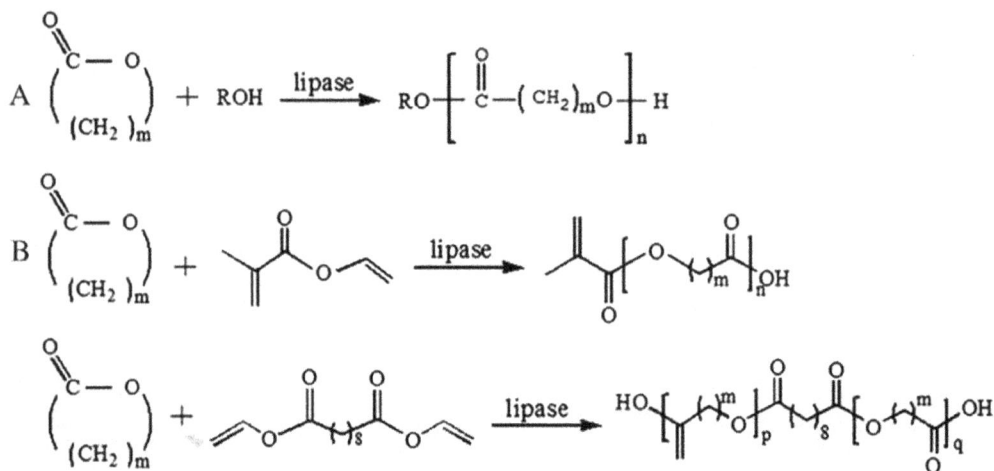

SCHEME 6.7 Enzymatic production of end-functionalized polyesters by the initiator (a) and terminator (b) methods [41].

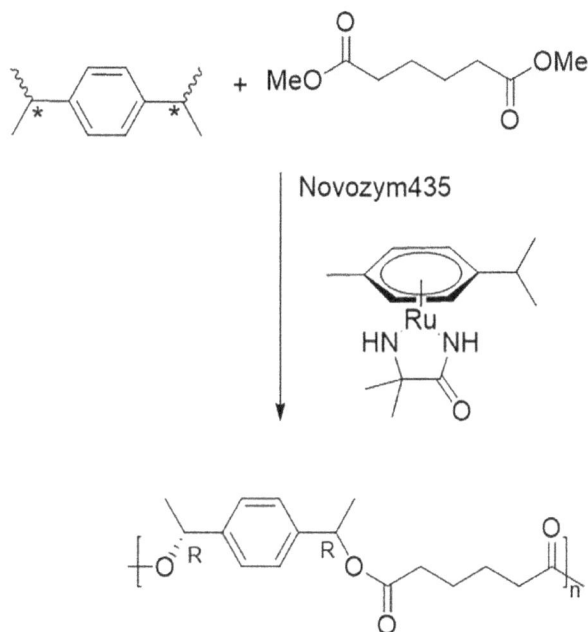

SCHEME 6.8 Chiral polyesters constructed by one-pot dynamic kinetic resolution polymerization [42].

In the pharmaceutical industry, the demand for enantiomerically pure compounds has become increasingly important. As such, lipase-catalyzed dynamic kinetic resolution polymerization protocols have been established to produce enantiopure, polymeric materials. For example, enzymatic polymerization and metal-catalyzed racemization have been found to work concurrently to synthesize polymers with high optical purity and significant molecular weight, Scheme 6.8 [42]. The examples given in this section highlight the significance of chemo-enzymatic polymerization using lipases as a powerful technique for the preparation of complex biodegradable and biocompatible polymers which can be used in the pharmaceutical industry.

6.3 APPLICATION OF BIODEGRADABLE POLYMERS

6.3.1 APPLICATION OF BIODEGRADABLE POLYMERS MEDICAL DEVICES

In the past century, considerable headway has been made in the advance of BPs, as they are favored in the engineering of therapeutic devices, temporary implants, and 3D scaffolds for tissue engineering as well. These polymers are exceptional biomaterials for usage in biomedical applications due to their improved biocompatibility, low toxicity, and chemical control properties. This section discusses various polymers (Figure 6.3) and their applications in the creation of medical devices.

6.3.1.1 Polyglycolide (PGA)

PGA and PLA belong to the PHA family, and together with their PLGA copolymer composed of PLA and PGA units, are BPs widely used in biological applications. PLA is a natural aliphatic polyester made from natural resources such as wheat, and due to its superior reproducibility, biocompatibility, and biodegradability, it is commonly used for surgical implants, tissue cultures, and absorbable surgical sutures, wound closure, controlled-release system, and prosthesis [43]. Despite having structural similarities to PGA, PLA has different physicochemical properties since it possesses additional methyl groups in the repeating unit. Moreover, a PLA scaffold can last for longer periods before it loses mechanical integrity, it, therefore, has been seen to offer better application

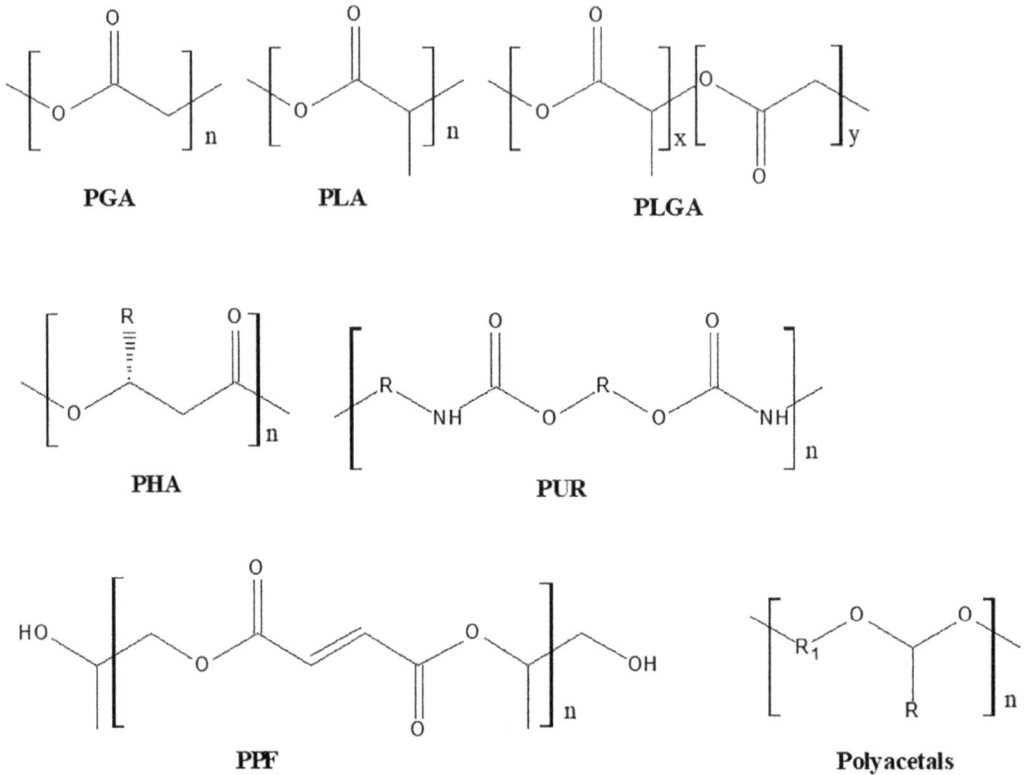

FIGURE 6.3 Chemical structures of commonly used biodegradable polymers in medical devices.

toward orthopedic fixation devices [44]. Another important characteristic to note is that PLA comes in different forms since it possesses chiral molecules, poly(L-lactic acid) (PLLA), poly(D-lactic acid) (PDLA), poly(D,L-lactic acid) (PDLLA) – a racemic mixture of PLLA and PDLA, and *meso*-poly(lactic acid) [45]. Only PLLA and PDLLA have been proven to be useful in biomedical research and have been intensively studied. Functionalization of these BPs through the advancement in chemistry has resulted in highly effective multifunctional medical devices.

6.3.1.2 Poly(Lactide-*co*-Glycolide) (PLGA)

PLGA is the best-studied BP for medical applications, it is extensively employed in sutures, medical devices, drug carriers, and tissue engineering. Depending on the composition of the PLGA used and the interaction between the payload and the polymer, the release profile of the active ingredient or protein may vary. PLGA is also widely used in the manufacture of various biomedical devices such as grafts, sutures, implants, and prostheses. In addition to drug delivery applications, PLGA is also one of the key polymers in the development of medical drug binding devices and tissue engineering applications. For example, few nanoparticle PLGA stents can be used to control the release of cardiovascular drugs. PLGA can be combined with stents for drug delivery applications and has great potential in this field. In cosmetics, the penetration rate of various assets can be improved by including them in PLGA [46]. Furthermore, functionalization of medical devices from PLGA-based nanofibers is an active area of research. Fibers in nanoscale diameter are being electrospun into yarn that can be applied in designing various medical devices. Such studies include one reported by Rouhollahi et al., who reported electrospun nanofibers with diameters ranging from 450 to 1170 nm and loaded them with silver nanoparticles to form a nanofibrous yarn. From the studies, it was found that by the addition of PGA to PLGA polymer, the mechanical strength of the yarn increased from 36.6 MPa and 0.9 GPa to 51.3 MPa and 1.9 GPa, respectively. Moreover, knotting the sutures increased the mechanical strength of the yarn while impregnation with silver nitrate enhanced wound healing due to its ability to prevent wound infections [47]. In another similar study by Bae et al., electrospun nanofibers were utilized for surgical connection of blood vessels, anastomosis, transplantation, and reconstructive surgical procedures with the PLGA-based nanofibers. The electrospun of the nanofibers was composed of PLGA, poly(ethylene oxide) (PEO), and the positively charged copolymer, poly(lactide-co-glycolide)-graft-polyethylenimine (PgP) to enhance electrostatic attraction and impregnated the fibers with heparin. The diameters of the nanofibers were found to be 48.0 ± 6.38 nm for PLGA/PEO, 51.9 ± 7.93 nm for (PLGA/PEO/ PgP$_1$, and 43.4 ± 5.11 nm for PLGA/PEO/PgP$_{3.7}$, nanofibers (Figure 6.4), while the yarns from the nanofibers were found to have diameters of yarns with the average diameter of 93.00 ± 5.93 μm, 105.56 ± 18.67 μm, and 92.74 ± 20.11 μm PLGA/PEO, PLGA/PEO/PgP$_1$, and PLGA/PEO/ PgP$_{3.7}$, respectively. The yarns containing PgP had significantly higher tensile strength and higher heparin loading slower release than those without the PgP grafting [48]. The results were in line with other functionalized PLGA-based nanofibers [49–52]. The continued development in PLGA chemistry is opening up possibilities for blends and "tailor-made" compounds that can be employed to produce advanced biomedical devices.

6.3.1.3 Polyhydroxyalkanoates (PHA)

PHA is a kind of natural biodegradable polyester, produced by microbes. They possess excellent biocompatibility and biodegradability and produce nontoxic degradation by-products, making them ideal for tissue engineering, and implantable device replacement, including sutures, patches, slings, orthotics, stents, stents, and adhesive barriers [53]. PHA can also be used as therapeutic support for tissues and organs, as a mechanical barrier to protect organs, nerves, and tendons from scar tissue. The piezoelectric property of P (3HB) promotes neuron regeneration and is also conducive to tissue regeneration [54]. The applicability of P (4HB) as a tissue matrix for heart valves, vascular plasters, sutures, orthopedic implants, stents, and local drug delivery systems has also been studied.

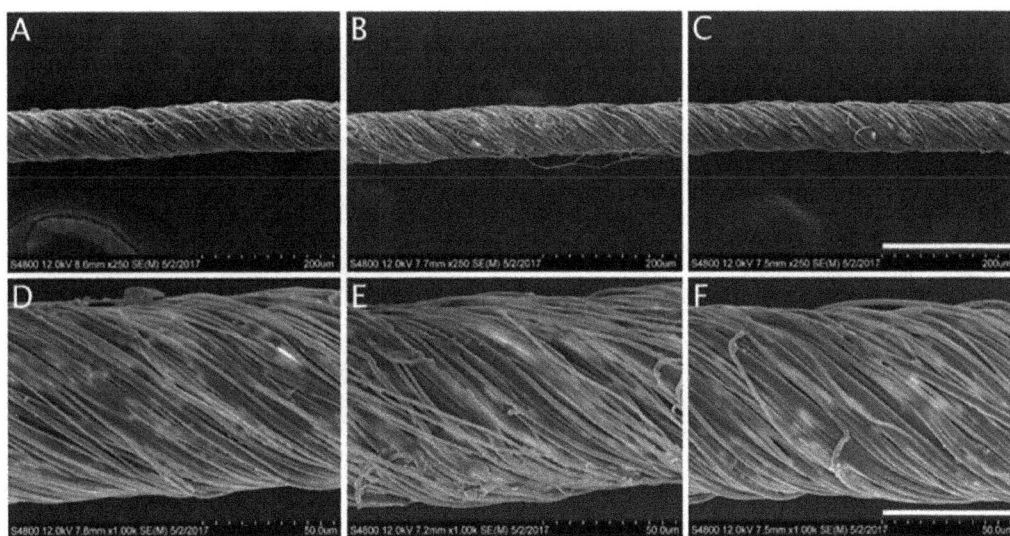

FIGURE 6.4 Surface characterization of the electrospun nanofibers yarns with different compositions. (a) and (d): PLGA/PEO; (b) and (e): PLGA/PEO/PgP1 and (c and f); PLGA/PEO/PgP3.7. (Reproduced with permission from Ref. [48]. Copyright (2018) American Chemical Society.)

6.3.1.4 Polyurethane (PUR)

Due to its strength, durability, biocompatibility, and biostability, polyurethanes are the first choice for medical devices and have a compelling potential as a basis for tissue regeneration. They are versatile in their mechanical and physicochemical properties. In addition, a unique two-component polyurethane injection system PolyNova® (PolyNovo Biomaterials Pvt. Ltd.) based on lysine diisocyanate was developed for orthopedic applications. Compared with standard bone cement, it has sufficient mechanical support and higher bonding strength for arthroscopic applications in liquid form at normal body temperature in situ [55]. In addition, it supports convenient cell adhesion and proliferation. Due to its strength, durability, biocompatibility, and higher biostability, polyurethanes have been utilized as inert material in heart valves, catheters, prostheses, and vascular grafts.

Moreover, Pus can be employed to develop shape memory polymers (SMPs) that are polymeric materials possessing shape memory effect via 3D printing. Wang et al. fabricated water-based biodegradable SMPs-PU as the leading component for 3D-printed bone scaffolds [56]. The novel scaffold exhibited excellent shape memory properties as shown in Figure 6.5. The scaffold was loaded with human mesenchymal stem cells (hMSCs) and superparamagnetic iron oxide nanoparticles (FeNPs) and evaluated for borne regeneration. The scaffold with FeNPs had 2.7-fold increase osteogenesis than scaffolds without FeNPs (Figure 6.5). Moreover, the release of FeNPs from the scaffold was consistent for 14 days [56]. The versatility of PUR makes them excellent biomaterials for various applications in the biomedical field.

6.3.1.5 Poly(Propylene Fumarate) (PPF)

PPF possesses high strength with a unique capability to cross-link unsaturated bonds in its backbone. Due to its ability to cross-link, PPF degradation depends on its molecular weight, a cross-linking agent, and cross-link density. PPF is an injection that solidifies when cross-linked; therefore, it is used biomedically in bone defect filling and long-term storage of ophthalmic drugs [57]. Additionally, in osteogenic tissue engineering, PPF is often mixed with ceramics (such as hydroxyapatite or aluminoxane) to create a stronger, more biologically active framework. Recent research activity has focused on the use of PPF to fill irregular-shaped bone defects, such as ossicles or mandibular defects [58]. In both cases, PPF-based stents can build structures that are not available with non-cross-linked, degradable polymers.

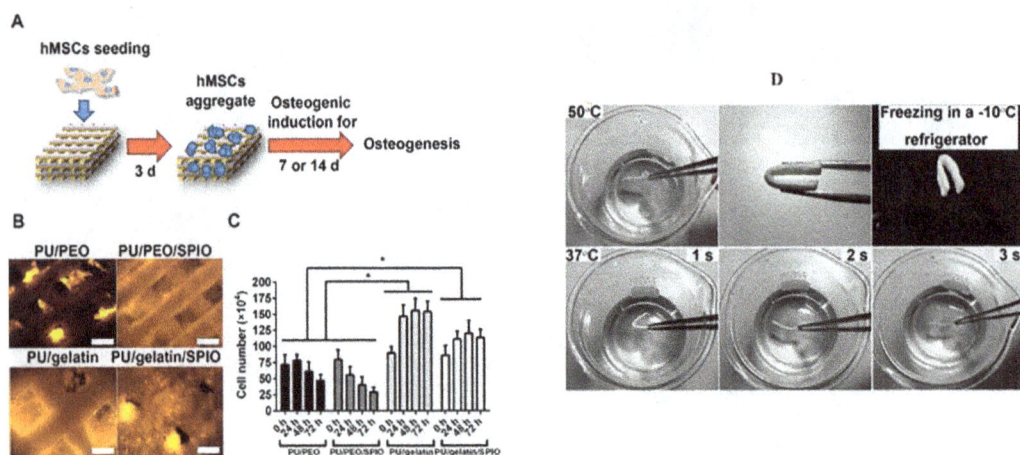

FIGURE 6.5   Design and characterization of 3DP shape memory PU scaffolds seeded with hMSCs for osteogenesis to enhance bone regeneration. (a) Preparation and seeding procedure. (b) Fluorescent images of the scaffold after 24 h culture. (c) Proliferation in PU-polyethylene oxide (PU/PEO), and PU gelatin scaffolds with and without superparamagnetic iron oxide nanoparticles during culture for 72 h. (d) Shape memory properties. (Reproduced with permission from Ref. [56]. Copyright (2018) American Chemical Society.)

6.3.1.6 Polyacetals

Polyacetals are degradable polymers and are generally divided into two subcategories: polyacetals and polyketals. Both of these polymers have been widely utilized in biomedical research since their decomposition products do not contain any carboxylic acids, which leads to a significantly mild pH microclimate, and their decomposition is acid-catalyzed [59]. A microenvironment with a milder pH value can deliver acidic and hydrolysis-sensitive payloads. For most implants, the use of polyacetals is restricted because they usually cannot be synthesized with a molecular weight high enough to confer mechanical strength requirements. One exception is DelrinVR (polyoxymethylene), which is a formaldehyde homopolymer that can be polymerized with high molecular weight through acid or anion catalysis. In the late 1960s, DelrinVR-based implants were used as oblique disc valves to repair defective heart valves. These implants were found to swell when used, and other materials have since replaced DelrinVR in artificial valves [60].

6.3.2 Applications of Biodegradable Polymers in Drug Delivery

The recent advancement in BPs represents a revolution in the biomedical field that has led to a significant biotechnological improvement in drug delivery encouraged by interdisciplinary research for effective drug delivery carriers [61, 62]. The biodegradable polymeric-based drug delivery platforms have been shown to overcome the limitation of biostable delivery systems by avoiding bioaccumulation [61, 63, 64] and programmed degradations [65] thereby improving the safety and efficacy of the delivered therapeutic payloads [65]. By taking the advantage of localized biochemical alterations in erratic disease states such as pH, reactive oxygen species, enzymatic activity, and reductive state, smart stimuli-responsive polymeric-based drug delivery systems have been introduced allowing for more precise drug therapy by controlling the rate and the site of drug release [66, 67] (Figure 6.6). The section below will discuss applications of BPs in controlled-release and targeted delivery systems.

6.3.2.1 Applications of Biodegradable Polymers in Controlled-release Delivery Systems

Mechanism of drug release from polymeric reservoir devices involved diffusion-controlled, drug-polymer affinity, and degradation of the polymeric matrix [61, 68]. Earlier polymeric controlled drug delivery systems were based on non-BPs, including polyurethanes, silicone rubber, and poly(ethylene-*co*-vinyl acetate) (PEVA) in which release of the payload was triggered by diffusion. However, the non-biodegradability of such biostable systems has necessitated the search for alternative BPs. This has led to the emergence of a new class of polymeric materials for drug delivery applications. Poly(hydroxyl acids) such as poly(lactic acid) (PLA), ploy(glycolic acid) (PGA), and their copolymer PLGA which were originally developed for sutures in the 1960s and 1970s [69]

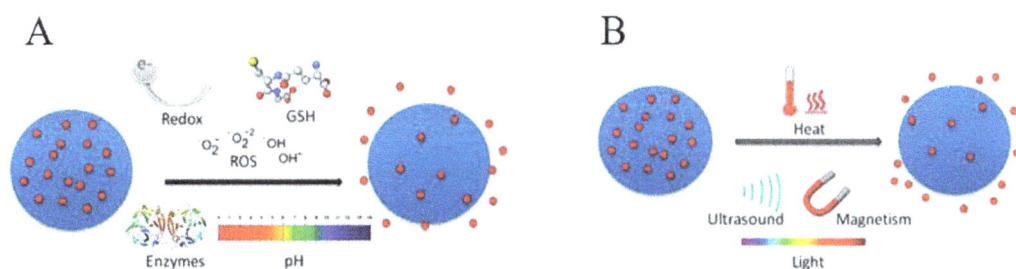

FIGURE 6.6 Design of (a) endogenous and (b) exogenous stimuli-responsive biodegradable polymeric nanoparticles. (Adapted with permission from Ref. [61]. Copyright (2016) American Chemical Society.)

have also been adopted by drug delivery scientists owing to their favorable biodegradability and safety profile.

BPs can be engineered into various drug delivery systems such as core-shell structures, solid nanoparticles, polyplexes, and polymeric micelles. These nanoscale delivery drugs morphed from BPs have shown to have attractive traits such as controlled release and extended releases of drugs which can result in reduced dosage frequency which leads to reduced pill burden and increased compliance and better drug safety. The application of BPs in drug delivery has resulted in several drugs being introduced in the market and several under clinical trials as illustrated in Table 6.3 [70].

6.3.2.2 Application of Biodegradable Polymers in Targeted Delivery

Advancement in synthetic chemistry and formulation science of BP has resulted in drug delivery systems that undergo programmed degradation for targeted drug release. Targeted delivery can be passive or active targeting. Passive targeting is based on taking the advantage of the drug carrier's characteristics to establish a physical interaction with the diseased tissue microenvironment (blood flow, lymphatic drainage, etc.), while active targeting can be achieved by conjugating tissue/receptor-specific ligands (antibodies, peptides, macromolecules, etc.) and stimuli-responsive ligands for spatial accumulation of nanoparticles and release of the drug in tissues of interest, thus avoiding nonspecific toxicity in healthy cells [72]. BP-based nanoparticles (80–150 nm) exploit the enhanced permeability and retention (EPR) effect to achieve passive targeting intrinsically. The anticancer activity of many clinically available nanoparticles (Doxil, Caelyx, and Abraxane) can be partially attributed to the passive targeting of the EPR effect [73]. Various formulations have been designed to exploit inflammatory microenvironment properties accompanied by a variety of diseases, including inflammatory bowel disease, inflammatory rheumatoid arthritis, bacterial infections, and tumors, and promote the accumulation of polymeric nanoparticles in such regions via passive diffusion [73].

Active targeting, on the other hand, provides a more precise and accurate accumulation of drugs at the targeted site [74]. Biodegradable polymeric nanoparticles can be modified and/or functionalized to afford the ability to target-specific pathological regions [75]. Functionalization using small molecules ligands such as folic acid and macromolecules, including antibodies, peptides, sugars,

TABLE 6.3
Selected Clinical Trials of Biodegradable Polymeric-Based Nanoparticles for Anticancer Drug Delivery

Name	Carrier	Drug	Indications	Status
Genexol-PM	PEG-PLA	PTX	Breast, lung, and ovarian cancer	Approved (South Korea)
NK 911	PEG-PAsp-DOX conjugate	DOX	Pancreatic and colorectal cancers	Phase II (Japan)
NK 105	PEG-PAPB	PTX	Stomach cancer	Phase II (Japan)
NK 012	PEG-PGlu-SN-38 conjugate	SN-38	Breast cancer	Phase II (Japan and USA)
NC 6004	PEG-PGlu	Cisplatin	Advance solid tumors	Phase I/II (UK)
NC 4016	PEG-PGlu	Oxaliplatin	Bowel cancer	Phase I (Japan)
BIND-014	PEG-PDLLA or PEGPLGA	Docetaxel	Advanced solid tumors and metastatic cancers	Phase I (USA)

Source: Adapted with permission from Ref. [71]. Copyright (2012) Elsevier.

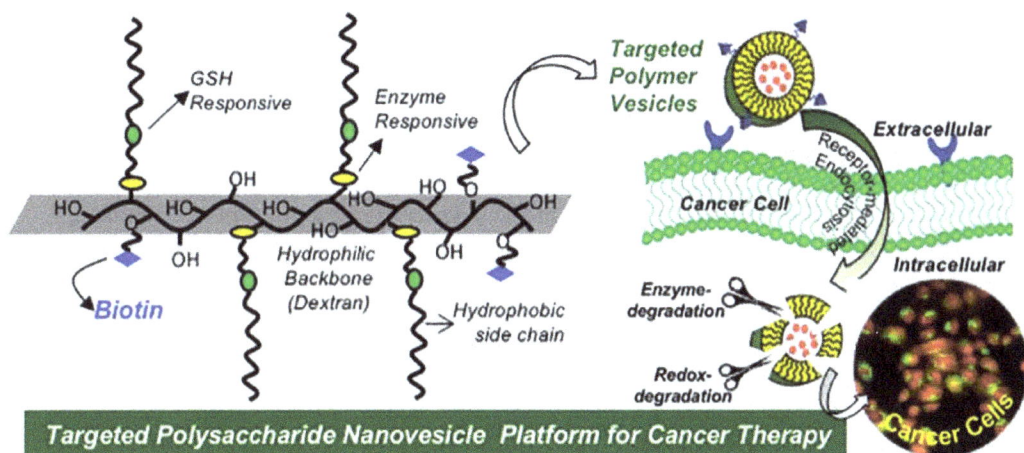

FIGURE 6.7   Modifications of BPs with responsive ligands for targeted delivery.

Reproduced with permission from Ref. [78]. Copyright (2018) American Chemical Society.

and lectins, acquired them affinity toward specific receptors/markers overexpressed on diseased tissues or organs and avoid nonspecific accumulation [72, 76]. Many anticancer molecules, including bevacizumab (anti-VEGF, Avastin®), trastuzumab (anti-ERBB2, Herceptin®), humanized anti-αvβ3 antibody (Abegrin), and etaracizumab, have been conjugated onto the surfaces of a variety of biodegradable polymeric drug delivery nanosystems in an attempt to enhance their localization and accumulation in the tumor region [77].

Apart from active targeting, stimuli-responsive polymeric-based delivery systems have also emerged extensively. These are designed to release their cargo in response to various physical, chemical, and biological cues. Such cues disrupt the nanocarrier structure resulting in the release of the payload in a specific pathological area [61]. Method of drug loading is the prime determinant of the mechanism of drug release from the stimuli-responsive polymeric delivery system, for example, if the drug is entrapped within the core of the polymeric nanoparticles, release can be promoted by structural deformities such as deshedding of polymer's surface layers, polymer degradation, and charge switching; on the other hand, in drug-polymer conjugates, a release is usually triggered by breaking of the bond between drug molecule and the polymer backbone. Both exogenous (heat, electrical, ultrasound, light, or induction) and endogenous (pH, elevated enzyme levels, and reactive oxygen species) stimuli have been employed for the design and fabrication of stimuli-responsive BP-based delivery systems. Such modifications of BPs have resulted in multifunctional drug delivery systems (Figure 6.7).

6.4   FUTURE PERSPECTIVES OF BIODEGRADABLE POLYMERS

Currently, numerous biodegradable polymeric materials have been assessed and utilized in medical devices and tissue engineering. Furthermore, with the advancement of polymer chemistry, many BPs have been useful to support effective therapy. The emergence of combined polymers is expected to surge in the quest for the development of novel biomaterials with outstanding properties. Moreover, advances in artificial intelligence and nanotechnology are expected to form "smart" scaffolds that can be used in medical devices and tissue engineering. Furthermore, with a strong collaborative team of materials scientists, engineers, and doctors, BP usage is expected to grow and to be further developed into varied applications to enhance therapeutics. The advance of 3D printing technology will also bring about new perspectives in medical devices and tissue engineering applications, where BPs can be used to build up or support organ function for the desired period. Both natural and synthetic BPs have been developed using 3D printing for medical devices where fibers and nanoparticles are combined with these polymers for the fabrication of smart materials with varied bioactivities as well as biocompatibility. Lastly, the advance of nanoscience will enable biomaterial researchers to design functional scaffolds that have biomimetic potential as well as regulate cells to fast-track tissue regeneration. Therefore, these functionalized scaffolds can be made into smart materials used in cell signaling. These biomaterials should therefore be seen as important templates to improve the area of tissue engineering and medical device construction for effective therapy.

REFERENCES

1. I. Vroman, L. Tighzert, Biodegradable polymers, Materials, 2 (2009) 307–344.
2. K. Dhaliwal, P. Dosanjh, Biodegradable polymers and their role in drug delivery systems, Biomedical Journal of Scientific & Technical Research, 11 (2018) 8315–8320.
3. Y. Ikada, H. Tsuji, Biodegradable polyesters for medical and ecological applications, Macromolecular Rapid Communications, 21 (2000) 117–132.
4. K. Modjarrad, 1 – Introduction, in: K. Modjarrad, S. Ebnesajjad (Eds.) Handbook of Polymer Applications in Medicine and Medical Devices, William Andrew Publishing, Oxford, 2014, pp. 1–7.
5. S. Daniel, Biodegradable Polymeric Materials for Medicinal Applications, in: S. Thomas, P. Balakrishnan (Eds.) Green Composites, Springer Singapore, Singapore, 2021, pp. 351–372.
6. M. Hino, O. Ishiko, K.I. Honda, T. Yamane, K. Ohta, T. Takubo, N. Tatsumi, Transmission of symptomatic parvovirus B19 infection by fibrin sealant used during surgery, British Journal of Haematology, 108 (2000) 194–195.
7. J.-M. Lü, X. Wang, C. Marin-Muller, H. Wang, P.H. Lin, Q. Yao, C. Chen, Current advances in research and clinical applications of PLGA-based nanotechnology, Expert Review of Molecular Diagnostics, 9 (2009) 325–341.
8. E. Marin, M.I. Briceño, C. Caballero-George, Critical evaluation of biodegradable polymers used in nanodrugs, International Journal of Nanomedicine, 8 (2013) 3071.
9. H. Yan, W. Jiang, Y. Zhang, Y. Liu, B. Wang, L. Yang, L. Deng, G.K. Singh, J. Pan, Novel multibiotin grafted poly(lactic acid) and its self-assembling nanoparticles capable of binding to streptavidin, International Journal of Nanomedicine, 7 (2012) 457.
10. L.S. Nair, C.T. Laurencin, Polymers as biomaterials for tissue engineering and controlled drug delivery, in: K. Lee, D. Kaplan (Eds.) Tissue Engineering I. Advances in Biochemical/Engineering Biotechnology, Springer, Berlin, Heidelberg, 102, 2005, pp. 47–90.
11. H. Tian, Z. Tang, X. Zhuang, X. Chen, X. Jing, Biodegradable synthetic polymers: preparation, functionalization and biomedical application, Progress in Polymer Science, 37 (2012) 237–280.
12. Á.J. Leite, J. Mano, Biomedical applications of natural-based polymers combined with bioactive glass nanoparticles, Journal of Materials Chemistry B, 5 (2017) 4555–4568.
13. S.I. Jeong, B.-S. Kim, S.W. Kang, J.H. Kwon, Y.M. Lee, S.H. Kim, Y.H. Kim, In vivo biocompatibilty and degradation behavior of elastic poly(L-lactide-co-ε-caprolactone) scaffolds, Biomaterials, 25 (2004) 5939–5946.
14. B. Guo, P.X. Ma, Synthetic biodegradable functional polymers for tissue engineering: a brief review, Science China Chemistry, 57 (2014) 490–500.
15. J. Hodde, Naturally occurring scaffolds for soft tissue repair and regeneration, Tissue Engineering, 8 (2002) 295–308.
16. M. Borowska, M. Glinka, N. Filipowicz, A. Terebieniec, P. Szarlej, A. Kot-Wasik, J. Kucińska-Lipka, Polymer biodegradable coatings as active substance release systems for urological applications, Monatshefte für Chemie-Chemical Monthly, 150 (2019) 1697–1702.
17. B. Singh, T. Garg, A.K. Goyal, G. Rath, Recent advancements in the cardiovascular drug carriers, Artificial Cells, Nanomedicine, and Biotechnology, 44 (2016) 216–225.
18. S.S. Panchal, D.V. Vasava, Biodegradable polymeric materials: synthetic approach, ACS Omega, 5 (2020) 4370–4379.
19. S.H. Zeng, P.P. Duan, M.X. Shen, Y.J. Xue, Z.Y. Wang, Preparation and degradation mechanisms of biodegradable polymer: a review, IOP Conference Series: Materials Science and Engineering, 137 (2016) 012003.
20. M. Okada, Chemical syntheses of biodegradable polymers, Progress in Polymer Science, 27 (2002) 87–133.

21. A.T. Jensen, W.S. Neto, G.R. Ferreira, A.F. Glenn, R. Gambetta, S.B. Gonçalves, L.F. Valadares, F. Machado, 8 – Synthesis of polymer/inorganic hybrids through heterophase polymerizations, in: P.M. Visakh, G. Markovic, D. Pasquini (Eds.) Recent Developments in Polymer Macro, Micro and Nano Blends, Woodhead Publishing, Duxford, United Kingdom, 2017, pp. 207–235.
22. M.Y. Kariduraganavar, A.A. Kittur, R.R. Kamble, Chapter 1 – Polymer synthesis and processing, in: S.G. Kumbar, C.T. Laurencin, M. Deng (Eds.) Natural and Synthetic Biomedical Polymers, Elsevier, Oxford, 2014, pp. 1–31.
23. H.R. Allcock, F.W. Lampe, Contemporary Polymer Chemistry, Prentice Hall, Englewood Cliffs, NJ, 1990.
24. A. Díaz, R. Katsarava, J. Puiggalí, Synthesis, properties and applications of biodegradable polymers derived from diols and dicarboxylic acids: from polyesters to poly(ester amide)s, International Journal of Molecular Sciences, 15 (2014) 7064–7123.
25. W.-Y. Lu, M.-W. Hsiao, S.C.N. Hsu, W.-T. Peng, Y.-J. Chang, Y.-C. Tsou, T.-Y. Wu, Y.-C. Lai, Y. Chen, H.-Y. Chen, Synthesis, characterization and catalytic activity of lithium and sodium iminophenoxide complexes towards ring-opening polymerization of l-lactide, Dalton Transactions, 41 (2012) 3659–3667.
26. C.S. Hege, S.M. Schiller, Non-toxic catalysts for ring-opening polymerizations of biodegradable polymers at room temperature for biohybrid materials, Green Chemistry, 16 (2014) 1410–1416.
27. W.A. Munzeiwa, V.O. Nyamori, B. Omondi, N,O-Amino-phenolate Mg(II) and Zn(II) Schiff base complexes: synthesis and application in ring-opening polymerization of ε-caprolactone and lactides, Inorganica Chimica Acta, 487 (2019) 264–274.
28. P.M. Schäfer, S. Herres-Pawlis, Robust guanidine metal catalysts for the ring-opening polymerization of lactide under industrially relevant conditions, ChemPlusChem, 85 (2020) 1044–1052.
29. O. Santoro, X. Zhang, C. Redshaw, Synthesis of biodegradable polymers: a review on the use of Schiff-base metal complexes as catalysts for the ring opening polymerization (ROP) of cyclic esters, Catalysts, 10 (2020) 800.
30. M.P. Arrieta, J. López, E. Rayón, A. Jiménez, Disintegrability under composting conditions of plasticized PLA–PHB blends, Polymer Degradation and Stability, 108 (2014) 307–318.
31. R.D. Saini, Biodegradable polymers, International Journal of Applied Chemistry, 13 (2017) 18.
32. Q. Qi, B.H.A. Rehm, Polyhydroxybutyrate biosynthesis in *Caulobacter crescentus*: molecular characterization of the polyhydroxybutyrate synthase, Microbiology, 147 (2001) 3353–3358.
33. Y. Poirier, C. Somerville, L.A. Schechtman, M.M. Satkowski, I. Noda, Synthesis of high-molecular-weight poly([r]-(-)-3-hydroxybutyrate) in transgenic *Arabidopsis thaliana* plant cells, International Journal of Biological Macromolecules, 17 (1995) 7–12.
34. Y.C. Yao, X.Y. Zhan, J. Zhang, X.H. Zou, Z.H. Wang, Y.C. Xiong, J. Chen, G.Q. Chen, A specific drug targeting system based on polyhydroxyalkanoate granule binding protein PhaP fused with targeted cell ligands, Biomaterials, 29 (2008) 4823–4830.
35. Y. Liu, L. Song, N. Feng, W. Jiang, Y. Jin, X. Li, Recent advances in the synthesis of biodegradable polyesters by sustainable polymerization: lipase-catalyzed polymerization, RSC Advances, 10 (2020) 36230–36240.
36. Q.-M. Gu, W.W. Maslanka, H.N. Cheng, Enzyme-catalyzed polyamides and their derivatives, in: Polymer Biocatalysis and Biomaterials II, American Chemical Society: Washington, DC, 2008, pp. 309–319.
37. X. Zhang, M. Cai, Z. Zhong, R. Zhuo, A water-soluble polycarbonate with dimethylamino pendant groups prepared by enzyme-catalyzed ring-opening polymerization, Macromolecular Rapid Communications, 33 (2012) 693–697.
38. W.-X. Wu, Lipase-catalyzed synthesis and post-polymerization modification of new fully bio-based poly(hexamethylene γ-ketopimelate) and poly(hexamethylene γ-ketopimelate-*co*-hexamethylene adipate) copolyesters, e-Polymers, 20 (2020) 214–225.
39. A. Heise, A.R.A. Palmans, Hydrolases in polymer chemistry: chemoenzymatic approaches to polymeric materials, in: A.R.A. Palmans, A. Heise (Eds.) Enzymatic Polymerisation, Springer Berlin Heidelberg, Berlin, Heidelberg, 2011, pp. 79–113.
40. S. Kobayashi, Recent developments in lipase-catalyzed synthesis of polyesters, Macromolecular Rapid Communications, 30 (2009) 237–266.
41. Y. Yang, J. Zhang, D. Wu, Z. Xing, Y. Zhou, W. Shi, Q. Li, Chemoenzymatic synthesis of polymeric materials using lipases as catalysts: a review, Biotechnology Advances, 32 (2014) 642–651.
42. I. Hilker, G. Rabani, G.K.M. Verzijl, A.R.A. Palmans, A. Heise, Chiral polyesters by dynamic kinetic resolution polymerization, Angewandte Chemie International Edition, 45 (2006) 2130–2132.

43. D. Kim, K. Shin, S.G. Kwon, T. Hyeon, Synthesis and biomedical applications of multifunctional nanoparticles, Advanced Materials, 30 (2018) 1802309.
44. P. Saini, M. Arora, M.N.V.R. Kumar, Poly(lactic acid) blends in biomedical applications, Advanced Drug Delivery Reviews, 107 (2016) 47–59.
45. E. Balla, V. Daniilidis, G. Karlioti, T. Kalamas, M. Stefanidou, N.D. Bikiaris, A. Vlachopoulos, I. Koumentakou, D.N. Bikiaris, Poly(lactic acid): a versatile biobased polymer for the future with multifunctional properties—from monomer synthesis, polymerization techniques and molecular weight increase to PLA applications, Polymers, 13 (2021) 1822.
46. O. Gherasim, G. Popescu-Pelin, P. Florian, M. Icriverzi, A. Roseanu, V. Mitran, A. Cimpean, G. Socol, Bioactive ibuprofen-loaded PLGA coatings for multifunctional surface modification of medical devices, Polymers, 13 (2021) 1413.
47. F. Rouhollahi, S.A. Hosseini, F. Alihosseini, A. Allafchian, F. Haghighat, Investigation on the biodegradability and antibacterial properties of nanohybrid suture based on silver incorporated PGA-PLGA nanofibers, Fibers and Polymers, 19 (2018) 2056–2065.
48. S. Bae, M.J. DiBalsi, N. Meilinger, C. Zhang, E. Beal, G. Korneva, R.O. Brown, K.G. Kornev, J.S. Lee, Heparin-eluting electrospun nanofiber yarns for antithrombotic vascular sutures, ACS Applied Materials & Interfaces, 10 (2018) 8426–8435.
49. J. Cai, X. Xie, D. Li, L. Wang, J. Jiang, X. Mo, J. Zhao, A novel knitted scaffold made of microfiber/nanofiber core–sheath yarns for tendon tissue engineering, Biomaterials Science, 8 (2020) 4413–4425.
50. S. Wu, Y. Qi, W. Shi, M. Kuss, S. Chen, B. Duan, Electrospun conductive nanofiber yarns for accelerating mesenchymal stem cells differentiation and maturation into Schwann cell-like cells under a combination of electrical stimulation and chemical induction, Acta Biomaterialia, (2020) 1–14.
51. M.A. Ghavimi, A.B. Shahabadi, S. Jarolmasjed, M.Y. Memar, S.M. Dizaj, S. Sharifi, Nanofibrous asymmetric collagen/curcumin membrane containing aspirin-loaded PLGA nanoparticles for guided bone regeneration, Scientific Reports, 10 (2020) 1–15.
52. A.S. Richard, R.S. Verma, Bioactive nano yarns as surgical sutures for wound healing, Materials Science and Engineering: C, 128 (2021) 112334.
53. G.-Q. Chen, J. Zhang, Microbial polyhydroxyalkanoates as medical implant biomaterials, Artificial Cells, Nanomedicine, and Biotechnology, 46 (2018) 1–18.
54. C. Ning, Z. Zhou, G. Tan, Y. Zhu, C. Mao, Electroactive polymers for tissue regeneration: developments and perspectives, Progress in Polymer Science, 81 (2018) 144–162.
55. I.C. Bonzani, R. Adhikari, S. Houshyar, R. Mayadunne, P. Gunatillake, M.M. Stevens, Synthesis of two-component injectable polyurethanes for bone tissue engineering, Biomaterials, 28 (2007) 423–433.
56. Y.-J. Wang, U.-S. Jeng, S.-H. Hsu, Biodegradable water-based polyurethane shape memory elastomers for bone tissue engineering, ACS Biomaterials Science & Engineering, 4 (2018) 1397–1406.
57. C.W. Kim, R. Talac, L. Lu, M.J. Moore, B.L. Currier, M.J. Yaszemski, Characterization of porous injectable poly-(propylene fumarate)-based bone graft substitute, Journal of Biomedical Materials Research Part A, 85A (2008) 1114–1119.
58. A.S. Mistry, A.G. Mikos, J.A. Jansen, Degradation and biocompatibility of a poly(propylene fumarate)-based/alumoxane nanocomposite for bone tissue engineering, Journal of Biomedical Materials Research Part A, 83A (2007) 940–953.
59. B.D. Ulery, L.S. Nair, C.T. Laurencin, Biomedical applications of biodegradable polymers, Journal of Polymer Science Part B: Polymer Physics, 49 (2011) 832–864.
60. V.O. Björk, A new tilting disc valve prosthesis, Scandinavian Journal of Thoracic and Cardiovascular Surgery, 3 (1969) 1–10.
61. N. Kamaly, B. Yameen, J. Wu, O.C. Farokhzad, Degradable controlled-release polymers and polymeric nanoparticles: mechanisms of controlling drug release, Chemical Reviews, 116 (2016) 2602–2663.
62. G. Tiwari, R. Tiwari, B. Sriwastawa, L. Bhati, S. Pandey, P. Pandey, S.K. Bannerjee, Drug delivery systems: an updated review, International Journal of Pharmaceutical Investigation, 2 (2012) 2.
63. A. Zaffaroni, Systems for controlled drug delivery, Medicinal Research Reviews, 1 (1981) 373–386.
64. A.R. Kirtane, S.M. Kalscheuer, J. Panyam, Exploiting nanotechnology to overcome tumor drug resistance: challenges and opportunities, Advanced Drug Delivery Reviews, 65 (2013) 1731–1747.
65. Y.K. Sung, S.W. Kim, Recent advances in polymeric drug delivery systems, Biomaterials Research, 24 (2020) 1–12.
66. S.S. Das, P. Bharadwaj, M. Bilal, M. Barani, A. Rahdar, P. Taboada, S. Bungau, G.Z. Kyzas, Stimuli-responsive polymeric nanocarriers for drug delivery, imaging, and theragnosis, Polymers, 12 (2020) 1397.
67. S. Guragain, B.P. Bastakoti, V. Malgras, K. Nakashima, Y. Yamauchi, Multi-stimuli-responsive polymeric materials, Chemistry—A European Journal, 21 (2015) 13164–13174.

68. C.J. Kearney, D.J. Mooney, Macroscale delivery systems for molecular and cellular payloads, Nature Materials, 12 (2013) 1004–1017.
69. R.L. Kronenthal, Biodegradable polymers in medicine and surgery, in: R.L. Kronenthal, Z. Oser, E. Martin (Eds.) Polymers in Medicine and Surgery, Springer US, Boston, MA, 1975, pp. 119–137.
70. J. Karlsson, H.J. Vaughan, J.J. Green, Biodegradable polymeric nanoparticles for therapeutic cancer treatments, Annual Review of Chemical and Biomolecular Engineering, 9 (2018) 105–127.
71. C. Deng, Y. Jiang, R. Cheng, F. Meng, Z. Zhong, Biodegradable polymeric micelles for targeted and controlled anticancer drug delivery: promises, progress and prospects, Nano Today, 7 (2012) 467–480.
72. B.S. Pattni, V.P. Torchilin, Targeted drug delivery systems: strategies and challenges, in: P.V. Devarajan, S. Jain (Eds.) Targeted Drug Delivery: Concepts and Design, Springer International Publishing, Cham, 2015, pp. 3–38.
73. A. Gagliardi, E. Giuliano, V. Eeda, M. Fresta, S. Bulotta, V. Awasthi, D. Cosco, Biodegradable polymeric nanoparticles for drug delivery to solid tumors, Frontiers in Pharmacology, 12 (2021) 17.
74. A. Gonda, N. Zhao, J.V. Shah, H.R. Calvelli, H. Kantamneni, N.L. Francis, V. Ganapathy, Engineering tumor-targeting nanoparticles as vehicles for precision nanomedicine, Med One, 4 (2019) e190021.
75. K. Li, B. Liu, Polymer-encapsulated organic nanoparticles for fluorescence and photoacoustic imaging, Chemical Society Reviews, 43 (2014) 6570–6597.
76. A. Wicki, D. Witzigmann, V. Balasubramanian, J. Huwyler, Nanomedicine in cancer therapy: challenges, opportunities, and clinical applications, Journal of Controlled Release, 200 (2015) 138–157.
77. F. Danhier, O. Feron, V. Préat, To exploit the tumor microenvironment: passive and active tumor targeting of nanocarriers for anti-cancer drug delivery, Journal of Controlled Release, 148 (2010) 135–146.
78. N.U. Deshpande, M. Jayakannan, Biotin-tagged polysaccharide vesicular nanocarriers for receptor-mediated anticancer drug delivery in cancer cells, Biomacromolecules, 19 (2018) 3572–3585.

7 Hybrid Conductive Polymers
Synthesis, Properties, and Multifunctionality

Daeik Jang,[1] Hammad R. Khalid,[2,3]
S.Z. Farooq,[1] H.N. Yoon,[1] and H.K. Lee[1]

[1]Department of Civil and Environmental Engineering, Korea Advanced Institute of Science and Technology (KAIST), Daejeon, South Korea

[2]Civil and Environmental Engineering Department, King Fahd University of Petroleum and Minerals (KFUPM), Dhahran, Saudi Arabia

[3]Interdesciplinary Research Center for Construction and Building Materials, King Fahd University of Petroleum and Minerals, Dhahran, Saudi Arabia

CONTENTS

7.1 INTRODUCTION

Conductive polymeric composites (CPCs) that are synthesized with a polymeric matrix and electrically conductive fillers have been highlighted in various applications such as human health monitoring, wearable sensors, flexible devices, and electrical heating composites [1–3]. Many types of electrically conductive fillers, e.g., carbon nanotubes (CNTs), carbon fiber (CF), graphene nanopowder (GNP), or nanowires (NWs) have been embedded into a polymer matrix, and they can form the conductive networks [1–3]. As the conductive fillers are embedded in the polymeric composites, the electrically conductive fillers-based conductive networks can be formed in the polymeric composites. Then, it can ensure the conductivity of the composites [1–3]. Meanwhile, the distances

DOI: 10.1201/9781003205418-7

between the embedded-conductive fillers change as the strain/stress is applied to the CPCs, imply-ing the piezoresistive properties of the CPCs [3]. It is generally known that the electrical conductiv-ity increased dramatically at some point as the embedded conductive filler increased, indicating the percolation threshold [1–3]. As the conductive filler content embedded in the CPCs is in the range of the percolation threshold, the conductivity of the CPCs can change dramatically when the strain/stress is applied to the CPCs [1–3]. Based on these phenomena, the CPCs can be utilized in various sensing applications. In addition, the high electrical conductivity of the CPCs leads to the multifunctionality of the CPCs such as wearable sensors, flexible devices, and electrical heating composites.

In recent years, it has been reported that the CPCs consisted of only one type of electrically con-ductive filler sometimes show the drawbacks in their functional performances during their service life [1–3]. For these reasons, many researchers have focused on the hybrid CPCs fabricated with multiple types of electrically conductive fillers. Furthermore, many studies have focused on the synergistic effects of multiple conductive fillers on the sensing performances, multifunctionality, and stability enhancement compared to the conventional CPCs consisted of one type of conductive filler [2]. Meanwhile, the dispersion of electrically conductive fillers in the polymeric matrix has been considered as the main obstacle to fabricate the CPCs since the high van der Waals attraction of the conductive fillers causes the agglomeration, leading to the poor dispersions in the polymer matrix [1–3]. Various novel dispersion methods have been proposed by many researchers to solve the dispersion obstacles, and these methods can improve the electrical characteristics and sensing performances of the CPCs.

This chapter provides the synthesis and properties of the hybrid CPCs incorporating multiple conductive fillers that can be utilized in various applications. First, the various electrically conduc-tive fillers are categorized, and the main utilized dispersion methods for improving the dispersion state are introduced. Second, the development of the synthesis of CPCs with various electrically conductive fillers is summarized. In addition, the synergistic effects of multiple conductive filler utilization on the sensing performances enhancement and their multifunctionality of the hybrid CPCs are also covered.

7.2 CLASSIFICATION OF ELECTRICALLY CONDUCTIVE FILLERS AND THEIR DISPERSION METHODS

The electrically conductive fillers can be categorized in different ways according to their chemical components, behaviors, and particle size. Each electrically conductive filler has distinct properties; thus, it is necessary to understand the advantages and drawbacks of the electrically conductive fillers. Carbon-based conductive fillers such as CNTs, CF, and GNP have been considered as pos-sible candidates for favorable conductive fillers in polymeric composites due to their high electrical conductivity [4]. In addition, wire-typed fillers including AgNWs, AuNWs, and CuNWs also have been used to induce the conductivity in polymeric composites due to their outstanding electrical conductivity with a small amount [5]. As the electrically conductive fillers are utilized in fabricat-ing the CPCs, high van der Waals attraction can induce the agglomeration of the conductive fillers [1–5]. van der Waals attraction indicates the repulsions between the molecular and surface of each conductive filler, and it causes difficulties for dispersing the conductive fillers [1–5]. Therefore, it should be considered the proper dispersion method to mitigate the van der Waals attraction for ensuring the favorable dispersion of the conductive fillers.

Several dispersion methods (e.g., ultrasonication, polycarboxylate-typed superplasticizer, poly-meric dispersant) have been introduced to improve the dispersion of the conductive fillers in poly-meric composites. The ultrasonication which is regarded as a common dispersion method can generate a physical vibration and it can induce the vibration of conductive fillers [2–3]. The dis-tances between the conductive fillers can increase as the conductive fillers are possibly vibrated, improving the dispersion. As the polycarboxylate-typed superplasticizer is utilized, it can wrap the conductive fillers and increase the distances between each filler, mitigating the van der Waals attraction [2–3]. The utilized spherical polymeric dispersant such as polysodium 4-styrenesulfonate (PSS) and silica fume can be inserted into the gap of the conductive fillers, indicating the ball-bearing effects, thereby improved the dispersion of conductive fillers. For these reasons, in recent years, most of the researchers have utilized more than one dispersion method for ensuring the dispersion of the conductive fillers due to the synergistic effect of the dispersion methods [2–3].

7.3 RECENT DEVELOPMENT IN THE SYNTHESIS OF CPCs WITH ONE TYPE OF ELECTRICALLY CONDUCTIVE FILLER

The properties of the electrically conductive fillers-embedded polymeric composites depend strongly on the dispersion state of embedded-electrical conductive fillers. As described in the above section, carbon-based electrically conductive fillers tend to form the agglomerates due to their high van der Waals attraction [1–5]. Thus, it can be seen that the various fabrication methods are adopted to synthesize the CPCs containing carbon-based conductive fillers. The selection of the fabrication method is majorly influenced by the host polymer type. Even there are various fabrication methods to synthesize the CPCs, the purpose of the proposed methods is to achieve the uniform dispersion state of the utilized conductive fillers inside the polymer matrix. This section discusses the widely adopted synthesis techniques for the fabrication of CPCs.

7.3.1 MELT PROCESSING

For insoluble and thermoplastic polymers (e.g., polypropylene, polyethylene, polystyrene, and polyamide-6), melt processing is the most widely adopted fabrication method for thermoplastic polymers. The receptor thermoplastic matrix polymer is melted when the polymer is heated, and it forms a viscous liquid. The molten flow exerts shear stresses on the conductive fillers-based agglomerates. As the shear stress increases the cohesive strength of the agglomerate, the agglomerates can be broken down into smaller agglomerate or individual entities [6, 7]. During the melt-mixing process, the wetting of conductive fillers, infiltration of molten polymer within the agglomerate, and redistribution of the conductive fillers within the polymer matrix occur simultaneously [8]. The infiltration of the polymer in the agglomerate generally depends on the porosity of the agglomerates and the viscosity of the melt [9]. The CPCs are being fabricated on the industrial scales using the sigma mixer. In a sigma mixer, two different hoppers are used to separately load the polymer granule and conductive fillers inside the barrel, which are caught and pushed forward by continuous rotating screws and melt during their forward motion due to externally provided heat and shearing of material between the screw and barrel [8]. The molten mixture can be homogenized by the time it reaches the outlet. The final oozed-out composites are thus cooled and form pellets for further use as an input of compression or injection molding. Several studies have adopted melt-processing as the fabrication technique using CNTs, GNP, and graphene oxide (GO) as conductive fillers.

7.3.2 SOLUTION PROCESSING

The most commonly practiced technique to fabricate the CPCs is solution processing, in which the electrically conductive fillers are first dispersed in an aqueous or organic solvent by ultrasonication, and then they are mixed with the polymer solution by mechanically shear mixing with a magnetic stirring. And, lastly, the composites can be cast by the evaporation of solvent [9, 10]. However, it should be noted that the homogeneity of CPC is dependent on the dispersion characteristics of the conductive fillers. Carbon-based nano-fillers are generally dispersed through sonicating the filler suspensions. However, the sonication for dispersing the electrically conductive fillers can shorten the lengths, to avoid this shortcoming, literature has suggested using the surfactants for CNTs dispersion or functionalization of CNTs [11]. The selection of the solvent is mainly dependent on the solubility of the polymer matrix. It is worth mentioning that this technique can only be implemented for the polymers that are soluble in the solvent and has been widely used for incorporating carbon-based conductive fillers in a variety of polymers including polyimides, PDMS, polycarbonate, and poly-methyl methacrylate.

7.3.3   IN-SITU POLYMERIZATION

For the polymer with thermally unstable and/or insoluble polymers, in-situ polymerization is a suitable fabrication technique. In this method, electrically conductive fillers are first dispersed in the monomer followed by the in-situ polymerization [12–14]. Literature has reported that the better initial dispersion of the conductive fillers in the monomer greatly influences the overall properties of hardened CPC. For the case of CNTs, researchers have reported better results for functionalized CNTs, that exhibit better dispersion characteristics, in comparison to the pristine CNTs [15, 16]. And, for GNP and GO the monomer is intercalated between their layers, and then polymerization separates the layers, which helps in achieving better dispersion [17, 18]. GO facilitates the intercalation with ease as compared to GNP because of its larger interlayer spacing [19]. In-situ polymerization has widely been used for fabricating the CPCs with polycarbonate, polymethyl methacrylate, polyurethane, and water-soluble polyvinyl alcohol as a host matrix.

7.3.4   ELECTRO-SPINNING

Apart from the dispersion state of embedded-conductive fillers, the orientation of the fillers also affects the overall properties of the CPCs [20]. In this regime, without compromising the structural integrity of conductive fillers, electrospinning offers the advantage of orienting the fillers in the direction parallel to the fiber axis due to large shear forces. Fibrous polymeric composites with conductive fillers of diameter ranging from microns to few nanometers can be fabricated using this technique [21]. The general electrospinning process includes preparing conductive filler and polymer solutions separately and then mixing them in one syringe [20]. The stable dispersion of the conductive fillers can be achieved by using amphiphilic surfactants such as a natural polysaccharide Gum Arabic and sodium dodecyl sulfate [15, 16]. These agents can absorb onto the surface of hydrophobic conductive fillers, thus weakening the interparticle attraction forces. An electric field is applied to the syringe containing solutions and collector, which is placed at some distance from the needle of the syringe, leading to the formation of cone-shaped droplets [21]. The conductive filaments eject from the needle when electric forces with a critical value of 5–20 kV overcome the surface tension of the polymer solution [22]. It has been reported that the various aspects including operating parameters (e.g., applied voltage, distance between the needle and collector, and feeding rate) and solution properties (e.g., viscosity, surface tension, and conductivity) can influence the characteristics of fabricated CPCs and their morphology [20, 21].

In the above-described methods, no direct linkage between the polymer matrix and embedded-conductive fillers is observed due to the relative sparseness of the usable functional groups on most carbon-based conductive fillers; thus, they lie under the category of non-covalent dispersion methods. For the covalent route, the conductive fillers are required to be functionalized. During this modification, carbon atoms are changed from sp2 to sp3 hybridization. For example, the CNTs are generally functionalized through oxidation or using different acids and their combinations, including H_2SO_4, HNO_3, and HCl. By treating CNTs with acids, its caps are removed due to the formation of carboxylic acid and hydroxyl groups at the ends. Nonetheless, the chemical functionalization may affect the aspect ratio of CNTs and the varying hybridization state results in bandgap widening thus influences the conductivity [15]. On the other hand, GO inherently contains several reactive functional groups that can be attached to a broad range of polymers following grafting-to or grafting-from approaches [23]. The selection between the explained approaches depends on the polymer being used. In general, the grafting-to approach may decrease the grafting density of polymer chains to the platelet surface and resultantly affect the dispersion of the polymer-grafted platelets [24, 25].

7.3.5   3D PRINTING TECHNOLOGY

In recent years, 3D printing (3DP) technology has been adopted to fabricate polymer composites [26]. However, the conventional polymeric composites fabricated with 3DP technology have various drawbacks such as lack of strengths and difficulty in combining with conductive fillers, leading to a low possibility of using in fabricating CPCs [26]. To solve such problems, many researchers have proposed novel printing methods for fabricating CPCs. First, fused deposition modeling (FDM) printers are introduced for fabricating polymeric composites. The thermoplastic polymers including polycarbonates (PC), acrylonitrile butadiene styrene (ABS), and polyactic acid (PLA) which have low melting temperature are commonly used in FDM printers. However, the main drawback of the FDM printing method is that the polymer matrix has to be in filament formation for extrusion, and it can lead to the poor dispersion of additional fillers in the polymeric composites [27]. In addition, the polymer matrix used in FDM printers is limited to thermoplastic polymers with favorable viscosity. Thus, the FDM printers with various advantages such as simplicity, high speed, and low cost are limited to some applications, and it is expected that FDM printers can be utilized in various applications as the disadvantages and limitations are improved. Second, powder-liquid 3DP technology that was proposed at MIT in 1993 has been proposed for manufacturing CPCs. The main advantages of 3DP technology are the flexibility of selecting polymer matrix and processing environment at room temperature [26]. Any polymer matrix in powder type can be used in a 3DP printer, and the advantages including low cost, simplicity of manufacturing method improved their applicability of using in fabricating CPCs; however, the disadvantages such as binder contamination and clogging of binder jet are needed to be improved. Except for the FDM and 3DP technology, there are many different kinds of 3DP technology such as stereolithography (SLA), selective laser sintering (SLS), digital light processing (DLP), and liquid deposition modeling (LDM) which have their advantages for manufacturing CPCs.

7.4   SYNERGISTIC EFFECTS OF MULTIPLE CONDUCTIVE FILLERS ON CPCs FABRICATION

It is worth mentioning that the synergistic effects of multiple conductive fillers with different dimensionality may lead to an improved physical property of the CPCs. The synergistic effect is termed when the combination of two or more species yield better results as compared to a summation of results obtained by the individual entity. Literature has shown that incorporating CNTs along with other conductive fillers can lower the percolation threshold, meaning the improved electrical properties of the CPCs. This phenomenon is attributed to the bridging-interactions between the conductive fillers inclusion of varying length scales, and that 0-D round particles may prevent clustering and help to achieve the better dispersion of fibrous conductive fillers inclusion [28–31]. This combination of different conductive fillers helps to achieve multifunctionality, reduce the composite cost, and have better physical properties. These multifunctional characteristics include piezoresistive sensing, electrical-heating, electromagnetic interference (EMI) shielding, and monitoring sensors for biosignals. The performances of these multifunctional CPCs are highly dependent on the electrical characteristics that are influenced by the percolation demand. It is reported that a better electric pathway can be achieved if different scale fillers are used in combination because it would reinforce and help to capture the response of the composite at varying length scales. Researchers have widely studied the combined effect of CNTs with carbon black (CB), GNP, and graphite [28–30].

The combined utilization of CNTs with CF has yielded better electrical sensing performances and helped to achieve percolation at lower content. The CNTs form an auxiliary conductive network and serve as a conductive bridge by interconnecting by flowing electrons between the proximately separated microfibers [32, 33]. In another study, the compound effect of CNTs and CB is studied in

TABLE 7.1

Fabrication Method for CPCs with Different Polymers and Conductive Fillers

Polymer	Filler	Synthesis Method	Reference
Nylon 6,6	CNTs	Electrospinning	[34]
Polyvinyl alcohol	CNTs	Electrospinning	[35, 36]
Polyacrylonitrile	CNTs	Electrospinning	[37, 38]
Polyacrylonitrile	CF	Electrospinning	[39]
Polycarbonate	CNT	Electrospinning	[40]
Polyamic acid	CNTs	Electrospinning	[41]
Polyvinyl chloride	GNP	In situ polymerization	[42]
Polypropylene	GNP	Melt processing	[43]
Polystyrene	Graphene	Solution processing	[44]
Polyvinyl alcohol	Graphene	Solution processing	[45]
Pitch	CF	Melt processing	[46]
Polyethylene terephthalate	CNT	Melt processing	[47]
Polymethyl methacrylate	CNT	Grafting-from	[48]

the polyvinylidene fluoride (PVDF) matrix, and it is concluded that CB complements better CNTs dispersion and bridges the gap between CNTs thus lower the percolation demand and sensing performance. It has also been shown that the synergistic effect of conductive fillers also improves the sensitivity and durability characteristics of the CPCs. Table 7.1 shows the fabrication method for CPCs with different polymers and conductive fillers.

7.5 MULTIFUNCTIONALITY OF CPCs INCORPORATING MULTIPLE ELECTRICALLY CONDUCTIVE FILLERS

In recent years, the CPCs incorporating multiple electrically conductive fillers that have outstanding electrical conductivity have been utilized in various applications such as human motion monitoring and wearable sensors, flexible devices, and electrical heating composites. In addition, the CPCs with multiple electrically conductive fillers can exhibit multifunctional properties. Thus, this section covers the various applications using the CPCs with multiple electrically conductive fillers and their multifunctional properties.

7.5.1 WEARABLE SENSORS

As the electrically conductive fillers are embedded into the polymeric composites, the conductive networks can be formed in the CPCs [1–3]. The distances between the conductive filler in CPCs can be changed as the stress/strain is applied to the CPCs, and it can change the conductive networks [1–3]. The electrical conductivity of the CPCs also can be changed, which indicates the piezoresistive mechanism [1, 2]. Based on the piezoresistive mechanism, the relationship between the applied stress/strain and electrical conductivity variation can be found. Thus, the CPCs can be utilized in stress/strain-based piezoresistive sensors according to the above relationship. Many studies related to the stress/strain-based piezoresistive sensors using the CPCs are trying to attach the sensors to the human bodies to monitor human movements [1, 2]. The motions caused by the human body movements can cause the stress/strain to change the electrical conductivity of the CPCs, showing the potential of using human motion monitoring sensors [49]. Meanwhile, the high sensitivity and repeatability have to be improved to utilize the CPCs in such applications. According to the previous literature, however, the original positions of the conductive fillers can be changed during their service life, leading to the redistribution of the conductive filler-based conductive networks. Thus, it can degrade the sensing performances including the sensitivity and repeatability. In this regard, many studies have been given to improve such sensing performances using the synergistic effects of multiple electrically conductive fillers utilizations [49]. The utilizations of the conductive fillers with different sizes can mitigate the redistributions of the conductive networks, improving not only the sensitivity but also repeatability of the sensing performances during their service life [49].

7.5.2 ELECTRICAL-HEATING COMPOSITES

The electrical conductive fillers-based conductive networks can be formed in CPCs, which can generate heat as the voltage is applied to the CPCs [3]. This phenomenon can be explained by Joule's heating mechanism, indicating the electrical energy can be converted into heat energy [3]. Thus, the temperature of CPCs increase as the input power is applied to the CPCs, and the heat-generation capability according to Joule's heating mechanism is proportional to the applied input power and the electrical conductivity of the CPCs [3]. Many studies have shown the electrical-heating characteristics of the CPCs incorporating various electrically conductive fillers since the CPCs for electrical heating have been highlighted in various applications such as micro-heaters, thermo-electrical devices, and bio-therapy devices [3]. According to the previous literature, the electrical-heating capability can be degraded during their service life. It can be deduced from the electrically conductive fillers-based conductive network can be disturbed due to the thermal expansion of the conductive fillers as the input power is applied to the CPCs [3]. Thus, many studies have tried to include two types of conductive fillers due to their synergistic effects of the two types of conductive fillers can mitigate the disturbances of conductive networks to improve the electrical-heating capability (i.e., temperature increase, repeatability, and durability) [3]. Specifically, in recent years, some researchers have sought to improve the heat-storage capability for improving the heating efficiency of the electrical-heating CPCs. Jang et al. [3] investigated the effects of silica aerogel inclusion on the electrical-heating characteristics of the CNTs-incorporated CPCs, showing the silica aerogel inclusion can improve not only the repeatability and/or durability, but also the heat-storage capability related to the heating efficiency of the electrical-heating CPCs.

7.5.3 ELECTROMAGNETIC SHIELDING COMPOSITES

According to the previous literature, the electromagnetic wave interference (EMI) shielding capability can be improved as the electrical conductivity of the CPCs increased [50]. For these reasons, many studies have been given to utilize the CPCs with high electrical conductivity as the EMI shielding composites [50]. In addition, the hybrid CPCs incorporating two types of conductive fillers can improve the EMI shielding capability compared to the CPCs with a single type of conductive filler [50]. These phenomena can be explained by the formation of conductive networks that can improve as the conductive fillers with multi-scale are included [50]. In addition, incorporating two types of conductive fillers can improve both electrical and polarization losses that are the main factors in improving the EMI shielding capability of the CPCs.

7.6 FUTURE RECOMMENDATIONS

The hybrid CPCs composed of multiple conductive fillers have been synthesized for their high applicability in various applications. Due to some drawbacks of the conventional hybrid CPCs for improvements, some points for future research are recommended as below:

1. The dispersion of multiple conductive fillers in the polymeric matrix affects the electrical characteristics and their functional performances. Thus, it is necessary to have a favorable synthesis method for dispersing each conductive filler in the polymeric matrix.

2. The synergistic effects of different conductive fillers on the improvements of functional performances are related to their applications. Therefore, the effects of multiple conductive fillers with different dimensions on the functional performances (i.e., piezoresistive-sensing, electrical heating, and EMI shielding) need to be more accurately investigated.

3. The more accurate manufacturing method for CPCs is required to have stable properties in electrical conductivity and their functional performances, and 3DP technology can thus be a possible solution for manufacturing CPCs more accurately.

7.7 CONCLUSIONS

The hybrid CPCs incorporating electrically conductive fillers have high potential in various applications. During the fabricating of the CPCs, the proper method for dispersing the conductive fillers in the polymer matrix is necessary to be considered. In addition, favorable amounts of conductive fillers are needed according to their applications. The hybrid CPCs utilizing two types of conductive fillers can lead to the synergistic effects of the different fillers on the improvements of performances, multifunctionality, and stability enhancement conventional to the CPCs with one type of conductive filler.

REFERENCES

1. Khalid H R, Choudhry I, Jang D, Abbas N, Haider M S and Lee H K (2021) Facile synthesis of sprayed CNTs layer-embedded stretchable sensors with controllable sensitivity *Polymers (Basel)* **13** 1–6
2. Jang D I, Yoon H N, Nam I W and Lee H K (2020) Effect of carbonyl iron powder incorporation on the piezoresistive sensing characteristics of CNT-based polymeric sensor *Compos. Struct.* **244** 112260
3. Jang D, Yoon H N, Seo J, Park S, Kil T and Lee H K (2021) Improved electric heating characteristics of CNT-embedded polymeric composites with an addition of silica aerogel *Compos. Sci. Technol.* **212** 108866
4. Matos M A S, Tagarielli V L and Pinho S T (2020) On the electrical conductivity of composites with a polymeric matrix and a non-uniform concentration of carbon nanotubes *Compos. Sci. Technol.* **188** 108003
5. Amjadi M, Pichitpajongkit A, Lee S, Ryu S and Park I (2014) Highly stretchable and sensitive strain sensor based on silver nanowire-elastomer nanocomposite *ACS Nano* **8** 5154–63
6. Eow, J.S., Ghadiri, M., and Sharif, A. O (2007) Electro-hydrodynamic separation of aqueous drops from flowing viscous oil *J. Pet. Sci. Eng.* **55** 146–55
7. Hartley P A and Parfitt G D (1985) Dispersion of powders in liquids. 1. The contribution of the van der Waals force to the cohesiveness of carbon black powders *Langmuir* **1** 651–7
8. Banerjee J and Dutta K (2019) Melt-mixed carbon nanotubes/polymer nanocomposites *Polym. Compos.* **40** 4473–88
9. Ma P C, Siddiqui N A, Marom G and Kim J K (2010) Dispersion and functionalization of carbon nanotubes for polymer-based nanocomposites: A review *Compos. Part A Appl. Sci. Manuf.* **41** 1345–67
10. Thostenson E T, Li C and Chou T W (2005) Nanocomposites in context *Compos. Sci. Technol.* **65** 491–516
11. Barrau S, Demont P, Perez E, Peigney A, Laurent C and Lacabanne C (2003) Effect of palmitic acid on the electrical conductivity of carbon nanotubes-epoxy resin composites *Macromolecules* **36** 9678–80
12. Potts J R, Dreyer D R, Bielawski C W and Ruoff R S (2011) Graphene-based polymer nanocomposites *Polymer (Guildf)* **52** 5–25
13. Wang X, Hu Y, Song L, Yang H, Xing W and Lu H (2011) In situ polymerization of graphene nanosheets and polyurethane with enhanced mechanical and thermal properties *J. Mater. Chem.* **21** 4222–7
14. Mrah L and Meghabar R (2021) In situ polymerization of styrene–clay nanocomposites and their properties *Polym. Bull.* **78** 3509–26
15. Mallakpour S and Soltanian S (2016) Surface functionalization of carbon nanotubes: Fabrication and applications *RSC Adv.* **6** 109916–35
16. Sun Y P, Fu K, Lin Y and Huang W (2002) Functionalized carbon nanotubes: Properties and applications *Acc. Chem. Res.* **35** 1096–104

17. Dresselhaus M S and Dresselhaus G (2002) Intercalation compounds of graphite *Adv. Phys.* **51** 1–186
18. Shioyama H (2000) Interactions of two chemical species in the interlayer spacing of graphite *Synth. Met.* **114** 1–15
19. Chen G, Wu D, Weng W and Wu C (2003) Exfoliation of graphite flake and its nanocomposites *carbon* **41** 619–21
20. Fong H, Liu W, Wang C S and Vaia R A (2001) Generation of electrospun fibers of nylon 6 and nylon 6-montmorillonite nanocomposite *Polymer (Guildf)* **43** 775–80
21. Yeo L Y and Friend J R (2006) Electrospinning carbon nanotube polymer composite nanofibers *J. Exp. Nanosci.* **1** 177–209
22. Yang Y, Jia Z, Liu J, Li Q, Hou L, Wang L and Guan Z (2008) Effect of electric field distribution uniformity on electrospinning *J. Appl. Phys.* **103** (10) 104307
23. Lee S H, Dreyer D R, An J, Velamakanni A, Piner R D, Park S, Zhu Y, Kim S O, Bielawski C W and Ruoff R S (2010) Polymer brushes via controlled, surface-initiated atom transfer radical polymerization (ATRP) from graphene oxide *Macromol. Rapid Commun.* **31** 281–8
24. Akcora P, Kumar S K, Moll J, Lewis S, Schadler L S, Li Y, Benicewicz B C, Sandy A, Narayanan S, Ilavsky J, Thiyagarajan P, Colby R H and Douglas J F (2010) "Gel-like" mechanical reinforcement in polymer nanocomposite melts *Macromolecules* **43** 1003–10
25. Coleman J N, Khan U and Gun'ko Y K (2006) Mechanical reinforcement of polymers using carbon nanotubes *Adv. Mater.* **18** 689–706
26. Wang X, Jiang M, Zhou Z, Gou J and Hui D (2017) 3D printing of polymer matrix composites: A review and prospective *Compos. Part B Eng.* **110** 442–58
27. Sood A K, Ohdar R K and Mahapatra S S (2010) Parametric appraisal of mechanical property of fused deposition modelling processed parts *Mater. Des.* **31** 287–95
28. Paszkiewicz S, Szymczyk A, Sui X M, Wagner H D, Linares A, Ezquerra T A and Rosłaniec Z (2015) Synergetic effect of single-walled carbon nanotubes (SWCNT) and graphene nanoplatelets (GNP) in electrically conductive PTT-block-PTMO hybrid nanocomposites prepared by in situ polymerization *Compos. Sci. Technol.* **118** 72–7
29. Ma P C, Liu M Y, Zhang H, Wang S Q, Wang R, Wang K, Wong Y K, Tang B Z, Hong S H, Paik K W and Kim J K (2009) Enhanced electrical conductivity of nanocomposites containing hybrid fillers of carbon nanotubes and carbon black *ACS Appl. Mater. Interfaces* **1** 1090–6
30. Socher R, Krause B, Hermasch S, Wursche R and Pötschke P (2011) Electrical and thermal properties of polyamide 12 composites with hybrid fillers systems of multiwalled carbon nanotubes and carbon black *Compos. Sci. Technol.* **71** 1053–9
31. Kim T, Park J, Sohn J, Cho D and Jeon S (2016) Bioinspired, highly stretchable, and conductive dry adhesives based on 1D-2D hybrid carbon nanocomposites for all-in-one ECG electrodes *ACS Nano* **10** 4770–8
32. Zakaria M R, Md Akil H, Abdul Kudus M H, Ullah F, Javed F and Nosbi N (2019) Hybrid carbon fiber-carbon nanotubes reinforced polymer composites: A review *Compos. Part B Eng.* **176** 107313
33. Jang M G, Ryu S C, Juhn K J, Kim S K and Kim W N (2019) Effects of carbon fiber modification with multiwall CNT on the electrical conductivity and EMI shielding effectiveness of polycarbonate/carbon fiber/CNT composites *J. Appl. Polym. Sci.* **136** 1–9
34. Jeong J S, Jeon S Y, Lee T Y, Park J H, Shin J H, Alegaonkar P S, Berdinsky A S and Yoo J B (2006) Fabrication of MWNTs/nylon conductive composite nanofibers by electrospinning *Diam. Relat. Mater.* **15** 1839–43
35. Naebe M, Lin T, Tian W, Dai L and Wang X (2007) Effects of MWNT nanofillers on structures and properties of PVA electrospun nanofibres *Nanotechnology* **18** (22) 225605
36. Naebe M, Lin T, Staiger M P, Dai L and Wang X (2008) Electrospun single-walled carbon nanotube/polyvinyl alcohol composite nanofibers: Structure-property relationships *Nanotechnology* **19** (30) 305702
37. Ko F, Gogotsi Y, Ali A, Naguib N, Ye H, Yang G, Li C and Willis P (2003) Electrospinning of continuous carbon nanotube-filled nanofiber yarns *Adv. Mater.* **15** 1161–5
38. Kedem S, Schmidt J, Paz Y and Cohen Y (2005) Composite polymer nanofibers with carbon nanotubes and titanium dioxide particles *Langmuir* **21** 5600–4
39. Park S J and Im S H (2008) Electrochemical behaviors of PAN/Ag-based carbon nanofibers by electrospinning. *Bull. Korean Chem. Soc.* **29** (4) 777–81
40. Kim G M, Michler G H and Pötschke P (2005) Deformation processes of ultrahigh porous multiwalled carbon nanotubes/polycarbonate composite fibers prepared by electrospinning *Polymer (Guildf)* **46** 7346–51

41. Keikhaei S, Mohammadalizadeh Z, Karbasi S and Salimi A (2019) Evaluation of the effects of β-tricalcium phosphate on physical, mechanical and biological properties of poly(3-hydroxybutyrate)/chitosan electrospun scaffold for cartilage tissue engineering applications *Mater. Technol.* **34** (10) 615–25.

42. Vadukumpully S, Paul J, Mahanta N and Valiyaveettil S (2011) Flexible conductive graphene/poly(vinyl chloride) composite thin films with high mechanical strength and thermal stability *Carbon NY* **49** 198–205

43. Bafana A P, Yan X, Wei X, Patel M, Guo Z, Wei S and Wujcik E K (2017) Polypropylene nanocomposites reinforced with low weight percent graphene nanoplatelets *Compos. Part B Eng.* **109** 101–7

44. Stankovich S, Dikin D A, Dommett G H B, Kohlhaas K M, Zimney E J, Stach E A, Piner R D, Nguyen S B T and Ruoff R S (2006) Graphene-based composite materials *Nature* **442** 282–6

45. Xu Y, Hong W, Bai H, Li C and Shi G (2009) Strong and ductile poly(vinyl alcohol)/graphene oxide composite films with a layered structure *Carbon NY* **47** 3538–43

46. Andrews R, Jacques D, Rao A M, Rantell T, Derbyshire F, Chen Y, Chen J and Haddon R C (1999) Nanotube composite carbon fibers *Appl. Phys. Lett.* **75** 1329–31

47. Li Z, Luo G, Wei F and Huang Y (2006) Microstructure of carbon nanotubes/PET conductive composites fibers and their properties *Compos. Sci. Technol.* **66** 1022–9

48. Kong H, Gao C and Yan D (2004) Functionalization of multiwalled carbon nanotubes by atom transfer radical polymerization and defunctionalization of the products *Macromolecules* **37** 4022–30

49. Jang D, Farooq S Z, Yoon H N and Khalid H R (2021) Design of a highly flexible and sensitive multifunctional polymeric sensor incorporating CNTs and carbonyl iron powder *Compos. Sci. Technol.* **207** 108725

50. Zhao X, Xu L, Chen Q, Peng Q, Yang M, Zhao W, Lin Z, Xu F, Li Y and He X (2019) Highly conductive multifunctional rGO/CNT hybrid sponge for electromagnetic wave shielding and strain sensor *Adv. Mater. Technol.* **4** 1–9

8 Electrically Conductive Polymers and Composites for Biomedical Applications

Paloma Patra, Aswathi Thomas, and Aravind Kumar Rengan
Department of Biomedical Engineering, Indian Institute
of Technology Hyderabad, Hyderabad, India

CONTENTS

8.1 INTRODUCTION

Shirakawa, MacDiarmid, and Heeger in 1977 were able to change our fundamentally orthodox view about organic polymers from being "insulators" to "electrically semi-conductive," for which they received the Nobel Prize in Chemistry twenty-three years later. Their serendipitous discovery led to extensive research in the field of conducting polymers (CPs), revealing that they exhibit unique optical and electrical properties matching that of inorganic semiconductors. A CP may be described as "any π-conjugated polymer—having a backbone with alternating single and double (or triple) covalent bonds—that can transport charges, independently of their intrinsic conductivities (conductor or semiconductor) and charge transport characteristics." Current flow may be feasible in such a conjugated system because the electrons are only weakly bound. There is a robust localized sigma bond in the conjugated system and a less strongly localized weak pi bond. This permits the delocalization of electrons across the entire polymer backbone, allowing them to be used by many atoms, making them "jump" throughout the system. However, for electron flow to take place, the polymer must be doped to make it conductive. "Doping" may be carried out by either the addition (reduction) of electrons or by their removal (oxidation) from the polymer. Iodine can be used as a dopant for carrying out oxidation doping. Upon doping, an electron is attracted by iodine from the weak pi bonds, thus making the electrons move across the polymer molecule; this is when current flows. An orderly arrangement of molecules in a tightly packed system curbs the vacant space available for the electrons to jump, thus improving conductivity. Manipulating specific properties such as chemical modification of the polymer chain, the extent of doping, merging with other polymers, and the type of dopant used may alter the conductivity of these polymers [1]. A dopant ion overcomes the energy gap that arises in the electronic spectrum due to alternate bonds, making the conjugated system unstable. These dopant ions offer a charge by introducing additional electrons, neutralizing the unstable polymer backbone [2, 3]. CPs can be doped with both p- and n-type dopants using a

DOI: 10.1201/9781003205418-8

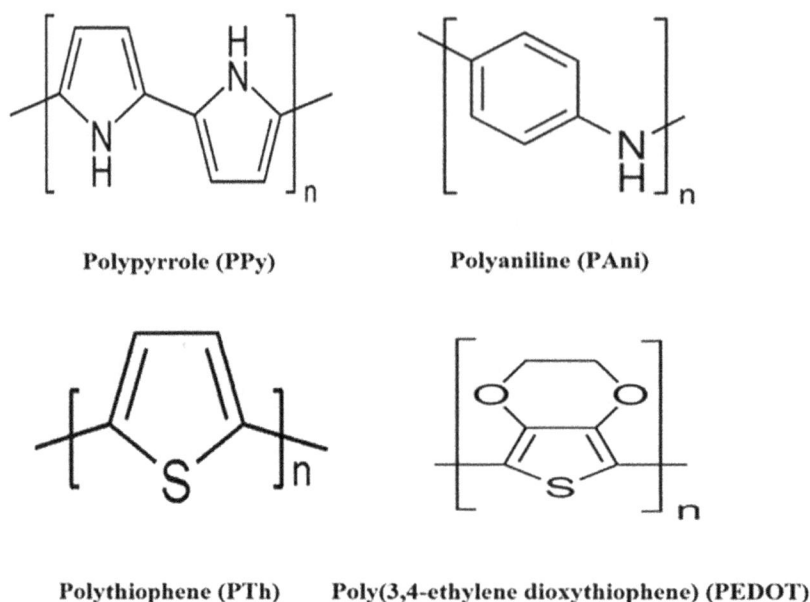

FIGURE 8.1 Structures of common CPs investigated for biomedical applications.

variety of molecules, such as small salt ions (Cl–, Br–, or NO3–), and larger dopants such as hyaluronic acid (HA), peptides, or polymers [4, 5].

The most promising CPs to be used in biomedical applications are polypyrrole (PPy), polyaniline (PAni), polythiophene (PTh), and poly(3,4-ethylenedioxythiophene) (PEDOT) (Figure 8.1). The electrical and optical properties of CPs are comparable to those of metals and semiconductors, with the added advantage of having simpler synthesis methods than conventional polymers. Ever since the 1980s, when it was revealed that such polymers are compatible with a broad spectrum of biological molecules, their biomedical applications have increased significantly. These "smart materials" have found potential usage in a range of biomedical applications, including modulation of cell activity (skeletal, bone, cardiac, and nerve), tissue engineering, biosensor coatings, electrical stimuli-responsive drug and gene delivery systems, artificial muscles, and diagnostic applications due to their excellent biocompatibility [6]. These polymers have attracted considerable attention in optical imaging modalities like fluorescence, chemi/bioluminescence, and photoacoustic imaging due to their remarkable photostability, high fluorescence brightness, and sharp NIR absorbance peaks, ease of modification, and low cytotoxicity. The hydrophilicity can be introduced to CPs by attaching a charged side chain, enabling it to have light-harvesting properties and signal amplification effect providing its potential application in biomedical engineering as a diagnostic agent [7]. With the emergence of CPs in the field of nanomedicine, diagnosis of various diseases can be made possible at the molecular and cellular level with superior therapeutic outcomes. Most research has concentrated on *in vitro* tests and techniques to improve biocompatibility to better understand how these polymers interact with biological tissues. The primary issue in utilizing these CPs for clinically effective biomedical devices, implants, or diagnostic agents has been designing them for suitable optoelectronic and mechanical properties and ease of processing and biocompatibility. The development of CP composites by blending with conducting nanoparticles or non-CPs has been one of the recent approaches to enhance the physicochemical properties and biocompatibility and overcome the existing limitations of CPs. The primary focus of this book chapter is to provide a general overview of the common CPs and composites, their relevant properties, and their biomedical applications, including their applications in various imaging modalities.

8.2 COMMON CONDUCTIVE POLYMERS, THEIR GENERAL PROPERTIES, AND BIOMEDICAL APPLICATIONS

Chemical or electrochemical synthesis methods can be used to make CPs. Chemical methods usually follow condensation polymerization or addition polymerization. Although chemical synthesis permits various methods to synthesize a range of CPs and also allows for scale-up, electrochemical synthesis is very simple and is thus the most widely utilized method [8, 9]. The electrochemical synthesis method offers certain advantages such as concurrent entrapment of molecules and doping is possible during synthesis, however, the CPs obtained cannot be easily modified covalently post-synthesis, also removal from the electrodes is difficult [4]. On the other hand, chemical synthesis provides additional choices for covalently modifying the CP backbone and allows for post-synthesis covalent modification, albeit it is a more difficult approach [4]. Another important difference between both methods is that chemical polymerization usually produces powdery thick films whereas in electrochemical synthesis usually very thin films are produced [4, 10]. Moreover, electrochemical synthesis is confined to conditions where the monomer may be oxidized to form reactive radical ion intermediates for polymerization upon application of voltage [4]. The most commonly utilized CPs (e.g. PPy, PTh, PAni, PEDOT) can be polymerized using both chemical and electrochemical synthesis methods; however, chemical polymerization is the only route in the case of some novel CPs having modified monomers [11]. Table 8.1 presents the general properties of the most commonly used CPs.

8.2.1 POLYPYRROLE

Doped PPy offers many advantages as it is simple to produce, can be surface modified easily, and also has high electrical conductivity. The above factors have made it the most thoroughly

TABLE 8.1

General Properties and Limitations of Commonly Used CPs in Biomedical Applications

Polymer	Conductivity (S cm^{-1})	Dopant Type	Properties	Constraints	Ref.
Polypyrrole	10–7.5×10^3	P	Easy to prepare and modify the surface. High electrical conductivity.	Insoluble, brittle, and rigid.	[12, 13–15]
Polyaniline	30–200	n, p	Cost-effective, stable to environmental degradation, wide variations in structure.	Nonbiodegradable, limited solubility, difficult to process.	[12, 14, 15]
Polythiophene	10–10^3	P	Appreciable optical properties, high conductivity, easy to prepare.	Difficult to process.	[12, 13–15]
Poly(3,4-ethylene dioxythiophene)	0.4–400	n, p	Stable toward electrochemical and environmental degradation, transparent semiconducting material.	Restricted solubility.	[16–18]

investigated CP for biomedical applications [5, 10]. PPy has been shown to facilitate cell growth and adherence in various types of cells and also exhibits excellent stability against environmental degradation. Drug delivery, tissue engineering, bioactuators, and biosensors are a few of the areas where the role of PPy has been investigated. At ambient temperature, PPy may be readily synthesized in large amounts in many common organic solvents, as well as in water. PPy films can have conductivities of up to ~103 S cm^{-1} depending on the kind and quantity of dopant used. A drawback of pure PPy is that once prepared, it is difficult to process further as its molecular structure renders it insoluble, crystalline, brittle, and rigid, thus making the unmodified form of PPy unfit for tissue engineering applications [19].

Schmidt and his colleagues described the method of preparation and physicochemical characterization of a PPy derivative – poly(1-(2-carboxyethyl)pyrrole) (PPyCOOH), that possesses a chemical moiety that could be modified at the N-site of a polymer chain with biological molecules, thus improving the biomaterial–tissue interface interactions and supporting intended tissue responses [20]. Thereafter, human umbilical vascular endothelial cells (HUVECs) were cultured on PPyCOOH films that were surface-modified with the Arg-Gly-Asp (RGD) cell adhesion motif which displayed enhanced spreading and attachment on the films. In research by Richardson and team, PPy-coated electrodes were utilized to transmit charge and neurotrophins to spiral ganglion neurons (SGNs) to minimize the deterioration of SGNs linked to prolonged cochlear implant usage [21]. Cochlear implant electrodes were coated with an electrically conductive polypyrrole/paratoluene sulfonate containing neurotrophin-3 (PPy/pTS/NT3). In vivo experiments on guinea pigs revealed that the cochlear implant may be used to administer neurotrophic drugs to SGNs in a secure and controllable mode over a short time, along with electrical stimulation for improved SGN conservation following hearing loss.

Several studies on the biocompatibility of PPy have been published. Martin and his group used electrochemical polymerization to co-deposit PPy and a synthetic peptide on an electrode surface [22]. In vitro soaking studies were used to assess the stability of PPy/peptide coatings, and their influence on brain tissue response and neural recording was investigated in vivo. The electrodes were implanted and analyzed in vivo for a maximum of three weeks. The findings revealed that PPy/peptide coating enhanced neuron attachment and excellent recordings could be obtained from coated locations with neurons attached. In one of the *in vivo* studies, a PPy-silicone tube was fabricated electrochemically and utilized to span a 10-mm sciatic nerve gap in rats [23]. Electrophysiological and histological methods were used to examine the regenerated tissues 24 weeks following the procedure. There was no detection of acute or subacute toxicity, pyrogen, hemolysis, allergens, or mutagenesis in the PPy extraction solution; however, there was the presence of mild inflammation.

In conclusion, despite PPy's appealing characteristics and findings, in vivo research on the compound has been restricted and primarily centered on brief toxicity assessment. More in vivo investigations are needed to validate the feasibility of PPy as a biomaterial, given its disadvantages such as low solubility and stiffness.

8.2.2 POLYANILINE (PAni)

PAni is the second most explored CP, with numerous benefits, including a wide range of structural forms, excellent environmental stability, cost-effectiveness, and the ability to transition between conductive and resistive states electrically via the doping/de-doping process. The completely oxidized pernigraniline base, half-oxidized emeraldine base, and reduced leucoemeraldine base are the three different forms PAni takes depending on its oxidation state. The most robust and conductive form is PAni emeraldine. PAni is likewise challenging to process since it has low solubility in most solvents [19].

In a study reported by Humpolicek and team [25], both the nonconducting PAni (emeraldine base) and its conducting form (PAni hydrochloride) were evaluated for their biocompatibility to skin irritation, skin sensitization, and cytotoxicity. In vivo skin irritation and sensitization, tests were conducted, whereas in vitro cytotoxicity experiments were done on immortalized non-tumorigenic keratinocyte and human hepatocellular carcinoma cell lines. Both PAni hydrochloride and PAni base have good biocompatibility qualities when it comes to dermal irritation and sensitization, according to the findings. However, both polymers were cytotoxic, with PAni hydrochloride having greater cytotoxicity than PAni base. Moreover, purification of the polymer by the reprotonation/deprotonation cycle resulted in a substantial decrease in cytotoxicity, indicating that low-molecular-weight reaction leftovers or by-products, instead of PAni alone, are more likely to be responsible for the detected cytotoxicity.

The primary limitation for using PAni and its derivatives for biomedical applications appears from its inadequate cytocompatibility, poor processability, lack of flexibility, and non-biodegradability. PAni has, however, been studied for use in biological applications, including biosensors, neurological probes, controlled drug release, and tissue engineering, with encouraging results [24].

8.2.3 POLYTHIOPHENE (PTh) AND DERIVATIVES

PTh exhibit characteristics that are similar to, and sometimes even better than, PPy. Electroactive scaffolds for cell culture, biosensors, and neural probes have all been investigated using polythiophene and its derivatives. PEDOT (poly(3,4-ethylenedioxythiophene)) is the most effective PTh derivative because of its greater electrical conductivity and chemical stability, allowing it to be used in biomedical applications. PEDOT has just recently been explored in comparison to PPy and PAni. PEDOT's biocompatibility has been thoroughly documented [19]. PEDOT can be fabricated into various structures such as nanofilms, nanofiber mats, and nanorod arrays. Mattoli and colleagues reported a modified supporting layer method to create free-standing conductive ultrathin nanofilms fabricated using PEDOT and polystyrene sulfonic acid (PSS). The authors indicated that the PEDOT:PSS nanofilms could be modified, folded, and unfolded in water several magnitudes without cracking, disaggregation, or loss of conductive properties, indicating that they could be used in sensing and actuation as well as biomedical applications, such as smart substrates for cell culturing and stimulation [25].

Carmena and her team investigated the use of PEDOT (PSS-doped)-coated microelectrodes as cortical neural prosthetics. When comparing PEDOT coated Pt–Ir electrodes to Pt–Ir electrodes, in vivo chronic testing of microelectrode arrays implanted in rat cortex indicated that the former had better signal-to-noise recordings and improved charge injection [26]. PEDOT nanofiber mats were manufactured by electrospinning coupled with in situ interfacial polymerization using FeCl3 as an oxidant in the research reported by Feng and his group. The PEDOT nanofiber mats had high mechanical characteristics and flexibility, as well as the electrical conductivity of 7.8 ± 0.4 S cm^{-1} and biocompatibility that was comparable to tissue culture plates [27]. PEDOT:PSS coatings integrating dopamine were synthesized on platinum electrodes in a work by Sui and coworkers and their electrochemical characteristics and dopamine transport capabilities were investigated in vitro and in vivo. The PEDOT:PSS/dopamine-coated electrodes were implanted into the brain striatum region of rats for in vivo investigations. The findings showed that when platinum electrodes were electrically stimulated, the PEDOT:PSS/dopamine coatings lowered electrode impedances, enhanced charge storage capacities, and released substantial amounts of dopamine. These findings suggested that PEDOT:PSS/dopamine-coated implanted electrodes might be useful to treat illnesses characterized by dopamine deficiencies, such as Parkinson's disease [28].

The addition of physiologically active dopants gives CPs the characteristics of a multi-stimulus responsive material, making them more appealing as biomaterials for biomedical applications. The capacity of nerve growth factor (NGF) to bring forth particular biological associations with neurons has been tested using it as a co-dopant in the electrochemical deposition of conductive polymers, PPy and PEDOT. PC12 (rat pheochromocytoma) cells attached to the NGF-modified substrate and propagated neurites on both PPy and PEDOT in these tests, showing that the NGF within the polymer film is physiologically active. This method may be used to develop biomedical materials that can be stimulated biologically and electrically [29, 30].

In conclusion, CPs offer many appealing characteristics, including excellent stability, adequate electrical conductivity, and the capacity to incorporate and release biomolecules. They can also be altered for electrical, chemical, physical, and biocompatibility characteristics to make them more suitable for a particular purpose. However, because CPs are frequently fragile and hard to control, and the use of bigger dopants might exacerbate this impact, their usage in biomedical applications is restricted. One approach to address a CP's limitations is to combine it with another polymer in the form of a blend or composite to integrate the materials' beneficial characteristics. CP composites can offer enhanced solubility and improved mechanical characteristics required for a range of biomedical applications without sacrificing their conductivity or other features, as will be discussed in the following sections.

8.3   CONDUCTIVE POLYMER COMPOSITES – COMPOSITES/ BLENDS BASED ON CONJUGATED CPs

Creating composites or blends of CPs with other polymers that have superior mechanical characteristics for the desired use is an efficient method to increase their mechanical properties. CPs such as PPy and PAni have also been investigated as conductive fillers, particularly with biopolymers, to overcome their poor processability and also to confer conductive properties to the insulating polymers [31, 32]. CP composites with better mechanical characteristics may also be made by doping with big molecules. However, because of the existence of insulating molecules, these interactions may create interference with electron conjugation within the CP [4]. Ma and colleagues created synthetic nerve conduits by dip-coating them with a PPy/poly(D,L-lactic acid) (PDLLA) composite solution made from PPy emulsion polymerization in PDLLA solution. The oxidative polymerization was started using an aqueous FeCl3 solution [33]. PC12 cells were utilized to test in vitro cytocompatibility, and after being stimulated with 100 mV for 2 hours, they produced even more and longer neurites on composite conduits than on PDLLA conduits. The 5% PPy/PDLLA composite was also utilized to prepare nerve conduits that bridged a 10-mm gap in a rodent sciatic nerve. The rats with the PPy/PDLLA conduits demonstrated functional restoration that was comparable to the gold-standard autologous nerve transplant and considerably better than the PDLLA conduits at 6 months. The scientists speculated that such a conduit might be employed to regenerate nerve tissue while avoiding the disadvantages of utilizing an autologous transplant, including minimal donor pool, donor site morbidity, numerous surgical sites, and probable size misfit.

Schmidt and coworkers created biomaterials for tissue engineering and wound healing by constructing conductive PPy composites with the physiologically active polysaccharide HA as the dopant [31]. These conducting, HA-incorporated PPy films sustained HA on their surfaces in vitro for many days and increased vascularization in vivo, making them intriguing tissue engineering and wound-healing candidates that are favored from both electrical stimulation and enhanced vascularization. Merging the properties of a CP, such as PPy, with the characteristics of an elastomer, such as polyurethane (PU), may result in a composite with electrical attributes and enhanced biocompatibility and mechanical robustness. Broda et al used an in situ polymerization of PPy with FeCl3 as an oxidant in a PU emulsion to create a set of electrically conducting PPy nanoparticles and PU composites with various ratios [34]. The polymerization produced a composite with a PU core interwoven with a web of PPy nanoparticles that conducts electricity. The electrical conductivity of the composites improved as the mass ratio of PPy to PU increased. Furthermore, the rigidity of the composite improved as the mass ratio of PPy to PU increased, but the maximum elongation dropped. The PPy–PU composites showed elastomeric and conductivity characteristics, as well as being cytocompatible with C2C12 myoblast cells. Degradable polymers featuring conductivity and electroactivity have also received a lot of interest lately. Electrically conducting degradable polymers have been shown to enhance cell adhesion and proliferation, and they might be deployed as scaffold materials for neuronal, cardiac, and bone tissue regeneration [11].

PAni has also been used to construct electroactive hydrogels, which are polymeric blends that merge the responsive properties of electroactive polymers with the highly hydrophilic hydrogels in an aqueous environment that is friendly to biological molecules like polypeptides, enzymes, antibodies, and DNA [35]. Wallace and colleagues described the preparation of a single-component CP hydrogel to be used as a tissue engineering scaffold. Covalent cross-linking of the polymer poly(3-thiopheneacetic acid) with 1,1′-carbonyldiimidazole resulted in the construction of hydrogels [36]. The hydrogels demonstrated significant swelling characteristics (with swelling ratios up to 850%), and the networks' mechanical properties were reported to be equivalent to those of muscle tissue. At physiological pH, hydrogels were shown to be electroactive and conductive. It was observed that fibroblast and myoblast cells grown on hydrogel substrates were adhered and proliferated.

8.4 OPTICAL PROPERTIES OF CONDUCTING POLYMERS AND BIOMEDICAL IMAGING APPLICATIONS

As the current diagnostic materials such as fluorescent dyes and metallic nanoparticles and nanocomposites are insufficient to meet all the desired properties of a diagnostic agent, it is necessary to develop an advanced imaging agent that has improved qualities in photostability, functionalization, targeting, particle stability, and biocompatibility. The CPs are found extremely photostable which allows the in vitro and in vivo emission signals detection over a long observation time. This helps in improved emission signal strength and easy detection of targeted cells and tissues [37]. Owing to their tunable optical properties, these materials are well investigated in the field of diagnosis. Most of the CPs are hydrophobic and thus leading to a rapid cellular uptake of the particle. [38] This enhanced particle uptake, improved fluorescence/optical properties, and low cytotoxicity under both in vitro and in vivo conditions have exposed the potential application of CPs in bioimaging [39, 40]. The presence of a π-conjugated electron system in the organic semiconducting material aids in the tuning of its photoluminescent properties. The emission properties of these CPs can be altered and improved by modifying the copolymers. A wide range of emission signals across the entire visible spectrum to near-infrared (NIR) can be achieved by changing the copolymer composition [41]. This tunability in the electronic bands has introduced CPs as photoacoustic contrast agents in various image-guided therapies (photothermal and photodynamic therapies) [42, 43]. The CPs are well considered in the biomedical research fields such as biosensing due to their explicit electronic and optical properties. The modification on CdS nanostructure, having a known photoluminescent property, with polymer PANI has resulted in an improved and enhanced emission signal strength when compared to one-dimensional nanostructure (nanowires) prepared out of bare CdS compound [44]. Various such polymers derivatives can be prepared by familiar synthesis routes such as electrochemical oxidative polymerization, resulting in an improved optical property by changing its optical contrast, bandgap and switching time, etc. [44, 45]. The targeting of the diagnostic agent is another challenging issue. The CPs can be targeted by functionalizing them with biological recognition entities. Various studies based on functionalizing the CPs with proteins, peptides, antibodies, etc. proved its potential application in targeting the cells in pathological tissue. The functionalized materials were found biocompatible and were further reported as an efficient targeted-imaging agent of cancer and tumor cells [37, 46, 47].

Photothermal therapy is the most successful non-invasive cancer treatment and a highly efficacious PTT always requires safe and potent PTT agents. To accomplish this, Li and the team prepared a 30-nm-sized Pdots having 65% PTCE which had excellent photostability. The PDOTS comprises CPs and DPP with various thiophene derivatives (monothiophene, thienothiophene, bithiophene, and benzodithiophene) synthesized via Stille coupling reaction. To produce 30-nm CPs NPs, a nanoprecipitation method was used. They were determined to be a viable contender for photothermal applications after in vitro and in vivo investigation utilizing a 4T1 tumor-bearing mouse model [48].

Chang and coworkers used a simple nanoprecipitation technique to create theranostic PDOTS constructed of BDT (benzodithiophenedione-based polymer). The PDOTS had excellent cytocompatibility and stability, as well as enhanced optical characteristics. Using MCF-7 cells and tumor model in vitro and in vivo studies, it was demonstrated that it may be used as a dual functional PAI/PTT. Even a low dose of these nanomaterials enabled excellent therapeutic benefits on the in vivo model (MCF-7 cancer cell-bearing tumor model [49]. For the first time, Men and his colleagues created metabolizable and highly NIR-II absorbing Pdots for PAI-guided PTT. The ultrasmall (4 nm) Pdots were produced utilizing the nano-reprecipitation technique and were made of D-A π-conjugated polymer (DPP-BTzTD). These PDOTS have good biocompatibility, photostability, and strong photoacoustic signals, as well as a high PTCE of 53 percent. Intravenously administered PDOTS on 4T1 tumor-bearing mice model demonstrated excellent PTT under a 1064-nm laser irradiation setting with 0.5 W cm^{-2} and showed fast clearance from the body. The pilot research sets the path for future clinical trials by demonstrating their efficacy [50].

8.5   CONCLUSIONS

The goal of this chapter was to analyze the current state of knowledge on the potential of CPs for biomedical applications, based on numerous researches published in the literature over the previous decade. The majority of the early researches focused on determining whether well-known CPs were suitable for biological purposes. Researchers have been exploring different chemical modification approaches, as well as combining with conducting nanoparticles and non-CPs, to address the constraints of CPs, such as limited processability, poor mechanical characteristics, and biocompatibility. The significance of choosing the right conducting particles, as well as using the right blending procedures, seems to be the way to develop effective composites that might be used in biomedical implants. While this technique may aid with processability, mechanical qualities, and biocompatibility, it will come at the expense of electrical conductivity, which may limit the spectrum of applications. CPs are promising materials for filling material needs in medical implants, particularly those utilized in neuronal stimulation and sensing. Tissue engineering is yet another field where these materials might be beneficial, namely as substrates for tissue regeneration where electrical conductivity can help cells grow faster. Because of their excellent optoelectronic characteristics, semi-CP-based theranostic nanoparticles are predicted to have a bright future in phototherapy, particularly for PTT and PDT as mono and combination therapy. There has been significant progress in the development of novel materials for phototherapy and immuno/radiotherapy. The biggest concerns, however, such as the long-term toxicity of such theranostic materials, remain unsolved. Semiconducting polymer-based research will undoubtedly continue to expand, but scientists must also consider their drawbacks to make them appropriate for real-world applications in cancer therapy. Future research on their safety, long-term buildup and elimination, will be extremely useful in moving CPs from the laboratory to the market.

REFERENCES

1. Joy, N., Gopalan, G. P., Eldho, J., & Francis, R. (2016). Conducting polymers: biomedical applications. *Biomedical Applications of Polymeric Materials and Composites*, 37–89.
2. Chiang, C. K., Fincher Jr, C. R., Park, Y. W., Heeger, A. J., Shirakawa, H., Louis, E. J., Gau, S.C., & MacDiarmid, A. G. (1977). Electrical conductivity in doped polyacetylene. *Physical Review Letters*, *39*(17), 1098–1101.
3. Bolto, B. A., McNeill, R., & Weiss, D. E. (1963). Electronic conduction in polymers. III. Electronic properties of polypyrrole. *Australian Journal of Chemistry*, *16*(6), 1090–1103.
4. Guimard, N. K., Gomez, N., & Schmidt, C. E. (2007). Conducting polymers in biomedical engineering. *Progress in Polymer Science*, *32*(8–9), 876–921.
5. Bendrea, A. D., Cianga, L., & Cianga, I. (2011). Progress in the field of conducting polymers for tissue engineering applications. *Journal of Biomaterials Applications*, *26*(1), 3–84.

6. Jagur-Grodzinski, J. (2012). Biomedical applications of electrically conductive polymeric systems. *e-Polymers*, *12*(1), 722–740.

7. Gu, Y., Li, J., Xie, A., Zhang, K., Jiao, Y., & Dong, W. (2018). Superfine palladium nanocrystals on a polyphenylene framework for photocatalysis. *Catalysis Science & Technology*, *8*(20), 5201–5206.

8. Diaz, A. F., Kanazawa, K. K., & Gardini, G. P. (1979). Electrochemical polymerization of pyrrole. *Journal of the Chemical Society, Chemical Communications*, (14), 635–636.

9. Kaneto, K., Yoshino, K., & Inuishi, Y. (1983). Electrical and optical properties of polythiophene prepared by electrochemical polymerization. *Solid State Communications*, *46*(5), 389–391.

10. Huang, Z. B., Yin, G. F., Liao, X. M., & Gu, J. W. (2014). Conducting polypyrrole in tissue engineering applications. *Frontiers of Materials Science*, *8*(1), 39–45.

11. Guo, B., Glavas, L., & Albertsson, A. C. (2013). Biodegradable and electrically conducting polymers for biomedical applications. *Progress in Polymer Science*, *38*(9), 1263–1286.

12. Balint, R., Cassidy, N. J., & Cartmell, S. H. (2014). Conductive polymers: Towards a smart biomaterial for tissue engineering. *Acta Biomaterialia*, *10*(6), 2341–2353.

13. Bredas, J. L., & Street, G. B. (1985). Polarons, bipolarons, and solitons in conducting polymers. *Accounts of Chemical Research*, *18*(10), 309–315.

14. Dai, L. (2004). Conducting polymers. *Intelligent Macromolecules for Smart Devices*, 41–80.

15. Dai, L. (1999). Conjugated and fullerene-containing polymers for electronic and photonic applications: advanced syntheses and microlithographic fabrications. *Journal of Macromolecular Science: Polymer Reviews*, 39, 273–387.

16. Gustafsson, H., Kvarnström, C., & Ivaska, A. (2008). Comparative study of n-doping and p-doping of poly (3, 4-ethylenedioxythiophene) electrosynthesised on aluminium. *Thin Solid Films*, *517*(2), 474–478.

17. Groenendaal, L., Jonas, F., Freitag, D., Pielartzik, H., & Reynolds, J. R. (2000). Poly (3,4-ethylenedioxy-thiophene) and its derivatives: past, present, and future. *Advanced Materials*, *12*(7), 481–494.

18. Ouyang, J., Chu, C. W., Chen, F. C., Xu, Q., & Yang, Y. (2005). High-conductivity poly (3, 4-ethylene-dioxythiophene): poly (styrene sulfonate) film and its application in polymer optoelectronic devices. *Advanced Functional Materials*, *15*(2), 203–208.

19. Kaur, G., Adhikari, R., Cass, P., Bown, M., & Gunatillake, P. (2015). Electrically conductive polymers and composites for biomedical applications. *Rsc Advances*, *5*(47), 37553–37567.

20. Lee, J. W., Serna, F., Nickels, J., & Schmidt, C. E. (2006). Carboxylic acid-functionalized conductive polypyrrole as a bioactive platform for cell adhesion. *Biomacromolecules*, *7*(6), 1692–1695.

21. Richardson, R. T., Wise, A. K., Thompson, B. C., Flynn, B. O., Atkinson, P. J., Fretwell, N. J., … O'Leary, S. J. (2009). Polypyrrole-coated electrodes for the delivery of charge and neurotrophins to cochlear neurons. *Biomaterials*, *30*(13), 2614–2624.

22. Cui, X., Wiler, J., Dzaman, M., Altschuler, R. A., & Martin, D. C. (2003). In vivo studies of polypyrrole/peptide coated neural probes. *Biomaterials*, *24*(5), 777–787.

23. Wang, X., Gu, X., Yuan, C., Chen, S., Zhang, P., Zhang, T., Yao, J., Chen, F., & Chen, G. (2004). Evaluation of biocompatibility of polypyrrole in vitro and in vivo. *Journal of Biomedical Materials Research Part A*, *68*(3), 411–422.

24. Humpolicek, P., Kasparkova, V., Saha, P., & Stejskal, J. (2012). Biocompatibility of polyaniline. *Synthetic Metals*, *162*(7–8), 722–727.

25. Greco, F., Zucca, A., Taccola, S., Menciassi, A., Fujie, T., Haniuda, H., Takeoka, S., Dario, P., & Mattoli, V. (2011). Ultra-thin conductive free-standing PEDOT/PSS nanofilms. *Soft Matter*, *7*(22), 10642–10650.

26. Venkatraman, S., Hendricks, J., King, Z. A., Sereno, A. J., Richardson-Burns, S., Martin, D., & Carmena, J. M. (2011). In vitro and in vivo evaluation of PEDOT microelectrodes for neural stimulation and recording. *IEEE Transactions on Neural Systems and Rehabilitation Engineering*, *19*(3), 307–316.

27. Jin, L., Wang, T., Feng, Z. Q., Leach, M. K., Wu, J., Mo, S., & Jiang, Q. (2013). A facile approach for the fabrication of core–shell PEDOT nanofiber mats with superior mechanical properties and biocompatibility. *Journal of Materials Chemistry B*, *1*(13), 1818–1825.

28. Sui, L., Song, X. J., Ren, J., Cai, W. J., Ju, L. H., Wang, Y., Wang, L.Y., & Chen, M. (2014). In vitro and in vivo evaluation of poly (3, 4-ethylenedioxythiophene)/poly (styrene sulfonate)/dopamine-coated electrodes for dopamine delivery. *Journal of Biomedical Materials Research Part A*, *102*(6), 1681–1696.

29. Kim, D. H., Richardson-Burns, S. M., Hendricks, J. L., Sequera, C., & Martin, D. C. (2007). Effect of immobilized nerve growth factor on conductive polymers: electrical properties and cellular response. *Advanced Functional Materials*, *17*(1), 79–86.

30. Gomez, N., & Schmidt, C. E. (2007). Nerve growth factor-immobilized polypyrrole: Bioactive electrically conducting polymer for enhanced neurite extension. *Journal of Biomedical Materials Research Part A*, *81*(1), 135–149.

31. Collier, J. H., Camp, J. P., Hudson, T. W., & Schmidt, C. E. (2000). Synthesis and characterization of polypyrrole–hyaluronic acid composite biomaterials for tissue engineering applications. *Journal of Biomedical Materials Research, 50*(4), 574–584.

32. Takano, T., Mikazuki, A., & Kobayashi, T. (2014). Conductive polypyrrole composite films prepared using wet cast technique with a pyrrole–cellulose acetate solution. *Polymer Engineering & Science, 54*(1), 78–84.

33. Xu, H., Holzwarth, J. M., Yan, Y., Xu, P., Zheng, H., Yin, Y., Li, S., & Ma, P. X. (2014). Conductive PPY/PDLLA conduit for peripheral nerve regeneration. *Biomaterials, 35*(1), 225–235.

34. Broda, C. R., Lee, J. Y., Sirivisoot, S., Schmidt, C. E., & Harrison, B. S. (2011). A chemically polymerized electrically conducting composite of polypyrrole nanoparticles and polyurethane for tissue engineering. *Journal of Biomedical Materials Research Part A, 98*(4), 509–516.

35. Guiseppi-Elie, A. (2010). Electroconductive hydrogels: synthesis, characterization and biomedical applications. *Biomaterials, 31*(10), 2701–2716.

36. Mawad, D., Stewart, E., Officer, D. L., Romeo, T., Wagner, P., Wagner, K., & Wallace, G. G. (2012). A single component conducting polymer hydrogel as a scaffold for tissue engineering. *Advanced Functional Materials, 22*(13), 2692–2699.

37. Anwar, N., Rix, A., Lederle, W., & Kuehne, A. J. (2015). RGD-decorated conjugated polymer particles as fluorescent biomedical probes prepared by Sonogashira dispersion polymerization. *Chemical Communications, 51*(45), 9358–9361.

38. Fernando, L. P., Kandel, P. K., Yu, J., McNeill, J., Ackroyd, P. C., & Christensen, K. A. (2010). Mechanism of cellular uptake of highly fluorescent conjugated polymer nanoparticles. *Biomacromolecules, 11*(10), 2675–2682.

39. Wu, C., & Chiu, D. T. (2013). Highly fluorescent semiconducting polymer dots for biology and medicine. *Angewandte Chemie International Edition, 52*(11), 3086–3109.

40. Chan, Y. H., & Wu, P. J. (2015). Semiconducting polymer nanoparticles as fluorescent probes for biological imaging and sensing. *Particle & Particle Systems Characterization, 32*(1), 11–28.

41. Palner, M., Pu, K., Shao, S., & Rao, J. (2015). Semiconducting polymer nanoparticles with persistent near-infrared luminescence for in vivo optical imaging. *Angewandte Chemie, 127*(39), 11639–11642.

42. Pu, K., Shuhendler, A. J., Jokerst, J. V., Mei, J., Gambhir, S. S., Bao, Z., & Rao, J. (2014). Semiconducting polymer nanoparticles as photoacoustic molecular imaging probes in living mice. *Nature Nanotechnology, 9*(3), 233–239.

43. Li, K., & Liu, B. (2014). Polymer-encapsulated organic nanoparticles for fluorescence and photoacoustic imaging. *Chemical Society Reviews, 43*(18), 6570–6597.

44. Xi, Y., Zhou, J., Guo, H., Cai, C., & Lin, Z. (2005). Enhanced photoluminescence in core-sheath CdS–PANI coaxial nanocables: A charge transfer mechanism. *Chemical Physics Letters, 412*(1–3), 60–64.

45. Turac, E., Sahmetlioglu, E., Toppare, L., & Yuruk, H. (2010). Synthesis, characterization and opto-electrochemical properties of poly (2,5-di (thiophen-2-yl) 1-(4-(thiophen-3-yl) phenyl)-1*H*-pyrrole-co-EDOT). *Designed Monomers and Polymers, 13*(3), 261–275.

46. Feng, L., Liu, L., Lv, F., Bazan, G. C., & Wang, S. (2014). Preparation and biofunctionalization of multicolor conjugated polymer nanoparticles for imaging and detection of tumor cells. *Advanced Materials, 26*(23), 3926–3930.

47. Ahmed, E., Morton, S. W., Hammond, P. T., & Swager, T. M. (2013). Fluorescent multiblock π-conjugated polymer nanoparticles for in vivo tumor targeting. *Advanced Materials, 25*(32), 4504–4510.

48. Li, S., Wang, X., Hu, R., Chen, H., Li, M., Wang, J., Wang, Y., Liu, L., Lv, F., Liang, X. J., & Wang, S. (2016). Near-infrared (NIR)-absorbing conjugated polymer dots as highly effective photothermal materials for in vivo cancer therapy. *Chemistry of Materials, 28*(23), 8669–8675.

49. Chang, K., Gao, D., Qi, Q., Liu, Y., & Yuan, Z. (2019). Engineering biocompatible benzodithiophene-based polymer dots with tunable absorptions as high-efficiency theranostic agents for multiscale photoacoustic imaging-guided photothermal therapy. *Biomaterials Science, 7*(4), 1486–1492.

50. Men, X., Wang, F., Chen, H., Liu, Y., Men, X., Yuan, Y., Zhang, Z., Gao, D., Wu, C., & Yuan, Z. (2020). Ultrasmall semiconducting polymer dots with rapid clearance for second near-infrared photoacoustic imaging and photothermal cancer therapy. *Advanced Functional Materials, 30*(24), 1909673.

9 Conducting Polymers for Gene Delivery

Aykut Arif Topcu,[1] Erdoğan Özgür,[2] and Adil Denizli[2]

[1]Medical Laboratory Program, Vocational Scholl of Health Service, Aksaray University, Aksaray, Turkey

[2]Chemistry Department, Hacettepe University, Beytepe, Ankara, Turkey

CONTENTS

9.1 INTRODUCTION

Conducting polymers (CPs) have received a lot of attention in various research areas ranging from biomedical applications to electronics due to their compatibility, electrical, and optical properties [1, 2]. The conjugated backbone of CPs consists of an alternating single (σ) and double (π) bonds, and these π-conjugated structures of CPs are responsible for their electric and optical features similar to those of inorganic semiconductors and metals [3] displaying electrical conductivity values between 10^{-3} and 10^3 Ω^{-1} cm^{-1} (Figure 9.1). Their chemical and physical properties can easily be tailored by chemical functionalization and/or doping processes which impart functional characteristics to be used in distinct biomedical applications [5]. During the doping process, the formation of positive and negative polarons/bipolarons (radicals) is facilitated, and counterions balance the charge of the polymer [6]. The extent of this process mainly increases the electrical conductivity of the CPs by many orders of magnitude. The removal of electrons from the valence band leads to the formation of positively charged holes (p-doping), while the insertion of electrons to the conduction band generates a negative charge (n-doping). Depending on the chemical structure of the polymer, the electrical pathway and the characteristic conductivity in CPs mainly rely on the free movement of delocalized electrons in the conjugated π-system or the electron hopping between the redox-active sites in redox polymers [7]. Polyacetylene (PA) is first fabricated CP doped with bromine and this invention resulted in Nobel Prize in 2000 and opened a new class of materials to be explored [8, 9].

DOI: 10.1201/9781003205418-9

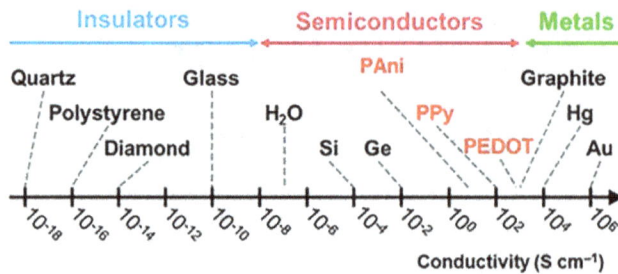

FIGURE 9.1 A scale of material's conductivity. (Adapted with permission from Ref. [4]. Copyright (2021) Elsevier.)

Nowadys, the conjugated organic polymers, for instance, PA, polyaniline (PANI), polypyr-role (PPy), poly (p-phenylene vinylene), and polythiophene (PTH), are synthesized with vari-ous approaches, including catalytic-polymerization, electrochemical polymerization, chemical oxidation, and direct oxidation methods [1]. The choice of the preparation method, the changing of the experimental conditions like the pH of the media, the reaction temperature, the reac-tion time, the use of suitable dopants, and the dopant concentrations are of great importance for enhancing their conductivities [3]. Moreover, their syntheses are simple, quick, and inexpen-sive; so, CPs are favorable materials utilized in various fields such as photocatalytic applications, energy storage studies, biomedical researchers including biosensor applications, tissue engineer-ing, and drug delivery studies [3, 5, 8]. One interesting feature of CPs is that their capability to entrap molecules (such as drugs, proteins, and enzymes) or release these molecules on demand. The release rate can easily be adjusted by the applied electrical stimulation resulting in expansion and contraction movements in the polymer backbone [9]. Furthermore, these conjugated organic polymers can enhance the sensitivity, speed, stability of the sensing devices and their interfaces with biological tissues [2]. From this point of view, CPs hold a great promise for biomedical stud-ies [10], and herein, the different CPs, their synthesis methods, and their gene delivery studies were summarized in depth.

9.2 GENE DELIVERY

Gene therapy is an alternative method that inserts the gene into a patient's cell to treat or recover various diseases such as cancers, inherited disorders, viral infections instead of using drugs or operations [11]. Viral vectors including DNA, RNA, oncolytic viral vectors, and nonviral vectors, e.g., liposomes, polymers are utilized in gene delivery systems [11, 12]. In clinical trials, the non-viral vector-based gene delivery systems have distinct advantages such as low-immune response, controllability, ease of modification, and simplicity compared to the viral-based delivery systems [12–16]. During the nonviral-based gene therapy studies, the physicochemical methods including needle injection, electroporation, DNA injection are applied to introduce the genetic materials such as plasmid DNA, siRNA into the cell through the cell membrane [11, 17]. Polymers and liposomes are generally used as nanovesicles to transfer and facilitate the genetic materials into the cells [11]. These nanocarriers are capable of interacting with DNA or RNA with electrostatic interactions (Figure 9.2) and can enter the cell via endocytosis [11, 12, 17]. Furthermore, by using the nonviral vectors, it is possible to eliminate the size limitation of DNA packaging and the spe-cific cell targeting can be achieved by ligand modifications [18]. For these reasons, the nonviral vectors are promising nanocarriers for gene delivery studies owing to some advantages above men-tioned. In this context, the applicability of various CPs including polythiophene, its derivatives, polydiacetylene, polypyrrole, poly(phenylene ethylene), polyfluorene, polyaniline for gene delivery applications is discussed.

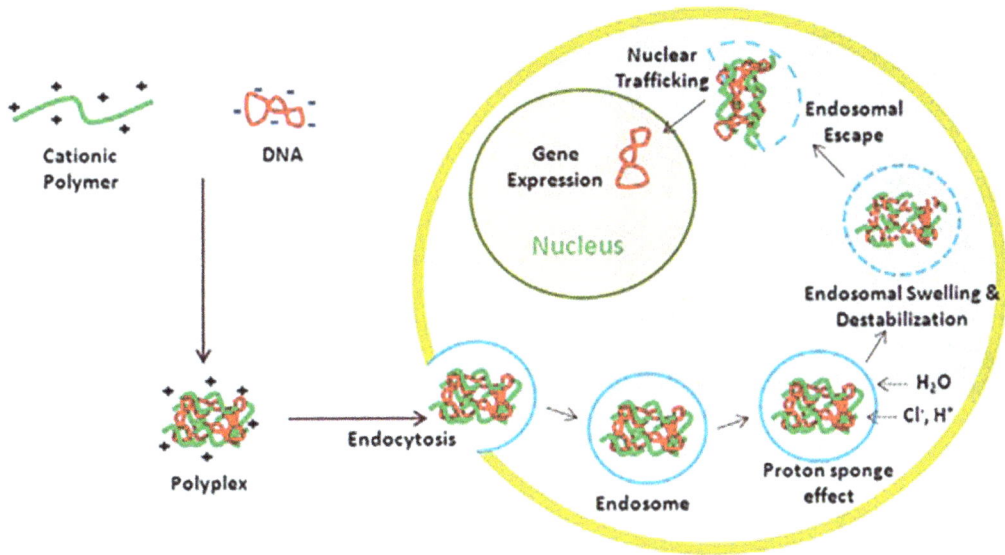

FIGURE 9.2 Gene transfection mechanism via cationic polymers. (Adapted with permission from Ref. [17]. Copyright (2021) Elsevier.)

9.2.1 Polythiophene (PTH)

The water solubility, remarkable biocompatibility, easy accessibility, and optical property of PTH make as a promising material for cell studies, tissue engineering applications, and the cationic forms of PTH are of great importance for gene delivery studies as well [19]. In the early 1980s, this conjugated polymer was chemically synthesized via Yamamoto and Lin-Dudek routes [11], recently, electropolymerization [20], oxidative synthesis [21], temple-assisted synthesis [20], and green synthesis [22] approaches have been used to synthesize for PTH and its derivatives. Zhang et al. [19] prepared cationic PTH onto the star-shaped degradable polyaspartate (SP)/plasmid to enhance the gene delivery performance of the carrier, reduce the toxicity, save the consumption of the delivery material, and understand the relationship between the cationic PTH and transfection process (Figure 9.3). Ethyl thiophene-3-acetate (TEt) and ethyl thiophene-3-manolate (TMEt) monomers were chosen and the linear and hyperbranched PTH esters were prepared using $FeCl_3$ oxidative polymerization method. After that, cationic PTH was produced via aminolysis, acidification, and purification steps, respectively. According to the experimental findings, the addition of cationic PTH that enhanced stability and endolysosomal permeability resulted in increasing gene delivery performance of the delivery material by moderate photosensitization of oxygen.

Small interfering RNA (siRNA)-based therapy is used for the treatment of inherited or acquired diseases by gene suppression, but naked siRNA (unprotected oligonucleotides) has a short half-life *in vivo* because of endonuclease degradation and is rapidly removed from the body due to the small size [23–25]. For its therapeutic effect, siRNA should escape the endosome and enter the cytosol [26]; in this regard, the need for siRNA carrier is in high demand for clinical studies. A cationic siRNA carrier was prepared by the combination of low-molecular-weight polyethylenimine (PEI-18K) and PTH because of the cell membrane penetrating capability of PTH and biocompatibility of PEI-18K [23]. PTH was synthesized via the chemical oxidative coupling reaction approach, following that, PEI-18K was coupled with PTH to prepare siRNA carrier (PEI-*co*-polythiophenes copolymers). The increasing density of PEI could increase siRNA retardation capacity and the siRNA copolymer nanocarrier could effectively deliver siRNA resulted in a knockdown of the target protein expression. Furthermore, the hydrophobic character of PTH could positively affect the siRNA delivery performance.

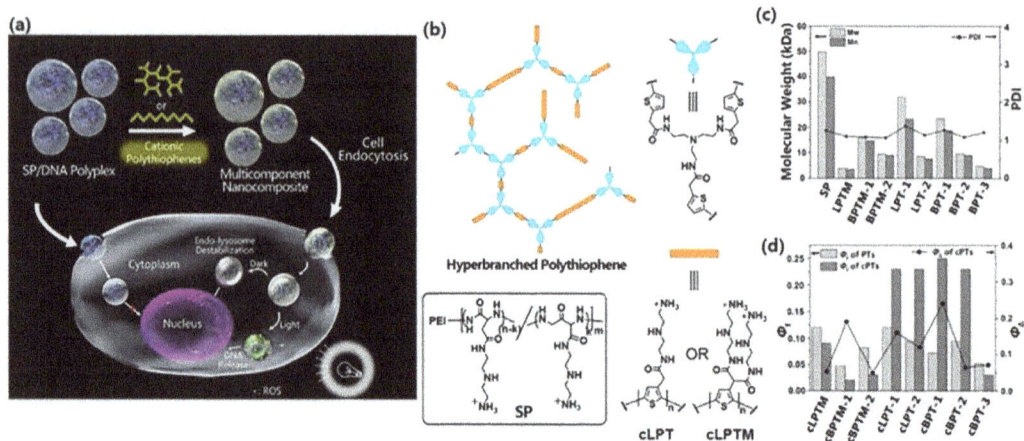

FIGURE 9.3 (a) Schematic illustrations of the mechanism for cPT-enhanced endolysosome escape in gene delivery to cells. (b) Chemical structures of linear and hyperbranched polythiophenes, and SP. (c) GPC-based molecular weights and PDIs for PTs and SP. (d) Fluorescence quantum yields of PTs and singlet oxygen quantum yields of cPTs. (Adapted with permission from Ref. [19]. Copyright (2021) American Chemical Society.)

In the next study, the conjugated nanocarrier and photodynamic therapy (PDT) were combined to siRNA delivery into MDA-MB-231 cancer cells [27]. The researchers used phosphonium groups to increase the transfection efficiency and decrease cell toxicity. The phosphonium-based-PTH conjugated polymers were prepared with Kumada Catalyst-Transfer Condensative Polymerization (KCTCP) method to control the molecular weights of PTH polymers (11.5–53 kDa). Afterward, the conjugated polymers containing phosphonium groups were synthesized using trimethyl phosphine. The PTH1 conjugated polymer did not generate O_2 and was not able to complex siRNA due to the higher molecular weight (53 kDa) than the other polymers, PHT2 (19 kDa) and PHT3 (11.5 kDa). The sizes of PTH2/siRNA and PTH3/siRNA complexes were found as 82 and 92 nm, respectively, and these sizes were suitable for endocytosis. Furthermore, the gene silencing effects of the nanocarriers (PHT2 and PHT3) were found over 30% due to their cationic and amphipathic properties.

DNA binding capability and gene delivery performance of the regioregular PTH nanocarrier were investigated by Zhang et al. [28]. The regioregular PTH nanocarriers containing ester side chains (P3ETs) were prepared using the KCTP method; thereafter, the cationic P3ETs were synthesized via aminolysis reaction. Then, DNA delivery performance into tumor cells of the cationic nanocarriers was investigated. In the light of the experimental findings, the nanocarriers with different molecular weights were obtained via aminolysis reaction and the DNA binding capability of the regioregular nanocarrier was higher than the nonregular nanocarrier. Moreover, the photochemical internalization (PCI) effect enhanced the gene delivery performance of the nanocarriers.

In one interesting study, Gautier et al. [29] combined the DNA biochip technology with the controlled electrochemical gene delivery to design a dual function system for both diagnostic and therapeutic applications. For this purpose, they fabricated a label-free DNA-sensing platform based on a synthetic 5′-amino-terminated single-stranded DNA probe immobilized on an electropolymerized polythiophene copolymer film through the activated arylsulfonamide terminal groups. The hybridization of the complementary target was accomplished on the DNA-modified CP matrix. This surface which can be oxidized and reduced with an ionic exchange at the film/solution interface acted as a transducer to detect DNA hybridization by electrochemical impedance spectroscopy and quartz crystal microbalance techniques. Significant changes were observed in the impedance spectra, which was a sign of the success of both the electrochemical probe fabrication and the hybridization processes. Impedimetric measurements revealed that the polymer matrix contained negative charges in its backbone after the DNA hybridization due to a superficial p-doping process with the aid of phosphate groups. After the hybridization, the DNA-modified substrates were subjected to electrochemical treatment for the release of this grafted DNA sequence species controllably (between −0.5 and −2.4 V versus saturated calomel electrode) by repetitive cycles. This led to partial cathodic degradation of the copolymer by electrochemical cleavage, which resulted in the controlled delivery of DNA. The released fragments of DNA were checked to confirm the retainment of the bioactivity by electrophoresis successfully and it was demonstrated that the CP functionalized system was appealing for DNA detection and controlled gene delivery purposes in a physiological environment.

Later on, the same group [30] studied a specifically designed CP obtained from the electrooxidation of cyclopentadithiophene (CPDT) units, for the controlled electrochemical delivery of DNA. Here, the electrochemical probe was constructed by the electropolymerized copolymer film (CPDT-co-M1) matrix on platinum electrodes. The electronic properties of the CP were tunable in which the cathodic cleavage of the SN bond was available by electrochemical treatment. The 5′-amino-terminated single-stranded DNA probe was accumulated covalently on the polymer via a sulfonyl chloride as linker group and hybridization was performed with the 675-base target DNA that was complementary to the probe. Voltammetric studies showed that under external electrical stimulus by successive scanning of the potential between −0.5 and −1.7 V, the aryl sulfonamide moieties were reduced in physiological media, which confirmed the bielectronic cleavage of the SN bond. The delivery profile was monitored using quartz crystal microbalance and the loss of mass was correlated with the amount of liberated DNA in a pulsatile way. The released species amplified by PCR were identified as long sequences of DNA strands by electrophoresis and the bioactivity was preserved after the electrochemical delivery process. The proposed "on demand-electrochemical DNA delivery strategy" appeared to be a promising sequence-specific therapeutic tool to detect genetic disorders and to deliver the genes at specific locations in physiological media. Furthermore, the polymer matrix exhibited a reusable character by only a new exposure to N-chlorosuccinimide after cleavage.

9.2.2 Polydiacetylene (PDA)

PDA is a member of the CPs family with chromatic and nonlinear optical properties and is formed by UV irradiation or ¥ irritation via 1.4-polymerization of diacetylene monomers [31]. The freshly prepared PDA is blue in color and non-fluorescent, but the change of the environmental conditions such as pH, temperature, ligand-receptor interactions, electrical stress, and mechanical stress leads to change its blue color to red [31]. The change of its color is generally irreversible and can be observed with the naked eye, UV/vis absorption, fluorescence spectroscopy; so, PDAs could be employed in biosensing applications by observing this phase transition [31]. Moreover, the good biocompatibility and the mechanical properties of PDA make this conjugated polymer to be utilized in wide applications including drug delivery systems, bioimaging, and tissue engineering studies [31].

Wang et al. [32] prepared PDA vesicles to visualize the membrane affinity of gene vectors with the aid of discoloration of PDA vesicles. The nanocarriers were prepared with a film dispersion method, the surface potential compositions of which were changing using different phospholipids, e.g., 1,2-dimyristoyl-sn-glycero-3-phospho-(1′-rac-glycerol) (DMPG), 1,2-dimyristoyl-sn-glycero-3-phospho-choline (DMPC), and stearamide (SA). After the optimization of the surface potential of PDA vesicles, the visualization performance of PDA vesicles was compared with three commercial gene vectors, Lipofectamine 2000 (Lipo), Entranster TM-H4000 (Entr), and polyethylenimine (PEI) by observing the color changes. Moreover, in vitro gene transfection efficiency of the commercial vectors was investigated in human normal cell (L02) and hepatocellular carcinoma (HepG2) cell lines via the reporter gene method. The experimental results supported that PDA vesicles would be a screening potential for drug delivery studies.

In the next research article, siRNA binding ability, siRNA transport capability, and siRNA delivery performances of the cationic PDA micelles containing ammonium groups (Figure 9.4)

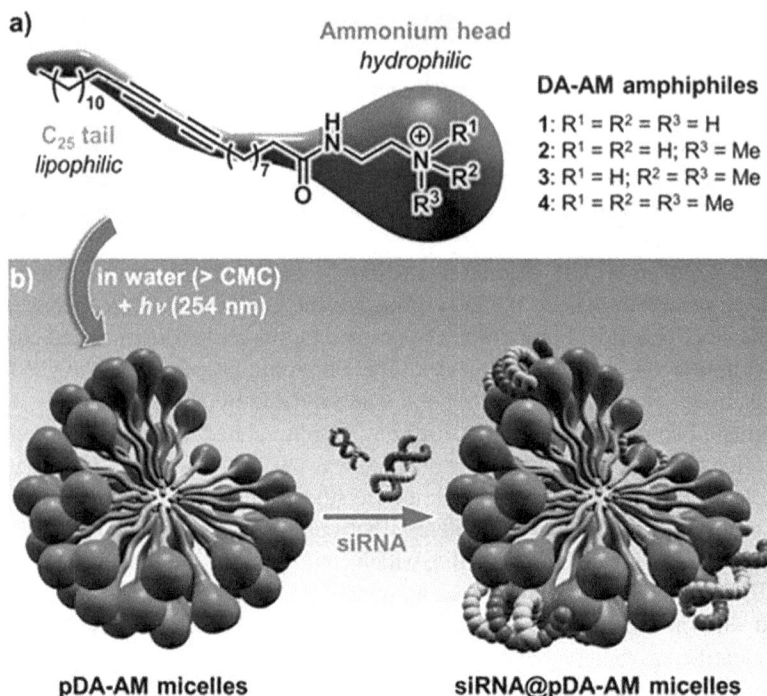

FIGURE 9.4 (a) Structure of the four DA-AM amphiphiles; (b) micelle assembly, photopolymerization, and complexation with siRNA. (Adapted with the permission from Ref. [33]. Copyright (2021) Royal Chemical Society.)

were investigated [33]. For that aim, the commercial pentocosa-10,12-diynoic acid was used in the starting point of producing four different cationic PDA micelles via the topochemical polymerization method. After that, the transfection efficiencies of four cationic PDA micelles (pDA-AM 1, pDA-AM 2, pDA-AM 3, pDA-AM 4) were compared with the reference lipidic carrier system (Lipofectamine RNAi Max). The siRNA transfection capabilities of pDA-AM 1 (containing primary amines) and pDA-AM 2 (containing secondary amines) were higher than pDA-AM 3 (containing tertiary amines) and pDA-AM 4 (containing quaternary amines). The pDA-AM 2 had the highest transfection efficiency and the lowest cytotoxicity.

In another study, pH-responsive PDA micelles were synthesized via UV irradiation to be used as a siRNA nanocarrier [34]. The researchers selected a cationic monomer with histidine amino acid (containing hydrophilic head group) to prepare siRNA nanocarrier and aimed to enhance siRNA delivery performance of the nanocarrier. The polymerized micelles including histidine (His-PMs, NH_2-PMs) and nonpolymerized micelles including (Hist-NPMs, NH_2-NPMs) were synthesized. The histidine groups enhanced siRNA delivery performance of the nanocarrier through the "proton sponge" effect thanks to the imidazole group of histidine. The polymerization of PDA micelles resulted in decreasing the toxicity, and the nonpolymerized micelles were almost three times more toxic than the polymerized micelles. Moreover, the dialyzed polymeric micelles had the lowest toxicity performances than the nondialyzed nanocarrier.

Core cross-linked micelles and shell cross-linked micelles are prepared using copolymers and have a potential for drug delivery studies; from this point of view, Morin et al. [35] prepared ammonium-based cationic PDA micelles for DNA delivery. An amount of 10,12-pentacosadiyonic acid (PCDA), a commercial monomer, was photopolymerized to synthesize two amphiphile (M_4 and M_8) compounds. The polymerized micelles (M4 and M8) and nonpolymerized micelles (M4NP and M8NP) were mixed with pCMV-Luciferase (cPMV-Luc) solution to form the lipoplexes (45 nm) and their transfection efficiencies were evaluated into HeLa cells. After that, Luciferase gene expression level was examined and compared with a commercial Jetsi-ENDO kit. The increasing of photopolymerization time resulted in increasing the diameter of the micelles; so, the polymerization was taken place an hour. The polymerized micelles had higher transfection efficiency and lower toxicity than the nonpolymerized micelles, and *in vitro* transfection efficiency results were highly satisfactory as compared with the commercial kit.

9.2.3 POLYPYRROLE (PPY)

The unique features of PPy such as chemical stability, *in vivo* and *in vitro* biocompatibility, high conductivity, ease of fabrication, and modification make this conjugated polymer a promising material to be utilized in wide applications including drug delivery studies [36]. The electrical conductivity of PPy materials can be achieved by using negatively charged dopants that enable conduction of charge along the polymer backbone upon the electrical stimulation and the higher conductivities (up to ~10^3 S cm^{-1}) can be attained depending on the type and amount of dopants [37]. The electroresponsive behavior of PPy backbone is used for this distinctive purpose in which loading and releasing can be achieved by electrical switching between the oxidative and reductive state leading to a change in polymer's charge, conductivity, and volume. So, the molecules have been entrapped or doped into the PPy film and released through a voltage-controlled approach, or either PPy was employed as an active element to trigger the release from other polymers [37].

The poly electrode multilayers (PEMs) were prepared and their DNA releasing performances were tried to promote by using an external electric field [38]. Before the deposition of PEI and DNA onto PPy films, PPy films were prepared using ammonium persulfate and PPy at 4°C for 20 min. After the PPy film formation, the surface of which was first coated with PEI and then, PEI functionalized surface was treated with DNA/PEI solution to form the PEMs. The vertically electrical fields significantly increased DNA releasing, whereas the horizontal electric field did not show the same delivery performance. Moreover, the increase of the voltages and treatment time resulted in enhanced DNA and PEI delivery. The surface morphologies of PEMs were changed by applying the different electrical fields and treatment time; additionally, the electrical fields also promoted DNA and PEI release and led to an increased N/P ratio. The transfection efficiency results showed that the higher voltage and longer treating time could enhance DNA and PEI delivery from PEMs.

Despite its potential to be utilized for delivery systems, practical application of PPy for drug and gene delivery remains a challenge because of low loading efficiency and rapid release of the encapsulated molecule. Considerable improvements can be achieved via increasing ion-doping capacity through the formation of a high surface area. Another major limitation is the lack of some functional groups in its structure for surface modification and unmodified PPy is not suitable for bioapplications. Moreover, PPy has inherent hydrophobic nature and also has been shown to have insufficient biocompatibility for cell adhesion [39]. A detailed investigation for gene delivery showed that previous efforts dealt with PPy synthesis by oxidative polymerization for the incorporation of DNA, or RNA, but did not investigate the *in vivo* or *in vitro* controlled release of these molecules. Electrochemical synthesis is another approach for convenient synthesis of PPy for the entrapment of nucleic acids within the polymer network through a simple, one-step, well-controlled route. In this technique (Figure 9.5), potential is cycled repetitively in the predetermined electroactive range, which causes an oxidative reaction to deposit PPy on the working electrodes placed in a solution containing monomer and a doping anionic biomolecule [39].

Accordingly, Wang and Jiang first time studied the incorporation of nucleic acid dopants in PPy and investigated the progressing growth profile during the film deposition and ion exchange properties by electrochemical quartz crystal microbalance (EQCM) [40]. Electropolymeric growth profiles indicated that the anionic 20–30-meter-long oligonucleotides (ODNs), as well as ssDNA, served as counterions during polymer growth by maintaining the electrical neutrality without using additional electrolytes and were irreversibly incorporated in the polymeric film. PPy/ODN film was

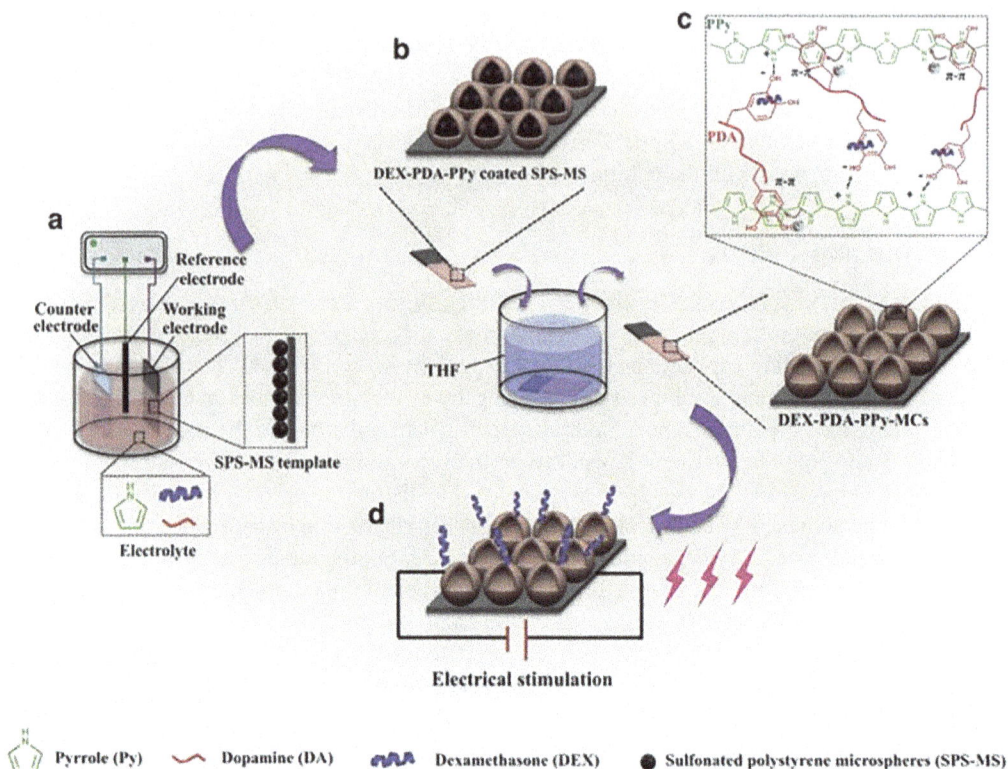

FIGURE 9.5 Schematic representation of the preparation process of DEX-loaded PDA-PPy-MCs. (a) Electrochemical deposition process. (b) Removing the SPS-MS template by THF etching, (c) PDA and PPy interact with each other through hydrogen bonding and π-π interaction. (d) Drug release by ES. (Adapted with permission from Ref. [39]. Copyright (2021) Springer Nature.)

able to exchange the ODNs with the anion of the electrolyte. However, such an irreversible entrapment of nucleic acids in PPy hindered its use for gene delivery applications using electrochemical release/undoping protocol. Therefore, PPy/ODN composite film was suggested to be utilized as a potential bioactive interface toward the development of new genoelectronic devices for genetic analysis purposes.

9.2.4 POLY (PHENYLENE ETHYLENE) (PPE)

PPE is a widely studied conjugated polymer for diverse applications because this conjugate polymer is easily synthesized and its different structural forms could be enriched by modifying its side chains [41]. Additionally, this conjugated polymer is capable of inducing reactive oxygen species (ROS) generation [42]. PPEs are generally synthesized by Sonogashira-type polycondensation reaction via a step-growth mechanism with the copolymerization of 1,4-dihaloarene and 1,4-diethynylbenzene using Pd-catalyzer [41]. The molecular weight, the polydispersity, and the end-group functionality of PPEs could be controlled by this reaction that boosts the optical and electrical properties of PPEs [41]. The first research study of this section was reported by Tu et al. [42] and four cationic PPEs (P-O-3, P-C-3, PIM-2, and PIM-4) with different side chains and charge densities were prepared to increase the gene delivery performance of branched PEI (BPEI) by inducing photochemical internalization (PCI) effect. The monomers M-O-3 and M-C-3 were synthesized by using 1,4-dimethoxybenzene and 1,4-phenylene diacetic acid, respectively. Thereafter, the polycationic-branched side chains of P-O-3 and P-C-3 polymers were synthesized via Sonogashira reaction with the copolymerization of para-substituted diiodobenzene derivatives (M-O-3 and M-C-3) and 1,4-diethynylbenzene. After the coupling reaction, the polycationic end groups of PPEs were obtained by hydrolysis of the Boc-protected amine groups. The other cationic polymers PIM-4 and PIM-2 containing two imidazolium units, which were located on different sides of phenylene units, were synthesized via Pd catalyst.

After the polymerization of these cationic PPEs, a tiny amount of the desired PPE solution was added to BPEI/DNA complex to form the polyplex, and photoactivation was used for the surface coating of the polyplex. The adsorption light of P-C-3 was found at 387 nm, whereas the adsorption waves of the other polymers were found longer due to the electron-donating of alkoxy groups on their backbones and the aggregation of the PPEs resulted in longer wavelengths. The singlet oxygen quantum yields of PIM-2 were higher than the other conjugated polymers, and the hydrodynamic radius of polyplexes was found to be at the same ratios; however, significant zeta-potential changes were observed. The polymeric shell prevented DNA from ROS attack and the conjugated polymers with PCL effects could enhance the DNA delivery performance of BPEI and transfection efficiency.

The cationic poly(p-phenylene vinylene) (PPV) conjugated polymer was used in siRNA delivery into the living cells [43]. The cationic PPV was exposed to white light to endosomal disruption to enhance the siRNA delivery performance of the nanocarrier. The electrostatic interactions between PPV and siRNA made the polyplex into a compact form and resulted in a smaller size. The endosomal disruption capability of the conjugated polymer toward HeLa cells was investigated with and without white light irritation and was visualized using a labeled agent. Furthermore, the biosafety, the transfection efficiency, and *in vitro* gene silencing ability of polyplex were examined using MMT assay, flow cytometry, and Luciferase assay, respectively. According to experimental results, the polyplex was more biocompatible than PEI25kD as a classic gene vector, white light irradiation could enhance the siRNA delivery performance of polyplex, disrupt the endosomal membrane, and boost gene silencing ability of the polyplex. The transfection efficiency of the polyplex was higher than PEI25kD and Lipofectamine 2000. Additionally, the polyplex is capable of imaging siRNA delivery due to its high fluorescence property.

The fluorescent property of positively charged CPs makes them favorable materials for *in vitro* siRNA delivery studies [44]. For this purpose, the siRNA delivery performance of the amine-containing PPE nanoparticles (47–52 nm) was visualized by fluorescent microscopic imaging [44]. The interactions between the carrier and siRNA resulted in increasing the hydrodynamic radius of PPE nanoparticles, and these results were in accordance with the other study findings. The fluorescent intensity of the nanocarrier was gradually decreased by increasing the amount of siLamin (lamin A/C gene) concentrations. Moreover, the complexation efficiency was affected by changing the nitrogen (N) to phosphate (P) (N/P) ratios. The siRNA delivery performance of the nanocarrier was evaluated toward A549 (human carcinoma epithelial cells) by visualizing the fluorescent signals and the experimental results were highly satisfactory for siRNA delivery *in vitro*. Although the escape mechanism of the siRNA/PPE nanoparticle complex is uncertain, probably, the increase of pH led to endosomal disruption and boosted the releasing siRNA/PPE complexes to the cytosol. The transfection efficiency of siRNA nanocarrier was examined the expression level of actin B protein and HeLa cells were incubated for transfection solution containing the nanocarrier, actin B, and siGLO (an indicator) for 24 h. After the incubation of transfection solution, the expression level of actin B was decreased (nearly 94%) without toxic effect while, the expression level of a control protein, tubulin was not changed.

The usability of siRNA delivery and gene silencing of the capability of polyvalent nanocarriers based on dendritic polyethylene-cationic poly(p-phenylene ethylene) (DPE-PPE⁺) were examined by Zhang et al. [45]. The neutral nanocarriers (DPE-PPE) were synthesized with palladium/copper-catalyzed Sonogashira cross-coupling reaction after that, and the nanocarriers were quaternized with bromoethane to obtain water-soluble cationic DPE-PPE⁺. The other water-soluble nanocarriers containing folic acid (FA) (DPE-PPE-FA⁺) were prepared via FA-Br mediated quaternization and their RNA protection abilities, cytotoxicities, and cell transfection capabilities were investigated.

The cytotoxicity results showed that the attachment of FA did not have a remarkable effect on cell viabilities; but, both of the nanocarriers with low cytotoxicities had a potential for siRNA delivery *in vitro*. The carriers protected siRNA toward RNaseA enzyme over 90 min at 37°C and according to flow cytometry results, siRNA delivery performance of DPE-PPE-FA⁺ into HeLa cells was better than DPE-PPE⁺ carrier. Furthermore, the cellular uptake of DPE-PPE-FA⁺ was higher than DPE-PPE⁺, and these nanocarriers could be used as imaging probes for siRNA delivery like quantum dots (QDs), but the carriers showed lower toxicity than QDs. The gene silencing of the carriers was evaluated with Western Blotting, and their gene silencing results were compared with the two commercial transfection reagents. The gene silencing efficiency of the carriers was found almost 80% in HeLa cells and the nanocarriers exhibited the same gene silencing efficiency with the commercial reagents.

The electroporation method is utilized in siRNA delivery or gene transfer into the protoplast by applying the electric field, which enhances the plasma membrane permeability and allows the penetration of nucleic acid into the protoplast [46]. However, the high voltage applied results in a significant loss of viable protoplasts depending on the source of the protoplasts. So, the researchers offered an alternative approach for siRNA delivery using conjugated nanocarriers with many advantages above-mentioned [46]. The amine-containing PPE nanoparticles (CPNs) as siRNA nanocarriers (60–80 nm) were synthesized with palladium/copper catalysts and incubated with *NtCesA-1*/siRNA. After that, the gene knockdown efficiency of CPNs was investigated toward tobacco BY-2 plant protoplasts. According to the experimental results, CNPs with 5–25 µM concentrations were not toxic whereas, over 50 µM CNPs concentrations could be harmful to BY-2 protoplast cell viabilities. The CNPs were capable of penetrating to BY-2 protoplasts and SiGLO Red could penetrate to the protoplasts after the incubation with the CNPs. Therefore, the CNPs had a potential for siRNA delivery into plant protoplast without significant loss of protoplast viability. Moreover, siRNA delivery could be monitored in real-time by CNPs, and these conjugated nano vehicles could be alternatively used for siRNA delivery among the other gene transfection methods.

9.2.5 POLYFLUORENE (PF)

PF is another conjugated polymer, the cationic homopolymers and copolymers of which are synthesized via Suzuki coupling polymerization, Heck, and Sonogashira coupling reactions using Pd-catalyzed [47]. The PF derivatives are synthesized by the copolymerization of the fluorene monomer and comonomers, e.g., benzene, benzothiadiazole, thiophene via Pd-catalyzed coupling reactions and are potentially used in various applications [47]. For example, cationic PF derivatives are utilized as optical probes for DNA detection due to light-harvesting properties and show optical amplification via Foerster resonance energy transfer (FRET) [48].

Feng et al. [49] prepared cationic poly(fluorenylene phenylene) (PFPL) nanoparticles via Suzuki coupling reaction for gene delivery and cell imaging studies. For those purposes, the nanoparticles were modified with lipid thanks to biocompatibility to facilitate the entrance of PFPL nanoparticles into the cytoplasm.

The cell imaging performance with the gene transfection efficiency of the nanocarrier (nanocarrier and pCX-EGFP plasmid complex) into lung cancer cells (A549) was evaluated. The PFPL nanoparticles with 50 nm showed bright fluorescence with a quantum yield of 6.3% and remarkable photostability. The PFPL nanocarriers exhibited low toxicity and could easily enter the cytoplasm via endocytosis. The gene delivery performance of the nanocarrier was highly satisfactory; hence, the fluorescent nanoparticle could be potentially used in gene delivery studies as a nonviral vector and cell imaging studies as a probe.

Chemiluminescence (CL) is a detection method for DNA hybridization and measures the specific luminescence activity of the labels linked to the DNA probe for the detection of the target DNA [48]. Various labels such as enzymes, metal nanoparticles, and the conjugated polymers are utilized in DNA hybridization studies [48]. Cationic PY derivatives are used as optical probes due to the light-harvesting property and show optical amplification via FRET. Hence, the researchers synthesized poly[(9,9-di(3,3-N,N-trimethyl ammonium)propylfluorenyl-2,7,-diyl)-alt-co-(1,4-phenylene)] dibromide (PFP$^+$) and aimed to develop CL-based detection method for DNA hybridization thanks to the enhancement effect of PFP$^+$ on luminol-H_2O_2 CL system. Additionally, the enhancement effect of PFP$^+$ on luminol-H_2O_2 can be controlled by adding DNAs, and the electrostatic interactions between DNA and PFP$^+$ result in decreasing the affinity of PFP$^+$ to luminol dianion and peroxide ion. Therefore, the hybridization of probe DNA and target DNA is investigated by the enhancement effect of PFP$^+$. The electrostatic interactions between PFP$^+$ and luminol dianion and the peroxide ion were responsible for enhancing CL intensity. The binding affinity of PFP$^+$ toward dsDNA was higher than ssDNA and the lengths of the base could influence DNA hybridization. The increase of the base lengths resulted in decreasing CL intensity.

9.2.6 POLYANILINE (PANI)

PANI is one of the favorable CPs with some features such as high stability, ease of synthesis, electroactivity, electric conductivity, tunable conducting, and optical properties [50]. PANI displays good *in vitro* and *in vivo* biocompatibility; hence, it has been studied for various biomedical applications. Generally, the chemical oxidation method is used for the synthesis of PANI in the presence of doping acids and oxidizing agents [3]. The formation of green color in the reaction medium is recognized as evidence of polymer film generation. PANI polymer backbone possesses both quinoid and benzenoid rings in different ratios, which results in different oxidation states. According to the oxidation state, it exists in a variety of forms: the fully reduced leucoemeraldine, the fully oxidized pernigraniline, and the partially oxidized and reduced emeraldine form. The latter, protonated emeraldine salt (ES), is the most important form of PANI. It is stable end electrically conductive, but when the pH ≥4, deprotonation leads to electrically nonconducting emeraldine base form. Therefore, the use PANI at the physiological pH for biomedical purposes is restricted.

To address this issue and preserve the electroactivity of PANI in physiological media, Gandla et al. [50] used inherently fluorescent carbon nanospheres as dopant agents for in-situ oxidative polymerization of PANI. The effect of the doping process on the obtained PANI composite was evaluated electrochemically. Cyclic voltammetry and electrochemical impedance spectroscopy results confirmed that the electroactivity of the polymer film was retained and exhibited stability in the cell culture medium at pH 7.4. The possible reason was the effective hindering of the dopant carbon nanospheres against the deprotonation of conducting ES of PANI at higher pH values. Accordingly, the obtained PANI composite, hence, provided a powerful strategy for its cellular adhesion and uptake of adipose-derived stem cells due to its small size (400–500 nm) and partial positive charge, which ensured permeability through the cell membrane into the cytoplasm and was supported by cellular imaging. Due to their highly migratory abilities, stem cells have been distinguished as efficient delivery vehicles to be exploited in gene therapy for neurodegenerative diseases [51]. Therefore, this research represented the capability of the composite to be promising to deliver various negatively charged species such as DNAs, RNAs, anionic drugs, and green fluorescent proteins into cells.

Since the implementation of PANI in a neutral pH environment restricts its redox activity and biological application, self-doped PANI or PANI doped with negatively charged polyelectrolytes have been reported to exhibit redox-active properties at physiological pH [52]. Therefore, the processing of doped PANI at a molecular level via the use of electrostatic interaction and complexation has been reported to be highly desirable for many applications.

In their work, Recksiedler et al. [53] used poly(anilineboronic acid) (PABA)/ribonucleic acid (RNA) multilayer films via the layer-by-layer (LbL) deposition method. Chemical synthesis of PABA in the presence of excess d-fructose and NaF resulted in the cationic nature of partially doped PABA and promoted multilayer assembly as well as extended electroactivity of PABA to neutral and alkaline media. The formation of these multilayer films and incorporation of RNA in the film structure was based on the specific interaction between boronic acid groups of the PABA with diol moieties and primary amines present in RNA molecules yielding an anionic boronate complex. A three-step procedure was followed and alternate deposition of PABA and RNA layers was repeated to build multilayer films on (indium tin oxide) ITO substrates. Deposited films were evaluated with UV-vis absorption spectroscopy, X-ray photoelectron spectroscopy, ellipsometry, and the results supported the reproducible formation of bilayers on the ITO and silicon substrate. RNA used as polyelectrolyte for multilayer formation also acted as a dopant for PABA and the resulting PABA/RNA multilayer films exhibited redox-activity at neutral pH. Voltammetric results proved that cycling the potential applied to a multilayer between −0.2 and +1.4 V caused successive oxidation and reduction of the conducting PABA polymer which induced the release of RNA from the multilayer. This behavior offered a potential for administering controlled release of RNA to targeted sites which is also optimal for *in vivo* application of different biomedical devices.

9.3   CONCLUSION

Gene therapy is an alternative treatment method for various diseases such as cancer, inherited disorders, viral infections without using drugs or operations. During the gene therapy-based treatment methods, the viral and nonviral vectors are employed in inserting the gene into a patient's cell and the choice of the vectors is of great importance to prevent some unwanted complications. From a point of view, the nonviral vectors with some good features especially low-immune response are favorable nanocarriers for gene therapy studies. CPs are highly important polymers for wide application areas ranging from bioelectronics to biomedical studies because of significant properties such as electric conductivity, biocompatibility, and optical property. Furthermore, the CPs can entrap the molecules such as enzymes, drugs, proteins, and release these molecules by applying electrical stimulation. So, the manipulation of the electric stimulation plays a key role in gene delivery studied. In this chapter, the gene delivery performance of the most preferred CPs, their synthesis, their interactions with DNA or siRNA, their entry into the cells, their gene silencing capabilities, their biocompatibilities, and the electric stimulation on their gene delivery performances were highlighted. Additionally, their gene delivery performances and their toxicities were discussed and compared with the commercial products.

REFERENCES

1. K. Namsheer, S. Rout, Conducting polymers: a comprehensive review on recent advances in synthesis, properties and applications, RSC Adv. (2021) 11: 5659–5697.
2. Y. Park, J. Jung, M. Chang, Research progress on conducting polymer-based biomedical applications, Appl. Sci. (2019) 9: 1070: 1–20.
3. T.H. Le, Y. Kim, H. Yoon, Electrical and electrochemical properties of conducting polymers, Polymers (2017) 9:150: 1–32.
4. A.P. Jou, L.J. Valle, C. Alemân, Drug delivery systems based on intrinsically conducting polymers, J. Control. Release, (2019) 309: 244–264.
5. S. Nambiar, J.T.W. Yeow, Conductive polymer-based sensors for biomedical applications. Biosens. Bioelectron. (2011) 26: 1825–1832.
6. J. Torras, J. Casanovas, C. Alemân, Reviewing extrapolation procedures of the electronic properties on the π-conjugated polymer unit, J. Phys. Chem. A (2012) 116: 7571–7583.
7. G. Inzelt, Conducting polymers: a new era in electrochemistry, 2nd, 2012, Springer, Berlin Heidelberg, Germany.
8. C.I. Awuzie, Conducting polymers, Mater. Today: Proc. (2017) 4: 5721–5726.
9. H. Shirakawa, The Discovery of polyacetylene film: the dawning of an era of conducting Polymers (Nobel Lecture), Angew. Chem. Int. Ed. (2001) 40: 2574–2580.
10. Rivers, T.J.; Hudson, T.W.; Schmidt, C.E. Synthesis of a novel, biodegradable electrically conducting polymer for biomedical applications. Adv. Funct. Mater. (2002) 12: 33–37.
11. Y.K. Sung, S.W. Kim, Recent advances in the development of gene delivery systems, Biomater. Res. (2019) 23 (8): 11–17.

12. Y.K. Sung S.W. Kim, The practical application of gene vectors in cancer therapy. Integrative Cancer Sci. Therap. (2018) 5 (5):1–5.
13. S. Li et al., Cationic poly(p-phenylene vinylene) materials as a multifunctional platform for light-enhanced siRNA delivery, Chem. Asian J. (2016) 11: 2686–2689.
14. H. Yin, R. L. Kanasty, A. A. Eltoukhy, A. J. Vegas, J. R. Dorkin, D. G. Anderson, Non-viral vectors for gene-based therapy, Nat. Rev. Genet. (2014) 15: 541.
15. M. Elsabahy, A. Nazarali, M. Foldvari, Non-viral nucleic acid delivery: key challenges and future directions, Curr. Drug Delivery (2011) 8: 235.
16. Y. He, Y. Nie, G. Cheng, L. Xie, Y. Shen, Z. Gu, Viral mimicking ternary polyplexes: A reduction-controlled hierarchical unpacking vector for gene delivery, Adv. Mater. (2014) 26: 1534.
17. V. Gajbhiye, S. Gong, Lectin functionalized nanocarriers for gene delivery, Biotechnol. Adv. (2013) 31 (5): 552–562.
18. D.J. Glover, H.J. Lipps, D.A. Jans, Towards safe, non-viral therapeutic gene expression in humans. Nat. Rev. Genet. (2005) 6: 299–310.
19. Y. Zhang, X. Li, T. Wu, J. Sun, X. Wang, L. Cao, F. Feng, Cationic polythiophenes as gene delivery enhancer, ACS Appl. Mater. (2017) 9: 16735–16740.
20. B. R. D. Mccullough, The chemistry of conducting polythiophenes, Adv. Mater., (1998) 10: 93–116.
21. M. Faisal, F.A. Harraz, M. Jalalah, M. Alsaiari, S.A. Al-Sayari, M.S. Al-Assiri, Polythiophene doped ZnO nanostructures synthesized by modified sol-gel and oxidative polymerization for efficient photo-degradation of methylene blue and gemifloxacin antibiotic, Mater. Today Commun. (2020) 24: 101048.
22. J. Chen, J. Zhu, N. Wang, J. Feng, W. Yan, Hydrophilic polythiophene/SiO2 composite for adsorption engineering: Green synthesis in aqueous medium and its synergistic and specific adsorption for heavy metals from wastewater, Chem. Eng. J. (2019) 260: 1486–1497.
23. P. He, K. Hagiwara, H. Chong, H.H. Yu, Y. Ho, Low-molecular-weight polyethyleneimine grafted polythiophene for efficient siRNA delivery, Biomed. Res. Int. (2015) 1–9.
24. N.P. Gabrielson, H. Lu, L. Yin, K.H. Kim, J. Cheng, A cell penetrating helical polymer for siRNA delivery to mammalian cells, Mol. Ther. (2012) 20 (8): 1599–1609.
25. D.C. Forbes, N.A. Peppas, Polycationic nanoparticles for siRNA delivery: comparing ARGET ATRP and UV-initiated formulations, ACS Nano (2014) 8 (3): 2908–2917.
26. R.U. Svenson, M.R. Shey, Z.K. Ballas, J.R. Dorkin, M. Goldberg, A. Akinc, R. Langer, D.G. Anderson, D. Bumcrot, M.D. Henry, Assessing siRNA pharmacodynamics in a luciferase-expressing mouse, Mol. Ther. (2008) 16 (12): 1995–2001.
27. L. Lichon, C. Kotras, B. Myrzakhmetov, P. Arnoux, M. Daurat, C. Nguyen, D. Durand, K. Boucmella, L.M.A. Ali, J.O. Durand, S Richeter, C. Frochot, M.G. Bobo, M. Surin, S. Clement, Polythiophenes with cationic phosphonium groups as vectors for imaging, siRNA delivery, and photodynamic therapy, Nanomaterials (2020) 10: 1–15.
28. C. Zhang, J. Ji, X. Shi, X. Zheng, X. Wang, F. Feng, Synthesis of structurally defined cationic polythiophenes for DNA binding and gene delivery, ACS Appl. Mater Interfaces (2018) 10: 4519–4529.
29. C. Gautier, C. Cougnon, J.F. Piland, N. Casse, B. Chenais, M. Laulier, Detection and modelling of DNA hybridzation by EIS measurements. Mention of a polythiophene matrix suitable for electrochemically controlled gene delivery. Biosens. Bioelectron. (2007) 22 (9-10): 2025–2031.
30. C. Gautier, C. Cougnon, J.F. Piland, N. Casse, B. Chenais, A poly(cyclopentadithiophene) matrix suitable for electrochemically controlled DNA delivery, Anal. Chem. (2007) 79: 7920–7923.
31. F. Fang, F. Meng, L. Luo, Recent advances on polydiacetylene-based smart materials for biomedical applications, Mater. Chem. Front. (2020) 4:1089–1104.
32. J.W. Wang, F. Zheng, H. Chen, Y. Ding, X.H. Xia, Rapidly visualizing the membrane affinity of gene vectors using polydiacetylene-based allochroic vesicles, ACS Sens. (2019) 4: 977–983.
33. M.D. Hoang, M. Vandamme, G. Kratassiouk, G. Pinna, E. Gravel, Tuning the cationic interface of simple polydiacetylene micelles to improve siRNA delivery at the cellular level, Nanoscale Adv. (2019) 1: 4331–4338.
34. M. Ripol et al., pH-responsive nanometric polydiacetylenic micelles allow for efficient intracellular siRNA delivery, ACS Appl. Mater. Interfaces (2016) 8: 30665–30670.
35. E. Morin, M. Nothisen, A. Wagner, J.S. Remy, Cationic polydiacetylene micelles for gene delevery, Bioconjugate Chem. (2011) 22: 1916–1923.
36. R. Balint, N.J. Cassidy, S.H. Cartmell, Conductive polymers: Towards a smart biomaterial for tissue engineering, Acta Biomater. (2014) 10: 2341–2353.
37. G. Kaur, R. Adhikari, P. Cass, M. Bown, P. Gunatillake, Electrically conductive polymers and composites for biomedical applications, RSC Adv. (2015) 5: 37553–37567.

38. Y.C. Cheng, S.L. Guo, K.D. Chung, W.W. Hu, Electrical field-assisted gene delivery from polyelectrolyte multilayers, Polymers (2020) 12 (133): 1–14.

39. C. Xie, P. Li, L. Han, Z. Wang, T. Zhou, W. Deng, K. Wang, X. Lu, Electroresponsive and cell-affinitive polydopamine/polypyrrole composite microcapsules with a dual-function of on-demand drug delivery and cell stimulation for electrical therapy, NPS Asia Mater. (2017) 358e: (9): 1–9.

40. J. Wang, M. Yiang, Toward genoelectronics: nucleic acid doped conducting polymers. Langmuir (2000) 16 (5): 2269–2274

41. P. Jagadesan, K.S. Schanze, Poly(phenylene ethynylene) conjugated polyelectrolytes synthesized via chain-growth polymerization, Macromolecules (2019) 52: 3845–3851.

42. T. Wu et al., Remarkable amplication of polyethylenimine-mediated gene delivery using cationic poly(phenylene ethylene)s as photosensitizers, ACS Appl. Mater. Interfaces (2018) 10: 24421–24430.

43. S. Li, H. Yuan, H. Chen, X. Wang, P. Zhang, F. Lv, L. Liu, S. Wang, Cationic poly(p-phenylene vinylene) materials as a multifunctional platform for light-enhanced siRNA delivery, Chem. Asian. J. (2016) 11 (19): 2686–2689.

44. J.H. More, E. Mendez, Y. Kim, A. Kaur, Conjugated polymer nanoparticles for small interfering RNA delivery, Chem. Commun. (2011) 47: 8370–8372.

45. L. Zhang, Q.H. Yin, J.M. Li, H.Y. Huang, Q. Zu, Z.W. Mao, Functionalization of dendritic polyethylene with cationic poly(p-phenylene ethynylene) enables efficient siRNA delivery for gene silencing, J. Mater. Chem. B. (2013) 1: 2245–2251.

46. A.T. Silva, A. Nguyen, C. Ye, J. Verchot, J.H. Moon, Conjugated polymer nanoparticles for effective siRNA delivery to tobacco BY-2 protoplasts, BMC Plant Biology, (2010) 10 (291): 1–14.

47. Y. Wang, B. Liu, Cationic water-solunle polyfluorene homopolymers and copolymers: synthesis, characterization and their applications in DNA sensing, Curr. Org. Chem. (2011) 15: 446–464.

48. M. Liu, B. Li, Detection of DNA hybridization using a cationic polyfluorene polymer as an enhancer of luminol chemiluminescence, Microchim. Acta (2016) 183: 897–903.

49. X. Feng, Y. Tang, X. Duan, L. Liu, S. Wang, Lipid-modified conjugated polymer nanoparticles for cell imaging and transfection, J. Mater. Chem. (2010) 20: 1312–1316.

50. D. Gandla, C. Putta, S. Ghost, B.K. Hazra, Carbon sphere-polyaniline composite: a fluorescent scaffold for proliferation of adipose derived stem cells and its cellular uptake, Chem. Sel. (2016) 1(12): 3063–3070.

51. F.J. Müller, E.Y. Synder, J.F. Loring, Gene therapy: can neural stem cells deliver? Nat. Rev. Neurosci. (2006) 7 (1): 75–84.

52. A. Kundu, S. Nandi, A.K. Nandi, Nucleic acid based polymer and nanoparticle conjugates: synthesis, Progr. Mater. Sci. (2017) 88: 136–185.

53. C.L. Recksiedler, B.A. Deore, M.S. Freund, A novel layer-by-layer approach for the fabrication of conducting polymer/RNA multilayer films for controlled release, Langmuir (2006) 22: 2811–2815.

10 Conducting Polymers for Regenerative Medicine

Merve Çalışır,[1] *Nilay Bereli,*[1] *İbrahim Vargel,*[2]
and Adil Denizli[1]

[1]Chemistry Department, Hacettepe University, Beytepe, Ankara, Turkey

[2]Deparment of Plastic, Reconstructive and Aesthetic Surgery, Hacettepe University, Sıhhıye, Ankara, Turkey

CONTENTS

10.1 INTRODUCTION

Regenerative medicine is an important and exciting branch of medical science, fundamentally seeks to heal the damaged organs due to many different reasons (trauma, aging, etc.) or to replace them in the closest way to the original [1]. Regenerative medicine supports the body's repair and regeneration, eclipsing the traditional treatment process and disease management. The aforementioned traumas left the patient with irreversible disabilities and deficiencies and the disadvantages such as incompatible tissues in organ donations, the very few donors, and the patient's lifetime use of immunosuppressive drugs after the donation led the scientific community to investigate the innovative and promising field of regenerative medicine [2–5]. Indeed, the success of treatments with regenerative medicine in both chronic and acute conditions is increasing day by day [6, 7]. Historically, the term regenerative medicine was first coined by William Haseltine at a conference in 1999 [8]. Although the term was first used there, the idea of "tissue creation", which is no longer a utopia, and its studies in the medical world has a precedent. In natural life, also it is possible to encounter examples of regenerative medicine, such as the regeneration of missing body parts by lizards and starfish. Thanks to the desire of imitating nature and past experiences of organ transplants, future medicine has started to take shape and regenerative medicine has been one step closer to the implementation of much faster and more effective treatments. Regenerative medicine delivery strategies are open to many combinations, such as structural and functional change, healing, or the inclusion of cells to promote healing [9]. Since all these strategies will also differ in the regions where the repair will be made, an idea about the working range of the field can be obtained. Although the healing ability of some underdeveloped creatures in nature is not in humans, regenerative medicine treatments can be activated with the innate healing response of the human body as well [10].

Polymers consisting of very large molecules or macromolecules composed of many repeating subunits were considered insulating materials before the production of conducting polymers [11]. After the production of organic conducting polymers, it was understood that they have distinctive electrical and optical properties. Conjugated carbon chains contain highly delocalized, polarized, and electron-dense π bonds, creating the aforementioned unique optical and electrical characteristics [12]. Typical conducting polymers include polyacetylene, polyaniline, polypyrrole,

DOI: 10.1201/9781003205418-10

polythiophene, poly(para-phenylene), poly(phenylenevinylene), and polyfuran [13]. Although the discovery of conducting polymers dates back to the 60s, they did not become popular at that time as the full potential of these materials was not realized. After the dramatic increase in conductivity was observed when iodine was added to polyacetylene, the first inherently conducting polymer was recognized [14]. After these and many other discoveries, conducting polymers began to attract more attention, and their stability, synthesis, and conductivity also began to be studied in detail for use in different research areas. Two main ways are followed in the synthesis of conducting polymers: chemical polymerization and electrochemical polymerization [15]. Both methods have their advantages and disadvantages and the choice is made according to the application type. The advantages of chemical polymerization include large-scale production and ease of covalent backbone modification. On the other hand, thin films cannot be produced and the synthesis step is quite complex. In electrochemical polymerization, thin films can be produced and the synthesis step is easy, but there are difficulties in removing the film from the electrode surface and modifying the bulk conducting polymer [14]. In the choice of the synthesis method, the aims and suitability of the envisaged study are taken into consideration and the demanding medical field especially favors conducting polymer in various applications.

In medical applications, the convenience of conducting polymers regarding easy synthesis and processing has come to the fore. Especially in sensor studies, its compatibility with biological molecules has been greatly benefited. Studies have proven that conducting polymers are also popular in terms of regulation of cellular activities such as cell migration, DNA synthesis, and protein secretion [16–19]. Conducting polymers that can mimic the extracellular matrix and are biodegradable and in tissue engineering, they are also advantageous because of easily removed. In these studies, in which nerve, bone, muscle, and heart cells respond to electrical stimuli, the biocompatibility of conducting polymers and the success of capturing and releasing biological molecules play a major role. Its successful integration into natural systems is also why conducting polymers are preferred in medical fields such as tissue engineering scaffolds, neural probes, drug delivery devices, and bioactuators. Considering all these advantages, in this chapter, the studies of conducting polymers in the field of regenerative medicine are presented in detail and it is aimed to guide to understand the true potentials of conducting polymers.

10.2 CONDUCTING POLYMER APPLICATIONS FOR REGENERATIVE MEDICINE

Thanks to the advanced level of today's technology, both the quality of life and the average human life span have been greatly extended. However, although much progress has been made in the treatment of diseases at the point of medication, surgeons have difficulties in the repair and restoration of skeletal and bone defects, especially after acute traumas. Conducting polymers are likely to be used in these treatments, thanks to their biocompatibility. In the study by Ragunathan and colleagues, nano-hydroxyapatite combined with natural polymer starch was used for exactly these purposes [20]. Hydroxyapatite resembles the mineral phase of the bone and is characterized as biocompatible, bioactive, and osteoconductive. They showed that the skeleton sample they synthesized was not toxic by cytotoxicity test, and they supported the mechanical strength of their scaffolds with polyvinyl alcohol.

The ease of modification of conducting polymers is becoming the cornerstone of many studies. Lee et al. provide a detailed comparison of conducting polymers designed for use in biointerfacing electronics, including composites, conducting hydrogels, and electrochemical deposition, with natural conducting polymers [21]. On behalf of regenerative medicine, they created a guide by giving examples for the use of conducting polymers in research such as tissue regeneration, neural recording, and stimulation. In Figure 10.1, the parameters that affect the conductivity of the conducting polymers are investigated. The parameters they looked at in their studies were molecular weights of

Dopant size

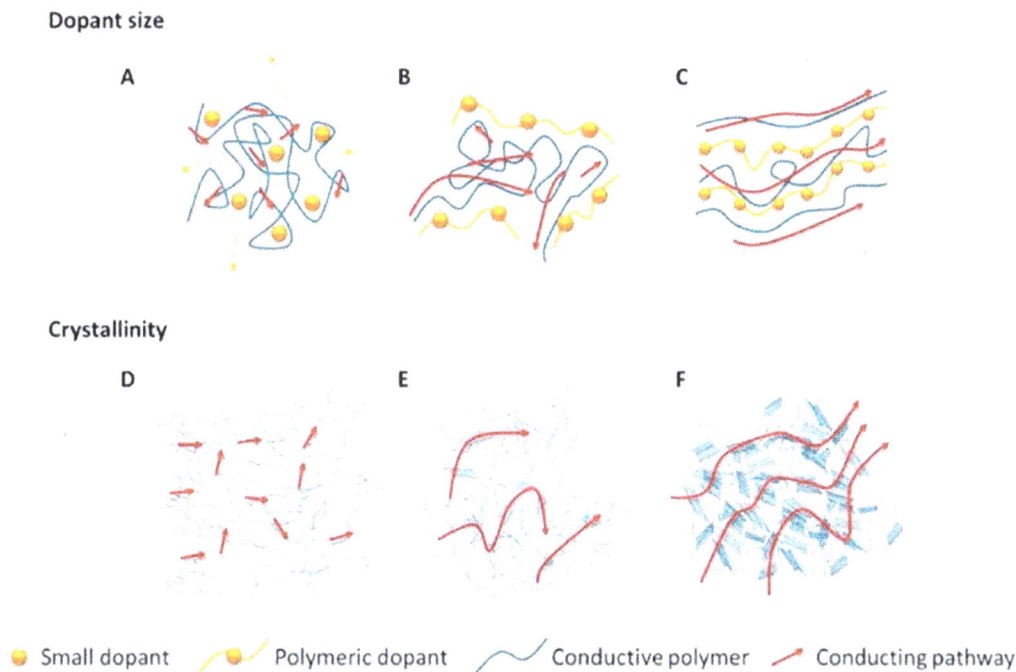

Crystallinity

Small dopant   Polymeric dopant   Conductive polymer   Conducting pathway

FIGURE 10.1 Schematics of molecular structures of CPs and parameters on which their conductivity is affected. (A) Small molecules, (B) oligomeric, (C) polymeric dopants, (D) amorphous, (E) low crystalline, and (F) high crystalline conjugated domains of the conductive structures. (Adapted with permission from Ref. [21]. Copyright (2021) Elsevier.)

conjugated backbones, π-π stacked crystalline domain sizes, interactions with p-type doping counter ions, protonic transporting pathways, electron transfer through hydrogen bonds, and molecular ordering. Today, the increasing need for organs has increased the need for studies in the field of tissue engineering. Tissue engineering is a branch of science that aims to create artificial tissues and organs to be transplanted into patients using stem cells and biomaterials in general one of the areas where conducting polymers are used most in regenerative medicine. A prospering study in tissue engineering in regenerative medicine was presented by Asadi et al. [22]. In their study, they synthesized conducting and soft polypyrrole/alginate hydrogels by chemically polymerizing polypyrrole in ionic cross-linked alginate hydrogel networks.

The outstanding work of Li et al. is an example of tissue engineering as well [23]. They succeeded in synthesizing the material that can be used in tissue engineering by functionalizing polyaniline, a conducting polymer, together with gelatin. Gelatin fibers containing electrospun polyaniline were characterized using scanning electron microscopy, electrical conductivity measurement, mechanical tensile testing, and differential scanning calorimetry. To observe the developments in cell growth, rat cardiac myoblast cells were cultured on fiber-coated coverslips and positive results were obtained. In Figure 10.2, scanning electron microscopy images of the combinations of polyalanine gelatin fibers they synthesized at different rates are shown.

Study using conducting polymers in tissue engineering was presented by Ateh et al. by synthesizing various charged polypyrrole films, including proteins and polysaccharides, on gold-coated polycarbonate coverslips [24]. The epidermis is the outermost layer of the skin and consists mainly of cells called "Keratinocytes". Its thickness varies depending on the part of the body, age, and gender. In their study, which showed that keratinocyte viability is charge-dependent, they concluded that optimized polypyrrole films adequately promote keratinocyte growth. As

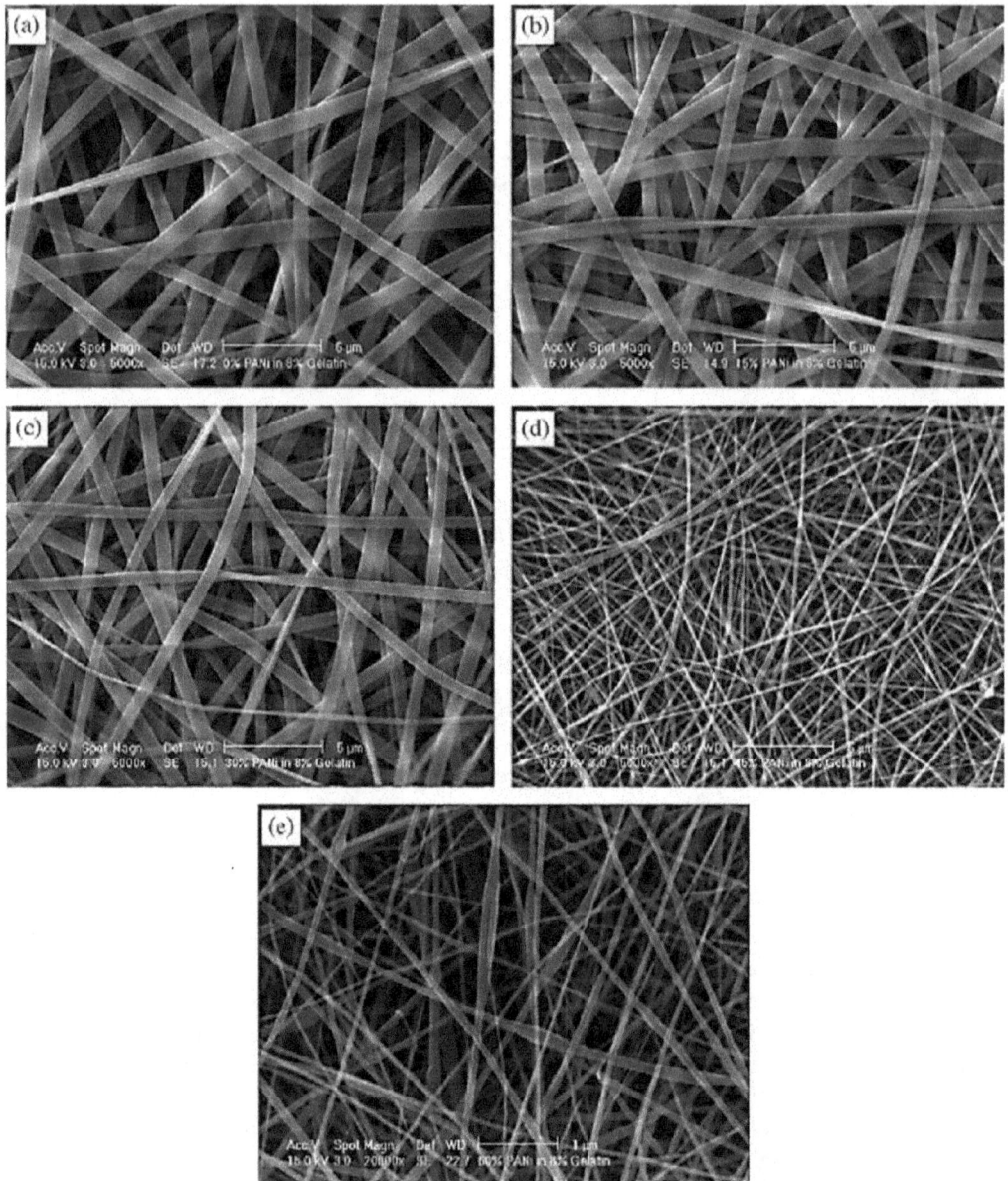

FIGURE 10.2 Scanning electron microscopy images of the combinations of polyalanine gelatin fibers (a) with ratio of (b) 15:85, (c) 30:70, (d) 45:55, and (e) 60:40. (Adapted with permission from Ref. [23]. Copyright (2021) Elsevier.)

evidence of how diverse the integration of conducting polymers into tissue engineering can be, the work of Dias Junior et al. can be presented [25]. The semi-interpenetrating polymeric scaffolds they designed based on polylactic acid and polyhydroxyethylmethacrylate may be a suitable alternative for repairing traumatic injuries of cartilage tissue. In Figure 10.3, there is a detailed display of the polylactic acid-containing scaffold formed after the reaction of polyhydroxyethylmethacrylate and di-tert-butyl peroxide.

A scaffold was produced in the absence of solvents to indicate the lack of biological interaction of biomaterials with surrounding tissues and demonstrated the absence of toxicity. The potential use

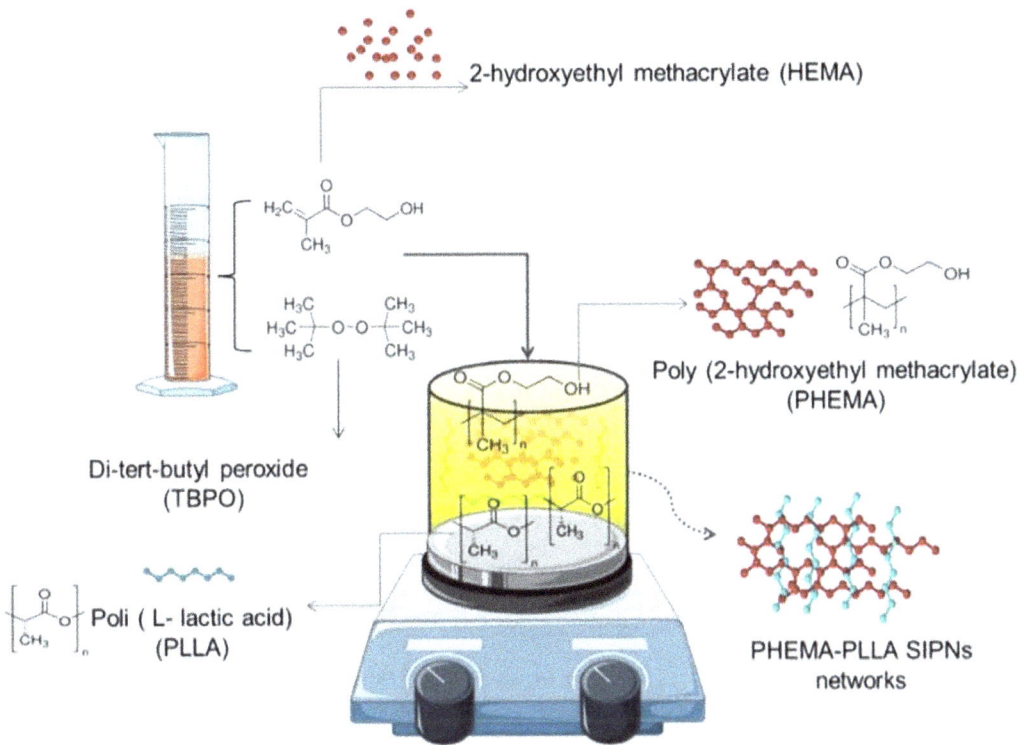

FIGURE 10.3 Schematic representation of conducting PHEMA-PLLA SIPNs scaffolds. (Adapted with permission from Ref. [25]. Copyright (2021) Elsevier.)

of the material in tissue engineering was thus proposed. In vitro studies with human bone marrow-derived mesenchymal stem cells (hMSCs) reveal that cell adhesion and growth are promoted in polypyrrole/alginate hydrogels. It has been published in the literature as a useful study to examine the effects of hydrogels in subcutaneous implantation and the effects of electrical and mechanical signals on stem cells and/or neural cells. As mentioned earlier, conducting polymers are also used in the regulation of cell and tissue functions because they can successfully mimic in vivo electrical stimulation. Bipolar electrochemistry aims to apply electrical stimulation systems in a noncontact mode in medical treatments. In the study of Qin et al., the development of a conducting polymer-based bipolar electrostimulation (BPES) system for living cells was adopted as a principle [26]. Conducting polypyrrole films were synthesized with different ingredients to demonstrate recoverable electrochemical activity, and a prototype was designed that provides wireless and programmable cell stimulation. The biggest proven advantage of wireless stimulation is that it enhances cell proliferation and differentiation. Bipolar conducting polymer electrodes have declared treatments in the regenerative part of medical applications even more effective.

As mentioned before, the physical therapy method in which electrical stimulation is used in regenerative medicine is tissue healing. The main purpose of regenerative medicine, the repair, and regeneration of damaged tissues, including muscle, bone, skin, nerves, tendons, and ligaments, can be carried to a further extent by electrical stimulation [27]. Another benefit of electrical stimulation is that it can function in multiple tissues. The conducting polymers used are successful in transmitting the stimulus applied from the skin electrodes to the desired tissue, which allows handling the healing in the desired place [28]. A better understanding of the molecular mechanisms of conducting polymers is essential for the widespread use of regenerative medicine procedures. In Figure 10.4, a broad perspective about the working range of the conducting polymers is presented.

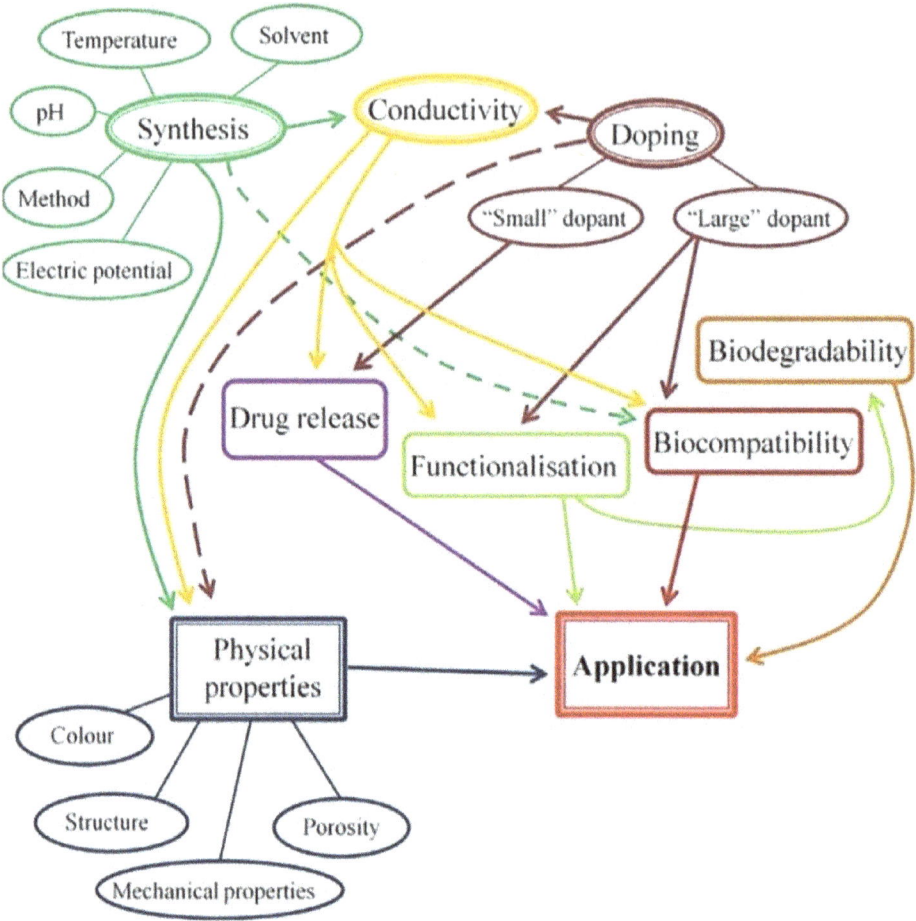

FIGURE 10.4 An illustration of the wide working range of conducting polymers [28]. (Adapted with permission from Ref. [28]. Copyright (2021) Elsevier.)

Huang et al.'s study is a good example of how conducting polymers promote nerve regeneration when integrated into electrical stimulation [29]. They created a local electrical environment by using polypirol/chitosan scaffold in their study in which they went through major nerve damage. Nerve regeneration and functional recovery were observed, thanks to the conducting scaffold, on which they applied intermittent 3 V, 20 Hz electrical stimulation. In addition, axonal regeneration and remyelination of regenerated axons were shown to be significantly increased. Their work is very encouraging to develop regenerative approaches. Nanotechnological developments are also promising for all kinds of medical applications. The inclusion of nanotechnology in regenerative medicine procedures will increase the sensitivity and precision of studies and encourage multidisciplinary development. Specifically, the study using nano conducting polymers synthesized by Shi et al. is important to support this combination [30]. Although nanostructured materials are successful in mimicking the extracellular matrix, they cannot provide the required strength. In their study, Shi et al. synthesized composite hydrogels by adding polypyrrole nanoparticles to nanoporous cellulose gels and converted them into aerogels by drying with CO_2, thereby obtaining the required mechanical strength. The addition of conducting polypyrrole nanoparticles increased the adhesion and proliferation of cells, and also attached and elongated the cells' longer neurites. All these results show that the material they synthesized is a valuable study in terms of nerve regeneration.

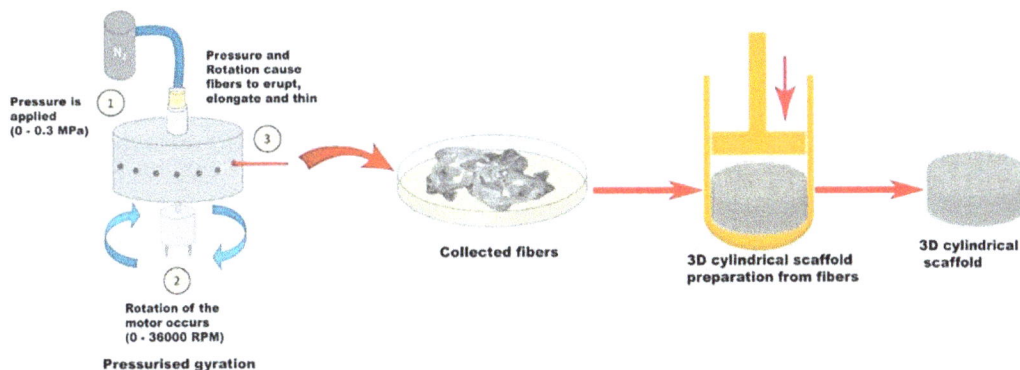

FIGURE 10.5 The illustration of the proposed pressurized gyration setup for a three-dimension polymer composite scaffold design. (Adapted with permission from Ref. [35]. Copyright (2021) Elsevier.)

Bone regeneration is an area that needs much improvement, especially in surgical interventions after serious accidents [31]. Bone grafting has been published in the literature as a method frequently used in defects of large bones and spinal injuries. Autografts used as bone grafts bring with them parameters such as donor site morbidity and amount of available bone that make treatment inadequate [32]. To overcome these problems, the use of human bone morphogenetic protein (BMP)-2 has started to be preferred especially in spine treatments [33]. The study by Bal et al. aims to contribute to effective bone regeneration with BMP-2 by using polylactic acid-polyethylene glycol (PLA-PEG) and an osteogenic nano-hydroxyapatite composite material to get rid of the side effects of BMP-2 [34]. They demonstrated the release of BMP-2 in the system they designed with the tissue-compatible PLA-PEG conducting material, provided efficient bone regeneration with low-dose BMP-2, and were able to bring an alternative to the standard graft approach.

It is also possible to use conductive polymer composite materials for bone regeneration. Polycaprolactone, montmorillonite nanoclay, and nano-hydroxyapatite-clay polymer composite fibers prepared by Kundu and colleagues are just examples of the studies given [35]. It has been reported that these composite polymers synthesized in a pressurized spinning assembly increase bone growth, cell viability, and proliferation. In Figure 10.5, the schematic representation of the pressurized gyration setup for three-dimension polymer composite scaffold design in three main stages is presented. After the gyration process, fibers are collected and shaped into three-dimensional cylindrical scaffolds. In addition, it has been shown that it has successful biocompatibility and that cells can develop and differentiate on fiber scaffolds, just like in natural processes.

Three-dimensional ceramic conductors are also good candidates for use in bone regeneration. The desired tissue scaffold can be created thanks to ceramic composites that can successfully transmit electrical impulses to the injured area. Poly(3,4 ethylenedioxythiophene) and poly(4-styrene sulfonate) conductive ceramic polymer produced by Tayebi et al. are one of the studies in this field [36]. Characterizations of this highly biocompatible material have also been made using nuclear magnetic resonance, in vitro degradation, as well as thermal and mechanical analysis. Thanks to this specific polymer content, the required mechanical strength and stability are also gained. The electrical signal between cells increased proportionally with increasing the concentration of conductive polymer in the scaffold.

When blood flow to part of your brain is reduced or interrupted, a stroke occurs that prevents brain tissue from getting oxygen and nutrients, and brain cells begin to die within minutes. Temporary or permanent loss of functions occurs in the area where the stroke occurred. Stroke (stroke) is a medical emergency and immediate treatment is of paramount importance. Early intervention can reduce brain damage and other complications. Effective treatments can also help prevent disability from stroke. There are two types of stroke: a blocked artery (ischemic stroke) or a blood vessel leaking or bursting (hemorrhagic stroke) [37]. Some people may suffer a stroke, known as a transient ischemic attack (TIA) that does not cause permanent symptoms and causes a temporary interruption in blood flow to the brain. This is the most common type of stroke. The brain's blood vessels become narrowed or blocked, causing severely reduced blood flow (ischemia). Clogged or narrowed blood vessels are caused by fat particles accumulating in the blood vessels, or by blood clots or other debris circulating in the bloodstream and lodged in the blood vessels in the brain. A hemorrhagic stroke occurs when a blood vessel in the brain leaks or ruptures. Brain hemorrhages can be caused by many conditions that affect your blood vessels. The treatment of stroke, which is a tough health problem for patients and caregivers, physiologically and psychologically, currently includes the most acute systematic treatments aimed at clot relieving. Poststroke brain tissue examination and possible treatments have not been implemented and remain in theory, and new ideas in medicine are needed to improve functional recovery after stroke [38]. It is noteworthy that conductive polymers can be very useful in poststroke recovery strategies that need to be established in the field of regenerative medicine. The ability to repair poststroke deformations and manipulate the environment makes conducting polymers stand out in this area. Some of the studies carried out in this context deserve to be presented in detail and it is necessary to see the opportunities that conductive polymers will provide in this field from a wide window.

Neuropsychiatric conditions with different clinical manifestations occur in a significant portion of poststroke cases. It is important to know these neuropsychiatric conditions, which often put clinicians in a difficult situation and cause difficulties in diagnosis and treatment because early diagnosis and treatment can be lifesaving. Functional changes after stroke include acute injury, poststroke inflammatory response, and neural repair. In an acute injury, the nutrients and oxygen needed by the brain cannot be supplied sufficiently and this causes cell death. Neural cell membranes become unstable, creating an environment for excitotoxicity, and after a series of complex events, inflammatory processes occur [39]. These processes increase free radical release, which hinders healing [40]. It has been determined that stroke is also an initiating risk factor for dementia. In addition, it has been reported that the rate of depression in the poststroke period is between 20 and 50% and harms functional recovery. Considering all these problems, it is inevitable to say that one of the areas where regenerative medicine is most needed is to heal poststroke damage.

Finding biomaterial compatible with nervous systems is the first step in the development of treatment and conducting polymers is the beginning of ways to solve compatibility in this respect. The biocompatibility of Polypyrrole is very suitable for supporting neural interactions as mentioned in previous treatments. In addition to the advantages mentioned, the fact that polypyrrole is a flexible and biodegradable material is the reason why it is at the center of many regenerative medicine studies. The study by George et al. focused specifically on the biocompatibility of polypyrrole in nervous systems [41]. Dissociated primary cerebral cortical cells were cultured on polypyrrole samples doped with polystyrene sulfonate or sodium dodecyl benzenesulfonate, and further modifications were attempted to capture different surface morphology. Neural networks sprawled on polypyrrole in a near-nature manner. Implants that were also integrated into the rat cerebral cortex were compared with Teflon implants. The results showed that the produced polypyrrole samples were a more compatible material than Teflon.

There is no currently approved medical treatment for the relief of poststroke complications. Developing stem cell therapies is presented as a promising area for poststroke recovery [42]. At this point, a good understanding of the poststroke environment is of vital importance. A biocompatible conductive scaffold is essential for the delivery of stem cells to the area that needs healing. Pires and colleagues specifically targeted the differentiation of neural stem cells by preparing a compatible environment for electrical stimulation with the polystyrene sulfonate-based conducting polymer they synthesized [43]. The desired cell elongation was achieved thanks to the material that proved to be noncytotoxic. Producing a novel material that they can present as part of neural recovery,

Pires et al. have also introduced a stable material that is easily processed from solution to interface with biological systems, particularly the neural stem cell.

Applying electrical stimulation to biological systems, as mentioned in previous studies, can be preferred as a working method in healing brain damage. Conductive core-sheath nanofibers prepared by Xie et al. are candidates for creating a favorable environment for the growth of nerve cells [44]. It has been proven in vitro that dorsal root ganglia adhere well to conductive core-sheath nanofibers and produce neurites along the surface. In these treatments, where cell manipulation is also very important, they were able to grow the neurites in the direction they wanted. Xie et al., who achieved a noticeable neurite length compared to the control groups to which they did not apply electrical stimulation, have brought an excellent material to medicine in nerve tissue engineering in the way of repairing the damage that occurs after a stroke.

Another neural stem cell study using conducting polymers to improve brain damage was presented by Wang et al. [45]. To heal the damaged tissue, the scaffold must be able to control both the proliferation and differentiation of neural stem cells. Achieving this control is possible with three-dimensional structures. Taking this advantage, Wang et al. prepared the poly(3,4 ethylene dioxythiophene)/chitosan/gelatin scaffold via in situ interface polymerization with a nanostructured PEDOT layer assembled on the channel surface of the porous Cs/Gel scaffold. Neural stem cells have successfully adhered to this three-dimensional conductive material. In conclusion, the designed scaffold not only promoted the adhesion and proliferation of neural stem cells but also enhanced their differentiation into neurons and astrocytes with higher protein and gene expression. The combined use of conducting polymers can also yield successful results for both skeletal muscle tissue engineering and stem cell evaluations. In a case study by Mohamadali and colleagues, stem cells were investigated on polyaniline/polyacrylonitrile copolymer conductive material [46]. After proving the biocompatibility of the copolymers, they electrospinned the material and modified it with oxygen plasma. The compatibility and viability of mouse fibroblast cells and mesenchymal stem cells revealed that the cells proliferated and adhered well on the scaffold, thus promoting nanofiber cellular growth.

In older patients, migration of neural stem cells is limited. This obstacle also creates difficulties in the treatment of nerve injuries because large brain size requires cells to travel longer distances. Therefore, increasing the migration distance of stem cells in the brain in adult patients is a sought-after parameter in treatments. In this context, Feng et al. presented a promising study that could direct the action of stem cells in the brain [47]. In their strategy for the transfer of currents in the brain, they directed the transplanted stem cells through the conducting medium and studied the symptoms for three weeks to four months. They aimed for a more controlled and accurate recovery with the systems that they achieved the electrical direction they wanted, and enviable results are obtained.

Supporting regenerative medicine's stem cell treatments in post-traumatic recovery with conducting polymers has brought along many different studies and has been a glimmer of hope for patients in determining cell fate. The last but last least work to be given to these studies was done by Stewart et al., and it will undoubtedly not be the last study [48]. Investigating conducting polymers that determine the behavior of progenitor cells, researchers have designed electrostimulation, a system that can promote neuronal differentiation of murine embryonic stem cells. The vital importance of the control of electrical stimulation in nervous system recovery has already been mentioned in previous studies. They investigated the effect of polypyrrole containing the anionic dopant dodecylbenzenesulfonate to differentiate new human neural stem cells. Electrical stimulation of polypyrrole/dodecylbenzenesulfonate caused stem cells to predominantly β-III tubulin expressing neurons and lower induction of glial fibrillary acidic protein-expressing glial cells. Electrical stimulation has also been shown to contribute to the elongation and branching of nerve cells. Cells on the conductive polypyrrole/dodecylbenzenesulfonate film achieved the desired distribution. Thanks to his work, the literature has incorporated another convenient and useful method for regenerative medicine.

10.3   CONCLUSION AND FUTURE ASPECTS

In this compiled chapter, it is aimed to shed light on future applications by explaining some pioneering studies in the field of regenerative medicine of conducting polymers. The definition of regenerative medicine takes the title of "regeneration" because it can heal the damaged tissues and organs, thereby improving body functions. Some serious injuries or chronic diseases can cause significant damage to the functional tissues of the body. In such cases, the body may fail to renew itself. Regenerative medicine also constitutes an important newly developed field in medicine for the regeneration of organ and/or tissue functions for these patients. New medical needs in the treatment of many diseases are among the important factors for the faster development of technologies in the field of regenerative medicine. Regenerative medicine studies to date have made great strides in controlling cell behavior, designing scaffold materials for tissue repair, and various cell manipulation techniques. The design of materials that can be successfully placed in the anatomy of patients with biocompatibility is one of the indispensable needs of regenerative medicine. Considering the material selection, conductive polymers have many advantages to meet this need. Conductive organic polymers are strong candidates for developing medical materials. The conjugated backbone in the conducting polymer gives it the ability to conduct electrons, unlike other polymers. In addition, being easily synthesized and modified are among the desired advantages for medical material design. This feature is especially important for providing control applications on cell proliferation and differentiation.

For future studies, conducting polymers have many more developmental stages concerning regenerative medicine applications. One of them is that it should play an important role not only in the treatment of damage after the problem develops but also in the field of preventive medicine to block problems from occurring. In this respect, the development of new cell sources for transplantation, which is a frequent obstacle in past practices, can be counted. Another improvement strategy could be to adopt patient-specific approaches to the development of bioengineered materials. Regenerative medicine methods powered by precision medicine applications will be a great contribution to the success of treatments from tissue design to cell formation. Adding information to treatments about how the patient's age, disease state, and microbiome affect regeneration is crucial in this regard. Increasing the biocompatibility further in the current studies is also important for future studies. Creating an environment of regeneration within the patient can significantly improve the outcomes of regenerative medicine strategies overall [49, 50].

REFERENCES

1. Mason, C., & Dunnill, P. (2008). A brief definition of regenerative medicine. Regenerative Medicine, 3(1), 1–5. https://doi.org/10.2217/17460751.3.1.1
2. Haseltine, W. A. (2003). Regenerative medicine: A future healing art. The Brookings Review, 21(1), 38. https://doi.org/10.2307/20081088
3. Jaklenec, A., Stamp, A., Deweerd, E., Sherwin, A., & Langer, R. (2012). Progress in the tissue engineering and stem cell industry "Are we there yet?". Tissue Engineering Part B: Reviews, 18(3), 155–166. https://doi.org/10.1089/ten.teb.2011.0553
4. Cohen, P., Hunsberger, J. G., & Atala, A. (2019). Regenerative medicine manufacturing—Challenges and opportunities. Atala A., Lanza R., Mikos T., Nerem R. (Eds) Principles of Regenerative Medicine, 1367–1376. Academic Press. San Diego, CA. https://doi.org/10.1016/b978-0-12-809880-6.00078-3
5. Fisher, M. B., & Mauck, R. L. (2013). Tissue engineering and regenerative medicine: Recent innovations and the transition to translation. Tissue Engineering Part B: Reviews, 19(1), 1–13. https://doi.org/10.1089/ten.teb.2012.0723
6. Falanga, V., & Sabolinski, M. (1999). A bilayered living skin construct (APLIGRAF®) accelerates complete closure of hard-to-heal venous ulcers. Wound Repair and Regeneration, 7(4), 201–207. https://doi.org/10.1046/j.1524-475x.1999.00201.x
7. Song, J. J., & Ott, H. C. (2011). Organ engineering based on decellularized matrix scaffolds. Trends in Molecular Medicine, 17(8), 424–432. https://doi.org/10.1016/j.molmed.2011.03.005

8. Mao, A. S., & Mooney, D. J. (2015). Regenerative medicine: Current therapies and future directions. Proceedings of the National Academy of Sciences of the United States of America, 112(47), 14452–14459. https://doi.org/10.1073/pnas.1508520112

9. Jagur-Grodzinski, J. (2006). Polymers for tissue engineering, medical devices, and regenerative medicine. Concise general review of recent studies. Polymers for Advanced Technologies, 17(6), 395–418. https://doi.org/10.1002/pat.729

10. Sipehia, R., Martucci, G., Barbarosie, M., & Wu, C. (1993). Enhanced attachment and growth of human endothelial cells derived from umbilical veins on ammonia plasma modified surfaces of PTFE and EPTFE synthetic vascular Graft biomaterials. Biomaterials, Artificial Cells and Immobilization Biotechnology, 21(4), 455–468. https://doi.org/10.3109/10731199309117651

11. Nezakati, T., Seifalian, A., Tan, A., & Seifalian, A. M. (2018). Conducting polymers: Opportunities and challenges in biomedical applications. Chemical Reviews, 118(14), 6766–6843. https://doi.org/10.1021/acs.chemrev.6b00275

12. Shirakawa, H., Louis, E. J., MacDiarmid, A. G., Chiang, C. K., & Heeger, A. J. (1977). Synthesis of electrically conducting organic polymers: Halogen derivatives of polyacetylene, (CH) X. Journal of the Chemical Society, Chemical Communications, (16), 578. https://doi.org/10.1039/c39770000578

13. Namsheer, N., & Rout, C. S. (2021). Conducting polymers: A comprehensive review on recent advances in synthesis, properties and applications. RSC Advances, 11(10), 5659–5697. https://doi.org/10.1039/d0ra07800j

14. Guimard, N. K., Gomez, N., & Schmidt, C. E. (2007). Conducting polymers in biomedical engineering. Progress in Polymer Science, 32(8–9), 876–921. https://doi.org/10.1016/j.progpolymsci.2007.05.012

15. Walton, D. J., & Davis, F. J. (2004). The synthesis of conducting polymers based on heterocyclic compounds. Davis, F. J. (Ed) Polymer Chemistry. Oxford University Press. Oxford. https://doi.org/10.1093/oso/9780198503095.003.0011

16. Foulds, N. C., & Lowe, C. R. (1986). Enzyme entrapment in electrically conducting polymers. Immobilisation of glucose oxidase in polypyrrole and its application in amperometric glucose sensors. Journal of the Chemical Society, Faraday Transactions 1: Physical Chemistry in Condensed Phases, 82(4), 1259. https://doi.org/10.1039/f19868201259

17. Umana, M., & Waller, J. (1986). Protein-modified electrodes. The glucose oxidase/polypyrrole system. Analytical Chemistry, 58(14), 2979–2983. https://doi.org/10.1021/ac00127a018

18. Wong, J. Y., Langer, R., & Ingber, D. E. (1994). Electrically conducting polymers can noninvasively control the shape and growth of mammalian cells. Proceedings of the National Academy of Sciences of the United States of America, 91(8), 3201–3204. https://doi.org/10.1073/pnas.91.8.3201

19. Shi, G., Rouabhia, M., Wang, Z., Dao, L. H., & Zhang, Z. (2004). A novel electrically conducting and biodegradable composite made of polypyrrole nanoparticles and polylactide. Biomaterials, 25(13), 2477–2488. https://doi.org/10.1016/j.biomaterials.2003.09.032

20. Ragunathan, S., Govindasamy, G., Raghul, D., Karuppaswamy, M., & VijayachandraTogo, R. (2020). Hydroxyapatite reinforced natural polymer scaffold for bone tissue regeneration. Materials Today: Proceedings, 23, 111–118. https://doi.org/10.1016/j.matpr.2019.07.712

21. Lee, S., Ozlu, B., Eom, T., Martin, D. C., & Shim, B. S. (2020). Electrically conducting polymers for bio-interfacing electronics: From neural and cardiac interfaces to bone and artificial tissue biomaterials. Biosensors and Bioelectronics, 170, 112620. https://doi.org/10.1016/j.bios.2020.112620

22. Asadi, N., Del Bakhshayesh, A. R., Davaran, S., & Akbarzadeh, A. (2020). Common biocompatible polymeric materials for tissue engineering and regenerative medicine. Materials Chemistry and Physics, 242, 122528. https://doi.org/10.1016/j.matchemphys.2019.122528

23. Li, M., Guo, Y., Wei, Y., Macdiarmid, A., & Lelkes, P. (2006). Electrospinning polyaniline-contained gelatin nanofibers for tissue engineering applications. Biomaterials, 27(13), 2705–2715. https://doi.org/10.1016/j.biomaterials.2005.11.037

24. Ateh, D. D., Vadgama, P., & Navsaria, H. A. (2006). Culture of human Keratinocytes on polypyrrole-based conducting polymers. Tissue Engineering, 12(4), 645–655. https://doi.org/10.1089/ten.2006.12.645

25. Dias Junior, E. M., Dias, D. D., Rodrigues, A. P., Dias, C. G., Bastos, G. D., Oliveira, J. A., Maciel Filho, R., & Passos, M. F. (2021). SIPNs polymeric scaffold for use in cartilaginous tissue engineering: Physical-chemical evaluation and biological behavior. Materials Today Communications, 26, 102111. https://doi.org/10.1016/j.mtcomm.2021.102111

26. Qin, C., Yue, Z., Chao, Y., Forster, R. J., Maolmhuaidh, F. Ó., Huang, X., Beirne, S., Wallace, G. G., & Chen, J. (2020). Bipolar electroactive conducting polymers for wireless cell stimulation. Applied Materials Today, 21, 100804. https://doi.org/10.1016/j.apmt.2020.100804

27. Ferrigno, B., Bordett, R., Duraisamy, N., Moskow, J., Arul, M. R., Rudraiah, S., Nukavarapu, S. P., Vella, A. T., & Kumbar, S. G. (2020). Bioactive polymeric materials and electrical stimulation strategies for musculoskeletal tissue repair and regeneration. Bioactive Materials, 5(3), 468–485. https://doi.org/10.1016/j.bioactmat.2020.03.010

28. Balint, R., Cassidy, N. J., & Cartmell, S. H. (2014). Conducting polymers: Towards a smart biomaterial for tissue engineering. Acta Biomaterialia, 10(6), 2341–2353. https://doi.org/10.1016/j.actbio.2014.02.015

29. Huang, J., Lu, L., Zhang, J., Hu, X., Zhang, Y., Liang, W., Wu, S., & Luo, Z. (2012). Electrical stimulation to conducting scaffold promotes Axonal regeneration and Remyelination in a rat model of large nerve defect. PLoS One, 7(6), e39526. https://doi.org/10.1371/journal.pone.0039526

30. Shi, Z., Gao, H., Feng, J., Ding, B., Cao, X., Kuga, S., Wang, Y., Zhang, L., & Cai, J. (2014). In situ synthesis of robust conducting cellulose/Polypyrrole composite Aerogels and their potential application in nerve regeneration. Angewandte Chemie, 126(21), 5484–5488. https://doi.org/10.1002/ange.201402751

31. Orciani, M., Fini, M., Di Primio, R., & Mattioli-Belmonte, M. (2017). Biofabrication and bone tissue regeneration: Cell source, approaches, and challenges. Frontiers in Bioengineering and Biotechnology, 5(17). https://doi.org/10.3389/fbioe.2017.00017

32. Wang, W., & Yeung, K. W. (2017). Bone grafts and bone substitutes for bone defect management. Orthopedic Biomaterials, 2(4), 495–545. https://doi.org/10.1007/978-3-319-73664-8_18

33. Boerckel, J. D., Kolambkar, Y. M., Dupont, K. M., Uhrig, B. A., Phelps, E. A., Stevens, H. Y., García, A. J., & Guldberg, R. E. (2011). Effects of protein dose and delivery system on BMP-mediated bone regeneration. Biomaterials, 32(22), 5241–5251. https://doi.org/10.1016/j.biomaterials.2011.03.063

34. Bal, Z., Korkusuz, F., Ishiguro, H., Okada, R., Kushioka, J., Chijimatsu, R., Kodama, J., Tateiwa, D., Ukon, Y., Nakagawa, S., Dede, E. Ç., Gizer, M., Korkusuz, P., Yoshikawa, H., & Kaito, T. (2021). A novel nano-hydroxyapatite/synthetic polymer/bone morphogenetic protein-2 composite for efficient bone regeneration. The Spine Journal, 21(5), 865–873. https://doi.org/10.1016/j.spinee.2021.01.019

35. Kundu, K., Afshar, A., Katti, D. R., Edirisinghe, M., & Katti, K. S. (2021). Composite nanoclay-hydroxyapatite-polymer fiber scaffolds for bone tissue engineering manufactured using pressurized gyration. Composites Science and Technology, 202, 108598. https://doi.org/10.1016/j.compscitech.2020.108598

36. Tayebi, S. A., Yazdimamaghani, M., Walker, K. J., Eastman, M., Hatami-Marbini, H., Smith, B., Ricci, J. L., Madihally, S., & Vashaee, D. (2013). 3D conductive nanocomposite scaffold for bone tissue engineering. International Journal of Nanomedicine, 9, 167. https://doi.org/10.2147/ijn.s54668

37. Lindsberg, P. J., & Grau, A. J. (2003). Inflammation and infections as risk factors for ischemic stroke. Stroke, 34(10), 2518–2532. https://doi.org/10.1161/01.str.0000089015.51603.cc

38. Lipton, P. (1999). Ischemic cell death in brain neurons. Physiological Reviews, 79(4), 1431–1568. https://doi.org/10.1152/physrev.1999.79.4.1431

39. Moskowitz, M. A., Lo, E. H., & Iadecola, C. (2010). The science of stroke: Mechanisms in search of treatments. Neuron, 68(1), 161. https://doi.org/10.1016/j.neuron.2010.08.019

40. Arai, K., Jin, G., Navaratna, D., & Lo, E. H. (2009). Brain angiogenesis in developmental and pathological processes: Neurovascular injury and angiogenic recovery after stroke. FEBS Journal, 276(17), 4644–4652. https://doi.org/10.1111/j.1742-4658.2009.07176.x

41. George, P. M., Lyckman, A. W., LaVan, D. A., Hegde, A., Leung, Y., Avasare, R., Testa, C., Alexander, P. M., Langer, R., & Sur, M. (2005). Fabrication and biocompatibility of polypyrrole implants suitable for neural prosthetics. Biomaterials, 26(17), 3511–3519. https://doi.org/10.1016/j.biomaterials.2004.09.037

42. Hess, D. C., Wechsler, L. R., Clark, W. M., Savitz, S. I., Ford, G. A., Chiu, D., Yavagal, D. R., Uchino, K., Liebeskind, D. S., Auchus, A. P., Sen, S., Sila, C. A., Vest, J. D., & Mays, R. W. (2017). Safety and efficacy of multipotent adult progenitor cells in acute ischaemic stroke (Masters): A randomised, double-blind, placebo-controlled, phase 2 trial. The Lancet Neurology, 16(5), 360–368. https://doi.org/10.1016/s1474-4422(17)30046-7

43. Pires, F., Ferreira, Q., Rodrigues, C. A., Morgado, J., & Ferreira, F. C. (2015). Neural stem cell differentiation by electrical stimulation using a cross-linked PEDOT substrate: Expanding the use of biocompatible conjugated conductive polymers for neural tissue engineering. Biochimica et Biophysica Acta (BBA) – General Subjects, 1850(6), 1158–1168. https://doi.org/10.1016/j.bbagen.2015.01.020

44. Xie, J., MacEwan, M. R., Willerth, S. M., Li, X., Moran, D. W., Sakiyama-Elbert, S. E., & Xia, Y. (2009). Conductive core-sheath nanofibers and their potential application in neural tissue engineering. Advanced Functional Materials, 19(14), 2312–2318. https://doi.org/10.1002/adfm.200801904

45. Wang, S., Guan, S., Li, W., Ge, D., Xu, J., Sun, C., Liu, T., & Ma, X. (2018). 3D culture of neural stem cells within conductive PEDOT layer-assembled chitosan/gelatin scaffolds for neural tissue engineering. Materials Science and Engineering: C, 93, 890–901. https://doi.org/10.1016/j.msec.2018.08.054

46. Mohamadali, M., Irani, S., Soleimani, M., & Hosseinzadeh, S. (2017). PANi/PAN copolymer as scaffolds for the muscle cell-like differentiation of mesenchymal stem cells. Polymers for Advanced Technologies, 28(9), 1078–1087. https://doi.org/10.1002/pat.4000

47. Feng, J., Liu, J., Zhang, L., Jiang, J., Russell, M., Lyeth, B. G., Nolta, J. A., & Zhao, M. (2017). Electrical guidance of human stem cells in the rat brain. Stem Cell Reports, 9(1), 177–189. https://doi.org/10.1016/j.stemcr.2017.05.035

48. Stewart, E., Kobayashi, N. R., Higgins, M. J., Quigley, A. F., Jamali, S., Moulton, S. E., Kapsa, R. M., Wallace, G. G., & Crook, J. M. (2015). Electrical stimulation using conductive polymer Polypyrrole promotes differentiation of human neural stem cells: A biocompatible platform for translational neural tissue engineering. Tissue Engineering Part C: Methods, 21(4), 385–393. https://doi.org/10.1089/ten.tec.2014.0338

49. Oh, J., Lee, Y. D., & Wagers, A. J. (2014). Stem cell aging: Mechanisms, regulators and therapeutic opportunities. Nature Medicine, 20(8), 870–880. https://doi.org/10.1038/nm.3651

50. Eming, S. A., Martin, P., & Tomic-Canic, M. (2014). Wound repair and regeneration: Mechanisms, signaling, and translation. Science Translational Medicine, 6(265), 265–265. https://doi.org/10.1126/scitranslmed.3009337

11 Conducting Polymers as Efficient Materials for Tissue Engineering

Srijita Sen,[1] Trishna Bal,[1] Aditya Dev Rajora,[1] Shreya Sharma,[1] Shubha Rani Sharma,[2] and Neelima Sharma[1]

[1]Department of Pharmaceutical Sciences and Technology, Birla Institute of Technology, Mesra, Ranchi, India

[2]Department of Bioengineering, Birla Institute of Technology, Mesra, Ranchi, India

CONTENTS

11.1 INTRODUCTION

Smart electroactive biomaterials are a major requirement for tissue-engineered regenerative medicine (TERM) as well as in tissue engineering, as these materials can respond to the specific stimuli which are necessary for cell growth [1]. This group of electroactive biomaterials comprises different conducting polymers (CPs), semiconducting materials, etc. CPs are most advantageous as they have the desired electrical and optical properties and these properties can be tailored to be biocompatible, biodegradable to be utilized as an effective polymeric scaffold for tissue regeneration. The field of tissue engineering has a tremendous impact on the health-care system, still there are certain limitations for example, the availability of cells for seeding in the scaffolds for further processing. For overcoming this issue, stem cells and progenitors came into being. Polymers have become an important part of the day-to-day research encompassing drug delivery, tissue engineering, biomimetic devices, sensors, efficient filtration system, etc. But the role of a special class of polymers that are electroconductive or stimuli-sensitive is investigated more as they can target various tissues and organs based on their response toward a specific site. The electronically CPs also called as first category CP should possess desired biocompatibility as it is applied for tissue growth and should possess zero toxicity when tested both *in vitro* as well as *in vivo*. Different biodegradable biomaterials like ceramics, metals, polymers, alloy, composites, etc. are used for the preparation of scaffolds for the proliferation of cells in them to develop into tissues and later into organs and biodegradability is an essential criterion for such materials. Amongst all biomaterials, polymeric materials being chemically more stable have good processability and their low production cost make them favorable candidates for applications in different fields of engineering. But general polymers lack the property of conductance with respect to body stimuli and thus sometimes become target inefficient. With the advent of CPs, all the limitations of common polymers were alleviated. The CPs have multiple properties of generalized polymers as well as both electrical and optical properties matching with semiconductors and metals which make them appropriate for use in the fabrication of different biomedical devices. For tissue engineering, CPs must be capable of modulating the molecular and mechanical signals for tissue regeneration. The current chapter mainly focuses on the various aspects related to applications of preparative techniques of CP and their applications in tissue engineering, basics of tissue engineering, and different biocompatibility studies of CPs both *in vitro* and *in vivo*. These conducting materials possess vast potential in tissue regeneration and thus this chapter summarizes all the details on CPs and their applications in various fields of tissue engineering.

11.2   TISSUE ENGINEERING – RATIONALE AND BASIC CONCEPTS

Tissue engineering is a broad aspect that encompasses tissue growth and biomaterials. Biomaterials are used as backbone above which the cells start regenerating and thus proliferate to produce new tissues and further to organs. The field of tissue engineering is an amalgamation of multiple sciences incorporating physics, mechanical engineering, material science, polymers, basic chemistry, and biology. The use of biomaterials that are cell compatible gives an excellent environment for mimicking the extracellular matrix (ECM) of the cell and also provides sufficient adhesion for required growth [2]. From the very ancient period, it's a matter of concern for organ transplantation for alleviating serious health issues and in many cases, recipients do not get proper donors and suffer from severe complications related to organ rejection by the body. In this aspect, the use of biomimetic materials for tissue regeneration can be an effective solution. The term "tissue engineering" was formally conceived at the National Science Foundation workshop in 1988 for describing the use of different principles and methods from engineering and life sciences that are used for restoring the biological functions in pathological mammalian tissues [3]. Tissue engineering actively involves the use of 3D scaffolds which are mostly porous providing an ambient environment for tissue proliferation. The efficiency of these porous 3D scaffolds is significantly enhanced when they respond to specific biophysical stimuli [4]. These responses can be achieved by when scaffolds are prepared with stimuli-sensitive polymers. Generally, it is seen that the role of tissue engineering is more prevalent in the development or replacement of any injured tissue or rectifying congenital defects, but rarely these tissue-engineered products are used for any diseased condition [5]. Prosthetic devices are used to replace organs but they are not efficient in restoring the normal functions, such use of fabricated 3D porous biomimetic scaffolds can serve the smooth functioning of diseased organs as well as disease treatment which cannot be cured with generally available active pharmaceutical ingredients [5]. There has always been a crisis for organ replacement for serious patients, thus with the emergence of tissue engineering and TERM, the ray of hope has percolated in the field of medical science. In this aspect, natural and synthetic materials have been tried for the fabrication of different types of prosthetics to replace organs in the human body. From ancient time onwards from the seventh-century BC, different materials were tried for replacement of the lost body part [6]. With time, advanced materials are utilized with proper vigilance of their response toward cell compatibility. Thus, the concept of a scaffold that matches with tissue environment came into the picture.

Langer and Vacanti in 1993 [6] elaborated certain strategies for regeneration of new tissues *in vitro* as mentioned below:

1. Isolated cells or substitutes: In this strategy, isolated cells are devised for the healing of injured tissue and in this aspect, stem cells are utilized but they sometimes lack their fixability at the injured site.
2. Tissue-inducing substance: In this strategy, cell growth promoters like growth factors, small molecules are employed within the cells for their quick proliferation.
3. Cells placed or within the matrix: According to this strategy, the cells are to be placed in a scaffold along with the growth regulators. This strategy is the whole concept of tissue engineering.

By far, the most common way for achieving a tissue-engineered scaffold is by seeding the cells in the preformed porous scaffolds or seeding the cells during the fabrication of scaffolds. But post the seeding method is a bit tedious and not very efficient, whereas the seeding of cells during the fabrication of scaffold is more advantageous as there is higher entrapment efficiency. Another method of tissue engineering is the decellularization of ECM from allogenic or exogenic tissues to be used as a scaffold [5]. In addition, another method involves the preparation of cell sheets using temperature-responsive culture dishes known as cell sheet engineering. For fulfilling these methods, interpenetrating polymeric network (IPN) or hydrogels are most appropriate provided they are exclusively biocompatible [7]. Different polymers of natural or synthetic origin, as well as bioceramics and also various hybrid materials, are used for the scaffold preparation. Polysaccharide-based scaffolds are widely explored for tissue regeneration being highly biodegradable and biocompatible and also have the ability in mimicking the chemical composition of ECM [7]. The porous structure of the scaffold mainly the honeycomb structure is very advantageous as they are featherweight and flexible which provides compressive strength to its structure. Materials used for the preparation of scaffold must have sufficient mechanical strength and appropriate structural configurations so that they can be easily fitted to the specific anatomical site when implanted [7]. An ideal scaffold should have micro and macrostructural properties like scaffold porosity, pore size, pore shape, and interconnectivity, etc. for mimicking the ECM [4]. Moreover, the cells that are entrapped within the scaffold should be able to grow by themselves to proliferate to a new matrix without any hindrance [8, 9]. Different types of anatomically important proteins like collagen, fibrin, fibrinogen, elastin along with polysaccharides like alginates, hyaluronic acid, chitosan, cellulose form the basic constituents of ECM. With a complex structure, the ECM provides a basic backbone for cell regeneration with sufficient mechanical and biochemical support facilities [10, 11]. The most important requirement for any scaffold is that it should not have immunological reactions with the host body [12].

11.3 CPs AND THEIR ROLE IN TISSUE REGENERATION

The concept of CPs has unfolded a fresh chapter in organic polymers after the accidental discovery of polyacetylene (PA) doped with iodine. Alan J. Heeger, Alan G. MacDiarmid, and Hideki Shirakawa are the three icons behind the discovery and development of CPs and were awarded the Nobel Prize in chemistry 2000 for their great contribution. The mechanism of transformation from insulating material to a conducting one is the oxidation of the PA polymeric chain, where iodine takes away electrons leaving cations that can move under influence of an external electric field [13]. However, the principle of conductivity is not the same for all CPs. When there are so many variants of materials and polymers available in the market, CP has been embraced due to some extraordinary qualities like lightweight, resistance to corrosion or chemical attack, nonmetallic surface behavior, superior flexibility, economical, and a lot more which has been discussed in brief in a later portion of this chapter. CPs suffer from adequate durability and poor cyclic performance and the challenge is to improve the overall performance of the polymer by increasing or maintaining the conductivity. Good quality of CP with pertinent applicability can be prepared by electrochemical methods whereas good quantities of polymers are obtained by chemical synthesis. Recently environment-friendly catalysts like horseradish peroxidase [14], hematin, PEG-hematin [15], and few others have been experimented with to speed up the polymerization reaction by green approach. Once the polymers have been synthesized or modified, extensive evaluation of fundamental physical, chemical, mechanical properties along with other desired features is essential to understand the need for subsequent redevelopment for their novel applications. The primary necessity for CPs evaluation is to assess their conductive behavior. Moreover, to extend their application, several studies are done over the past few decades to develop microstructured and nanostructured materials, including gel matrix from CPs with controllable non-disperse morphology [16]. Progressive studies are going on to prepare multifunctional CP nanocomposite and blends of CPs as a part of nanoscale engineering.

The optimal properties to be possessed by a biomaterial for its utilization as a scaffold relies on physiological and morphological aspects, but to make them responsible for stimulating cell propagation, there arises a necessity for fabricating electrically conductive specially designed polymers doped with different functional groups involved in monitoring cell behavior and biological properties necessary for tissue regeneration [17]. These specially engineered polymers help in cell proliferation, cell differentiation, adhesion, and alignment by conduction of electrical charges. To increase the softness, CPs are blended with different biodegradable and flexible polymers like poly(lactic-*co*-glycolic acid) (PLGA), polycaprolactone (PCL), which also increases their resistivity as compared to the native cardiac tissues. Also, nanofibrous scaffolds prepared by blending polyaniline (PANI) with gelatin or polylactic acid (PLA) show electrical conductivity of 4.2×10^{-3} S/cm which is sufficient for cardiac tissue regeneration [18]. CPs blended with polymers in the form of aqueous hydrogels are efficient materials for tissue regeneration having desired flexibility, mechanical strength, and also excellent electrical conductivity. But in the presence of physiological pH and fluid content, they lack electrical conductivity as the charges on the CPs become neutral in the presence of physiological pH, and also CPs escape the network of the hydrogels, thereby further diminishing the electrical conductivity [19]. These can be remedied by fabricating single component CP hydrogel, but such single-component CP hydrogel lacks flexibility, hydrophilicity, and functional side chains [20]. Polythiophenes (PTs) are potential candidates for fabricating singe component CP hydrogels. These conducting hydrogels can be modified by varying the pore size, cross-linking density, and swelling ratio. The CPs can be utilized as actuators for artificial muscle applications based on their electrochemomechanical properties which result in their volume expansion and they operate at a very low voltage of 2–10 V. CPs applied as actuators for muscle are activated for faster response by incorporating nanofibers and nanotubes within them where the porosity of the whole system gets enhanced [21]. Intrinsically CPs (ICPs) possess the property of ion exchange. Different polymers under this category are like PA, PANI, polypyrrole (PPy), polycarbazole (PC), poly(3,4-ethylenedioxythiophene) (PEDOT), PT, polyphenazine (PPH), poly(*p*-phenylene) (PPP), poly(2,5-thienylenevinylene) (PTV), etc. and their derivatives. The presence of benzenoid or nonbenzenoid or heterocyclic structure in these polymers on partial oxidation or reduction state becomes electroactive. ICPs can be tuned to be stretchable by performing structural engineering by altering the chain length of polymer backbone or by side-chain branching, or by introducing soft blocks to the rigid oligomer or macromonomer structure The conductivity of extrinsically conductive polymers (ECPs) is due to externally added elements which sometimes support the enhancement of mechanical strength of the polymers and can also be furnished by embedding conductive fillers onto the flexible and stretchable polymeric matrix. Natural polymers can be harmonized to such a state which will create a new horizon in the sphere of biomedical technology. Development of ECPs matrix illicit special advantages over metals like platinum, gold, iridium, titanium, tungsten, stainless steel, etc., which suffers from mechanical incompatibility with delicate tissues. Widely investigated conductive fillers like multi-walled carbon nanotubes, silver nanowires, and poly(hydroxymethyl 3,4-ethylene dioxyphenylene) CP microspheres, when incorporated in the soft biocompatible matrix of poly(ε-decalactone) (EDL), a 3D percolation network was formed. *In vitro* assessment of cytocompatibility using ventral mesencephalon cells from embryonic rat brain tissue shows the ability of the network to reduce reactive astrocytes responsible for gliosis while maintaining sufficient neurons, which may positively influence coupling of a neuron to electrode [22]. Generally, higher extent of stretching or deformation allows the polymeric system to come into close contact with the biological environment/ tissues/effective surface more easily. Stretchable polymers are also called self-healing polymers due to their ability to restore initial mechanical as well as functional properties after damage or structural deformation by intrinsic/extrinsic mechanisms. In intrinsic mechanism, healing is aided by the polymer's own dynamic molecular features, chain entanglements, and chemical bonding (covalent, non-covalent interaction) in the presence of external action whereas a healing agent is encapsulated/embedded into the polymeric matrix in the external mechanism. Release of a healing agent from incorporated microencapsulate or polymer matrix is needed to trigger often by external stimuli, serve additional advantages. After mechanical deformation or microcracking, the percolation pathway for the charge transport gets altered in conducting matrix leading to enhanced electrical resistance which generates heat encouraging self-healing. Among them PEDOT: poly(styrene sulfonate), PANI, PPy are the most common [23]. These create an interest to evaluate their ideality in tissue engineering and wound management therapy. Still to achieve high conductivity in combination with high stretchability is a limitation that must be scrutinized for getting the best outcome. Versatile natural, semisynthetic, and synthetic biodegradable polymers like gelatin, heparin, collagen, chitosan, polyurethane, poly(L-lactide), poly(D,L-lactide), PCL, fumarate, poly(glycolic acid), poly(lactide-*co*-glycolide), poly(lactide-*co*-polycaprolactone), etc. have been explored to make blended CPs. Some of these blended systems have the potential to mimic the mechanical as well as electrical nature of ECM. These blended CPs have enlivened their applications in the different forms of hydrogels, fibers, films, coatings, etc. Composite emulsion of PPy/PDLLA (poly(D,L-lactic acid) has been transformed into a conduit system which after *in vivo* implantation favors neurite, axon regeneration with comparable conducting velocities to healthy nerves without any infective issues. As PPy degrade very slowly, the amount of it kept minimized [24]. However, these blended polymers are not completely removed from the body. So careful evaluation of percentage CPs remaining inside the body must be well evaluated along with toxicity prediction. So, CPs that are 100% degradable is the need of the hour. Although many new biodegradable CPs are synthesized in the last few years, still there is a need for more to be added to these spectra for their wide applicability in biomedical technology and regenerative medicine. Another important matter to be focused on is the solubility of these biodegradable polymers. Almost all of these degradable CPs are soluble in organic solvents that therefore arise the question of toxicity and biocompatibility. The only answer can be the development of water-soluble degradable CPs without disturbing their electro-optical properties and stability.

In the year 1960, scientists highlighted the importance of electrical conductivity on the regeneration of body parts like bone regrowth, wound healing, nerve function, cardiac cell functioning, etc. Thus, studies on the electrical conductivity of such polymers have gained importance in recent times. The CPs can be used for the preparation of cardiac patches, hydrogel, and 3D-printed constructs, etc. The CPs are used for the fabrication of conducting nanomaterial for mimicking the noninotropic structure of myocardium and thereby promoting electrical signal, excitability, and growth of cardiac cells and this electrical conductivity can be enhanced by the incorporation of conducting additives like carbon nanotubes, reduced graphene oxide, graphene, etc. Within the polymeric matrix [25], the CPs have been used as scaffolds for increasing the contraction ability of heart cells but due to their slow degradation *in vivo*, they exhibit certain incompatibilities related to inflammation, and thus surgical procedures are required for their removal. Thus, scientists are working on the fabrication of CPs that are degradable at both *in vitro* and *in vivo*. The CPs can be modified with the desired properties of tissues for enhanced tissue regeneration [26]. Some of the most widely used CPs like PANI, PPy, PT, and their derivatives are an excellent choice due to their biocompatibility, ease of synthesis, and easy modification of physicochemical properties [27]. In order to enhance the cellular response and regenerative potency, the CPs are conjugated with biologically active dopants like peptides which makes the CPs more electroactive functionally [28] and these modified CPs become like a vessel for these biologically active molecules which enhances their cell response and functionality [29]. The electrochemical oxidation or reduction potential of CPs results in the change in the local conformation of the binding proteins of the biotic or abiotic surface resulting in adhesion and migration of cells [30, 31].

11.4 APPLICATIONS OF CPs IN TISSUE ENGINEERING

11.4.1 NERVE TISSUE ENGINEERING

Neurons or nerve cells of the nervous system can transmit the signals at a wink of an eye by exciting the cells electrically. CPs like PPy and PANI have been developed and are being used as conductive scaffolds for boosting the growth as well as regeneration of nervous tissue. The conductive PCL/PPy polymer blend is considered to be conductive nanofibers with excellent cytocompatibility. They support the differentiation of PC12 cell-producing neurite outgrowth which clings on the conductive nanofibers. It was observed that the growth of the PC12 cells was widespread as compared to those of PCL devoid of PPy coating. Due to these exciting outcomes, the scientists declared that the conductive nanofiber scaffolds have immense application in the area of nerve tissue engineering [32]. It was seen that the PEDOT or chitosan or gelatin scaffolds demonstrate wonderful biocompatibility as well as appreciable enhanced neuron-like adhesion and proliferation of rat PC12 cells. They were also able to maintain cells in further active proliferation and neurite growth apart from upregulation of GAP43 and SYP protein and gene expression levels. Growth of neuronal cells and increase of the length of neurites were exhibited by electroactive tobacco mosaic virus (TMV)/PANI/PSS (polystyrene sulfonate) nanofibers [33]. It was observed that cells with neurites were larger in number than cells that were cultured on the TMV-derived nonconductive nanofibers. So these unique conductive nanofibers are very much instrumental in the formation of bipolar neural cell morphology. This suggests that the electroactivity, as well as topographical signals from these types of nanofibers, could support the differentiation as well as the growth of the neural cell. Due to the complexities of the anatomy of the nervous system, these types of tissue involving conductive biomaterials make their way in the area of nervous tissue repair. The property of electrical stimulation by the electric conductive polymers makes it a major attraction for the construction of the scaffolds for nerve tissue engineering. The most commonly used conductive polymers, PPy and PANI, because of their design and modifications, form the most appropriate scaffolds for the engineering of nerve tissue. PANI along with gelatin and PPy along with poly(ε-caprolactone) is the other scaffolds that are electrospun, prepared composites, conductive scaffolds, having electrical stimulation [34].

11.4.2 BONE REGENERATION

Electrically conducting materials because of their property of vulnerability to electrical, mechanical, optical as well as thermal phenomena could be used as a boundary with the external and the physiological environment. PPy as a conductive polymer is an excellent substrate for the attachment, proliferation, and differentiation of BMSC (bovine bone marrow stromal cells) and thus can be used as an interactive substrate for inducing differentiation in BMSC [35]. The conducting substrates can improve the attachment and propagation of the different bone cells like MC3T3-E1 cells, osteoblast-like SaOS-2 cells, C2C12 cells, and mesenchymal stem cells [36]. It was observed that the PEDOT: PSS scaffold which is ice-templated is porous and conductive having a median pore diameter above 50 μm facilitating the infiltration and matrix deposition within the void space of MC3T3-E1 cell. In addition, the scaffold also led to the increase in the expression of alkaline phosphatase (ALP), collagen type 1 alpha 1 (COL1), and runt-related transcription factor 2 (Runx2) in genes, as well as an increase in the mineralization of the ECM [37]. It also stimulates the deposition of osteocalcin of MC3T3-E1 cells. So, from the above observations, it was declared that conductive PEDOT: PSS scaffold can encourage the differentiation of osteogenic precursor cells (MC3T3-E1) into osteocalcin positively stained osteoblasts [38]. The 3D conductive composite scaffolds made up of PEDOT: PSS, gelatin as well as bioactive glass nanoparticles exhibited porosity of about 60%, and the pore size could be manipulated by altering the PEDOT: PSS content in the scaffold. It was witnessed that these scaffolds assisted the adhesion of human mesenchymal stem cells [39]. Also, with the increase in the concentration of the conductive polymer in the scaffold, the viability of cells can be increased. The above characteristics endowed to the cells can be credited to the improved microstructure of the scaffolds or improved electrical signaling amongst the cells.

11.4.3 As Electroconductive Biomaterial Scaffolds in Cardiac Tissue Engineering

The capacity of the human heart to regenerate is very less. This shortcoming may have led to a plethora of problems leading to various heart issues like chronic heart failure and subsequently myocardial infarction. Even though many supporting devices related to the heart have been conceived, yet the permanent cure of heart failure is whole heart transplantation. Various promising devices like the engineered scaffolds, implantable patches, and injectable hydrogels are used to reinstate the function of the heart, still, the property of ideal tissue regeneration is yet to be achieved [40]. Also, we witness a lot of mismatches between the present biomaterial with the natural tissue of the human body that may lead to rejection of these tissues. Now for obtaining adequate electrical conductivities to be used in cardiac regeneration, conductive fillers such as graphene, carbon nanotubes, metallic nanoparticles, MXenes, and CPs viz. PANI, PPy, and poly(3,4-ethylendioxythiophene) can be used [41]. A synchronized propagation of electrical signals through the cardiac cells results in excitation–contraction coupling of the heart. The contractibility of the cardiac muscles is synchronized by variations in the concentration of intracellular Ca ($[Ca^{2+}]_i$). The normal functioning of the heart needs sufficiently high $[Ca^{2+}]_i$ in the systole stage as well as low in diastole [42]. The conductive film of PPy/PCL can perk up the rate of propagation of calcium waves and shorten the calcium transient duration of the cardiomyocyte monolayers in comparison to PCL film. It was observed that more cardiomyocytes on the PPy/PCL film offered localization of the gap junction protein connexin-43 (CX43) along the margins. PLA/PANI conductive nanofibrous sheets were prepared by electrospinning with the same diameter of the nanofibers which exhibited elevated cell viability and encouraged the differentiation of H9c2 cardiomyoblasts [43]. The cell–cell interaction was improved by PLA/PANI nanofibrous sheets. These also enhanced the maturation of cardiac cells as well as the spontaneous beating of primary cardiomyocytes. These properties suggested that PLA/PANI conductive nanofibrous sheets possess a lot of potential in cardiac tissue engineering applications [44]. The negatively charged adhesive proteins could be attracted by the doped conductive PLGA/PANI nanofiber meshes thereby enhancing cell adhesion. It was witnessed that the cardiomyocytes on the nanofiber meshes could connect and form isolated cell clusters with elongated and aligned morphology along the main axis of the fibrous mesh [45]. The CX43 was also present in the clusters and synchronous beating of the cells was observed in each cluster present. A native heart mimicking electrical stimulation could be used to synchronize the beating rate of the heart.

11.4.4 Skeletal Muscle Tissue Engineering

Though the skeletal muscles exhibit a vigorous capability to regenerate, under certain brutal conditions, the function of the muscle may be lost forever. Differentiation and maturation of muscle precursor cells (the prefabrication of muscle tissue *in vitro*) on a functional scaffold forms the basis of skeletal muscle tissue engineering [46]. The conductive biomaterials like poly(L-lactide-co-ε-caprolactone) (PLCL)/PANI fibers are found to be compatible with the muscle cells. The myogenin expression was improved by PLCL/PANI fibers in comparison to that of the PLCL fibers [47]. They also increased the expression of troponin T and myosin heavy chain genes [48]. This signifies that the substrates that were electrically conductive help in the regulation of the myoblasts into myotube formation without additional electrical stimulation. It was observed that gelatin + camphor sulfonic acid (CSA) + PANI nanofibers could improve C_2C_{12} cell myogenic differentiation relative to those of gelatin or gelatin + CSA nanofibers [49]. The other novel properties exhibited by the conductive nanofiber group are enhanced intracellular organization, localization of both dihydropyridine receptor and ryanodine receptor, expression of genes associated with excitation-contraction coupling apparatus, calcium transients, and myotube contractibility [49]. The functionality of the formed myotubes could be further enhanced when myotubes were electrically stimulated, which showed more calcium transients and contractions with higher amplitude and regularity.

11.4.5 Skin Tissue Engineering

A skin forms a protective layer on the human body and saves it from all types of damage and microbial attack. Injury to the skin in different ways occurs, be it burns due to acid, flames, or injury to the skin in accidents. There are certain ways for replenishing the destroyed skin like skin grafting, but due to insufficient supply of natural skin, we have to depend upon various other man-made biomaterials. So many biomaterials have been synthesized or fabricated which has a great property of wound dressing and good antibacterial activity [50]. Through various researches, certain conductive materials have been found to support cellular behavior which includes keratinocytes as well as fibroblasts. Among the CPs, PANI that is a CP exhibits antibacterial property that makes them a promising therapeutic candidate in the area of wound healing [51]. A number of research activities were carried to examine the wound-healing capacity of conductive nanofiber composites based on poly(aniline-co-aminobenzene sulfonic acid), poly(vinyl alcohol), and chitosan oligosaccharide, etc. The Sprague Dawley (SD) rat was used for the study. A double full-thickness skin wound model was employed on the dorsum of SD rats [52]. It was observed that conductive nanofiber composite dressing showed lesser stimulating responses than the control group in the experiment. After treatment of about 15 days, it was observed that the conductive dressing composed of conducting biomaterial exhibited nearly absolute healing and improved collagen and granulation in comparison to the control group post-treatment of 15 days. This result signified the promising preposition of conductive nanofiber composites in wound healing [53]. It was also observed that a PANI/chitosan nanofiber mat in the ratio 1:3 exhibited an effect that was collaborative in enhancing the growth of both cells osteoblast and fibroblast. These cells attributed toward the appropriate level of hydrophilicity and conductivity contributing toward wound-healing applications [54]. An antibacterial and conductive injectable benzaldehyde-functionalized poly-(ethylene glycol)-co-poly(glycerol sebacate) PEGS-FA/quaternized chitosan-g-PANI hydrogels was developed which exhibited vital properties like enhanced mechanical properties, high-quality electroactivity, free-radical scavenging capability, antibacterial action, tackiness, conductivity as well as biocompatibility. A special hydrogel was developed which had a high cross-linker capacity and proved to be possessing outstanding *in vivo* blood clotting potential. It also worked for considerably improved wound healing *in vivo*. A full-thickness skin defect model was used as a control over those of nonconductive quaternized chitosan/PEGS-FA hydrogel and commercial dressing (Tegaderm film) [55]. These conductive biomaterials showed excellent properties in upregulating the expression of genes of growth factors, including vascular endothelial growth factors (VEGF), epidermal growth factor (EGF), and transforming growth factor-beta (TGF-β), and encouraged granulation tissue thickness and collagen deposition [56].

11.5 CONCLUSION AND FUTURE PERSPECTIVE

The CPs are a versatile class of materials having a wide range of applications right from utilization in the fabrication of piezoelectric devices to biomimetic scaffolds used for tissue regeneration. This class of materials possesses unique properties of electricity conduction as well as characteristic optical properties which can be maneuvered to attain the desired response when used as a polymeric scaffold for tissue engineering purposes. There is a wide range of CPs that are used but the most common ones frequently and successfully used are PANI, PEDOT, PPy, PLA, PPH, PPP, PTV, etc. and their derivatives. Although the CPs are highly advantageous over conventional polymers, yet they suffer from certain limitations like poor water solubility, poor biodegradability, low charge carrier mobility, reduced entrapment efficiency for bioactive moieties as well as drug, which are the major criteria for the scaffold to be used for tissue engineering. To combat such limitations, CPs are combined with a large quantity of biodegradable polymers which drastically enhances their biodegradability; moreover, the hydrophobicity is reduced by surface modification as well as functionalization, also doping with different dopants can modify the chemical and physical properties of CPs, thereby making them ambient for tissue regeneration. Despite such remedies employed to idealize these CPs for application in tissue engineering, still, there are lots to be explored in the fabrication of CPs nanocomposite, with high efficiency in terms of biodegradability and increased entrapment.

REFERENCES

1. Huang NF, Lee RJ, Li S (2010) Engineering of aligned skeletal muscle by micropatterning. Am J Transl Res 2(1): 43.
2. O'brien FJ (2011) Biomaterials & scaffolds for tissue engineering. Mater Today 14(3): 88–95.
3. Martin I, Wendt D, Heberer M (2004) The role of bioreactors in tissue engineering. Trends Biotechnol 22(2): 80–86.
4. Mhanna R, Hasan A (2017). Introduction to tissue engineering. In: Hasan A (Eds.) Tissue engineering for artificial organs: regenerative medicine, smart diagnostics and personalized medicine. Weinheim: 1–24.
5. Finch J (2011) The ancient origins of prosthetic medicine. Lancet 377(9765): 548–549.
6. Langer R, Vacanti JP (1993) Tissue engineering. Science 260(5110): 920–926.
7. Meyer U, Meyer T, Handschel J, Wiesmann HP (Eds.) (2009) Fundamentals of tissue engineering and regenerative medicine. Springer Science & Business Media, Berlin, Heidelberg.
8. Li Z, Xie MB, Li Y, Ma Y, Li JS, Dai FY (2016) Recent progress in tissue engineering and regenerative medicine. J Biomater Tissue Eng 6(10): 755–766.
9. Eltom A, Zhong G, Muhammad A (2019) Scaffold techniques and designs in tissue engineering functions and purposes: a review. Adv Mater Sci Eng 2019: 1–3.
10. Celikkin N, Rinoldi C, Costantini M, Trombetta M, Rainer A, Święszkowski W (2017) Naturally derived proteins and glycosaminoglycan scaffolds for tissue engineering applications. Mater Sci Eng C 78: 1277–1299.
11. Roseti L, Parisi V, Petretta M, Cavallo C, Desando G, Bartolotti I, Grigolo B (2017) Scaffolds for bone tissue engineering: state of the art and new perspectives. Mater Sci Eng C 78: 1246–1262.
12. Deb P, Deoghare AB, Borah A, Barua E, Lala SD (2018) Scaffold development using biomaterials: a review. Mater Today Proc 5(5): 12909–12919.
13. Shirakawa H, Louis E, MacDiarmid AG, Chiang CK, Heeger AJ (1977) Synthesis of electrically conducting organic polymers: halogen derivatives of polyacetylene, (CH)x. J Chem Soc Chem Commun 16: 578–580.
14. Alvarez S, Manolache S, Denes F (2003) Synthesis of polyaniline using horseradish peroxidase immobilized on plasma-functionalized polyethylene surfaces as initiator. J Appl Polym Sci 88(2): 369–379.
15. Bruno FF, Nagarajan R, Roy S, Kumar J, Samuelson LA (2003) Biomimetic synthesis of water soluble conductive polypyrrole and poly (3,4-ethylenedioxythiophene). J Macromol Sci A 40(12): 1327–1333.
16. Zhao F, Shi Y, Pan L, Yu G (2017) Multifunctional nanostructured conductive polymer gels: synthesis, properties, and applications. Acc Chem Res 50(7): 1734–1743.

17. Molino BZ, Fukuda J, Molino PJ, Wallace GG (2021) Redox polymers for tissue engineering. Front Med Technol 3:669763. doi: 10.3389/fmedt.2021.669763.

18. Park Y, Jung J, Chang M (2019) Research progress on conducting polymer-based biomedical applications. Appl Sci 9: 1070. doi: 10.3390/app9061070.

19. Lin J, Tang Q, Wu J, Li Q (2010) A multifunctional hydrogel with high-conductivity, pH-responsive, and release properties from polyacrylate/polyptrrole. J Appl Polym Sci 116: 1376–1383.

20. Mawad D, Lauto A, Wallace GG (2016) Conductive polymer hydrogels. In: Kalia S (Eds.) Polymeric hydrogels as smart biomaterials. Springer International Publishing, Cham: 19–44.

21. Molino BZ, Fukuda J, Molino PJ, Wallace GG (2021) Redox polymers for tissue engineering. Front Med Technol 3: 669763. doi: 10.3389/fmedt.2021.669763.

22. Krukiewicz K, Britton J, Więcławska D, Skorupa M, Fernandez J, Sarasua JR, Biggs MJ (2021) Electrical percolation in extrinsically conducting, poly (ε-decalactone) composite neural interface materials. Sci Rep 11(1): 1–10.

23. Nik Md NoordinKahar, NNF, Osman AF, Alosime E, Arsat N, Mohammad Azman NA, Syamsir A, Itam Z, Abdul Hamid ZA (2021) The versatility of polymeric materials as self-healing agents for various types of applications: a review. Polymers 13(8): 1194.

24. Xu H, Holzwarth JM, Yan Y, Xu P, Zheng H, Yin Y, Li S, Ma PX (2014) Conductive PPY/PDLLA conduit for peripheral nerve regeneration. Biomaterials 35(1): 225–235.

25. Mostafavi E, Medina-Cruz D, Kalantari K, Taymoori A, Soltantabar P, Webster TJ (2020) Electroconductive nanobiomaterials for tissue engineering and regenerative medicine. Bioelectricity 2(2): 120–149.

26. Saberi A, Jabbari F, Zarrintaj P, Saeb MR, Mozafari M (2019) Electrically conductive materials: opportunities and challenges in tissue engineering. Biomolecules 9(9): 48.

27. Guo B, Ma PX (2018) Conducting polymers for tissue engineering. Biomacromolecules 19(6): 1764–1782.

28. Higgins MJ, Molino PJ, Yue Z, Wallace GG (2012) Organic conducting polymer–protein interactions. Chem Mater 24(5): 828–839.

29. Wu B, Cao B, Taylor IM, Woeppel K, Cui XT (2019) Facile synthesis of a 3,4-ethylene-dioxythiophene (EDOT) derivative for ease of bio-functionalization of the conducting polymer PEDOT. Front Chem 7: 178. doi: 10.3389/fchem.2019.00178.

30. Bolin MH, Svennersten K, Nilsson D, Sawatdee A, Jager EWH, Richter-Dahlfors A, Berggren M (2009) Active control of epithelial cell-density gradients grown along the channel of an organic electrochemical transistor. Adv Mater 21(43): 4379–4382.

31. Wan AMD, Schur RM, Ober CK, Fischbach C, Gourdon D, Malliaras GG (2012) Electrical control of protein conformation. Adv Mater 24(18): 2501–2505.

32. Jin L, Feng ZQ, Zhu ML, Wang T, Leach MK, Jiang Q (2012) A novel fluffy conductive polypyrrole nano-layer coated PLLA fibrous scaffold for nerve tissue engineering. J Biomed Nanotechnol 8(5): 779–785.

33. Wu Y, Feng S, Zan X, Lin Y, Wang Q (2015) Aligned electroactive TMV nanofibers as enabling scaffold for neural tissue engineering. Biomacromolecules 16(11): 3466–3472.

34. Boni R, Ali A, Shavandi A, Clarkson AN (2018) Current and novel polymeric biomaterials for neural tissue engineering. J Biomed Sci 25(1): 1–21.

35. Xue Y, Dånmark S, Xing Z, Arvidson K, Albertsson AC, Hellem S, Finne-Wistrand A, Mustafa K (2010) Growth and differentiation of bone marrow stromal cells on biodegradable polymer scaffolds: an in vitro study. J Biomed Mater Res A 95(4): 1244–1251.

36. Coelho MJ, Fernandes MH (2000) Human bone cell cultures in biocompatibility testing. Part II: Effect of ascorbic acid, β-glycerophosphate and dexamethasone on osteoblastic differentiation. Biomaterials 21(11): 1095–1102.

37. Wang W, Hou Y, Martinez D, Kurniawan D, Chiang WH, Bartolo P (2020) Carbon nanomaterials for electro-active structures: a review. Polymers 12(12): 2946.

38. Shahini A, Yazdimamaghani M, Walker KJ, Eastman MA, Hatami-Marbini H, Smith BJ, Ricci JL, Madihally SV, Vashaee D, Tayebi L (2014) 3D conductive nanocomposite scaffold for bone tissue engineering. Int J Nanomed 9: 167–181.

39. Marzocchi M, Gualandi I, Calienni M, Zironi I, Scavetta E, Castellani G, Fraboni B (2015) Physical and electrochemical properties of PEDOT: PSS as a tool for controlling cell growth. ACS Appl Mater Interfaces 7: 17993–18003.

40. Solazzo M, O'Brien FJ, Nicolosi V, Monaghan MG (2019) The rationale and emergence of electroconductive biomaterial scaffolds in cardiac tissue engineering. APL Bioeng 3(4): c041501.

41. Roshandel M, Dorkoosh F (2021) Cardiac tissue engineering, biomaterial scaffolds, and their fabrication techniques. Polym Adv Technol 32: 2290–2305.

42. Eisner DA, Caldwell JL, Kistamás K, Trafford AW (2017) Calcium and excitation-contraction coupling in the heart. Circ Res 121(2): 181–195.

43. Spearman BS, Hodge AJ, Porter JL, Hardy JG, Davis ZD, Xu T, Zhang X, Schmidt CE, Hamilton MC, Lipke EA (2015) Conductive interpenetrating networks of polypyrrole and polycaprolactone encourage electrophysiological development of cardiac cells. Acta Biomater 28: 109–120.

44. Mohammadi Nasr S, Rabiee N, Hajebi S, Ahmadi S, Fatahi Y, Hosseini M, Bagherzadeh M, Ghadiri AM, Rabiee M, Jajarmi V, Webster TJ (2020) Biodegradable nanopolymers in cardiac tissue engineering: from concept towards nanomedicine. Int J Nanomedicine 15: 4205–4224.

45. Teplenin A, Krasheninnikova A, Agladze N, Sidoruk K, Agapova O, Agapov I, Bogush V, Agladze K (2015) Functional analysis of the engineered cardiac tissue grown on recombinant spidroin fiber meshes. PLoS One 10(3): e0121155.

46. Ostrovidov S, Hosseini V, Ahadian S, Fujie T, Parthiban SP, Ramalingam M, Bae H, Kaji H, Khademhosseini A (2014) Skeletal muscle tissue engineering: methods to form skeletal myotubes and their applications. Tissue Eng Part B Rev 20(5): 403–436.

47. Jun I, Jeong S, Shin H (2009) The stimulation of myoblast differentiation by electrically conductive sub-micron fibers. Biomaterials 30: 238–247.

48. Ford SJ, Chandra M (2012) The effects of slow skeletal troponin I expression in the murine myocardium are influenced by development-related shifts in myosin heavy chain isoform. J Physiol 590(23): 6047–6063.

49. Strovidov S, Ebrahimi M, Bae H, Nguyen HK, Salehi S, Kim SB, Kumatani A, Matsue T, Shi X, Nakajima K, Hidema S, Osanai M, Khademhosseini A (2017) Gelatin-polyaniline composite nanofibers enhanced excitation-contraction coupling system maturation in myotubes. ACS Appl Mater Interfaces 9(49): 42444–42458.

50. Mir M, Ali MN, Barakullah A, Gulzar A, Arshad M, Fatima S, Asad M (2018) Synthetic polymeric biomaterials for wound healing: a review. Prog Biomater 7(1): 1–21.

51. Korupalli C, Li H, Nguyen N, Mi FL, Chang Y, Lin YJ, Sung HW (2021) Conductive materials for healing wounds: their incorporation in electroactive wound dressings, characterization, and perspectives. Adv Healthcare Mater 10: 2001384.

52. Fadilah NIM, Rahman MBA, Yusof LM, Mustapha NM, Ahmad H (2021) The therapeutic effect and in vivo assessment of palmitoyl-GDPH on the wound healing process. Pharmaceutics 13:193.

53. Gao C, Zhang L, Wang J, Jin M, Tang Q, Chen Z, Cheng Y, Yang R, Zhao G (2021) Electrospun nanofibers promote wound healing; theories, techniques and perspectives. J Mater Chem B 9: 3106–3130.

54. Moutsatsou P, Coopman K, Georgiadou S (2017) Biocompatibility assessment of conducting PANI/chitosan nanofibers for wound healing applications. Polymers (Basel) 9(12): 687.

55. Zhao X, Wu H, Guo B, Dong R, Qiu Y, Ma P (2017) Antibacterial anti-oxidant electroactive injectable hydrogel as self-healing wound dressing with hemostasis and adhesiveness for cutaneous wound healing. Biomaterials 122: 34–47.

56. Yamakawa S, Hayashida K (2019) Advances in surgical applications of growth factors for wound healing. Burn Trauma 7: 1–13. 41038-019-0148.

12 Conducting Polymer-Based Nanomaterials for Tissue Engineering

Murugan Prasathkumar,[1] Chenthamara Dhrisya,[1] Salim Anisha,[1] Robert Becky,[1] and Subramaniam Sadhasivam[1,2]

[1]Biomaterials and Bioprocess Laboratory, Department of Microbial Biotechnology, Bharathiar University, Coimbatore, India

[2]Department of Extension and Career Guidance, Bharathiar University, Coimbatore, India

CONTENTS

DOI: 10.1201/9781003205418-12

12.1 INTRODUCTION

Biomaterials are generally implantable or inserted into the human body for reconstructing or repairing damaged tissues or organs. They are synthesized by diverse molecules to directly interact with human cells, organs, and tissues for diverse medical applications [1]. The biomaterials are designed by a variety of polymers, copolymers, composite, and different cell sources that are capable of playing the ultimate role in the promotion of cell and materials interaction [2]. Tissue engineering is a multidisciplinary and interdisciplinary research area that aims to develop novel functional biomaterials that reconstruct, improve, and maintain tissue or organ function by combining polymers, cells, and biological molecules. A variety of polymers are used to design effective tissue engineering biomaterials with improved physical and biological properties such as hydrophilicity, electrical conductivity, thermal stability, mechanical strength, biocompatibility, cellular response, antimicrobial, and tissue regeneration [3]. Among the polymers, conducting polymers (CPs) such as polypyrrole (PPy), polyaniline (PANI), and polythiophene (PT) exhibit electrical conductivity, improved cell proliferation, and differentiation, and also serve as a template for the regeneration of functional organs and tissues. CPs are blended with other flexible and degradable polymers such as poly(lactic acid) (PLA), polycaprolactone (PCL), poly(lactic-co-glycolic acid) (PLGA) to improve and adjust the biological and mechanical properties. Therefore, the CPs-based biomaterials are promising candidates for biomaterials for tissue engineering applications to regenerate or reconstruct damaged tissues or organs [4]. Further, this chapter aims to detail the efficacy of CPs in skin, bone, nerve, and cardiac tissue engineering applications.

12.2 CONDUCTING POLYMERS AND THEIR DERIVATIVES

CPs have achieved much significance due to their unique electrical and optical properties akin to those of inorganic semiconductors. For the past several decades, CPs (or "synthetic metals") was considered to be both insoluble and intractable. In 1977, Shirakawa et al. discovered metallic properties of polyacetylene and followed by this discovery a variety of important CPs have been investigated constantly, such as PT, PPy, poly(3,4-ethylenedioxythiophene) (PEDOT), PANI, *trans*-polyacetylene, and poly(*p*-phenylenevinylene) (PPV). Figure 12.1 shows the chemical structures of CPs and their derivatives. Generally, a CP contains alternating single (σ) and double (π) bonds, and these delocalized π-electrons render them intrinsic electrochemical, optical, and electrical properties. The physical properties of CPs are highly dependent on the parameters such as intra- and interchain interactions, degree of crystallinity, and conjugation length. One of the significant properties of CP is electrical conductivity by several orders of magnitude of doping and it ranges from 10^{-3} to 10^{3} S/cm for common doped CPs whereas intrinsic conjugated polymers without doping are in the range of 10^{-9}–10^{-6} S/cm. Once doped the conductivity of conjugated polymers observed an increase of six- to ninefolds. The conductivity of CP depends on the degree of ordering of the polymer main chain in the solid film and the doping degree. Thus, the doping degree is directly related to the charge carrier concentration on the conjugated polymer.

Generally, CPs have synthesized either by chemical or electrochemical methods. The chemical synthesis will take place through the oxidation or reduction of monomers and polymerization of corresponding monomers. The condensation polymerization occurs by the loss of simple molecules, often water or hydrochloric acid. The addition polymerization is based on chain growth (e.g. radical, cation, and anion polymerizations). The electrochemical polymerization is based on solvent, three-electrode configuration (counter, working, and reference electrodes) in a solution of the monomer, and an electrolyte (dopant). The morphology, polymer growth rate, and properties of polymer films deposited on the electrode are highly dependent on the electrolyte. The main

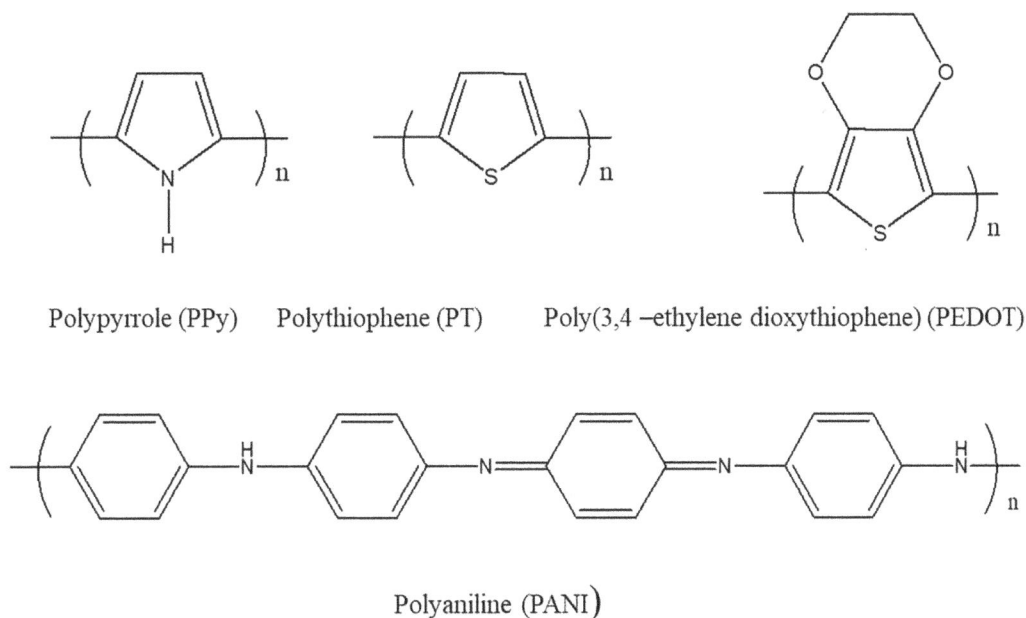

Polypyrrole (PPy) Polythiophene (PT) Poly(3,4 –ethylene dioxythiophene) (PEDOT)

Polyaniline (PANI)

FIGURE 12.1 Chemical structure of the important CPs.

difference between chemical and electrochemical polymerization is that thick films or powders can be synthesized with chemical polymerization whereas thin films (20 nm) are synthesized by electrochemical techniques. The standard CPs (PPy, PT, PEDOT, and PANI) can be polymerized by both techniques though, the majority of the novel CPs with modified monomers are only amenable to chemical polymerization [5].

12.3 CONDUCTING POLYMER–BASED BIOMATERIALS

Electrically responsive organs such as the heart, brain, and skeletal muscles and active cells (fibroblasts, neural crest cells, and myoblasts) have been widely explored by interfacing metallic or semiconductor electrodes to provide electrical stimulation. Hence, it received significant attention in regenerative medicine because of cellular signaling, migration, development, division, wound healing, and muscle contraction. Electroactive CPs are one of the smart biomaterials that permit the direct delivery of electrical, electrochemical, and electromechanical stimulation to cells. Especially, the conducting biomaterials have been exhibited to improve the differentiation and proliferation behavior of electrical stimuli-responsive cells, including muscle cells, neuron cells, and bone cells [6]. The addition of electroconductive materials is reported to enhance various cellular activities of electrically excitable cells such as fibroblasts, keratinocytes, nerve, bone, muscle, cardiac, and mesenchymal stem cells. The usage of CPs alone as a wound dressing material is not recommended because of their high fragility and lower processability. Due to these reasons, it is blended with non-CPs such as chitosan, PCL, polyurethane (PU), cellulose, and many others altering mechanical properties and electrical conductivity and used in electroactive wound dressings. Researchers have developed different types of conducting biomaterials, such as hydrogels, films, aerogels, scaffolds, and nanofibers [7]. The list of widely used CPs to develop conducting biomaterials for biomedical applications is represented in Table 12.1.

TABLE 12.1

The Properties and Biomedical Applications of CPs

S. No	Conducting Polymer	Synthesis	Conductivity	Properties	Applications
1.	Polypyrrole	Electrochemical and chemical synthesis	High conductivity (up to 160 S/cm) when doped with iodine	Electrical conductivity, improved cell proliferation, adhesion and regeneration, high surface area, flexible, lightweight	Drug delivery, tissue engineering, bio-, chemical-, immuno-, and DNA-based sensors, neural implants, bioactuator
2.	Polythiophenes		Good electrical conductivity and optical property	Electrical conductivity, improved cell proliferation, and regeneration	Neural implants
3.	Polyaniline		High conductivity up to 100 S/cm	Electrical conductivity, tunable morphology, redox properties, environmental stability, antimicrobial, improved cell proliferation regeneration, and adhesion	Chemosensor, biosensor, bioactuator, tissue engineering
4.	Poly(3,4-ethylenedioxythiophene)		Conductivity up to 210 S/cm	High charge capacity density and low electrode impedance	Neural implants, tissue engineering

12.3.1 Hydrogels

The CP hydrogels are fabricated by cross-linking the CPs by crystallization, chemical reaction, and aggregation to hold a large amount of water. Natural polysaccharides viz., alginate, cellulose, chitosan, starch, and their derivatives have been adopted to synthesize conducting hydrogels [8]. The CP hydrogel offers high electrical conductivity, interconnected three-dimensional (3D) networks, good mechanical properties, and specific surface area that facilitate ion diffusion compared to bulk materials. The polymerization and subsequent gelation properties of PPy, PANI, and PEDOT lead to the formation of CP hydrogels with outstanding properties without any additives. In addition, CPs exhibit different intermolecular interactions like hydrophobic interactions, crystallization forces, ionic interactions, hydrogen bonding, and π–π stacking interactions. These interactions are responsible for the formation of an attractive force between polymer chains and other surrounding molecules. Moreover, conducting hydrogels are characterized by redox activity, ionic conductivity, and mixed electronic conductivity. Recently, the conducting hydrogels have received great attention in biosensors, drug delivery systems, and tissue engineering applications due to their plasticity, porosity, mechanical integrity, electroactivity, biocompatibility, elasticity, and fine-tuning the preferred mechanical softness [9].

12.3.2 Films

Researchers have explored CP's unusual electronic properties for a range of biomedical applications as they exhibit high electron affinities, low ionization potential, and low energy optical transmission due to the presence of a conjugated π-electron backbone. These individual properties make CPs capable of applications as film materials. The CPs such as PPy, PANI, and PT are widely used to synthesis in the form of conducting film materials on the surface of an electrode either chemically or electrochemically due to their capability of proliferation, adhesion, and differentiation of diverse cells. These polymers can blend with other degradable materials such as PCL, PLA, and PLGA to adjust the mechanical properties [4]. For instance, PPy doped with *p*-toluene sulfonate can modulate the cellular response, whereas fibronectin-coated PPy film material can support cell growth, cell spreading, and control cell function. Similarly, PPy-coated hyaluronic acid bilayer film promoted the vascularization process [10]. PPy–cellulose biodegradable CP films have exhibited great potential in drug delivery systems and tissue engineering applications due to their porous structure, drug-releasing efficacy, 1.58 S/cm conductivity, high surface area, flexible and lightweight [4].

12.3.3 Aerogels

Aerogels are porous materials with attractive properties such as high specific surface area, low thermal conductivity, high pore volume, and high porosity. Their potential biomedical applications include wound care, drug delivery system, tissue engineering. Alginate, cellulose, gellan gum, and DNA have been widely used as a template for *in situ* polymerization with CPs. The nanoporous cellulose gels–PPy composite aerogel can provide sufficient electrical conductivity and no cytotoxic effects against nerve cells [11]. PANI aerogels are notable conductive polymeric biomaterials that have been received good attention due to their low cost, tunable morphology, redox properties, and environmental stability, but lack of strong chemical or physical interactions. The PANI cross-linked with pectin aerogel has exhibited considerable electrical conductivity, hierarchical pores, high mechanical integrity, and self-supported 3D nanoporous network structures with high surface areas [12].

12.3.4 Scaffolds

The usage of CPs in tissue engineering and wound management systems presents a novel possibility for improved tissue recovery, better antibacterial effects, and controlled drug delivery. Through their expanded electric conductivity, CPs can facilitate the utility of electrical stimulation directly to the injured tissue surface for faster regeneration [13]. The CP-based scaffolds or wound dressing materials provide the opportunity for electrical stimulation to be routinely delivered more uniformly to the injured tissue. Moreover, some researchers have reported that usage of CPs alone, even without the external electrical stimuli, resulted in improved organ reconstruction and antimicrobial effects. For example, PANI has shown potential antimicrobial effects against *Escherichia coli*, *Staphylococcus aureus*, and *Pseudomonas aeruginosa* [13]. Skin is the largest organ responsive to electrical signals and has a conductivity from 2.6 to 1×10^{-4} mS/cm, depending on its components. Studies have reported that the electroconductive scaffolds offer good conductivity to natural skin to promote wound healing and functional recovery.

12.3.5 Nanofibers

Nanofibers can be synthesized by different techniques, like electrospinning, biological synthesis, interfacial polymerization, phase separation, self-assembly, and electrically conductive materials. The main advantages of nanofibers are that they mimic the fibrous element of the natural extracellular matrix (ECM), maintain the moisture environment, absorb the exudates, act as a protective barrier against penetration of pathogens, and improve skin regeneration [14]. In recent years, conducting polymeric nanofibers has involved significant concern due to their inexpensive, flexible, superior optoelectronic properties, high electrical conductivity, high electrochemical activity, and large surface area. In addition, CP nanofibers offer unique properties like high charge carrier collection, possible surface modification, transport in the axial orientation, good flexibility in surface functionality, and excellent mechanical integrity relative to their microscale counterparts [15].

12.4 BIOMEDICAL APPLICATIONS OF CONDUCTING POLYMERS

The properties such as flexibility, non-toxicity, and availability make CPs an ideal candidate in various fields of science. Some of the commonly used biomedical applications of CPs include biosensors, neural implants, drug delivery systems, bioactuators, and tissue engineering (Figure 12.2).

12.4.1 BIOSENSORS

Biosensors are a multifaceted tool for monitoring a range of analytes in the sectors like environment, agriculture, healthcare, food, environment, and biosecurity. Pyrrole-based bio- or immunosensors have received significant attraction because they can function stably under ambient environments. Among the CPs, PPy synthesized by oxidative polymerization is biocompatible and protects electrodes from fouling and rarely causes any significant disturbance of the working environment. In some cases, they form permselective films to exclude endogenous electrochemically active interferents. Extensive research has been conducted with PPy for bio-, chemical-, immuno-, and DNA-based sensors development. The first commercial CP PANI is synthesized by the chemical or electrochemical oxidative polymerization of aniline. The conductive nature of PANI can be reduced by raising the pH level (>4) couple to couple with other CP, graphene, or carbon nanotubes (CNTs). PANI has been widely employed in various chemosensing and biosensing platforms for the detection of cancer biomarkers, infectious diseases, heavy metals, acrylamides, drugs, and sulfites [16].

12.4.2 NEURAL IMPLANTS

For the modulation of neural behavior in neuronal diseases, neural implants are used for the past few years. It stimulates the nervous system (motor or sensory function) with the support of implanted

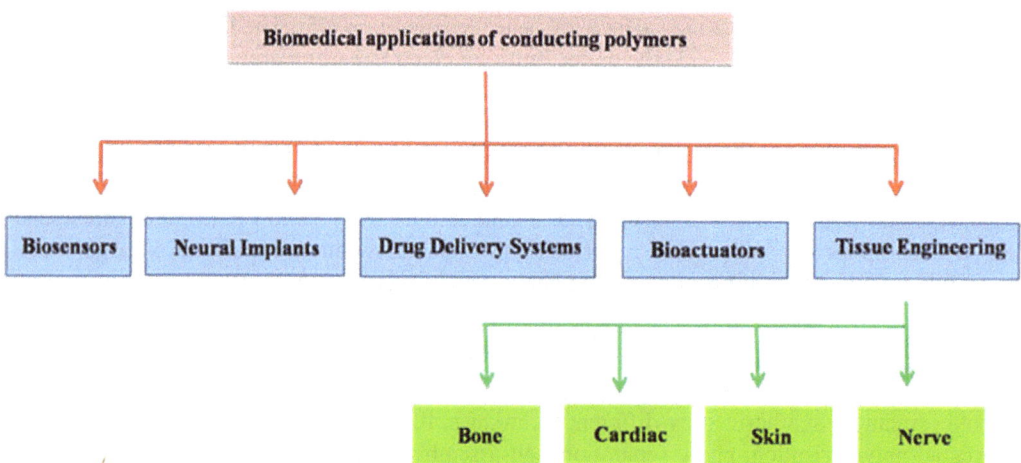

FIGURE 12.2 Biomedical applications of conducting polymers.

electrical circuitry. The success of a neural implant is based on the interface between the material and tissue. At present, CPs such as PPy and PT derivative PEDOT have been widely used to improve the properties of neural interfaces. CP coatings play a vital role in enhancing the charge transfer characteristics of conventional metal electrodes. Since conventional electrodes are not conducive to tissue integration, the biological interaction can eliminate the fluid gap through close contact among the electrode and tissue. The CPs modified with conventional gold, platinum, or alloys of these metals and iridium oxide electrodes reported a long-term performance of neural implants such as the vision prosthesis, cochlear implant for hearing loss, neural recording electrodes, and neural regeneration devices [17].

12.4.3 DRUG DELIVERY SYSTEMS

The drug delivery technique should assure the transport of the specific pharmaceutical compound to its target site to attain the desired therapeutic effect. Controlled-release drug delivery systems exhibit advantages over conventional therapies by maintaining drug concentrations at active levels while simultaneously improving patient adherence. The controlled-release implantable technologies are unsuitable for certain medications as they release the drug at an inflexible preprogrammed rate. However, a CP-based implantable system (PEDOT and *N*-methylpyrrole) would allow changes in drug release rate even after administration. The high charge capacity density and low electrode impedance make PEDOT an efficient candidate for a controlled release of the anti-inflammatory drug dexamethasone. For this, nanofibers loaded with dexamethasone were coated with PEDOT and then covered with an alginate hydrogel. This system was then assembled onto neural electrodes, with the intended application of reducing inflammation following implantation [18].

12.4.4 BIOACTUATORS

CP scaffolds are commonly used in bioactuator devices for the fabrication of artificial muscles. In the triple-layer arrangement, a nonconductive material is placed in between the layers of CP. As the current passes through the CP film, oxidation and reduction take place. Due to the inflow of dopant ions, the oxidized film undergoes expansion, whereas the expulsion of dopant ions in reduced film shrinks. This simultaneous expansion and contraction mimic the effect of muscles in biological systems. The properties such as high strength, electrically controllable, low voltage actuation, finite strain, ability to work with body fluids, and physiological temperature make CP actuator an ideal candidate for artificial muscles. PPy, PANI, and PPy–PANI composites and composites of these polymers with CNTs (e.g. PANI–CNTs and PANI–CNTs–PPy) are the well-studied actuators. Amidst this, PPy–PANI reported the highest work per cycle, which is advantageous for effective mechanical properties [19].

12.4.5 TISSUE ENGINEERING

The conductive nature of CPs allows cells or tissues cultured on them to be stimulated by electrical signals and hence used as bioactive scaffolds for tissue regeneration. But, the poor processability, as well as mechanical brittleness, limits the application. Thus, conductive polymeric composites have been developed using CPs and degradable polymers. Electrospinning is a method widely used for the formation of polymer nanofibers for various biomedical applications. The polymer nanofibers such as polydioxanone (PDO) and PCL are used in cardiac tissue replacement [20].

12.5 SKIN TISSUE ENGINEERING

Skin is the largest organ in the human body and covers a distance of 1.8–2.0 m^2 and is composed of epidermis, dermis, and hypodermis. The epidermis contains stem cells, and it will induce self-regeneration once the wound occurred. A severe wound or burns result in a chronic or nonhealing wound with delayed recovery, and the loss of full skin more than 4-cm diameter needs skin grafting. Surgical options are limited due to the lack of donors, and foreign grafts often exert immune rejection and infection. Hence, tissue-engineered skin substitutes are recommended to overcome the drawbacks. The skin was the first organ that went from laboratory research to patient care among the various engineered organs. Tissue engineering is rapidly evolving to find a better skin substitute that can be efficiently used for clinical applications [21].

The process of wound healing is complex and involves various stages such as hemostasis, inflammation, proliferation, and remodeling. The application of wound dressing facilitates maintaining moist conditions, promotes speedy re-epithelization and recovery. Advanced material formulation viz., films, foams, hydrogels, hydrocolloids, and hydrofibers were used to promote wound healing. Natural polymers, such as chitosan, fibrin, elastin, gelatin, and synthetic polymers (PCL and PLA), were generally used to fulfill the material fabrication. In current years, CPs are widely used as wound dressing materials and enable faster recovery due to their electrochemical properties [13]. The integration of CPs into wound dressing enables improved antibacterial activity and aids the controlled release of drugs.

12.5.1 Conducting Polymer–based Hydrogel for Wound Healing

Chitosan offers an excellent antibacterial and hemostatic activity; further, it renders many of its surface functional groups for cross-linking reactions. However, the antibacterial activity is limited at nonacidic conditions due to its low solubility. In the same way, CP such as PANI also showed poor solubility. A self-healing electroactive hydrogels composed of quaternized chitosan-*g*-PANI (QCSP) and benzaldehyde group functionalized poly(ethylene glycol)-*co*-poly(glycerol sebacate) (PEGS-FA) was reported to improve the *in vivo* wound healing. The as-prepared hydrogel was better than commercial wound dressing as it increased the vascular endothelial growth factor (VEGF), epidermal growth factor (EGF), transforming growth factor-β (TGF-β) values and also promotes granulation tissue thickness. The enhancement of growth factors may be due to the movement of electrical signals from conductive hydrogels to the wound site and the establishment of cellular metabolism. The electroactive compounds would act on the calcium/calmodulin pathway to increase the calcium level and thereby activate the cytoskeletal calmodulin to enhance cell proliferation [22]. Wound healing is also promoted via ionic interaction of neutrophils, macrophages, fibroblasts, and epidermal cells due to electronic stimulation of CPs. A thermo-responsive gelatin hydrogel with polyethyleneimine (PEI)–PPy nanocomplex has shown excellent photothermal performance and hyperthermic response. The incorporation of such nano-hydrogels resulted in complete wound closure in the Wistar rat model (wound size 20 mm) within 21 days (Figure 12.3) [23].

12.5.2 Conducting Polymer–based Scaffold for Wound Healing

Tissue engineering scaffolds offer a platform for cell growth, adhesion, and migration. A scaffold should offer ECM mimic environment as well as higher biocompatibility and biodegradability to support the growth of neural and muscle cells. Chitin is often used in tissue engineering applications because it can easily convert into many forms such as fibers, membranes, hydrogels, sponges, etc. Biocompatible chitin/PANI (Chi/PANI) scaffold made by the electrospinning method showed excellent potential for tissue engineering application. The study reports the electrical conductivity of nanofiber scaffold was >91% in comparison with random nanofibers. The cell viability of aligned scaffolds was 2.1 folds more than that of pure chitin [24]. The incorporation of PANI to the chitosan gelatin nanohydroxyapatite (nHA) scaffold has improved the water retention capacity, biodegradation, and biomineralization. The nontoxicity of the scaffold was confirmed

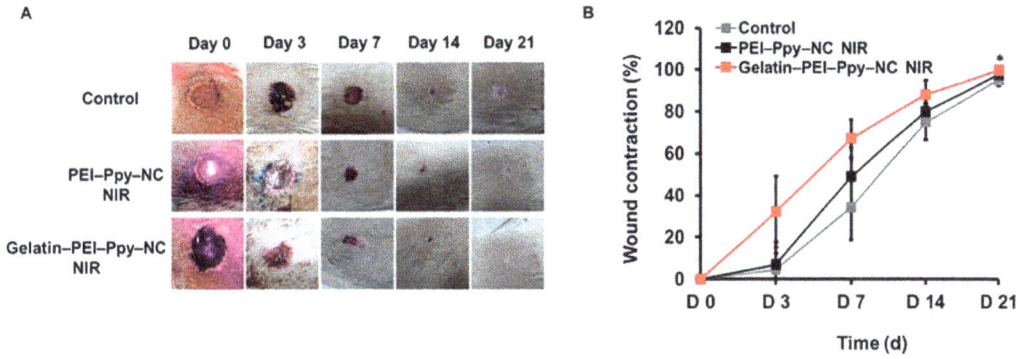

FIGURE 12.3 (a) Periodical wound healing evaluation (full-thickness wound in the Wistar rat model): macroscopic images of the wound site and wound area of the control and two experimental groups (polyethyleneimine–polypyrrole nanocomplex subject to near-infrared light [PEI-Ppy-NCNIR] and gelatin hydrogel containing polyethylenimine–polypyrrole nanocomplex subject to near-infrared light [Gelatin–PEIPpy–NC-NIR]) at different time points (day (d) 0, 3, 7, 14, and 21) (n = 3); (b) wound contraction (%) at various stages of wound healing and complete wound closure from day 0 to day 21 (n = 3) (p < 0.05). (Adapted with permission from Ref. [23]. Copyright (2018) Copyright the Authors, some rights reserved; exclusive licensee [MDPI]. Distributed under a Creative Commons Attribution License 4.0 (CC BY) https:// creativecommons.org/licenses/by/4.0/.)

in BLV-FLK and dental pulp stem cell (DPSC) cell lines [25]. A novel 3D, highly conductive, porous, and elastic scaffold was fabricated using polyester, PANI, and poly(ε-caprolactone). The as-synthesized scaffold has shown appropriate conductivity and mechanical properties that suit many tissue engineering applications [26].

12.5.3 Conducting Polymer–based Composite Biomaterials

Several studies reported that the incorporation of CPs also facilitated the controlled delivery of drugs. A polymeric film was prepared by sodium alginate/gelatin/hyaluronic acid/reduced graphene oxide (SAlg/Gel/HA/RGO) and the RGO acted as an electroactive material in the composite film. The wound-healing potential of the film was further improved by ibuprofen loading as it promotes wound-healing processes. The RGO-based film showed moderate release of ibuprofen i.e., 0.94-mg drug/g dry film after 48 h, whereas RGO lacking SAlg/Gel/HA film showed rapid release of ibuprofen (1.35 mg/g). The adequate level of water and oxygen permeability along with high biocompatibility supports SAlg/Gel/HA/RGO film in tissue engineering applications [27]. In another study, PU and CNTs have been cited as smart dressing materials for biomedical applications. The major drawback of this composite is that at high concentrations, CNTs cause direct deposition and cytotoxic effects. The cytotoxicity can be minimized by coating it with PPy since it lowers the direct contact of CNTs with the skin [13]. Most of the natural fibers are hydrophilic which is not ideal for adequate ECM protein adsorption that in turn regulates cell adhesion and spreading. To overcome this drawback and to improve cell adhesion, natural fibers can be coated with hydrophobic materials such as conductive materials. Silk fibroin (SF) polymer coated with PPy/PANI showed higher cell adhesion than SF fibers. Hence, hydrophilic and hydrophobic combined materials suggested its excellence in tissue engineering applications [28]. In addition to the development of biomaterial for wound healing, 3D-skin mapping technology also makes use of CPs. The conductive skin mapping technology is currently at the stage of evaluation to check its possibility in wound healing–related applications [13].

12.6   BONE TISSUE ENGINEERING

The bone structure is highly complex and can be viewed as an open-cell composite material consisting of ECM proteins, growth factors, and mineral calcium. 10% of the total bone volume consists of osteoblasts, osteocytes, and osteoclasts. Preosteoblasts are derived from mesenchymal stem cells and are induced by growth factors like bone morphogenetic proteins (BMPs), TGF-β, fibroblast growth factor (FGF), insulin-derived growth factor (IGF), platelet-derived growth factor (PDGF), and interleukins. The stages of osteoblasts differentiation are classified into (1) cell proliferation, (2) matrix maturation, and (3) matrix mineralization. During the proliferation stage, osteoblasts would secrete ECM protein that forms a non-mineralized bone matrix or osteoid that leads to a complex, stable cross-linked structure. Among the various biomedical applications, the most forthcoming conductive biomaterials are mainly concerned with bone tissue designing. The ideal biomaterials used for bone regeneration should offer good mechanical properties and mimic the natural bone ECM. Such biomaterials would provide ideal environmental conditions to support cell adhesion and differentiation.

12.6.1   BONE TISSUE ENGINEERING BIOMATERIALS

Bone tissue engineering applications generally use scaffolds, implants, hydrogels, and films. Electric stimulations have been widely used to treat orthopedics to cure fractures in bones. Electrospinning methods hold a special place in bone tissue engineering due to their ability to produce 3D fibrous structures. A bilayer hybrid scaffold was prepared by the electrospun method using gelatin (Gel) modified with calcium phosphate nanoparticles (SG5) and PCL modified with osteogenon (Osteo) drug. Then the CP PANI was deposited on the surface of the scaffold of PCL/ Osteo and Gel/SG5 by inkjet printing. These hybrid scaffolds showed improved bioactivity as well as osteoblast cell proliferation [29]. A nanofibrous scaffold was prepared by electrospinning using PU and PANI to improve the bioactive properties and hydrophilic nature. The resultant scaffold was further coated with poly(vinyl alcohol) (PVA) and 3-glycidoxypropyltrimethoxysilane (GPTMS) to enhance the biomineralization, cell proliferation, alkaline phosphatase (ALP), and osteogenic expression [30]. ALP plays a vital role in bone formation by excising phosphate groups of the macromolecules that in turn enhance the extracellular mineralization. An electrically conductive and injectable, interpenetrating polymer network (IPN) hydrogel has received substantial attention in recent times for its conductivity, pore size, swelling ratio, good mechanical properties, and biocompatibility.

The piezoelectricity of the bone tissue helps to convert electronic stimulus into mechanical energy. Hence, the piezoelectric smart bone biomaterials are widely used for tissue repair and regeneration. An electroactive composite scaffold was fabricated with the piezoelectric material, polyvinylidene fluoride (PVDF), and CP PANI at the extremely low-frequency electromagnetic field (ELF, 0–300 Hz) to study the osteogenic differentiation in DPSCs. The as-synthesized PVDF-PANI scaffold exhibited higher ALP activity, calcium content, osteogenic gene (Runx2, osteonectin, and osteocalcin), and osteocalcin protein expression. The study suggested that synergistic combination of low-frequency electromagnetic field and PVDF–PANI facilitates the regeneration process [31]. Artificial implants and orthopedic implantations often use titanium (Ti) and its alloy for higher biocompatibility, mechanical properties, and corrosion resistance. The major drawback of these implants is it causes bacterial infections since it is biologically inert and lacks non-integration with bone tissue. Hence, titanium nanotube (TNT) surfaces are modified in a multilayer pattern with gelatin- and chitosan-grafted PANI to overcome these drawbacks. The surface coating with PANI considerably improved antibacterial activity due to the presence of amine groups that will combine with negatively charged bacterial membrane and consequently results in the release of intracellular components such as DNA and RNA. PANI was grafted in chitosan and deposited over the surface of TNTs via layer-by-layer (LBL) techniques. The TNTs-LBL (PANI) showed higher cytocompatibility, antibacterial activity and also supported osteoblast adhesion, proliferation, and differentiation [32]. A nanocomposite (NC) film (Pd/PPy/RGO) prepared by PPy and palladium (Pd) nanoparticles with RGO has reported a higher antibacterial activity and enhanced Saos-2 osteo cells proliferation [33].

12.7 NERVE TISSUE ENGINEERING

The nervous system comprising the central nervous system (CNS) and peripheral nervous system (PNS) regulates the sensory and motor functions of the body. Neural injuries that manifest in stroke, trauma, and other neurological diseases emerge as a critical medical emergency affecting the social and occupational quality of life. The anatomical complexity of neurons and the nervous system presents distinctive challenges in treating neurodegenerative disorders clinically. In the current scenario, conventional approaches like autologous nerve grafting are used for treating the damaged neural tissues of PNS. But the major drawbacks are the mismatch, risk of infections, potential neuroma formation, and limited availability of autologous donor tissues. All these reasons made the CNS treatment highly complex. Neural tissue engineering is an effective way to attain neural recovery, especially with the use of nerve-compatible materials (hydrogel, scaffolds, and films) via electrical stimulation. The porous, elastic, and moist nature with the chemical and biological cues provide a microenvironment for the neural cells to attach, proliferate, and differentiate efficiently. This neural supportive biomaterials support the growth of neuritis, transmit biological cues to lead the axonal growth cone to the distal stump [34]. Advancement in biomaterial research with CPs for electroactive tissues has attained extensive interest in the recent year due to their beneficial outcomes.

12.7.1 EFFECT OF ELECTRICAL STIMULATION IN NERVE REGENERATION

Electrical stimulation is commonly used in regenerative medicine that effectively transmits electrical charges to manipulate the cell behaviors of electrically active neurites. The nervous system is a network of neurons that are electrically excitable and can evoke neurotransmitter signaling. The cell membrane of the neurons exists in resting potential, and any rise or fall of the electrical signal with a characteristic pattern results in an action potential. It subsequently changes the membrane potential to either positive or negative. Neurotransmission occurs along the axon in the event of series of action potentials [35]. Change in the potential and polarity generates a biological electric field of voltage gradient −10 to −90 mV across cell membranes. These electrical cues can induce growth, neuronal signaling, and communication networks, thereby regulating neuronal activity. Technically, any nerve injuries can block the bioelectrical transmission between neurons and lead to the loss of essential body functions. However, restoration of the electrical cues can guide the axonal growth of the severed neuron in the right direction and establish regeneration. The electrical stimuli may mimic the Ca^{2+} influx that produces a retrograde signal that initiates the cellular cascade mechanism for regeneration. This electrical surge can stimulate the neuron across scaffolds made of electrically active CPs. Conductive biomaterials can induce sufficient electrical stimulation and transmit the bioelectrical cues to induce neural regeneration for modulating the cellular behavior and mimic the electrical microenvironment for the neurons [36]. Such exogenous stimulation has presented improved nerve generation in both *in vitro* and *in vivo* conditions. The electrical stimulation can offer increased myelin formation, inhibit Schwann cell apoptosis, synaptic remodeling, and elevated expression of neural growth factors like brain-derived neurotrophic factor (BDNF), growth-associated protein 43 (GAP 43) to accelerate neural regeneration [35]. Suitable substrates in combination with electrical stimulation have demonstrated potential neurite outgrowth and regeneration in several studies. In such instances, researchers developed electrospun conductive PVA and PEDOT scaffolds to study neural cell differentiation upon electrical stimulation. Further, the electroconductive scaffolds presented the upregulation of neural markers viz., nestin, β-tubulin III, and enolase to improve neural recovery [37].

12.7.2 CONDUCTING POLYMERS AS A THERAPEUTIC SYSTEM FOR NEURAL INJURIES

The inherent electrical properties of CPs such as PPy, PANI, and PEDOT have shown distinctive features and provide an electrical niche for neural tissue engineering [36]. Amidst this, PPy is the consistent polymer in neural tissue engineering as they inherit good biocompatibility, conductivity, and tunable surface properties. They also promote excellent cell adhesion properties for multiple cells such as neuronal, glial, endothelial, and mesenchymal cells. PPy/alginate/nano chitosan composites were reported to study the neural cell adhesion on scaffolds. Increased concentration of PPy elevated the conductivity of the scaffold up to 10^{-4} S/cm which results in uniform distribution and well-ordered adherence of OLN 93 cells [38]. Similarly, nanofibers developed from PCL–PANI–CSA (camphor sulfonic acid) have shown improved expression of neural genes like DCX, MAP2, enhancing the differentiation profile of neural cells [39].

12.7.2.1 Conducting Polymers as Nerve Conduits

The polymer scaffolds are fabricated into nerve guidance conduits (NGCs) to treat peripheral nerve injury. Nerve conduits are biomaterial-based nerve guiding channels that are sutured to the proximal and distal terminals of the severed nerve, facilitating the regrowth of axons. A promising PPy-based conductive nerve conduit has shown the higher expression of neurotrophic factors when the neural progenitor stem cells are cultured *in vitro*. Likewise, improved myelination and axon regeneration was observed in the *in vivo* examination. Such functional cues represent improved peripheral nerve regeneration and enhanced functional recovery. Another study involved the fabrication of PPy/PLGA fiber-based nerve conduits on peripheral nerve generation has established a higher proliferation rate and PC12 cell attachment. The *in vivo* efficacy of the scaffold was assessed in the sciatic nerve transected rats and the PPy/PLGA conductive conduits had stimulated increased nerve outgrowth and extension and higher recovery of sciatic-injured nerves. It is explained that the electrical cues directed the regulated cell proliferation and neurite extension [40].

12.7.2.2 Conducting Polymers in the Delivery System for Nerve Regeneration

The neurotrophic factors like nerve growth factor (NGF), neurotrophin-3 (NT-3), BDNF, ciliary neurotrophic factor (CNTF), FGF, TGF-β, and glial-derived neurotrophic factor (GDNF) are commonly involved in neural regeneration. Encapsulating the growth factors/therapeutic drugs on the conductive scaffolds can stabilize the controlled release of the drug for constructive regeneration of the nerve cells. The PLGA/dextran/hyaluronic acid scaffold is a classic example of the encapsulation and delivery of BDNF at the site of spinal cord injury. The developed hydrogel showed excellent electrical conductivity mimicking the natural spinal cord and exhibited a stable release of BDNF. Moreover, histological results demonstrated the neural differentiation to neurons and inhibitory effects on astrocytes differentiation as astrocytes contribute to the prevention of axonal regeneration during injury [41]. Likewise, injectable gelatin/PANI-based hydrogels with bone marrow stromal cells (BMSCs) could act as a drug delivery system for Parkinson's disease. It also improved the expression of tyrosine hydroxylase positive (TH$^+$) dopaminergic neurons, BDNF, and GDNF [42]. In this way, electroactive materials transmit the electrical cues to the neural tissue and improve the regenerative process.

12.8 CARDIAC TISSUE ENGINEERING

The cardiomyocytes are the only cell type present in the heart and are responsible for the contraction and relaxation of heart muscles. The internal structural organization of cardiomyocytes contributes to the electrophysiological nature of the heart. Myocardial infarction (MI), the most common heart disorder known to be the primary cause of death worldwide, results in cardiac myocyte slippage, scar tissue formation, and electrical conductivity loss. Presently, no standardized therapies are available to repair or improve the myocardial damage. The existing drugs and treatments do not establish a complete recovery or regeneration of the damaged myocardium. Moreover, the restoration of mechanical and electrical integrity of the heart tissues is difficult to achieve. Therefore, framing novel strategies by understanding the cellular and electrophysiological properties of the cardiac tissues was necessary. The advent of cardiac tissue engineering has revolutionized the significant areas of medicine to overcome the complications of conventional transplantations and cardiac treatments.

The conductive biomaterials are an ideal platform for cardiomyocytes regeneration because of the electroconductivity. They contribute both the physical and chemical niche, mimic the native tissue environment, and facilitate cell attachment and migration. The electrically conductive PANI/poly(glycerol sebacate) (PGS) composite has served as a suitable platform to modulate the cell behavior via electrical stimulation and also provided enough mechanical support [43]. Development of poly(thiophene-3-acetic acid) (PTAA)/methacrylate aminated gelatin (MAAG)-based double-network hydrogel with electronic conductivity 10^{-4} S/cm was reported to support the proliferation of brown adipose-derived stem cells (BADSCs). BADSCs are the potential progenitor cells having the ability to differentiate into cardiomyocytes upon electrical stimulation. The resulting cardiomyocytes had displayed cardiac-specific protein, cardiac troponin T (cTnT), and sarcomeric α-actinin on the hydrogel surface [44]. Thus, in the therapeutics view, biomaterials made of CPs remain a promising strategy for cardiac tissue regeneration as they can evoke electrical and biological cues via electrical stimulation.

12.8.1 Conductive Biomaterials for Cardiac Tissue Engineering

Conductive nanomaterials are the first choice for culturing the engineered cardiac tissues. Direct delivery of cardiomyocytes into the injured cardiac muscles may not be an appropriate technique as 90% of the cells die after implantation. The introduction of the individual cells without a suitable microenvironment for them to adapt may expose them to oxidative stress and inflammatory reactions leading to cell apoptosis. Hence, regenerative therapy demands a substantial need to construct the microarchitecture by fabricating biomaterials efficiently to establish cardiac regeneration. The substrate that provides the relatable environment for cellular regeneration can be hydrogels, films, scaffolds, or cardiac patches. Hydrogels are broadly used as a mechanical support for the cardiac cells and also stimulate vascularization under certain conditions. Especially the conductive hydrogel can maintain mechanical stability and mimic the biological and electrical properties of tissues in the human body. A self-healing conductive hydrogel made up of chitosan-graft-aniline tetramer and dibenzaldehyde-terminated poly(ethylene glycol) (PEG-DA) was reported for cardiac repair [45]. The conductivity of the hydrogels is ~10^{-3} S/cm, which is relatively similar to native cardiac tissue, further the injectable nature of the hydrogel could help to retain more cardiomyocytes and promote their proliferation. In this way, the hydrogel acted as a stable cell delivery system for cardiac repair.

Further developments in cardiac tissue engineering have launched cardiac patches with appropriate bioactive molecules and scaffolds for heart regeneration. It predominantly focuses to support or replace the scarred tissues in cardiomyopathic patients. The 1-cm thick patch possesses features to mimic the cardiac environment and induce the repair of heart tissues. It is also important for the patches to coordinate with the electrical rhythm of the myocardium for efficient results. A gelatin-based gel foam coupled with a self-doped conductive polymer (poly-3-amino-4-methoxy benzoic acid, PAMB) is reported as a cardiac patch (PAMB-Gel patch) to repair an infarcted heart [46]. The bioengineered PAMB-Gel patch implanted in infarcted rat hearts has shown significant electrical activity in the fibrotic tissue and synchronizing cardiomyocytes (CM) contraction across the scar region, markedly reducing its susceptibility to cardiac arrhythmias. The echocardiogram scans revealed that the PAMB-Gel patch could effectively repair injured cardiac tissues and restore cardiac function.

12.8.2   Cardiac Tissue Engineering via Stem Cells

The construction of well-standardized scaffolds to support stem cell attachment and proliferation is indispensable in tissue engineering. Various stem cells like mesenchymal stem cells, embryonic stem cells (ESCs), induced pluripotent stem cells (iPSCs), and cardiac stem cells (CSCs) have played a vital role in cardiac regeneration. Electrospun PANI–contained gelatin fibers served as a novel conductive and biocompatible scaffold for the growth of cardiac myoblast cells in cardiac tissue engineering applications [47]. Similarly, a conductive polymer, PANI was tested for its suitability in promoting cell attachment and proliferation. The study concluded that PANI could be used as an electroconductive scaffold for various cardiac and neural tissue engineering applications. The development of new and unique smart materials presents endless opportunities in tissue engineering applications. The as-designed PPy polymer was doped with certain molecules to support and stimulate the steam cells on an engineered surface [48]. The surface properties viz., roughness, surface energy, and surface chemistry of the polymer dopant molecules (NaCl, chondroitin sulfate A sodium salt, dodecylbenzene sulfonic acid, dextran sulfate sodium salt, hyaluronic acid sodium salt, lithium perchlorate, poly(sodium 4-styrene sulfonate), and sodium para-toluenesulfonate) had a positive influence on the viability of cardiac progenitor cells. An electroactive scaffold that provides electrical, mechanical, and topographical cues for stem cell stimulation is a promising approach in cardiac tissue engineering. The PLGA fiber scaffold coated with conductive polymer PPy has offered the necessary electrical and mechanical stimulation to induce human pluripotent stem cells (iPS) is presented [49]. This study demonstrates the first application of PPy-based fiber scaffold for iPS with dynamic mechanical, actuation, and electrical properties.

12.9   CONCLUSION

The development of tailor-made materials to fulfill the needs of functionally limited individuals is the prime aim of tissue engineering. A notable example under the smart materials category is conducting polymers. The positive attributes of conductive polymers like versatility, intrinsic monomers, facile synthesis, electrical conductivity, tunable conductivity, flexibility, and effortless modification make CPs very attractive for different biomedical applications. Contrarily, CPs also possess weak mechanical properties, low processability, and low biocompatibility (*in vivo*). With the advent of recent progress in material research and regenerative medicine, substantial improvements in the properties and extended applications of CPs are discussed in this chapter. Various CP-based hydrogel, scaffold, composite film, implants utilized for bone, nerve, and cardiac regeneration are discussed. Applying electrical conductivity through CP biomaterials exhibited improved conductivity and sensitivity at the cellular level that fastens the regenerative capability of tissue or organs. However, it requires further understanding to study the cell behavior upon electrical stimulation to assess the effectiveness of the CPs over time. Likewise, long-term studies on the assessment of degradation byproducts from the electroactive materials are essential. The studies or work discussed in this chapter are *in vitro* and preliminary *in vivo* studies, a more systemic data should be required before taking this into clinical practices. Once the shortcomings mentioned would be resolved, the potential of CPs in clinical tissue engineering becomes a reality.

REFERENCES

1. Lazurko C, Harden S, Suuronen E, Alarcon E (2019) Biomaterials for Organ and Tissue Repair. Front. Young Minds 7:8.
2. De Bartolo L, Piscioneri A (2017) New Advanced Biomaterials for Tissue and Organ Regeneration/ Repair. Cells Tissues Organs 204:123–124.

3. Reddy M, Ponnamma D, Choudhary R, Sadasivuni KK (2021) A Comparative Review of Natural and Synthetic Biopolymer Composite Scaffolds. Polymers 13:1105.

4. Park Y, Jung J, Chang M (2019) Research Progress on Conducting Polymer-based Biomedical Applications. Appl. Sci. 9:1070.

5. Guimard NK, Gomez N, Schmidt CE (2007) Conducting Polymers in Biomedical Engineering. Prog. Polym. Sci. 32:876–921.

6. Dong R, Ma PX, Guo B (2020) Conductive Biomaterials for Muscle Tissue Engineering. Biomaterials 229:119584.

7. Guo B, Ma PX (2018) Conducting Polymers for Tissue Engineering. Biomacromolecules 19:1764–1782.

8. Sharma, K, Kumar, V, Kaith, BS, Kalia, S, Swart, HC (2017) Conducting Polymer Hydrogels and Their Applications. Kumar V, Kalia S, Swart H (eds.) Conducting Polymer Hybrids. Springer Cham. Switzerland.

9. Tomczykowa M, Plonska-Brzezinska ME (2019) Conducting Polymers, Hydrogels and Their Composites: Preparation, Properties and Bioapplications. Polymers 11:350.

10. Filimon A (2016) Perspectives of Conductive Polymers toward Smart Biomaterials for Tissue Engineering. Yilmaz F (ed.) Conducting Polymers. IntechOpen: London.

11. Shi Z, Gao H, Feng J, Ding B, Cao X, Kuga S, Wang Y, Zhang L, Cai J (2014) In Situ Synthesis of Robust Conductive Cellulose/Polypyrrole Composite Aerogels and Their Potential Application in Nerve Regeneration. Angew. Chem. Int. Ed. 53:5380–5384.

12. Zhao HB, Yuan L, Fu ZB, Wang CY, Yang X, Zhu JY, Qu J, Chen HB, Schiraldi DA (2016) Biomass-based Mechanically Strong and Electrically Conductive Polymer Aerogels and Their Application for Supercapacitors. ACS Appl. Mater. Interfaces 8:9917–9924.

13. Talikowska M, Fu X, Lisak G (2019) Application of Conducting Polymers to Wound Care and Skin Tissue Engineering: A Review. Biosens. Bioelectron. 135:50–63.

14. Bacakova L, Zikmundova M, Pajorova J, Broz A, Filova E, Blanquer A, Matejka R, Stepanovska J, Mikes P, Jencova V, Kostakova EK, Sinica A (2019) Nanofibrous Scaffolds for Skin Tissue Engineering and Wound Healing Based on Synthetic Polymers. Stoytcheva M, Zlatev R (eds.) Applications of Nanobiotechnology. IntechOpen: London.

15. Mirzaei A, Kumar V, Bonyani M, Majhi SM, Bang JH, Kim JY, Kim HW, Kim SS, Kim KH (2020) Conducting Polymer Nanofibers Based Sensors for Organic and Inorganic Gaseous Compounds. Asian J. Atmos. Environ. 14:85–104.

16. Luong JH, Narayan T, Solanki S, Malhotra BD (2020) Recent Advances of Conducting Polymers and Their Composites for Electrochemical Biosensing Applications. J. Funct. Biomater. 11:71.

17. Green RA, Lovell NH, Poole-Warren LA (2009) Cell Attachment Functionality of Bioactive Conducting Polymers for Neural Interfaces. Biomaterials 30:3637–3644.

18. Svirskis D, Travas-Sejdic J, Rodgers A, Garg S (2010) Electrochemically Controlled Drug Delivery Based on Intrinsically Conducting Polymers. J. Control. Release 146:6–15.

19. Ravichandran R, Sundarrajan S, Venugopal JR, Mukherjee S, Ramakrishna S (2010) Applications of Conducting Polymers and Their Issues in Biomedical Engineering. J. R. Soc. Interface 7:S559–S579.

20. Punjabi PB, Chauhan NPS, Jangid NK, Juneja P (2015) Conducting Polymers: Biodegradable Tissue Engineering. Encycl. Biomed. Polym. Polym. Biomater. 2016:1972–1981.

21. Vig K, Chaudhari A, Tripathi S, Dixit S, Sahu R, Pillai S, Dennis VA, Singh SR (2017) Advances in Skin Regeneration Using Tissue Engineering. Int. J. Mol. Sci. 18:789.

22. Zhao X, Wu H, Guo B, Dong R, Qiu Y, Ma PX (2017) Antibacterial Anti-Oxidant Electroactive Injectable Hydrogel as Self-Healing Wound Dressing with Hemostasis and Adhesiveness for Cutaneous Wound Healing. Biomaterials 122:34–47.

23. Satapathy MK, Nyambat B, Chiang CW, Chen CH, Wong PC, Ho PH, Jheng PR, Burnouf T, Tseng CL, Chuang EY (2018) A Gelatin Hydrogel-Containing Nano-Organic PEI–Ppy with a Photothermal Responsive Effect for Tissue Engineering Applications. Molecules 23:1256.

24. Gu BK, Park SJ, Kim CH (2018) Beneficial Effect of Aligned Nanofiber Scaffolds with Electrical Conductivity for the Directional Guide of Cells. J. Biomater. Sci. Polym. Ed. 29:1053–1065.

25. Farshi Azhar F, Olad A, Salehi R (2014) Fabrication and Characterization of Chitosan–Gelatin/Nanohydroxyapatite–Polyaniline Composite with Potential Application in Tissue Engineering Scaffolds. Des. Monomers Polym. 17:654–667.

26. Sarvari R, Massoumi B, Jaymand M, Beygi-Khosrowshahi Y, Abdollahi M (2016) Novel Three-Dimensional, Conducting, Biocompatible, Porous, and Elastic Polyaniline-based Scaffolds for Regenerative Therapies. RSC Adv. 6:19437–19451.

27. Aycan D, Selmi B, Kelel E, Yildirim T, Alemdar N (2019) Conductive Polymeric Film Loaded with Ibuprofen as a Wound Dressing Material. Eur. Polym. J. 121:109308.

28. Gh D, Kong D, Gautrot J, Vootla SK (2017) Fabrication and Characterization of Conductive Conjugated Polymer-coated Antheraea Mylitta Silk Fibroin Fibers for Biomedical Applications. Macromol. Biosci. 17:1600443.

29. Rajzer I, Rom M, Menaszek E, Pasierb P (2015) Conductive PANI Patterns on Electrospun PCL/ Gelatin Scaffolds Modified with Bioactive Particles for Bone Tissue Engineering. Mater. Lett. 138:60–63.

30. Ghorbani F, Zamanian A, Aidun A (2020) Conductive Electrospun Polyurethane-Polyaniline Scaffolds Coated with Poly (Vinyl Alcohol)-GPTMS under Oxygen Plasma Surface Modification. Mater. Today Commun. 22:100752.

31. Mirzaei A, Saburi E, Enderami SE, Barati Bagherabad M, Enderami SE, Chokami M, Moghadam AS, Salarinia R, Ardeshirylajimi A, Mansouri V, Soleimanifar F (2019) Synergistic Effects of Polyaniline and Pulsed Electromagnetic Field to Stem Cells Osteogenic Differentiation on Polyvinylidene Fluoride Scaffold. Artif. Cells Nanomed. Biotechnol. 47:3058–3066.

32. Yu Y, Tao B, Sun J, Liu L, Zheng H (2020) Fabrication of Chitosan-Graft-Polyaniline-based Multilayers on Ti Substrates for Enhancing Antibacterial Property and Improving Osteogenic Activity. Mater. Lett. 268:127420.

33. Murugesan B, Pandiyan N, Arumugam M, Sonamuthu J, Samayanan S, Yurong C, Juming Y, Mahalingam S (2020) Fabrication of Palladium Nanoparticles Anchored Polypyrrole Functionalized Reduced Graphene Oxide Nanocomposite for Antibiofilm Associated Orthopedic Tissue Engineering. Appl. Surf. Sci. 510:145403.

34. Ai J, Kiasat-Dolatabadi A, Ebrahimi-Barough S, Ai A, Lotfibakhshaiesh N, Norouzi-Javidan A, Saberi H, Arjmand B, Aghayan HR (2014) Polymeric Scaffolds in Neural Tissue Engineering: A Review. Arch. Neurosci. 1:15–20.

35. Moskow J, Ferrigno B, Mistry N, Jaiswal D, Bulsara K, Rudraiah S, Kumbar SG (2019) Bioengineering Approach for the Repair and Regeneration of Peripheral Nerve. Bioact. Mater. 4, 107–113.

36. Luo Y, Xue F, Liu K, Li B, Fu C, Ding J (2021) Physical and Biological Engineering of Polymer Scaffolds to Potentiate Repair of Spinal Cord Injury. Mater. Des. 201:109484.

37. Babaie A, Bakhshandeh B, Abedi A, Mohammadnejad J, Shabani I, Ardeshirylajimi A, Moosavi SR, Amini J, Tayebi L (2020) Synergistic Effects of Conductive PVA/PEDOT Electrospun Scaffolds and Electrical Stimulation for More Effective Neural Tissue Engineering. Eur. Polym. J. 140:110051.

38. Manzari-Tavakoli A, Tarasi R, Sedghi R, Moghimi A, Niknejad H (2020) Fabrication of Nanochitosan Incorporated Polypyrrole/Alginate Conducting Scaffold for Neural Tissue Engineering. Sci. Rep. 10:1–10.

39. Garrudo FF, Mikael PE, Rodrigues CA, Udangawa RW, Paradiso P, Chapman CA, Hoffman P, Colaço R, Cabral JMS, Morgado J, Linhardt RJ, Ferreira FC (2021) Polyaniline-Polycaprolactone Fibers for Neural Applications: Electroconductivity Enhanced by Pseudo-Doping. Mater. Sci. Eng. C 120:111680.

40. Jing W, Ao Q, Wang L, Huang Z, Cai Q, Chen G, Yang X, Zhong W (2018) Constructing Conductive Conduit with Conductive Fibrous Infilling for Peripheral Nerve Regeneration. Chem. Eng. J. 345:566–577.

41. Huang F, Chen T, Chang J, Zhang C, Liao F, Wu L, Wang W, Yin Z (2021) A Conductive Dual-Network Hydrogel Composed of Oxidized Dextran and Hyaluronic-Hydrazide as BDNF Delivery Systems for Potential Spinal Cord Injury Repair. Int. J. Biol. Macromol. 167:434–445.

42. Xue J, Liu Y, Darabi MA, Tu G, Huang L, Ying L, Xiao B, Wu Y, Xing M, Zhang L, Zhang L (2019) An Injectable Conductive Gelatin-PANI Hydrogel System Serves as a Promising Carrier to Deliver BMSCs for Parkinson's Disease Treatment. Mater. Sci. Eng. C. 100:584–597.

43. Qazi TH, Rai R, Dippold D, Roether JE, Schubert DW, Rosellini E, Barbani N, Boccaccini AR (2014) Development and Characterization of Novel Electrically Conductive PANI–PGS Composites for Cardiac Tissue Engineering Applications. Acta Biomater. 10:2434–2445.

44. Yang B, Yao F, Hao T, Fang W, Ye L, Zhang Y, Wang Y, Li J, Wang C (2016) Development of Electrically Conductive Double-Network Hydrogels via One-Step Facile Strategy for Cardiac Tissue Engineering. Adv. Healthc. Mater. 5:474–488.

45. Dong R, Zhao X, Guo B, Ma PX (2016) Self-Healing Conductive Injectable Hydrogels with Antibacterial Activity as Cell Delivery Carrier for Cardiac Cell Therapy. ACS Appl. Mater. Interfaces 8:17138–17150.

46. Chen S, Hsieh MH, Li SH, Wu J, Weisel RD, Chang Y, Sung HW, Li RK (2020) A Conductive Cell-Delivery Construct as a Bioengineered Patch that Can Improve Electrical Propagation and Synchronize Cardiomyocyte Contraction for Heart Repair. J. Control. Release 320, 73–82.

47. Li M, Guo Y, Wei Y, MacDiarmid AG, Lelkes PI (2006) Electrospinning Polyaniline-Contained Gelatin Nanofibers for Tissue Engineering Applications. Biomaterials 27:2705–2715.
48. Gelmi A, Ljunggren MK, Rafat M, Jager EWH (2014) Influence of Conductive Polymer Doping on the Viability of Cardiac Progenitor Cells. J. Mater. Chem. B 2:3860–3867.
49. Gelmi A, Cieslar-Pobuda A, de Muinck E, Los M, Rafat M, Jager EW (2016) Direct Mechanical Stimulation of Stem Cells: A Beating Electromechanically Active Scaffold for Cardiac Tissue Engineering. Adv. Healthc. Mater. 5:1471–1480.

13 Conducting Polymers for Neural Tissue Engineering

Zahra Allahyari,[1]# Shayan Gholizadeh,[1]# Hossein Derakhshankhah,[2] Katayoun Derakhshandeh,[3] Seyed Mohammad Amini,[4] and Hadi Samadian[5]

[1]Department of Biomedical Engineering, Rochester Institute of Technology, Rochester, New York, United States

[2]Pharmaceutical Sciences Research Center, Health Institute, Kermanshah University of Medical Sciences, Kermanshah, Iran

[3]Medicinal Plants and Natural Products Research Center, Hamadan University of Medical Sciences, Hamadan, Iran

[4]Radiation Biology Research Center, Iran University of Medical Sciences (IUMS), Tehran, Iran

[5]Research Center for Molecular Medicine, Hamadan University of Medical Sciences, Hamadan, Iran

#These authors contributed equally to this chapter.

CONTENTS

13.1 INTRODUCTION

The human nervous system is a complex functional unit that provides the crucial role of coordinating and controlling voluntary and involuntary bodily activities, vital for maintaining human body function [1–3]. Despite the maturity and highly organized nature of the nervous system, it is prone to systemic and peripheral damage due to a variety of diseases and trauma events, which are likely to happen at some point in the life span of a human being [2, 4, 5]. Traditional treatment options are mostly limited to corrective surgeries using autologous, allogeneic, or xenogeneic grafts with maximal limitations and complications [6]. Limitations and complications such as limited donor grafts, graft incompatibility and rejection, lack of original nerve functionality, ethical concerns, and, most importantly, the limited length and extent to which the grafts can support regeneration led to explorations of alternative methods for nerve replacement and regeneration [1, 2, 7]. One of the most investigated striking venues is nerve tissue engineering [8]. Since the inception of the tissue engineering field, neural regeneration has been one of the most highly sought topics to the inefficiency and limitations of established methods for nerve repair. Nerve tissue engineering is sought after either to obtain replacements or, in more recent work, nerve conduits using a new generation of biomaterials [5, 8]. These biomaterials often include structures made of conducting polymers, carbon nanotubes, metallic nanoparticles, and other conductive materials to either resemble the healthy tissue or to provide the conductivity for exogenous electrical stimulation as an alternative strategy for nerve repair [9]. The former is more of a passive approach while the latter is an active approach for nerve regeneration, both relying on tissue engineering methods to succeed [2].

Among the potential materials as conductive components, conducting polymers can be considered the most promising materials so far due to their higher biocompatibility as compared to materials such as carbon nanotubes, their high conductivity, their mechanical robustness, and straightforward synthesis and surface modification with biological molecules [1–3, 9]. The most used conducting polymers in neural tissue engineering include polypyrrole (PPy), polyaniline (PANI), as well as polythiophene, and derivatives (namely poly(3,4-ethylenedioxythiophene) [PEDOT]) [9]. A variety of fabrication and modification techniques are used to obtain conducting polymer structures in different forms such as composites and hydrogels, which will be discussed further in this chapter along with their limitations and future opportunities. Despite the numerous advances and promising work in this area of research, practical outcomes for improving human life are still not up to the initial expectations [9–11]. Whether this can be interpreted as further potential work to improve the area or can be considered as a downfall will be discussed. A fresh perspective will also be presented as to the current path of this subfield and the necessary steps it takes for a successful transition into clinical results.

13.2 CONDUCTING POLYMERIC CHOICES FOR NERVE TISSUE REPAIR

Conducting polymers are synthetic materials that can be fabricated using chemical or electrochemical approaches [2, 9, 12]. They possess conjugated double bonds in their backbone that render them with high electrical conductivity [2, 12]. Inception of conducting polymers was triggered by the discovery that when iodine vapor is used to oxidize polyacetylene, polyacetylene transitions from being a material with low electrical conductivity to one with millions of times more electrical conductivity [3, 10, 12]. Building upon this finding, subsequent research led to the development of polyacetylenes, a family that contains the most commonly used conducting polymers, including PANI and PPy [9, 12]. Due to their high conductivity, robust stability, and the ability to easily alter their properties, they became favorite target materials for neural tissue replacements and conduits [12].

However, a shift of focus away from conducting polymers in neural tissue engineering in recent years toward doping natural polymers with conductive constituents commands us to look back at material choices to understand how to utilize these polymers to the best of our ability to achieve the initial goal integrating conducting polymers in neural tissue engineering and this step back starts

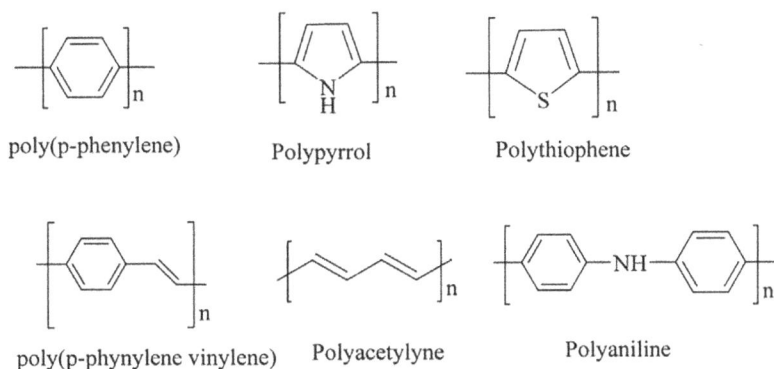

FIGURE 13.1   Chemical structures of commonly used conducting polymers. (Adapted with permission from Ref. [16]. Copyright (2021) Royal Society of Chemistry.)

with material choices [13–15]. PPy, PANI, and polythiophenes have been proved to hold the highest potential for this purpose [2, 9]. Chemical structures of commonly used conducting polymers, including these three polymers, are shown in Figure 13.1. Each of these conducting polymers provides unique properties, challenges, and opportunities that can affect the outcomes for neural tissue engineering applications, which will be discussed further.

13.2.1   POLYANILINE

PANI is an inexpensive conducting polymer with high electrical conductivity and antioxidant capability, and there are multiple variations of it with different degrees of oxidation [2, 12, 16]. It is used in multiple formats, most notably in hydrogel and nanofibrous forms. Similar to other conductive polymers, PANI exhibits no biodegradability, high rigidity, increased chance of inflammations or immunogenic reactions, and lackluster biocompatibility, collectively posing major challenges in its applications for neural tissue engineering [2, 9]. The predominant strategy for improving the properties of PANI structures for neural tissue engineering applications is the combination of PANI with more biocompatible, more biodegradable polymers such as poly(ε-caprolactone) (PCL) [2, 9]. Modification strategies include the addition of biological molecules and surface and bulk treatment, which will be discussed in the following sections [2, 9, 12].

In a pioneering breakthrough study of PANI-based biomaterials for tissue engineering applications, Li et al. used electrospinning to fabricate a nanofibrous structure made from a combination of PANI and gelatin for the first time [17]. The results indicated that doping of gelatin with a low percentage PANI results in a structure with high electrical conductivity and substantially smaller fiber diameter. They observed composition uniformity throughout the structure, biocompatibility, and suitable cell attachment and proliferation. Among more recent notable works on developing PANI-based biomaterials, a recent study was dedicated to the development and characterization of injectable hydrogel made of PANI-graft-gelatin, alginate, and polyethyleneimine for neural tissue engineering applications [18]. Karimi-Soflou et al. utilized a dual cross-linking strategy, relying on self-cross-linking of gelatin and periodate oxidized alginate in their structures. They observed non-toxicity, tunable structural properties with different composition ratios, and differentiation of P19 embryonic carcinoma cells toward neural cells increased with the application. Although the combination of PANI with more physiologically relevant and more biocompatible materials is not a new concept and, using different properties of PANI and the other composite components toward engineering the properties of the scaffold (in this case, gelation time, pore size, and swelling ratio), to optimize and achieve maximum neural regeneration is a vital step that should be considered as a necessity in future studies.

13.2.2 POLYPYRROLE

PPy is another conductive polymer with tunable electrical conductivity, facile electrochemical and chemical polymerization synthesis, high stability, and superior mechanical strength [7, 19]. PPy is the most investigated conducting polymer for neural tissue engineering purposes and there has been substantial progress made in this area of research using these [2, 9, 12]. However, it still entails the same issues associated with other conducting polymers such as PANI. It is brittle, has suboptimal biocompatibility on par with PANI, poor solubility, no biodegradation and as compared to PANI, it often exhibits lower and more inconsistent electrical conductivity [2, 12]. Similar to PANI, a variety of modifications and combinations with other materials are used to compensate for these deficiencies [2, 9, 16]. Yang et al. notably synthesized conductive PPy/alginate neural tissue engineering applications by polymerizing PPy in the more biocompatible naturally derived alginate hydrogel networks formed by ionic cross-linking [20]. This natural/synthetic combination approach has been a strategy adopted by many tissue engineering research groups. The addition of PPy led to an order of magnitude higher electrical conductivity, higher mechanical strength, enhancement of neural differentiation markers expression and exhibited minimal inflammatory response *in vivo* studies. Such platforms over the last two decades have provided an opportunity for neural regeneration using biomimetic and multifunctional tissue scaffolds.

13.2.3 POLYTHIOPHENE AND ITS DERIVATIVES

Polythiophene and its derivatives have been long used in tissue engineering applications for a variety of purposes due to their superior electrical properties. PEDOT has been the most widely used member of this family of materials for neural tissue engineering [2, 19, 21]. PEDOT exhibits electrical stability and as opposed to polythiophene and PPy which have α–β and β–β' couplings in its structure that contribute to unstable and diminished electrical properties, PEDOT is free of these coupling [2, 12]. It has superior electrical properties and it has been suggested that it shows mildly better biocompatibility. There are obstacles such as nonbiodegradability and high stiffness that should be considered when using these materials for neural tissue engineering [2, 9, 12]. A simple but resourceful approach has been provided by Xu et al., where they fabricated a carboxymethyl chitosan hydrogel network with a PEDOT layer as the secondary network using *in situ* chemical polymerization [22]. Carboxymethyl chitosan is biodegradable, biocompatible and has enough flexibility that compensates for the rigidity of PEDOT. They showed remarkable adhesion and proliferation of PC12 cells, laying a foundational work for exogenous electrical stimulation of neural cells for nerve tissue engineering applications.

13.3 FABRICATION AND MODIFICATION OF CONDUCTING POLYMER-BASED STRUCTURES FOR NEURAL TISSUE: STRATEGIES AND CONSIDERATIONS

Since the early pioneering works on neural tissue engineering in the 1990s, every possible fabrication strategy has been explored for neural tissue engineering biomaterials, including conducting polymer-based structures. Due to the unique electrical and geometrical considerations for neural regeneration and repair, nanofibrous scaffolds with high electrical conductivity have been among the favorite choices for scaffold material [2, 12, 23, 24]. Conducting polymers are the most frequently used electrically conductive component either solely or in combination with other materials [2]. Electrospinning, lyophilization, chemical or physical cross-linking, and, more recently, 3D printing have been proposed as being more suited methods for fabricating neural tissue engineering biomaterials [2, 6, 21, 23]. Each of these methods provides benefits for neural tissue engineering application, which will be briefly discussed in the following sections. Due to the synthetic nature of conducting polymers and their limited capacity for tissue integration, these materials are tailored to gain neural tissue biocompatibility [2, 18, 23]. A brief discussion will be provided on these modifications as well.

13.3.1 ELECTROSPINNING

Electrospinning is a versatile platform for the fabrication of nano- and microfibrous structures, enabled by establishing high-voltage polymer solutions from a nozzle and collecting the formed fibers at the collector electrode [24–26]. Before any attempt for developing conducting polymer-based scaffolds for neural tissue engineering, electrospinning was a well-established method for the fabrication of neural tissue scaffolds [27]. This was due to the opportunities in this method to produce structures which a high aspect ratio, high porosity, and large surface area together with the geometry which mimics the structure of healthy neural tissue [2, 24]. Such nonconductive scaffolds were shown to facilitate neural regeneration in different models [27].

The rise in the use of conducting polymers inspired works that combine electrospinning and conducting polymers, either by electrospinning a conducting polymer solution alone or by blending it with another polymer [2, 12, 28]. Earlier works on electrospinning conducting polymer solutions proved to be challenging due to poor solubility along with other complications mainly arising from strategies to combat this suboptimum solubility in organic solvents. The real progress was made upon research toward electrospinning polymer blends that include a conducting polymer [17, 23, 28]. For instance, Sadeghi et al. used electrospinning to fabricate nanofibers containing PCL, chitosan, and PPy, benefiting from suitable properties of each component for neural tissue regeneration to yield a material with superior biological, mechanical, and electrical properties [23]. Their results indicated improved cell attachment, spreading, proliferation as well as a significant increase in neurite extension in PC12 cells, indicating a promising approach for neural tissue engineering.

13.3.2 LYOPHILIZATION

Lyophilization, otherwise known as freeze-drying, is a multifunctional cost-effective fabrication strategy to create porosity in tissue engineering scaffolds [15, 21]. Its advantages, including highly controlled pore size, interconnecting pores with significant-high porosity, being porogen-free, and close resemblance to natural extracellular matrix (ECM), have led to a high potential for neural tissue engineering [15, 29–31]. Porous scaffolds with oriented pores are another possibility that can be realized using freeze-drying [31]. Baniasadi et al. used freeze-drying as part of their fabrication process to fabricate conductive scaffolds modified with PANI and graphene [32]. Incorporation of PANI and graphene with different concentrations in the chitosan/gelatin matrix was evaluated for neural tissue engineering applications. They found that even in low contents, the conductive scaffolds superiorly supported Schwann cell adhesion and proliferation as compared to nonconductive controls.

13.3.3 COVALENT AND IONIC CROSS-LINKING

Most tissue engineering scaffold applications, including neural tissue scaffolds, require desirable mechanical properties and structural integrity that conventional choices such as hydrogels cannot provide [15, 33]. Covalent and ionic cross-linking agents are suitable candidates which provide the opportunity for modifying the mechanical and biological properties of the scaffolds [15, 33]. However, covalent and ionic cross-linking agents entail limitations and challenges that should be considered for designing and fabrication neural tissue engineering scaffolds [33].

As for ionic cross-linkers, they can be reversible, temperature or pH dependant, and often with minimal cytotoxicity concerns, and in some cases such as β-glycerophosphate, they can have generic- or cell-specific positive effects on cell behavior and cell fate [15, 34]. However, the improvement in mechanical properties as a result of ionic cross-linking is limited in many cases and not sufficient for various neural tissue engineering applications [15]. On the other hand, covalent cross-linkers such as glutaraldehyde have more pronounced effects on scaffold mechanical properties, while they pose threats for cell and tissue biocompatibility of the scaffolds due to the possibility of unreacted cross-linkers [15, 35]. Choice of cross-linker agent is a delicate matter for neural tissue engineering applications and all potential benefits, limitations, and challenges of the chosen agent should be considering throughout the scaffold design and development process.

13.3.4 3D Printing

3D printing is a promising strategy for the development of neural tissue scaffolds and conduits which has received substantial attention in recent years due to its ability for creating complex porous geometries with high resolution down to micron or even nanoscale [4, 6]. 3D bioprinting for neural tissue involves layer-by-layer deposition of conducting polymer-based bioinks (spatial embedment of cells and biomaterials) to create functional neural tissue replacements or conduits. Bioprinting techniques include inkjet, stereolithography, and fused extrusion printing [4, 6, 28]. A notable work in this area was done by Heo et al. where they demonstrated that their PEDOT-modified 3D printed scaffolds had excellent electrochemical and mechanical properties and provided the excellent capability for neural differentiation for neural tissue engineering of dorsal root ganglion cells [21].

13.3.5 Modification Techniques

As previously mentioned, despite numerous properties of conducting polymers that resemble healthy neural tissue as well as being useful in providing neural regeneration cues, they lack adhesive ligands for cell adhesion often present in naturally derived biomaterials [2, 12]. Depending on the specific purpose that the conducting polymer-based structures should fulfill, different levels of biocompatibility are required. To achieve these higher levels of biocompatibility, modification strategies are pursued. This is of high importance for almost all tissue engineering applications, and when it comes to neural tissue engineering, modification of conducting polymers is even of more importance due to the sensitive and fragile nature of neural cells and their more challenging attachment [25, 36, 37]. Addition of biological molecules (often referred to as "decoration") and chemical and physical treatments are the two main routes with the most promising perspectives [2, 38]. Adopting either strategy entails consequences that will be addressed. Even if a combination of natural polymer and conducting polymer is used for improving biocompatibility, scaffold modification techniques should still be considered for neural tissue application as high biological requirements of neural tissue can be further satisfied with these modifications.

13.3.5.1 Addition of Biologically Active Molecules

The incorporation of biologically active compounds into the conducting polymer network is a simple and effective method to introduce cell adhesion sites to the structure [2, 12, 15]. These molecular components include a wide range of proteins, growth factors, ECM components as well as commercial ECM formulations such as Matrigel® [2, 3, 39]. Entrapment of biological molecules, adsorption, and covalent bonding are the three main methods used to embed biologically active compounds in conductive polymers [2, 12]. Entrapment of biological molecules involves mixing the molecules within the conducting polymer solution before electrochemical or chemical synthesis [12]. Complications and limitations associated with entrapment include a potential decrease in electrical conductivity, the limited concentration of biological molecules the structure can tolerate before the collapse, and a decrease in structural integrity as a result of decreasing the reaction sites for polymerization [2].

Adsorption involves exposing the conducting polymer network to the biological molecules solution and relies on static interactions between the two networks for achieving a more biocompatible conducting polymer network [12, 40, 41]. The limitation of this simple but effective method is the weak attachment of molecules to the structure and the possibility of release and subsequent compromise of the conducting polymer network conductivity [12, 41]. Covalent bonding follows the same logic but through covalent reaction and bonding of biological molecules and the conducting polymer [2, 12]. The other difference is that the bonding is more permanent and stronger as compared to adsorption, with the major drawback being the potential decrease in electrical conductivity as a result of the presence of these strong covalent bonds in the structure.

13.3.5.2 Surface Treatments

Considering the advantages and shortcomings of the addition of biologically active molecules, that strategy is not well-suited under all circumstances. When the decoration is no longer an option, chemical and physical treatments can be strategies with more complexity and potentially more promising outcomes [2, 12, 42, 43]. Surface and topographical modifications are the most widely used techniques for these purposes [2, 42]. Introducing surface and topographical changes often require chemical treatments and introducing physical topographies that lead to drastic changes in surface properties [44]. These changes can include imparting hydrophilicity to the surface and creating rough surface geometries and most of these changes lead to better cell attachment and higher scaffold-tissue integration. There can be an overlap between chemical/physical treatments and the addition of biologically active molecules since the addition of molecules can affect the surface properties.

13.4 APPLICATION OF CONDUCTING POLYMERS FOR NEURAL TISSUE ENGINEERING: DESIRABLE PROPERTIES AND DESIGN CONSIDERATIONS

The most important consideration for biomaterial design is the intended application [15, 45]. There are hundreds of scientific papers "exploring" a variety of scaffold biomaterials made of different combinations of conducting polymers with various materials with the goal of neural tissue engineering. Even though developing new biomaterials is always noteworthy and needed, there must come a time where biomaterials introduced as functional smart materials serve as functional biomaterials in practice for patients and find themselves in the clinic. Unfortunately, there is no evidence to suggest such a milestone is close to realization. If a road map was to be drawn for achieving this long-awaited goal, the first and most practical type of application would be active or passive electrical stimulation toward neural regeneration and repair. Such applications require specific physical and chemical requirements to be satisfied, where developing new conducting polymer-based biomaterials for the sole reason of new material development will not be able to satisfy.

13.4.1 Electrical Stimulation

Electrical stimulation for neural regeneration is the most widely used electrical cue in tissue engineering ever since its inception [15, 46]. This extended use is even more pronounced for neural tissue engineering and this can be attributed to the neurons being the most electrically excitable cells in the human body and the well-studied role of electrical signaling in the function, modulation, and regeneration of neurons and glial cells [47, 48]. Any artificial change in the electrical properties of neural cells can be referred to as neural electrical stimulation [2, 49]. Such change can lead to a wide range of changes in cell behavior and fate, including neural growth, proliferation, differentiation, and migration, as well as changing the permeability of the cell membrane [49]. As a result of artificial electrical cues, nerve cell membranes can go through the normal stages of rest, depolarization, and repolarization and can restore the normal function of neural cells [2]. Among the tools that facilitate or provide the opportunity for electrical stimulation of neural tissue, conducting polymers have been favorite choices due to their superior electrical properties which can help either deliver an external electrical cue or can on their own serve to guide and engineer a proper neural regeneration [2, 9]. A brief description of these two applications is provided in the following sections.

13.4.1.1 Exogenous Electrical Stimulation Delivery

Exogenous electrical stimulation is a straightforward and feasible method to excite neural cells [50–52]. This traditional format of electrical stimulation involves passing an electrical current through two metallic electrodes embedded in the distal and proximal end of the damaged site [53]. To this date, *in vitro* studies are being carried out using this simplistic but versatile method to investigate the effect of exogenous electrical stimulation on neural response. However, embedding such electrodes in the human body is not feasible in almost all settings, specifically not possible for neural tissue engineering application. The advent of tissue engineering provided the opportunity

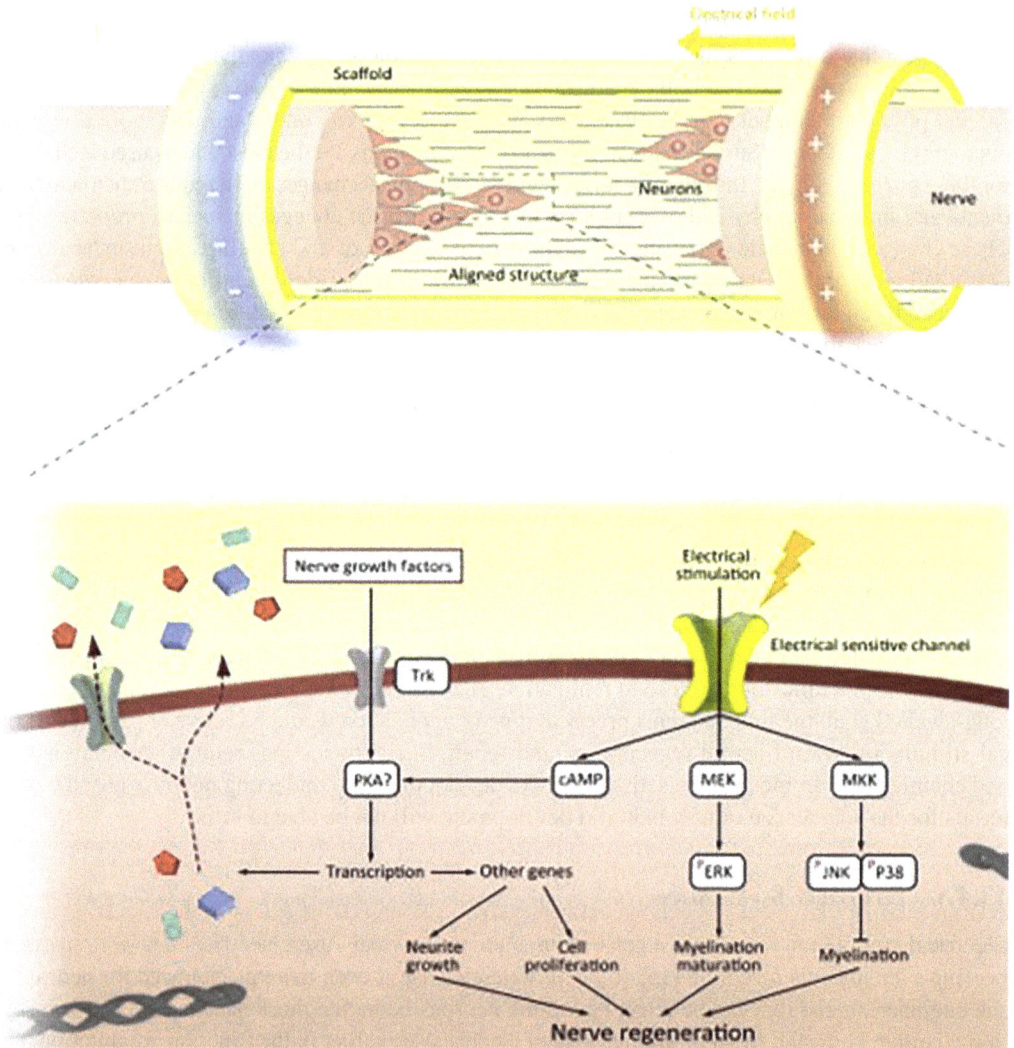

FIGURE 13.2   A possible mechanism for contribution of conducting polymer-enabled electrical stimulation to neural tissue regeneration. (Adapted with permission from Ref. [28]. Copyright (2020) Elsevier.)

for developing conductive tissue scaffolds that can serve as temporary housing of neural cells and neural tissue and then can be implanted in the damaged site. Another venue of research with neural conductive scaffolds is developing *in vitro* models of the neural tissue which can be used to optimize electrical stimulation parameters such as voltage, current, frequency, duty cycle, and wave shape [46, 54]. Most importantly, potential mechanisms for beneficiary effects of electrical stimulations, as proposed in Figure 13.2, can be explored in these platforms [28].

Conducting polymers are the perfect candidates for these applications due to the available robust methods of fabrication which can yield 2D or 3D networks with desirable properties [9, 12, 16]. Cells are simply seeded in the conducting polymer-based scaffold and by placing the scaffolds in the electrically conductive cell culture media, different forms of electrical simulation can be applied and investigated [46]. For peripheral nerve regeneration *in vivo* approaches, embedding a cell-laden conducting polymer-based scaffold in the damaged area and applying electrical stimulation either through direct contact or through the skin can also be an opportunity to facilitate the delivery of electrical stimulation to the otherwise delicate, untouched tissue. Due to their mechanical strength and their support for tissue regeneration, these scaffolds can play the additional traditional role of neural tissue engineering scaffolds [2, 9]. Their tailored biodegradability, when combined with other materials or otherwise modified, can ensure that neural tissue over time replaces the neural tissue scaffold, eliciting minimal inflammatory and potential long-term adverse effects of the scaffolds [2, 23].

13.4.1.2 Passive Electrical Cues

Despite the highly promising results from *in vitro* studies, translation of exogenous electrical stimulation and by extension, conducting polymer-based neural tissue-engineered scaffolds have been very slow and almost non-existent. Although this slow progress necessitates the current research for scaffold-mediated electrical stimulation, parallel interesting research has been carried out on using conducting polymer-based neural tissue-engineered scaffolds as an electrical cue which improves the electrical properties of injured or nonfunctional neural tissue and improve the capability for neural regeneration in the absence of any external electrical stimulation [2, 30, 55].

Although this line of studies might be perceived as conducting research on low-hanging fruits, it has shown such potential that justified scientific inquiry. Clear indications of this promising potential have been inadvertently shown in scaffold-mediated electrical stimulation where there are controls with no exogenous electrical stimulation and only with the presence of electrically conductive scaffolds, neural tissue regeneration was facilitated. For instance, Zhao et al. during presenting the data for their scaffold-based exogenous electrical stimulation have shown that although neurite ingrowth and stem cell infiltration was slightly inferior in conducting scaffold only sample compared to the scaffold with exogenous electrical stimulation, it still showed remarkable improvement in neural regeneration over nonconductive samples [28].

13.5 NEURAL CELL/TISSUE INTERACTIONS WITH CONDUCTING POLYMERS: CHALLENGE OR OPPORTUNITY?

It is a generally accepted fact that synthetic polymers, including the family of conducting polymers, are not capable of elucidating neural cellular and tissue response on par with naturally derived polymers for neural regeneration even with surface and structural modifications due to limited interaction capability with the target tissue [13]. This is more pronounced when they are not used as part of a composite structure with natural biological macromolecules [2, 12]. Natural polymers actively interact with damaged tissue, make up a structurally inseparable tissue complex, play their roles in tissue regeneration, and get eliminated from the body through biodegradation [15].

In the case of natural/synthetic composites, researchers also rely on the concept of biodegradation of natural polymer constituents and the capability of the body for the disposal of leftover synthetic components [15, 29, 32, 56]. However, there would be still concerns over how a precise body can eliminate the entire remaining conducting polymer to avoid complications in the regenerated tissue and whether the transport and elimination process leads to polymer accumulation in a secondary organ with devastating consequences [57]. A well-designed postimplantation elimination strategy or a substantially bulletproof approach for continued harmless presence is required for alleviating the aforementioned concerns. Given the sensitive nature of neural tissue, both these remedies require comprehensive *in vivo* research to test their potential success for neural tissue regeneration.

Despite these considerations and concerns, the weaker neural cell/tissue integration of conducting polymers is not necessarily a negative trait for all applications. This characteristic can be specifically useful when the "electrode" scaffold is used over an extended period for recurrent electrical stimulation, scaffold structural integrity is required, and it must be eventually taken out of the site through surgery. Conducting polymer-based structures can be perfect candidates for this scenario. For instance, exogenous electrical stimulation of neural tissue in the central nervous system is empowered by using a conducting polymer-based scaffold and in this case, the scaffold essentially serves to increase the electrical conductivity of the damaged tissue site, and it does not need to remain or integrate into the target neural tissue. Providing permanent physical cues through conducting polymer-based nerve conduits is also a promising avenue that hasn't been fully explored, which is a perfect fit due to the robust fabrication of aligned nanofibrous structures with conducting polymers.

13.6 FUTURE PERSPECTIVE

Conducting polymers have proved themselves to be highly suitable smart and versatile biomaterials for neural tissue engineering applications. The foundation for this area of research was built on high conductivity and another physical and chemical similarity of conducting polymers with neural tissue continued with their application as vehicles for exogenous electrical stimulation. Once dubbed as one of the most promising materials for neural regeneration, their transition into practical application in the clinic has been slowed down due to various aforementioned hurdles. These complications are further shifted in the wrong direction by developing strategies that yield positive *in vitro* outcomes, while their applicability for *in vivo* settings was limited. However, there is still a substantial interest and potential for conducting polymers. Passive and active electrical stimulation research requires a push toward clinical research through developing precise reproducible fabrication methods and substantial *in vivo* studies, as well as *in vitro* studies that yield a deeper understanding of the potential mechanics of conducting polymer-based neural tissue engineering.

Continuation of somewhat recent progress in conducting polymer-based nerve conduits is an area with high potential which requires further exploration. This line of research goes beyond the prosperous but simplistic earlier ideas that developing any biocompatible material that matches the high level of electrical conductivity of neural tissue will lead to desired neural regeneration. The new generation of these smart materials effectively can be a turning point for peripheral tissue engineering. Another idea that has been floated around but has not been thoroughly investigated is developing artificial neural replacements with conducting polymers-based structures. Due to the sensitive nature of neural tissue and the nervous system in general, there are many cases, including trauma or neurological disease, caused neural damage and degeneration, where the host tissue is unable to regenerate or heal itself even with the assistance of passive and active electrical stimulation, and nerve conduits. Neural replacements are required in these instances, and these replacements need to mimic the electrical, mechanical, and chemical properties of healthy neural tissue, and more importantly, they need to be functional. This is where conducting polymer-based neural tissue engineering can step in to provide biomaterials as active and smart tissue replacements with matching properties and functionality. Such outcomes are still away from realization and require a great amount of dedicated research.

In conclusion, any meaningful progress in neural tissue engineering with conducting polymers demands regeneration strategies that go beyond gathering preliminary *in vitro* and *in vivo* results. These strategies need to encompass the requirements for neural regeneration at every step of the regeneration and healing process, from the start point until full recovery and function restoration is achieved.

REFERENCES

1. B. Guo and P. X. Ma, "Conducting polymers for tissue engineering," *Biomacromolecules*, vol. 19, no. 6, pp. 1764–1782, 2018.
2. L. Ghasemi-Mobarakeh *et al.*, "Application of conductive polymers, scaffolds and electrical stimulation for nerve tissue engineering," *J. Tissue Eng. Regen. Med.*, vol. 5, no. 4, pp. e17–e35, 2011, doi: https://doi.org/10.1002/term.383.
3. T. Nezakati, A. Seifalian, A. Tan, and A. M. Seifalian, "Conductive Polymers: Opportunities and Challenges in Biomedical Applications," *Chem. Rev.*, vol. 118, no. 14, pp. 6766–6843, 2018, doi: 10.1021/acs.chemrev.6b00275.
4. X. Yu, T. Zhang, and Y. Li, "3D printing and bioprinting nerve conduits for neural tissue engineering," *Polymers (Basel).*, vol. 12, no. 8, pp. 1–27, 2020, doi: 10.3390/POLYM12081637.

5. P. Abdollahiyan, F. Oroojalian, and A. Mokhtarzadeh, "The triad of nanotechnology, cell signalling, and scaffold implantation for the successful repair of damaged organs: An overview on soft-tissue engineering," *J. Control. release*, 2021.

6. S. J. Lee *et al.*, "Advances in 3D Bioprinting for Neural Tissue Engineering," *Adv. Biosyst.*, vol. 2, no. 4, pp. 1–18, 2018, doi: 10.1002/adbi.201700213.

7. S. Farzamfar *et al.*, "Neural tissue regeneration by a gabapentin-loaded cellulose acetate/gelatin wet-electrospun scaffold," *Cellulose*, vol. 25, no. 2, pp. 1229–1238, 2018.

8. C. E. Schmidt and J. B. Leach, "Neural tissue engineering: strategies for repair and regeneration," *Annu. Rev. Biomed. Eng.*, vol. 5, no. 1, pp. 293–347, 2003.

9. H. Nekounam *et al.*, "Electroconductive Scaffolds for Tissue Regeneration: Current opportunities, pitfalls, and potential solutions," *Mater. Res. Bull.*, vol. 34, p. 111083, 2020.

10. A. Saberi, F. Jabbari, P. Zarrintaj, M. R. Saeb, and M. Mozafari, "Electrically Conductive Materials: Opportunities and Challenges in Tissue Engineering," *Biomolecules*, vol. 9, no. 9, p. 448, 2019.

11. S. Vandghanooni and M. Eskandani, "Electrically conductive biomaterials based on natural polysaccharides: Challenges and applications in tissue engineering," *Int. J. Biol. Macromol.*, vol. 141, pp. 636–662, 2019.

12. R. Balint, N. J. Cassidy, and S. H. Cartmell, "Conductive polymers: Towards a smart biomaterial for tissue engineering," *Acta Biomater.*, vol. 10, no. 6, pp. 2341–2353, 2014, doi: 10.1016/j.actbio.2014.02.015.

13. H. Samadian *et al.*, "Naturally occurring biological macromolecules-based hydrogels: Potential biomaterials for peripheral nerve regeneration," *International Journal of Biological Macromolecules*, vol. 154. Elsevier B.V., pp. 795–817, Jul. 01, 2020, doi: 10.1016/j.ijbiomac.2020.03.155.

14. H. Samadian, H. Maleki, Z. Allahyari, and M. Jaymand, "Natural polymers-based light-induced hydrogels: Promising biomaterials for biomedical applications," *Coord. Chem. Rev.*, vol. 420, p. 213432, 2020.

15. S. Gholizadeh *et al.*, "Preparation and characterization of novel functionalized multiwalled carbon nanotubes/chitosan/β-Glycerophosphate scaffolds for bone tissue engineering," *Int. J. Biol. Macromol.*, vol. 97, pp. 365–372, Apr. 2017, doi: 10.1016/J.IJBIOMAC.2016.12.086.

16. K. Namsheer and C. S. Rout, "Conducting polymers: a comprehensive review on recent advances in synthesis, properties and applications," *RSC Adv.*, vol. 11, no. 10, pp. 5659–5697, 2021.

17. M. Li, Y. Guo, Y. Wei, A. G. MacDiarmid, and P. I. Lelkes, "Electrospinning polyaniline-contained gelatin nanofibers for tissue engineering applications," *Biomaterials*, vol. 27, no. 13, pp. 2705–2715, 2006.

18. R. Karimi-Soflou, S. Nejati, and A. Karkhaneh, "Electroactive and antioxidant injectable in-situ forming hydrogels with tunable properties by polyethylenimine and polyaniline for nerve tissue engineering," *Colloids Surfaces B Biointerfaces*, vol. 199, no. January, p. 111565, 2021, doi: 10.1016/j.colsurfb.2021.111565.

19. G. Kaur, R. Adhikari, P. Cass, M. Bown, and P. Gunatillake, "Electrically conductive polymers and composites for biomedical applications," *RSC Adv.*, vol. 5, no. 47, pp. 37553–37567, 2015, doi: 10.1039/c5ra01851j.

20. S. Yang *et al.*, "Polypyrrole/alginate hybrid hydrogels: electrically conductive and soft biomaterials for human mesenchymal stem cell culture and potential neural tissue engineering applications," *Macromol. Biosci.*, vol. 16, no. 11, pp. 1653–1661, 2016.

21. D. N. Heo, S. J. Lee, R. Timsina, X. Qiu, N. J. Castro, and L. G. Zhang, "Development of 3D printable conductive hydrogel with crystallized PEDOT:PSS for neural tissue engineering," *Mater. Sci. Eng. C*, vol. 99, no. January, pp. 582–590, 2019, doi: 10.1016/j.msec.2019.02.008.

22. C. Xu *et al.*, "Biodegradable and electroconductive poly (3, 4-ethylenedioxythiophene)/carboxymethyl chitosan hydrogels for neural tissue engineering," *Mater. Sci. Eng. C*, vol. 84, pp. 32–43, 2018.

23. A. Sadeghi, F. Moztarzadeh, and J. Aghazadeh Mohandesi, "Investigating the effect of chitosan on hydrophilicity and bioactivity of conductive electrospun composite scaffold for neural tissue engineering," *Int. J. Biol. Macromol.*, vol. 121, pp. 625–632, 2019, doi: 10.1016/j.ijbiomac.2018.10.022.

24. H. Nekounam, Z. Allahyari, S. Gholizadeh, E. Mirzaei, M. A. Shokrgozar, and R. Faridi-Majidi, "Simple and robust fabrication and characterization of conductive carbonized nanofibers loaded with gold nanoparticles for bone tissue engineering applications," *Mater. Sci. Eng. C*, vol. 117, no. July, p. 111226, 2020, doi: 10.1016/j.msec.2020.111226.

25. J. Xie, M. R. MacEwan, A. G. Schwartz, and Y. Xia, "Electrospun nanofibers for neural tissue engineering," *Nanoscale*, vol. 2, no. 1, pp. 35–44, 2010, doi: 10.1039/b9nr00243j.

26. F. Imani, R. Karimi-Soflou, I. Shabani, and A. Karkhaneh, "PLA electrospun nanofibers modified with polypyrrole-grafted gelatin as bioactive electroconductive scaffold," *Polymer (Guildf).*, vol. 218, no. September 2020, p. 123487, 2021, doi: 10.1016/j.polymer.2021.123487.

27. F. Yang, R. Murugan, S. Wang, and S. Ramakrishna, "Electrospinning of nano/micro scale poly(l-lactic acid) aligned fibers and their potential in neural tissue engineering," *Biomaterials*, vol. 26, no. 15, pp. 2603–2610, 2005, doi: 10.1016/j.biomaterials.2004.06.051.

28. Y. Zhao, Y. Liang, S. Ding, K. Zhang, H. quan Mao, and Y. Yang, "Application of conductive PPy/SF composite scaffold and electrical stimulation for neural tissue engineering," *Biomaterials*, vol. 255, no. May, p. 120164, 2020, doi: 10.1016/j.biomaterials.2020.120164.

29. V. Raeisdasteh Hokmabad, S. Davaran, A. Ramazani, and R. Salehi, "Design and fabrication of porous biodegradable scaffolds: a strategy for tissue engineering," *J. Biomater. Sci. Polym. Ed.*, vol. 28, no. 16, pp. 1797–1825, 2017.

30. Y. Liang and J. C.-H. Goh, "Polypyrrole-Incorporated Conducting Constructs for Tissue Engineering Applications: A Review," *Bioelectricity*, vol. 2, no. 2, pp. 101–119, 2020, doi:10.1089/bioe.2020.0010.

31. V. Chiono and C. Tonda-Turo, "Trends in the design of nerve guidance channels in peripheral nerve tissue engineering," *Prog. Neurobiol.*, vol. 131, pp. 87–104, 2015, doi:10.1016/j.pneurobio.2015.06.001.

32. H. Baniasadi, A. Ramazani S.A., and S. Mashayekhan, "Fabrication and characterization of conductive chitosan/gelatin-based scaffolds for nerve tissue engineering," *Int. J. Biol. Macromol.*, vol. 74, pp. 360–366, 2015, doi: 10.1016/j.ijbiomac.2014.12.014.

33. G. S. Krishnakumar, S. Sampath, S. Muthusamy, and M. A. John, "Importance of crosslinking strategies in designing smart biomaterials for bone tissue engineering: A systematic review," *Mater. Sci. Eng. C*, vol. 96, no. May 2018, pp. 941–954, 2019, doi: 10.1016/j.msec.2018.11.081.

34. J. Maitra and V. K. Shukla, "Cross-linking in hydrogels-a review," *Am. J. Polym. Sci*, vol. 4, no. 2, pp. 25–31, 2014.

35. F. S. Tenório, T. L. do Amaral Montanheiro, A. M. I. dos Santos, M. dos Santos Silva, A. P. Lemes, and D. B. Tada, "Chitosan hydrogel covalently crosslinked by gold nanoparticle: Eliminating the use of toxic crosslinkers," *J. Appl. Polym. Sci.*, vol. 138, no. 6, p. 49819, 2021.

36. Y. Pooshidani, N. Zoghi, M. Rajabi, M. Haghbin Nazarpak, and Z. Hassannejad, "Fabrication and evaluation of porous and conductive nanofibrous scaffolds for nerve tissue engineering," *J. Mater. Sci. Mater. Med.*, vol. 32, no. 4, 2021, doi: 10.1007/s10856-021-06519-5.

37. R. Boni, A. Ali, A. Shavandi, and A. N. Clarkson, "Current and novel polymeric biomaterials for neural tissue engineering," *J. Biomed. Sci.*, vol. 25, no. 1, pp. 1–21, 2018.

38. B. Weng *et al.*, "Bio-interface of conducting polymer-based materials for neuroregeneration," *Adv. Mater. Interfaces*, vol. 2, no. 8, p. 1500059, 2015.

39. H. Kaisvuo, "3D Printed Conducting Polymer Microelectrode Arrays for Electrical Stimulation of Neural Tissues–A-Proof-of-Concept." 2019.

40. J. Y. Lee, "Electrically conducting polymer-based nanofibrous scaffolds for tissue engineering applications," *Polym. Rev.*, vol. 53, no. 3, pp. 443–459, 2013.

41. Z.-B. Huang, G.-F. Yin, X.-M. Liao, and J.-W. Gu, "Conducting polypyrrole in tissue engineering applications," *Front. Mater. Sci.*, vol. 8, no. 1, pp. 39–45, 2014.

42. C. Vallejo-Giraldo, A. Kelly, and M. J. P. Biggs, "Biofunctionalisation of electrically conducting polymers," *Drug Discov. Today*, vol. 19, no. 1, pp. 88–94, 2014.

43. R. Ravichandran, S. Sundarrajan, J. R. Venugopal, S. Mukherjee, and S. Ramakrishna, "Applications of conducting polymers and their issues in biomedical engineering," *J. R. Soc. Interface*, vol. 7, no. suppl_5, pp. S559–S579, 2010.

44. H. Ghasemi *et al.*, "Tissue stiffness contributes to YAP activation in bladder cancer patients undergoing transurethral resection," *Ann. N. Y. Acad. Sci.*, 2020.

45. G. M. Raghavendra, K. Varaprasad, and T. Jayaramudu, "Biomaterials: design, development and biomedical applications," in *Nanotechnology applications for tissue engineering*, William Andrew Publishing, Oxford, 2015, pp. 21–44.

46. Z. Allahyari *et al.*, "Optimization of electrical stimulation parameters for MG-63 cell proliferation on chitosan/functionalized multiwalled carbon nanotube films," *RSC Adv.*, vol. 6, no. 111, pp. 109902–109915, 2016.

47. J. Zhang *et al.*, "Conductive composite fiber with optimized alignment guides neural regeneration under electrical stimulation," *Adv. Healthc. Mater.*, vol. 10, no. 3, p. 2000604, 2021.

48. R. Zhu, Z. Sun, C. Li, S. Ramakrishna, K. Chiu, and L. He, "Electrical stimulation affects neural stem cell fate and function in vitro," *Exp. Neurol.*, vol. 319, p. 112963, 2019.

49. J. Wang, L. Tian, N. Chen, S. Ramakrishna, and X. Mo, "The cellular response of nerve cells on poly-l-lysine coated PLGA-MWCNTs aligned nanofibers under electrical stimulation," *Mater. Sci. Eng. C*, vol. 91, pp. 715–726, 2018.

50. M. Imaninezhad, K. Pemberton, F. Xu, K. Kalinowski, R. Bera, and S. P. Zustiak, "Directed and enhanced neurite outgrowth following exogenous electrical stimulation on carbon nanotube-hydrogel composites," *J. Neural Eng.*, vol. 15, no. 5, p. 56034, 2018.

51. M. R. Love, S. Palee, S. C. Chattipakorn, and N. Chattipakorn, "Effects of electrical stimulation on cell proliferation and apoptosis," *J. Cell. Physiol.*, vol. 233, no. 3, pp. 1860–1876, 2018.

52. S. Grossemy, P. P. Y. Chan, and P. M. Doran, "Electrical stimulation of cell growth and neurogenesis using conductive and nonconductive microfibrous scaffolds," *Integr. Biol.*, vol. 11, no. 6, pp. 264–279, 2019.

53. Y. Shan, H. Feng, and Z. Li, "Electrical stimulation for nervous system injury: research progress and prospects," *Acta Physico-Chimica Sin.*, vol. 36, no. 12, p. 2005038, 2020.

54. H. Cheng, Y. Huang, H. Yue, and Y. Fan, "Electrical Stimulation Promotes Stem Cell Neural Differentiation in Tissue Engineering," *Stem Cells Int.*, vol. 2021, 2021.

55. A. Casella, A. PANItch, and J. K. Leach, "Endogenous Electric Signaling as a Blueprint for Conductive Materials in Tissue Engineering," *Bioelectricity*, vol. 3, no. 1, pp. 27–41, 2021.

56. H. Baniasadi, S. A. Ahmad Ramazani, S. Mashayekhan, M. R. Farani, F. Ghaderinezhad, and M. Dabaghi, "Design, Fabrication, and Characterization of Novel Porous Conductive Scaffolds for Nerve Tissue Engineering," *Int. J. Polym. Mater. Polym. Biomater.*, vol. 64, no. 18, pp. 969–977, 2015, doi: 10.1080/00914037.2015.1038817.

57. N. Kamaly, B. Yameen, J. Wu, and O. C. Farokhzad, "Degradable controlled-release polymers and polymeric nanoparticles: mechanisms of controlling drug release," *Chem. Rev.*, vol. 116, no. 4, pp. 2602–2663, 2016.

14 Conducting Polymers for Ophthalmic Applications

Ismail Bal[1,2] *and Israfil Kucuk*[2]

[1]Vocational School of Health Services, Istanbul Okan University, Istanbul, Turkey

[2]Programme of Nanoscience and Nanoengineering, Institute of Nanotechnology, Gebze Technical University, Gebze, Turkey

CONTENTS

14.1 INTRODUCTION

The topic of ophthalmic drug distribution has become one of the most widely researched topics in recent years. Despite significant progress has been made in this area, new approaches are being incorporated. The whole drug carrier systems serve a couple of basic purposes: deliver the active ingredients to the target tissue(s), provide the therapeutic effect safely and effectively. Controlled drug release systems are being developed that allow adjustment of the release kinetics, time, and target of the therapeutic agent in the body, and thus enable drug therapy with increased efficacy and safety. Controlled drug release systems stand out in the treatment of many diseases, especially eye diseases, due to their biocompatibility and sterility. The ability to keep the drug at the therapeutic level for a longer time prevents the application of repetition of the dose as in the treatments performed with traditional methods, thus increasing the benefit that can be obtained from the drug. Besides, reducing the possible side effects of the drug by targeting it and keeping the drug at the effective level, and using fewer amounts of drugs helps to reduce the cost and toxic effect. Thanks to controlled drug release systems, treatment safety, treatment cost, and patient's quality of life are positively affected.

Drugs can be taken in a variety of ways such as swallowing, injection, or inhalation. Each method has its advantages and disadvantages, and not all methods are suitable for all drugs. In addition, this is also related to the applied tissue or organ to which the drug is administered. For example, drug delivery to the eye has different and more difficult limitations compared to other tissues. The physiological barriers present in the eye make it difficult to administer effective drugs. Recent advances in nanotechnology and polymeric drug delivery provide a new alternative route that can be effective in the treatment of ocular(ophthalmic) diseases. Especially the use of biodegradable polymers has led to advantageous results such as increased bioavailability and retention time. Drug-active ingredients can now be delivered to the deeper tissues of the eye using nanoscale drug delivery systems [1]. Among the benefits of using polymers to deliver ocular therapeutics to tissues, it is necessary to refer to their mucoadhesive properties in the cornea and conjunctiva region. This is significant in terms of extending the penetration of the drug and residence time on the corneal surface [2].

Besides polymers, the development of novel materials and designs has resulted in new drug delivery system alternatives. With the growth in material chemistry, micro and nanoplatforms have been designed for the development of controlled and smart drug release systems. Metal clusters and organic ligands enable the formation of 1D to 3D networks by self-assembly with large pore volumes and tunable host-guest interactions [3]. Metal-organic frameworks (MOFs) are a useful class of coordination polymers consisting of clusters or metal ions coordinated to organic ligands. Because of their facile synthesis, highly porous structure, biocompatibility, and high surface area, smart MOF-based nanocomposites are used in gas storage/separation, drug release, and biological imaging/sensing systems [4].

Advances in nanotechnology affect the biomedical fields as well. To solve the problems that traditional treatment methods can't handle and to enhance ocular bioavailability, a variety of ways are recommended. This chapter describes polymeric structures and MOFs used in ocular drug delivery (ODD) applications. Understanding ODD routes and methods require basic knowledge about eye anatomy. In the beginning, a basic overview of eye anatomy and physiology is available.

14.2 EYE ANATOMY AND OPHTHALMIC DRUG DELIVERY ROUTES

14.2.1 EYE ANATOMY

The eye can be split into two parts, called the anterior and posterior segments. The anterior segment includes the cornea, the iris, the ciliary body, the conjunctiva, the crystalline lens, and a chamber filled with aqueous humor [5]. The cornea is the outermost surface of the eye and consists of five layers, transparent to allow light refraction and transmission to the retina. In addition, it covers the eye surface to provide a protective layer. The iris is the colored circular part surrounding the pupil. The pupil changes its size in response to light levels, allowing the proper amount of light to reach the retina through muscular contraction and relaxation. The ciliary body produces the aqueous humor that has immunological and nutritional tasks as well as maintaining a 10–21 mmHg intraocular pressure (IOP) in a healthy eye. The conjunctiva is a clear mucous membrane that covers the front of the eye up to the cornea and the inner surfaces of the eyelids, responsible for tear production and preservation of the tear film. The human crystalline lens has a clear and biconvex shape that helps in focusing light onto the retina. The lens is flexible and can vary its shape and focusing power based on the distance between the object and the eye thanks to muscles called zonules.

The posterior segment has three layers as sclera, choroid, and retina, as well as a vitreous cavity filled with vitreous humor. The sclera is the external white shell of the eye and serves as a protective layer with the eyelids as well as helps in the preservation of the eye's form by giving mechanical support. The hydrophobicity and charge of the molecules are critical parameters for drug molecule penetration through the sclera. The high surface area and relatively permeable nature of the sclera make it suitable for researches in drug delivery to the posterior segment [6]. The choroid is the middle layer of the eyeball, with a dense vascular system and maintains the temperature of the eye medium and feeds the inner layer. The retina is the innermost layer of the eye and is primarily responsible for vision. The incoming light is transformed into an electrical signal by specialized neurons called photoreceptors in the retina (i.e., rods and cones) then delivered to the brain for image formation.

14.2.2 Ophthalmic Drug Delivery Routes

Eye diseases, which can be caused by a variety of factors, have a direct impact on a patient's visual acuity and quality of life. Visual impairment affects an estimated 2.2 billion people worldwide [7]. Many eye diseases affect the vision of the patient, including cataracts and glaucoma in the anterior segment while age-related macular degeneration (AMD), uveitis, and diabetic retinopathy in the posterior segment [8]. Tissue damage or irreversible visual impairment can occur in the case that those diseases are not treated effectively. This may have consequences that directly affect the patient's quality of life. In the treatment of ophthalmic infections caused by diseases or traumas, there have been some promising developments. However, despite their easily accessible nature, these treatment options are limited by physiological barriers in the eye. The eye is an outgrowth of the central nervous system embryologically. It has a lot in common with the brain in terms of anatomical and physiological features. Several membranes and barriers regulate the movement of fluids and solutes in the eye. Systemic administration for ocular diseases is frequently ineffective, owing to the blood-ocular barrier such as blood-aqueous (BAB) and blood-retina (BRB) barriers (Figure 14.1). Drug molecules are limited from systemic circulation to the anterior segment due to

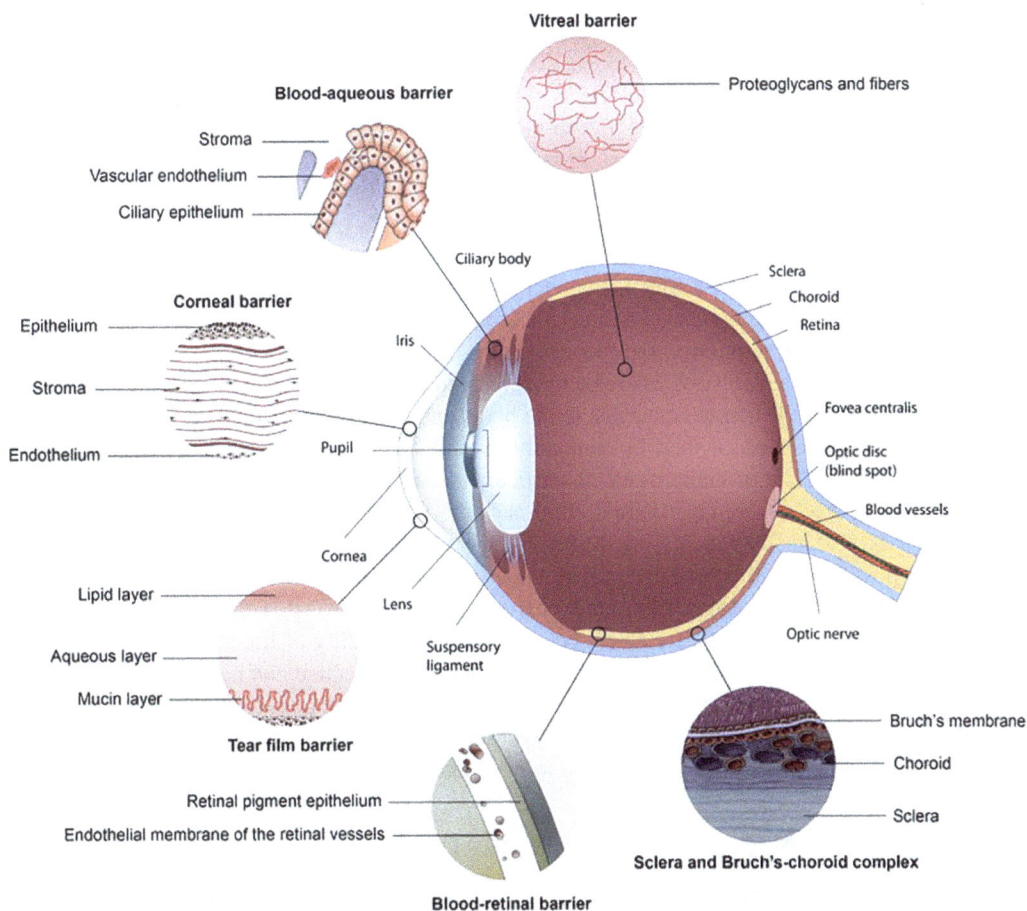

FIGURE 14.1 Physiological barriers in ODD. (Adapted with permission from Ref. [9]. Copyright (2018) Copyright The Authors, some rights reserved; exclusive licensee [Elsevier].)

the tight junctions of the iris vascular system and the ciliary body in the BAB. In BRB, similarly in BAB, delivery of drug molecules to the posterior segment is limited, because of having tight junctions of retinal pigment epithelium and retinal capillary endothelium cells [5].

Two main approaches can be identified in the treatment of ocular disorders: (a) delivery of drug molecules to the relevant tissues and (b) surgical procedure for repairing or replacing damaged tissue. Various methods are available for drug delivery to the eye. However, it has major deficiencies in the treatment of diseases that occur primarily in the posterior segment [10]. Intraocular (intracameral and intravitreal), topical, and systemic routes are the most common ocular drug delivery systems (ODDS). The anterior and posterior segments can both be treated with the intracameral approach. Applications to the anterior chamber, in particular, are conducted during surgery (i.e., cataract surgery). For the posterior segment, despite its relatively high bioavailability, some infections may develop such as retinal detachment or hemorrhage, and injections need to be repeated [11, 12]. The intravitreal route includes the injection of the drug directly in the vitreous humor, thus a significant way for the treatment of posterior segment diseases. Drug distribution in the vitreous is regulated by a variety of parameters, including the drug's molecular weight (MW), and is rarely uniform [6].

Topical ophthalmic applications include eye drops, ointments, and gels. Among the ocular ophthalmic applications, the most common are eye drops [5, 13]. Although they are preferred due to their ease of use, 1–5% of the administered dose can penetrate the anterior segment. The delivery of ocular therapeutics is hampered by dynamic barriers as well as ocular barriers. Due to the sudden increase in the tear fluid after the application of the eye drops, the majority of the eye drops are removed from the surface of the eye as a result of reflexive blinking. The efficacy of the remaining eye drops is also slightly reduced due to the diluting effect of the tears, resulting in low bioavailability of the eye drops in the range of 1–5% [14]. In addition, the barrier function of the cornea is another reason that reduces the effectiveness of the topical application. After the tear film and ocular mucosa have been passed, the next obstacle is the multilayered corneal structure with hydrophilic and lipophilic characteristics. The epithelium, the first layer of the five-layer cornea, is resistant to approximately 90% of hydrophilic drugs and about 10% of hydrophobic formulations due to its lipophilic nature. Furthermore the presence of tightly bound cell membrane regions, all of them are considered, makes it an effective barrier structure that prevents the transmission of foreign substances as well as therapeutic agents, thus poses a significant challenge for ODD [15].

Therefore, studies on promoting the bioavailability of the drug have been focused on in recent years. Controlled drug release systems are being developed that allow adjustment of the release kinetics, time, and target of the therapeutic agent in the body, and thus enable drug therapy with increased efficacy and safety. Because of their biocompatibility and sterility, controlled drug release systems stand out in the treatment of many diseases, particularly eye diseases. The ability to keep the drug at the therapeutic level for a longer period eliminates the need for dose repetition, as in traditional methods of treatment, thereby increasing the benefit that can be obtained from the drug. Furthermore, reducing the possible side effects of the drug by targeting it and keeping the drug at the effective level and using fewer amounts of drugs help to reduce the cost and toxic effect. Thanks to controlled drug release systems, treatment safety, treatment cost, and patient's quality of life are positively affected [16]. Sustained drug release systems allow the development of ways that can be effective in penetration through ocular barriers. As previously stated, the presence of strong barriers prevents ophthalmic drugs from diffusing or penetrating the eye. To address these issues, various strategies have been developed. The following sections focus on some strategies used for this purpose.

14.3   POLYMERS IN OPHTHALMIC APPLICATIONS

For any drug to be used for medical purposes such as the treatment or prevention of disease, its efficacy and safety must be proven. However, the barriers that protect the eye from pathogens also block drug molecules from passing through to the eye tissues. Considering the difficulties in delivering drugs to the target tissue in the eye, the use of new delivery routes such as polymers and nanosystems is regarded as a novel approach. Polymers can be used in medical applications such as advanced drug delivery and tissue engineering due to their biocompatible characteristics. The mucoadhesive nature of the polymers contributes to increased drug bioavailability by increasing contact time with the cornea and conjunctiva. As a result, they have a potential to improve drug release in the treatment of both anterior and posterior segment diseases [17]. Considering that rapid clearance of eye drops from the corneal surface results in low bioavailability, polymeric formulations make an indisputable contribution to the development of ocular therapeutics [18]. Biodegradable polymers, in particular, are frequently used in the development of therapeutic formulations. The fact that a polymer is biodegradable means that it is not toxic when broken down into smaller fragments, and drug release occurs as the polymer breaks down, which means degradation.

The biocompatible nature and easy availability of biopolymers make them ideal for utilization in drug delivery systems. They can be used in medical applications, particularly controlled drug release systems, due to their nontoxicity, non-immunogenicity, and good swelling capacity. Furthermore, the mucoadhesive properties of most natural polymers make them suitable for drug release studies on different mucous membranes [19]. The basic concept of mucoadhesion includes the binding of a drug-laden carrier to mucosal membrane structures. The mucoadhesion efficiency of the ocular therapeutics is affected by factors such as tear film content and physicochemical properties. At the same time, some critical points should be noticed in the selection of polymers, such as MW (high MW enhances adhesion), optimum polymer chain length (long enough to support penetration, short enough to facilitate diffusion), crosslinking (mucoadhesive strength decreases as the degree of crosslinking increases) as well as being nontoxic and biocompatible [20]. Mucoadhesive polymers exhibit adhesive properties through the interaction between the polymer chain and the mucin layer in the tear film. The interaction between the mucous layer and the drug form manipulates mucoadhesive drug delivery. Owing to this interaction, more time is provided for the drug to transmission into the tissues.

The mucoadhesion process can be summarized in three steps. In the first step, the biopolymer gets wet, then swells after it settles on the mucosal membrane. This process is governed by the wetting theory. The second step involves the interpenetration of polymer chains with the mucosal surface by forming physical bonds. This process is governed by a combination of adsorption and electronic theories. In the third and final step, the adhesion is reinforced by the covalent bonding of the polymer chains. This process is governed by a combination of electronic theory, adsorption theory, and diffusion theory. Most of biopolymers are known for their ability to adhere to tissues and have an ionized to known as a polyelectrolyte (PE). PEs are a subclass of polymers that have different properties from other polymers thanks to their ionic structure. PEs are beneficial in biomedical applications because of their ability to bind to oppositely charged surfaces or form complexes with oppositely charged polymers. Many natural PEs rely on electrostatic interactions with charged subcellular membranes or lipid bilayers of cellular to function biologically [21]. Some of the most widely known mucoadhesive polymers can be listed as chitosan (CS), alginate (ALG), carbomer, methylcellulose (MC), and hyaluronic acid (HA) [20].

Polyelectrolyte complexes (PEC) or multilayers formed by combining oppositely charged polymers without any chemical solvent have a wide range of functional structures. Changes in properties such as ionic strength, viscosity, and pH of the solution directly affect the behavior of PE. Since it is possible to optimize the properties of these structures, their use in the medical field is increasingly being investigated [22]. Polymers and drug complexes are being investigated as drug delivery systems for PEs. Natural polymers like CS, hypromellose phthalate, and HA have been studied for their mucoadhesive capabilities in the development of nano-PECs. PE matrices, in the form of hydrogels, coatings, or membranes, can be employed as controlled delivery vehicles. The formation of the PEC is based on the association of oppositely charged species such as PE-PE, PE-drug, or PE-surfactant. Thanks to van der Waals (vdW) forces, H-bonds, and dipole-dipole interactions, in the PE structure, temporary networks are formed without the need for crosslinking agents. The polymers that make up the PECs combine without losing their unique properties. It is well suited for use in biomedical applications due to its optimizable nature as well as being biodegradable and biocompatible [23].

14.4 NATURALLY OCCURRING POLYELECTROLYTES

PEs are naturally occurring linear or branched biopolymers that can dissociate when dissolved in a polar solvent (i.e., water) due to the presence of ionic groups. These molecules exhibit interesting behavior due to their chain and high charge characteristics. Although PEs are neutral molecules in their natural state, they become promising platforms for drug delivery after dissociation due to electrostatic interactions [23]. Figure 14.2 depicts how PEs are classified briefly.

The mucoadhesiveness of polyanions and polycations can occur through different processes. Polycation mucoadhesion is based on electrostatic interactions with negatively charged mucin while hydrogen bonds and vdW interactions are effective in the mucoadhesion mechanism in polyanions. PEs in which the macromolecule backbone contain both acidic (known as polyanion) and basic (called polycation) groups are called polyampholytes and are divided into three groups: annealed, quenched, and betaine (also referred to as zwitterionic) [22]. Generally, nontoxic and biocompatible nature of PECs based on polysaccharides and their derivatives makes them good

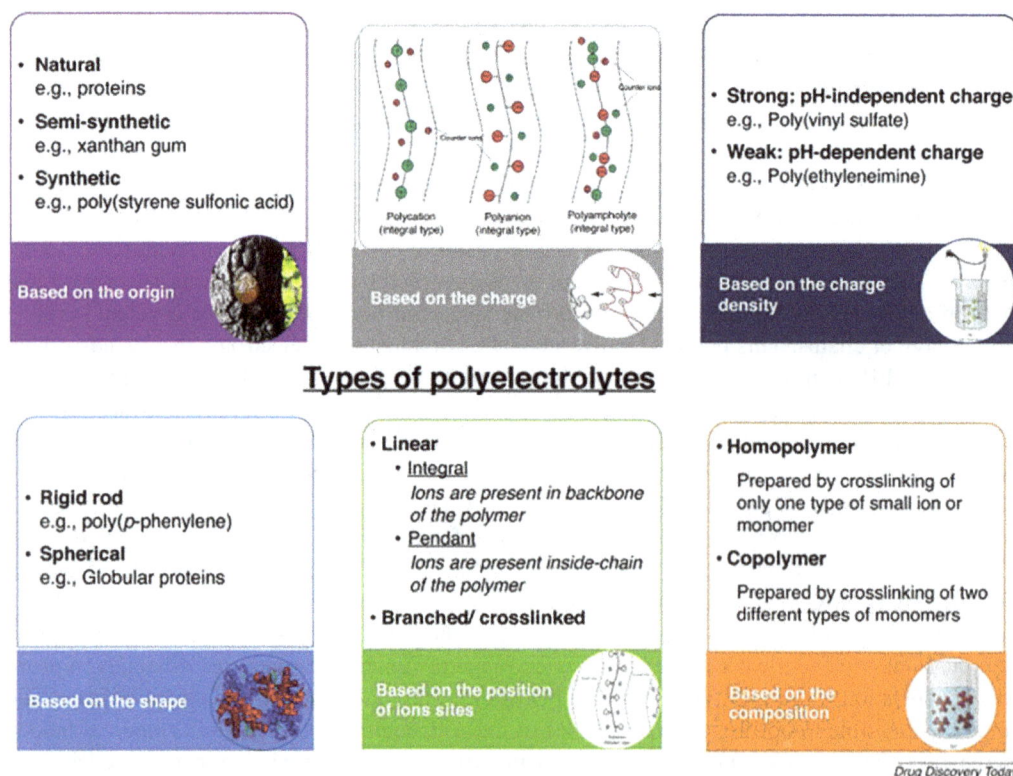

FIGURE 14.2 Classification of PEs based on the charge and position of ionic sites in the polymer chain. (Adapted with permission from Ref. [23]. Copyright (2018) Copyright The Authors, some rights reserved; exclusive licensee [Elsevier]).

candidates for use in medical applications. PEC examples can be employed in pharmaceutical applications, tissue engineering, wound dressing studies, and drug delivery systems [22]. The fact that PECs are biocompatible does not mean that they can be used directly as drug carriers. Some cytotoxicity issues may arise depending on the properties of the drug or nanoparticles. Therefore, a full understanding of the toxicological properties of the particles as well as cytotoxicity studies for PEC-based drug delivery systems is required [24]. Natural, semisynthetic, and synthetic polymers can be used to form PEs or preparing of mucoadhesive drug formulation. Some of the polymers in this field and their properties, as well as some usage examples from the literature, are included below.

14.4.1 Natural Polymers

14.4.1.1 Alginate (ALG)

ALGs are an anionic polymer containing 1,4-linked β-D-mannunoric acid (M) and α-L-glucuronic acid (G), typically derived from brown algae including *Ascophyllum nodosum*, *Laminaria hyperborea*, *Macrocystis pyrifera*. The M and G contents differ depending on the extraction source of the ALGs [25]. ALGs have been extensively researched and used for many biomedical applications due to their high biocompatibility, low toxicity, mucoadhesiveness, relatively low cost, and induced gelling properties [25, 26]. ALGs experience quick gelation by interaction with multivalent cations. Calcium chloride, which has high biocompatibility, is one of the commonly used divalent cations (Ca^{++}) that's responsible for fast gelation. The Ca^{++} particles are bind to G blocks of ALG and frame an egg-box-like structure. The characteristic properties of ALG hydrogels, such as porosity and, mechanical strength depend on the G/M ratio, the type of ionic crosslinker (divalent cations), the concentration, and viscosity of the ALG solution [27]. It is known that polymers with charge density exhibit good mucoadhesive properties. Mucoadhesive nature of ALG enhances the residence time of drugs on the surfaces, thus making it a good candidate for drug release studies in mucosal tissues [28].

14.4.1.2 Chitosan (CS)

CS is a natural polymer composed of glucosamine and *N*-acetyl glucosamine copolymers, known for its biocompatible, biodegradable, nontoxic, and mucoadhesive properties. The presence of free hydroxyl and amino groups allows easy adjustment of physicochemical properties. These features make CS a promising candidate for pharmaceutical and biomedical applications [29]. CS, a polymer derived from chitin, is a common polymer ingredient in ophthalmic formulations. It can interact with negative charges in the mucosa and conjunctiva due to its cationic nature. The mucoadhesive performance of CS varies depending on the pH of the medium. A remarkable increase in mucoadhesive property exhibits at slightly alkaline (e.g. in tear film) or neutral pH. It has been shown that the drug formulation increases the residence time on the corneal surface and increases penetration by opening the tight junctions in the epithelial tissue. Furthermore, some medical applications are available such as tissue engineering and therapeutic drug delivery. It is also approved by the FDA for use in wound dressing [22].

14.4.1.3 Hyaluronic Acid (HA)

HA is a non-sulfated glycosaminoglycan found in a variety of tissues throughout the human body, including the aqueous humor of the eye, synovial fluid of joints, skin, umbilical cord, connective, and epithelial tissues. It is a linear anionic polysaccharide composed of repeating D-glucuronic acid and *N*-acetyl-D-glucosamine disaccharide units [30]. Furthermore, it is a natural polysaccharide with a MW range ranging from 20,000 to several million Daltons, depending on the enzyme that catalyzes its synthesis, remarkable for its excellent biocompatibility, non-immune response, mucoadhesiveness, pseudoplastic, viscoelasticity behavior, and biodegradability [28]. HA has important physiological and biological functions in the human body, including maintaining the mechanical integrity of tissues and contributing to the preservation of viscoelasticity. Because of its biocompatibility, biodegradability, and mucoadhesive properties, it has been used in ocular therapeutic formulations [29]. It has been reported that topical application accelerates the healing process of the corneal epithelium and is used to treatment of dry-eye syndrome as well as prolong the retention time of many ocular active substances thanks to its mucoadhesive and viscosity-regulating properties [31].

14.4.1.4   Pectin

Pectin, a branched macromolecule with a high MW, is found naturally in most plant walls and is widely used in biomedicine. Pectin gels can be made from pectin solutions with a high pectin concentration and at low pH conditions. It can also be obtained by cross-linking with divalent or trivalent cations. Pectin gels are an alternative for extending drug residence time and increasing ocular absorption. Furthermore, due to the presence of lacrimal electrolytes, pectin is being evaluated for in situ gel formation [32]. Due to its biocompatible, nontoxic, and biodegradable nature, it is considered a good candidate for the development of drug delivery systems or tissue engineering.

14.4.2   Synthetic Polymers

14.4.2.1   Poly(Lactic-*co*-Glycolic Acid) (PLGA)

Poly(lactic-*co*-glycolic acid) (PLGA), a synthetic and biodegradable polymer, has been frequently studied in applications such as drug delivery and tissue engineering. It can be modified with other polymers to improve its properties and make it usable in other applications. In ODDSs, it is employed in formulations that improve drug permeability and drug residence time.

14.4.2.2   Poly(Ethylene Glycol) (PEG)

Poly(ethylene glycol) (PEG) is another polymer used in ocular formulations to increase the permeability of nanoparticles. It can be used alone or in combination with polymers such as PLGA and CS. The addition of PEG in a drug formulation has been shown to increase mucoadhesive properties [33].

The aforementioned polymers and more, play a role in improving the mucoadhesive properties of ocular formulations and PEC formation. Some of the studies in the literature can be exemplified as follows. PE multilayers were formed on hydrophobic silicon intraocular lens (IOL) by soaking the substrates in hydrated HA and CS solutions, respectively. It has been reported that the multilayer formed by the combination of the antifouling property of HA and the antibacterial properties of CS is successful in preventing postoperative endophthalmitis and proliferation of lens epithelial cells [34]. In another study, brinzolamide-laden CS/pectin PEC mucoadhesive nanocapsules were produced for the treatment of glaucoma, and drug release lasted for 8 hours. CS-pectin PEC has been shown to promote corneal permeability and show promising benefits in decreasing IOP when compared to commercially available eye drops [35]. An ion exchange controlled-release delivery system was tested from montmorillonite (Mt), a hydrated aluminum silicate with a two-dimensional layer structure, to be used in the treatment of glaucoma, and betaxolol hydrochloride (BH) was released. In vitro experiments showed good compliance with ocular tissues. In conclusion, this new approach is promising for the controlled release of BH drugs in the treatment of glaucoma [36].

14.5   METAL-ORGANIC FRAMEWORKS (MOFs)

MOFs are a type of crystalline material having a high porosity internal surface area with a metal cation and an organic linker. MOFs also have tunable structures, crystal character, porosity, and polymeric network flexibility [37]. Inorganic structural units and organic binders are the two basic components of MOFs. Transition metal ions or tiny metal clusters are common inorganic structural units, while organic linkers include a wide range of molecules such as dicarboxylates [38]. MOFs can be classified according to the organic ligands in their structure. Besides carboxylate ligands are the most common ones, many different types of ligands, such as phosphonates, sulfonates, imidazolate, pyridine, and porphyrin, are typically found in MOF structures. By selecting the suitable ligand and metal for the intended usage and desired frame structure, features and network structures of hybrid materials can be altered. Especially due to the nontoxic metals and ligands are used in the release of drugs and active biomolecules, they can be considered as alternative drug carriers. High loading capacity, biodegradability, and miscellaneous functionality make them great candidates in drug delivery applications. Another benefit is that by managing drug release with pH and temperature, it can be guided to the desired tissues while avoiding problems like overdosing and side effects [39].

14.5.1 Synthesis of MOFs

With the growing interest in MOF structures, traditional MOF production methods need to be developed and organized in a way that allows for larger scale production and shorter synthesis times. Various approaches can be used to manipulate the desired crystallinity, morphology, and particle size of MOFs. Temperature, additives, manufacturing methods, etc. all have a direct impact on the physical and chemical properties of materials. The following are some of the methods for producing MOFs:

i. **Solvothermal/Hydrothermal method:** The metal salt, organic ligand, and appropriate solvent (water, dimethylformamide, etc.) are placed in bottles (at a certain temperature and pressure), and the mixture is electrically heated. The method is known as hydrothermal if water is used as a solvent. Particle size and shape can be controlled, resulting in particles with good dispersibility and uniform size distribution. Changing parameters such as temperature, pH, stoichiometry, and reaction time can affect the properties of the resultant product. This technique typically necessitates a long reaction time, a high operating temperature, and high pressure.

ii. **Sonochemical synthesis:** Using high-energy ultrasound into the reaction mixture is known as a rapid, simple, and environmentally friendly method. Bubbles form and collapse in the solution when high-energy ultrasound interacts with liquids. Ultrasonic waves cause high temperatures and pressures during synthesis. Because of the rapid heating and cooling rates, fine crystal growth is possible.

iii. **Microwave-assisted synthesis:** Microwaves heat any material that contains electric charges by acting like high-frequency electric fields. The advantages of this method include shorter reaction times, lower polydispersity, and easy morphology control.

iv. **Mechanochemical synthesis:** It is an environmentally friendly method that allows the formation of MOF without the need for a solvent. The difficulties of solvent-based techniques are thereby avoided, and high yield products are obtained. The drawbacks of the method include the limited number of MOFs that can be formed and the fact that only a small amount of material can be obtained [40].

14.5.2 MOFs in Biomedical Applications

MOFs are well suited for biomedical applications such as drug delivery, bioimaging, and photodynamic therapy, due to their simple surface functionality and adjustable pore diameter and size [41]. Because of their large pore volumes, MOFs also are effective in drug encapsulation studies. The presence of controllable pore size makes MOFs suitable candidates for drug delivery systems. So, MOFs have been extensively studied in biomedical applications. Neglecting of size and surface chemistry in drug delivery systems makes it difficult to control properties such as drug diffusion, penetration, and targeting. Therefore, the impact of rational designs especially in ODDS is examined.

The drug release behavior of MOFs can be controlled by factors such as temperature and pH, as well as drug framework interactions (bonding, interactions, etc.). The easy designable ability of MOFs allows the creation of carriers capable of controlled release into the target tissue. The side effects associated with conventional drug delivery systems can be avoided by employing MOFs as drug carriers. Regardless of the field in which they are used, some key parameters are available to consider in MOF-based material design. For example, scale is an important parameter because of their higher cellular uptake, improved permeability, and retention effect; nanoscale MOFs can retain higher concentrations of drug at targeted sites. Particle size influences drug release kinetics, i.e., the smaller the particle the faster the release kinetics. Since some MOF structures contain various toxic molecules, their biomedical applications have been limited by their toxicity. According to the toxicity values of metals are taken as reference, it can be concluded that Fe, Zn, Zr, and Mn-based MOFs are more suitable for biomedical applications. The presence of high and regular porous structures, which can be prepared in accordance with the structure and nature of the drug, allows controlled and prolonged drug release with avoiding burst release [42].

14.5.3   MOFs in Ophthalmic Applications

As previously stated, conventional drug delivery methods, such as eye drops, have low bioavailability in the range of 1–5%. In short, nearly 5% of drugs can penetrate desired ocular tissues. Increasing the drug dose may be an option in this case, but the toxic effect and potential side effects will increase. The inability of elderly patients, in particular, to apply eye drops on their own or at the appropriate time reduces the effectiveness of eye drops [3]. Studies on the development of well-designed and long-term drug release systems have been conducted regularly, particularly in recent years. The growing area of interest examples includes the use of ocular systems as drug reservoirs (contact lenses or IOL), nanoparticles trapped in hydrogel networks, biodegradable film coatings, and so on. Although the utilizing of these technologies has opened the way to more specific and targeted delivery systems, the number of new ocular products hitting the market is delayed due to clinical trial failures, industrial scale-up issues, and formulation features. Thus, it requires a more feasible approach [43].

However, the loss in physical (poor gravimetric capacity for drug) and textural qualities of some of these materials is their principal limitation for possible application. In recent years, new drug delivery platforms composed of polymeric materials (solid or semi-solid inserts) for treating ocular diseases have gained in popularity. These systems are advantageous in that they improve the residence time on the ocular surface, increase drug penetration, and provide therapeutic drug release at lower doses. There are even some commercially available products that deliver ocular drugs. Ocusert was one of the first commercialized products that could release drugs for glaucoma treatment for 5–7 days. On the other hand, the lack of a regular 3D network within the polymeric matrices, limits overall drug uptake and, as a result, prevents controlled release.

In a healthy eye, eye fluid (aqueous humor) should be drainage smoothly. Sometimes drainage channels can become clogged or damaged and resulting in a rapid increase in IOP. Glaucoma, a second leading cause of blindness, is a disease that causes vision loss when IOP rises high enough to destroy the optic nerves that connect the eye and the brain. Incorporation into platforms or vehicles capable of controlled release at the therapeutic level, increasing bioavailability, and avoiding a burst of release, is one of the most efficient alternatives in the treatment of glaucoma. MOFs are used as an approach that combines improved drug uptake and controlled release due to their high surface area and porous structure, brimonidine tartrate which is one of the most widely used drugs in the treatment of glaucoma, is included in the MOF for the treatment of chronic glaucoma. Gandara-Loe et al. evaluated the release of brimonidine drugs from various MOFs, its efficacy in the treatment of glaucoma, and its toxic effect. HKUST-1 (Cu), UiO-66, UiO-67, and MIL-100(Fe) MOFs have been selected to define the effectiveness of glaucoma treatment as injected intraocular cavity. It's reported that UiO-67 and MIL-100 (Fe) allow for increased loading capacity to values greater than 50 wt%. UiO-67 in aqueous media with a structural amorphization can have a drug release time of 12 days or more. In vitro models based on cell cultures were preferred over in vivo models for toxicity testing due to advantages such as low cost, time efficiency, reliable results, and no ethical issues. Toxicity tests with the 661W cell line did not result in retinal photoreceptor cell damage, except HKUST-1. Based on these premises, MOFs can be attributed as good candidates for intraocular therapeutics [3].

Functional MOF-based polyurethane (PU) thin films were created for drug encapsulation and releases studies, and uptake/release performance of UiO-67@PU nanocomposites for brimonidine in the liquid phase was assessed. It was discovered that embedding MOF nanocrystals in a hydrophobic polymer like PU allows for high-quality crystallinity and increases their stability. Brimonidine adsorption is limited by partial access to embedded MOFs, but a 60-fold improvement is seen when the composite system is compared with pure PU film. These absorption values in the liquid phase have promising results in the design of polymeric inserts or for other ophthalmic disorders [3]. As a new drug carrier, NH2-MIL-88(Fe) MOF prepared by using solvothermal synthesis was tested. A metal ion (Fe) MOF with a 2-aminoterephthalic acid organic ligand has proven to be biocompatible [44]. In vitro, drug release experiments were conducted in PBS (pH 7.4). The anti-glaucoma drug brimonidine was encapsulated in the MOF (NH2-MIL-88(Fe)/Br), loaded brimonidine was 121.3 μg/mg, after 40% of the brimonidine drug was released in the first 30 minutes, it demonstrated a sustained drug release kinetic with a release rate of 5.2% per hour for up to 12 hours. While NH$_2$-MIL-88(Fe)/Br exhibited mucoadhesive properties in rabbit eyes, drug release occurred for up to 4 hours. In conclusion, high brimonidine concentrations in tear fluid demonstrated improved bioactivity and bioavailability results in decreasing IOP when compared to clinically approved eye drops. The carrier NH$_2$-MIL-88(Fe)/Br seems to be a promising candidate for topical application to the eye [44].

The ability of 2D nanosheet cyclodextrin-based MOFs (NS-CD-MOFs) to accommodate the anatomical structure of the eye was investigated. The particle size of the crystals was adjusted efficiently and the effect of particle size on ophthalmic drug release was investigated. The relationship between nucleation rate and crystal growth rate is critical in crystal growth and size adjustment. When the rate of crystal growth exceeds the rate of nucleation, large particles are formed. The particle size (from 60 to under 10 μm) was optimized by altering the reaction temperature (from 25 to 3°C) and using a crystal growth suppressor (acetone in this case). The relationship between nucleation rate and crystal growth rate is critical in crystal growth and size adjustment. The polymerization reaction of cross-linked CD-MOF (CL-CD-MOF) showed an increase in the stability of NS-MOF. Crosslinking allowed the particles to retain their size and shape, even though they lost their crystallinity during the process. The effect of particle geometry and size of the nanoporous material was compared using charged sheet-like DXM@CL-NS-MOF and 3D-cubic shape DXM@CL-CD-MOF. Dexamethasone release kinetics of both carriers were determined in vivo in tear fluid and aqueous humor. In terms of pre-corneal residence time and biocompatibility, the 2D-nano-sheet particles were stated to outperform the 3D-cubic structure and its commercial counterpart [43].

Infectious endophthalmitis is an immunological reaction that causes damage to optic tissues and nerves caused by bacteria such as *Pseudomonas aeruginosa*, *Staphylococcus epidermidis*, *Staphylococcus aureus* as well as viruses and fungi. Infectious endophthalmitis compromises the eye's physiological barrier, making it vulnerable to bacteria and damaging optic tissues. Although there are various methods for treating infections, MOFs are preferred because of their high drug loading capacity and low toxicity. UiO-66-NH2 is preferred because it releases drugs faster at lower pH. Furthermore, due to cell wall lysis, the degradation products of UiO-66-NH2 can inhibit bacterial growth. A lipopolysaccharide (LPS)-targeting polypeptide (PEP) and PEG are grafted onto UiO-66-NH2 nanoparticles (NPs) and toluidine blue (TB), attending as a photosensitizer (PS) loaded in it, to improve the targeting to the bacterial cell walls, resulting in TB@UiO-66-NH2@ PEP&PEG NPs (UTPP NPs). At pH 5.5, UTPP nanoparticles exhibit faster TB release behavior (compared to pH 7.4). It also demonstrated good biocompatibility with human retinal pigment epithelial cells, indicating that UiO-66-NH2 has potential as a theranostic nanoplatform [45].

14.6 CONCLUSION

The eye is such a vital organ and failure to treat eye diseases can lead to irreversible tissue damage or visual impairment. Although there are numerous ophthalmic formulations available in the treatment of ocular diseases, in most cases, they cannot provide the optimum treatment plan for the patient due to low bioavailability. Therefore, patients are exposed to frequent administration due to the side effects of drugs and low efficacy. However, recently developed nanosystems and biodegradable polymers offer new suggestions to overcome these problems. The properties of polymers such as biocompatibility, mucoadhesiveness, and biodegradation contain promising properties to provide treatment where conventional treatment methods have failed. In addition, PEC structures that combine two different polymers without losing their inherent properties are also used in ODD. The versatility of these systems may enable similar systems to be adapted to different diseases. In recent years, MOF-based materials have also been developed as drug delivery systems. The features of MOFs such as high surface area, functionality, and high drug loading capacity have attracted the attention of the scientific world. However, more detailed studies are needed on their toxicity, especially after long-term use. Although the researches carried out to date has yielded successful results, there are still aspects that need improvement. More interdisciplinary studies are needed to obtain more effective results.

REFERENCES

1. Lynch C, Kondiah PPD, Choonara YE, du Toit LC, Ally N, Pillay V (2019) Advances in biodegradable nano-sized polymer-based ocular drug delivery. *Polymers*, 11, 1371.
2. Ludwig A (2005) The use of mucoadhesive polymers in ocular drug delivery. *Adv Drug Delivery Rev*, 57, 1595–1639.
3. Gandara-Loe J, Souza BE, Missyul A, Giraldo G, Tan J-C, Silvestre-Albero J (2020) MOF-based polymeric nanocomposite films as potential materials for drug delivery devices in ocular therapeutics. *ACS Appl Mater Interfaces*, 12, 30189–30197.
4. Yang J, Yang Y-W (2020) Metal–organic frameworks for biomedical applications. *Small*, 16, 1906846.
5. Xu J, Xue Y, Hu G, Lin T, Gou J, Yin T, He H, Zhang Y, Tang X (2018) A comprehensive review on contact lens for ophthalmic drug delivery. *J Control Release*, 281, 97–118.
6. Gaudana R, Ananthula HK, Parenky A, Mitra AK (2010) Ocular drug delivery. *AAPS J*, 12(3) 348–360.
7. Demmin DL, Silverstein SM (2020) Visual impairment and mental health: Unmet needs and treatment options. *Clin Ophthalmol*, 14, 4229.
8. Torres-Luna C, Fan X, Domszy R, Hu N, Wang NS, Yang A (2020) Hydrogel-based ocular drug delivery systems for hydrophobic drugs. *Eur J Pharm Sci*, 154, 105503.
9. Huang D, Chen YS, Rupenthal ID (2018) Overcoming ocular drug delivery barriers through the use of physical forces. *Adv Drug Delivery Rev*, 126, 96–112.
10. Kirchhof S, Goepferich AM, Brandl FP (2015) Hydrogels in ophthalmic applications. *Eur J Pharm Biopharm*, 95, 227–238.
11. Bachu RD, Chowdhury P, Al-Saedi ZHF, Karla PK, Boddu SHS (2018) Ocular drug delivery barriers-role of nanocarriers in the treatment of anterior segment ocular diseases. *Pharmaceutics*, 10, 28.
12. Rodrigues EB, Grumann A J, Penha FM, Shiroma H, Rossi E, Meyer CH, Stefano V, Maia M, Magalhaes O J, Farah ME (2011) Effect of needle type and injection technique on pain level and vitreal reflux in intravitreal injection. *J Ocul Pharmacol Ther*, 27(2), 197–203.
13. Carvalho IM, Marques CS, Oliveira RS, Coelho PB, Costa PC, Ferreira DC (2015) Sustained drug release by contact lenses for glaucoma treatment – A review. *J Control Release*, 202, 76–82.
14. Jumelle C, Gholizadeh S, Annabi N, Dana R (2020) Advances and limitations of drug delivery systems formulated as eye drops. *J Control Release*, 321, 1–22.

15. Morrison PWJ, Khutoryanskiy VV. (2014) Advances in ophthalmic drug delivery. *Ther Deliv*, 5, 1297–1315.
16. Bhowmik D, Gopinath H, Pragati Kumar B, Duraivel S, Sampath Kumar KP (2012) Controlled release drug delivery systems. *Pharma Innovation*, 1(10).
17. Xu Q, Kambhampati SP, Kannan RM (2013) Nanotechnology approaches for ocular drug delivery. *Middle East Afr J Ophthalmol*, 20, 26.
18. Peterson GI, Dobrynin AV, Becker ML (2017) Biodegradable shape memory polymers in medicine. *Adv Healthc Mater*, 6, 1700694.
19. Bíró T, Aigner Z (2019) Current approaches to use cyclodextrins and mucoadhesive polymers in ocular drug delivery—a mini-review. *Sci Pharm*, 87, 15.
20. Mansuri S, Kesharwani P, Jain K, Tekade RK, Jain NK (2016) Mucoadhesion: A promising approach in drug delivery system. *React Funct Polym*, 100, 151–172.
21. Rosso F, Barbarisi A, Barbarisi M, Petillo O, Margarucci S, Calarco A, Peluso G (2003) New polyelectrolyte hydrogels for biomedical applications. *Mater Sci Eng C*, 23, 371–376.
22. Cazorla-Luna R, Martín-Illana A, Notario-Pérez F, Ruiz-Caro R, Veiga M-D (2021) Naturally Occurring polyelectrolytes and their use for the development of complex-based mucoadhesive drug delivery systems: An overview. *Polymers*, 13, 2241.
23. Meka VS, Sing MKG, Pichika MR, Nali SR, Kolapalli VRM, Kesharwani P (2017) A comprehensive review on polyelectrolyte complexes. *Drug Discovery Today*, 22, 1697–1706.
24. Shang L, Nienhaus K, Nienhaus GU (2014). Engineered nanoparticles interacting with cells: Size matters. *J Nanobiotechnol*, 12(1), 1–11.
25. Chowhan A, Giri TK (2020) Polysaccharide as renewable responsive biopolymer for in situ gel in the delivery of drug through ocular route. *Int J Biol Macromol*, 150, 559–572.
26. Lee KY, Mooney DJ (2012) Alginate: Properties and biomedical applications. *Prog Polym Sci*, 37, 106–126.
27. Augst AD, Kong HJ, Mooney DJ (2006) Alginate hydrogels as biomaterials. *Macromol Biosci*, 6, 623–633.
28. Mărțău GA, Mihai M, Vodnar DC (2019) The use of chitosan, alginate, and pectin in the biomedical and food sector—Biocompatibility, bioadhesiveness, and biodegradability. *Polymers*, 11, 1837.
29. Zamboulis A, Nanaki S, Michailidou G, Koumentakou I, Lazaridou M, Ainali NM, Xanthopoulou E, Bikiaris DN (2020) Chitosan and its derivatives for ocular delivery formulations: Recent advances and developments. *Polymers*, 12, 1519.
30. Vasvani S, Kulkarni P, Rawtani D (2020) Hyaluronic acid: A review on its biology, aspects of drug delivery, route of administrations and a special emphasis on its approved marketed products and recent clinical studies. *Int J Biol Macromol*, 151, 1012–1029.
31. Salzillo R, Schiraldi C, Corsuto L, D'Agostino A, Filosa R, De Rosa M, La Gatta A (2016) Optimization of hyaluronan-based eye drop formulations. *Carbohydr Polym*, 153, 275–283.
32. Noreen A, Nazli Z i. H, Akram J, Rasul I, Mansha A, Yaqoob N, Iqbal R, Tabasum S, Zuber M, Zia KM (2017) Pectins functionalized biomaterials; a new viable approach for biomedical applications: A review. *Int J Biol Macromol*, 101, 254–272.
33. Vasconcelos A, Vega E, Pérez Y, Gómara MJ, García ML, Haro I (2015) Conjugation of cell-penetrating peptides with poly(lactic-*co*-glycolic acid)-polyethylene glycol nanoparticles improves ocular drug delivery. *Int J Nanomedicine*, 10, 609.
34. Lin QK, Xu X, Wang Y, Wang B, Chen H (2016) Antiadhesive and antibacterial polysaccharide multilayer as IOL coating for prevention of postoperative infectious endophthalmitis. *Int J Polym Mater Polym Biomater*, 66(2), 97–104.
35. Dubey V, Mohan P, Dangi JS, Kesavan K (2020) Brinzolamide loaded chitosan-pectin mucoadhesive nanocapsules for management of glaucoma: Formulation, characterization and pharmacodynamic study. *Int J Biol Macromol*, 152, 1224–1232.
36. Li J, Tian S, Tao Q, Zhao Y, Gui R, Yang F, Zang L, Chen Y, Ping Q, Hou D (2018) Montmorillonite/chitosan nanoparticles as a novel controlled-release topical ophthalmic delivery system for the treatment of glaucoma. *Int J Nanomedicine*, 13, 3975.
37. Lustig WP, Mukherjee S, Rudd ND, Desai AV., Li J, Ghosh SK (2017) Metal–organic frameworks: Functional luminescent and photonic materials for sensing applications. *Chem Soc Rev*, 46, 3242–3285.
38. Shin J, Kim M, Cirera J, Chen S, Halder GJ, Yersak TA, Paesani F, Cohen SM, Meng YS (2015) MIL-101(Fe) as a lithium-ion battery electrode material: A relaxation and intercalation mechanism during lithium insertion. *J Mater Chem A*, 3, 4738–4744.
39. Lin W, Hu Q, Jiang K, Yang Y, Yang Y, Cui Y, Qian G (2016) A porphyrin-based metal–organic framework as a pH-responsive drug carrier. *J Solid State Chem*, 237, 307–312.

40. Safarifard V, Morsali A (2015) Applications of ultrasound to the synthesis of nanoscale metal–organic coordination polymers. *Coord Chem Rev*, 292, 1–14.
41. He S, Wu L, Li X, Sun H, Xiong T, Liu J, Huang C, Xu H, Sun H, Chen W, Gref R, Zhang J (2021) Metal-organic frameworks for advanced drug delivery. *Acta Pharm Sin B*, 11, 2362–2395.
42. Carrillo-Carrión C (2019) Nanoscale metal–organic frameworks as key players in the context of drug delivery: Evolution toward theranostic platforms. *Anal Bioanal Chem*, 412, 37–54.
43. Bello MG, Yang Y, Wang C, Wu L, Zhou P, Ding H, Ge X, Guo T, Wei L, Zhang J (2020) Facile synthesis and size control of 2D cyclodextrin-based metal–organic frameworks nanosheet for topical drug delivery. *Part Part Syst Charact*, 37, 2000147.
44. Kim SN, Park CG, Huh BK, Lee SH, Min CH, Lee YY, Kim YK, Park KH, Choy Y Bin (2018) Metal-organic frameworks, NH2-MIL-88(Fe), as carriers for ophthalmic delivery of brimonidine. *Acta Biomater*, 79, 344–353.
45. Jin Y, Wang Y, Yang J, Zhang H, Yang YW, Chen W, Jiang W, Qu J, Guo Y, Wang B (2020) An integrated theranostic nanomaterial for targeted photodynamic therapy of infectious endophthalmitis. *Cell Rep Phys Sci*, 1, 100173.

15 Conducting Polymers for Biomedical Imaging

Gaurav Kulkarni,[1,2] Krishna Dixit,[1] Jhansi Lakshmi Parimi,[1] Hema Bora,[1] Baisakhee Saha,[1] Soumen Das,[2] and Santanu Dhara[1]

[1]Biomaterials and Tissue Engineering Laboratory, School of Medical Science and Technology, Indian Institute of Technology Kharagpur, Kharagpur, India

[2]Bio-MEMS Laboratory, School of Medical Science and Technology, Indian Institute of Technology Kharagpur, Kharagpur, India

CONTENT

15.1 INTRODUCTION

Biomedical imaging embodies a range of noninvasive techniques to help researchers in unveiling the complex mechanism related to structural changes at cellular or tissue levels due to disease conditions [1]. Imaging modalities utilize electromagnetic ionizing or nonionizing radiations to obtain images based on the wavelength and energy of the photon. Specifically, photons of high frequency or short wavelengths are ionizing, which include gamma-rays, X-rays, and ultraviolet (UV) light [2]. However, the past UV range (after 400 nm) of the electromagnetic spectrum is followed by visible, infrared (IR), microwave, and ultrasonic waves having nonionizing photon energies. Visible light imaging techniques employ a visible range of the spectrum (400–700 nm) and are mainly used in research and diagnostics. Fluorescence labels are attached to intensify the microscopic structure of cell organelles and associated biological processes with enhanced specificity [3]. In addition, fluorescent labels have been also used as a biomarker to detect specific macromolecules in cells or tissues related to cancer [4]. Applications of optical imaging are largely expanding in flow cytometry to separate particular cell populations [5], immunohistochemistry [6], and various cytoskeleton staining [7]. For tumor detection, imaging modalities such as computed tomography (CT), positron emission tomography (PET), and magnetic resonance imaging (MRI) are widely practiced based on the tumor type and location [8]. Moreover, another well-known diagnostic modality is ultrasound where the images are produced using ultrasonic echoes received upon reflection with underlying tissues/organs [9]. Despite the progress in the aforementioned techniques in recent years, it has several drawbacks. The interpretation of the reports from the imaging modalities is mainly dependent on the physician's expertise. Also, factors such as the high equipment costs, lengthy scanning times, ionizing radiation exposure, and allergic reaction to external contrast agents may restrict the use of these modalities for diagnostic and imaging [10].

There has been a remarkable breakthrough in in vivo imaging, which provides more insights into complex biological reactions, and out of various techniques, optical imaging is a preferred choice [11]. In this, either endogenous [12] or exogenous [13] compounds are used to generate images with high contrast. The application areas of in vivo imaging using contrast agents include research and medical diagnostic using small animal models for cancer detection [14], guided surgery [15], photothermal therapy [16], and controlled drug release [17]. Sometimes, full-body imaging in animals utilizes UV or IR region to avoid interference from the tissue self-fluorescence. Moreover, because of the limited tissue penetration and the body's attenuation of visible light, the use of near-IR becomes a more appropriate choice [1]. In general, the selection of electromagnetic region for imaging is dependent on the tissue's properties to attenuate or absorb or scatter the light, which directly affects the penetrating depth [18]. Photoacoustic imaging (PAI) operates in the optical range and is based on the principle of formation of sound waves followed by light absorption in material. The major advantage of optoacoustic imaging is improved resolution owing to lower scattering and increased depth [19]. While several natural or synthetic chromophores have emerged as contrast agents for improving image quality, they presented restrictions related to penetration depth, sensitivity, and resolution [20]. Subsequently, a variety of materials were studied to increase the throughput in biomedical imaging for diagnostics, therapeutics, and research. Among the different types of emerging imaging agents, conjugated and conducting polymers (CPs) are investigated majorly considering their tunable emission and absorption spectra, high photostability, ease of synthesis into various forms, and applicability in fluorescence as well as PAI [21].

After the discovery of the very first electrically CP oxyacetylene (halogen doped), many second-generation conducting as well as conjugated polymeric materials were developed possessing excellent solubility and processing abilities. As a consequence of easy processability, different CPs were investigated as a material in photoluminescence devices, including light-emitting diodes (LEDs) [22]. The photo-conductive attributes of CPs arise after π-band excitations leading to π-π^* transitions with a specific energy gap depending upon the material's chemical structure [23]. Moreover, by using doping where electron or hole charge carriers were incorporated to get enhanced properties was also explored for tuning optical properties of CPs. The bandgap of π-π^* transition also controls the emission spectra, which in turn can be controlled by refinements in CPs structure as well as doping concentrations. CPs can be used to make fluorescent NPs (particle size <100 nm) due to their high quantum yield, and thus, they are one of the suitable candidates for in vivo fluorescence imaging in small animals [24]. In this context, cellular labeling for different cell lines using NPs of CPs was found to be promising where cytocompatibility of CPs was also studied. The stretching of absorption and emission spectra of CPs in the near-infrared (NIR) range can be accomplished by carefully choosing donor/receptor moieties [25]. NIR wavelength is an interesting range in a bioimaging perspective as it enables higher penetration into tissues with minimal auto-fluorescence. In addition, the lower energy of NIR optical waves is harmless for tissues.

Apart from in vivo and in vitro imaging, CPs are used extensively in biosensing for the colorimetric detection of various biomarkers. Here, the ligand conjugated CPs undergo conformational changes in their backbone after receptor-ligand binding [26]. A similar principle was used to detect other bio-analytes including bacteria [27], viruses [28], and toxins [29]. The modification of CPs for colorimetric detection requires functionalization and optimized synthesis routes to get high sensitivity. Specifically, for biological applications, the fluorescent probe should be water soluble. Although there have been a lot of efforts in synthesis approaches of CPs, the biodegradability and

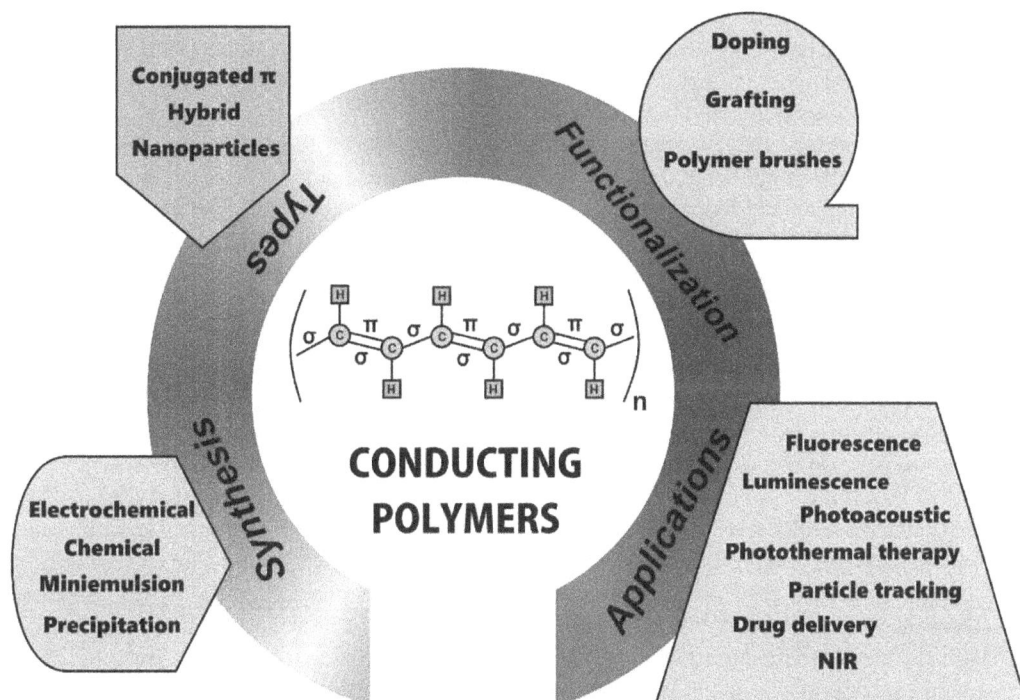

FIGURE 15.1 Schematics describing the role of conducting polymers in biomedical imaging.

their hydrophobic nature remain challenging [30]. Another aspect in fabricating NPs of CPs is that optical properties are mainly dependent on the particle size and the conjugation length. In this perspective, the development of CP-based composite materials was found to exhibit better imaging performance than its native material. With appropriate modifications, CPs find their applications in a variety of bioimaging such as cellular and immunofluorescent labeling, in vivo imaging, single-particle tracking [31], drug and gene delivery [32], PAI [33], cancer biomarker detection [1], photo-thermal therapy [34], and many more. The present chapter depicts a summary of CPs for biomedical imaging with a focus on synthesis, types, functionalization, and applications (Figure 15.1).

15.2 PROPERTIES OF CONDUCTING POLYMER-BASED NANOPARTICLES FOR BIOIMAGING

NPs of CPs are emerging in biomedicine for imaging and as therapeutics. The optoelectronic behavior of CPNPs can be predicted by their size, shape, and dispersity profile. Mostly, particles prepared in the 10–100-nm range are suitable for travel through the bloodstream, as they are large enough to avoid clearance by the renal system and adequately small to evade phagocytosis [35]. Various synthesis methods including nanoprecipitation, mini-emulsion, and self-assembly enable a good control over size as well as particle dispersion. Recently, imaging in biomedical research demands the synthesis of NPs having absorption in the NIR region [4]. Therefore, to take advantage of deeper tissue penetration, the introduction of push-pull or donor-receptor motif into CPs has been commonly practiced [36]. In addition, the polymeric side chains of hydrophobic CPs can be altered by the addition of charged moieties resulting in conjugated polyelectrolytes. The emission spectrum of NPs can be pushed into the first (650–950 nm) or second (950–1350 nm) NIR window based on the combination of the push-pull conjugated polymer [37]. In this aspect, particles of CPs were excited using two-photon microscopy, and it was found to display promising results [38]. Currently,

CP backbones such as polyaniline (PANi), polyacetylene (PA), polypyrrole (PPy), polythiophene (PT), poly(p-phenylene) (PPP), and poly(p-phenylene vinylene) (PPV) are often modified by incorporating another polymer or copolymer for imaging [36]. The most frequently studied polymers are based on fluorene, copolymerized together with quinoxaline, benzothiadiazole, and thiophene [36]. NPs prepared from CPs display identical characteristics as those of the bulk material and exhibit features that depend strongly on the solvent used. One drawback that must be taken into account is fluorescence quenching and reduction of quantum yield (QY) whenever polymer chains are tightly packed. One of the solutions to the problem of quenching was addressed by incorporating bulky side chains onto polymer backbone. Apart from this, introduction of nonconjugated polymer and varying monomer ratios were also investigated to increase QY of the probe [36].

Organic or inorganic NPs, bio-functionalized magnetic NPs, and carbon-based nanostructured materials like graphene and carbon nanotubes entrapped within CP matrix yield hybrid conjugated polymers (HCPs). HCPs based on PT, PPy, PANi, and their derivatives have generated interest since these nanocomposites are adorned with combined physicochemical properties of high electrical conductivity, thermal and mechanical stability, redox activity, biocompatibility, and bioactivity depending on their structures, compositions, and conditions applied during synthesis and dispersity [39, 40]. Biorecognition molecules like DNA, antibodies, and enzymes can easily bioconjugate with amino (–NH$_2$) and carboxyl group (–COOH) containing polymers. Thus, DNA, a thiolated molecule, can self-assemble on silver or gold NP decorated CP. HCP of PANi with other CPs, metallic NPs, graphene, or carbon nanotubes protects it from losing its conductivity above pH 4 and acts as electrode modifier in biosensing platforms for detection of infectious diseases, cancer biomarkers, drugs, phenolic, pesticides, sulfites, acrylamides, and heavy metals, etc. and as immobilization platforms for biomolecules and enzymatic reactions [41, 42]. For instance, coating PANi with quantum dots (QDs) was found to enhance the fluorescence by ~40 times with improved biocompatibility (Figure 15.2a) [43]. Properties of PPy NPs as a photothermal agent were explored due to their excellent NIR absorbance (Figure 15.2b) [44]. In another example (Figure 15.2c), PPy was combined with chitosan owing to the low cytotoxicity of PPy NPs [45]. The ability of tantalum oxide modified PPy NPs as a biomedical contrast agent was studied by Jin et al. and coworkers (Figure 15.2d) [46].

The issues of electrical conductivity, optical properties, solubility, etc. can be resolved to a great extent by incorporating functional groups and biomolecules into the CPs. The methods and advantages of functionalization are discussed below.

15.3 FUNCTIONALIZATION OF CPs

The potential bioimaging ability of CPs was one of the greatest efforts toward technology advancement. For guiding the NPs to the specific cells or organelle, a probe has to be engineered. This can be achieved by using different functionalization methods to impart properties such as solubility improvement, response to stimulus, and ability to modulate cell growth and differentiation. Above all, a covalent technique for engineering such a kind of agent was found more feasible. All these meritorious features were possible with the inclusion of bio-molecule attachment and polymer grating so on as depicted in Figure 15.3.

15.3.1 DOPING

Quite often dopant attachment to the CPs can be either adsorption or covalent bonding. Covalent binding to the CPs was made possible by using chlorine to prepare doped polypyrrole (PPyCl) for electrical and biomedical application and which is then functionalized with T59 peptide linked with cell adhesion molecule for better bio interfacing [47]. In another study, functionalization of CS doped PPy coupled with fibrillar collagen was presented for neural bio-interfacing. PPy doping promoted the cellular adhesion and differentiation along with fibrillar collagen self-assembling with improved electrode-implant interface [48]. Chen et al. covalently conjugated streptavidin with

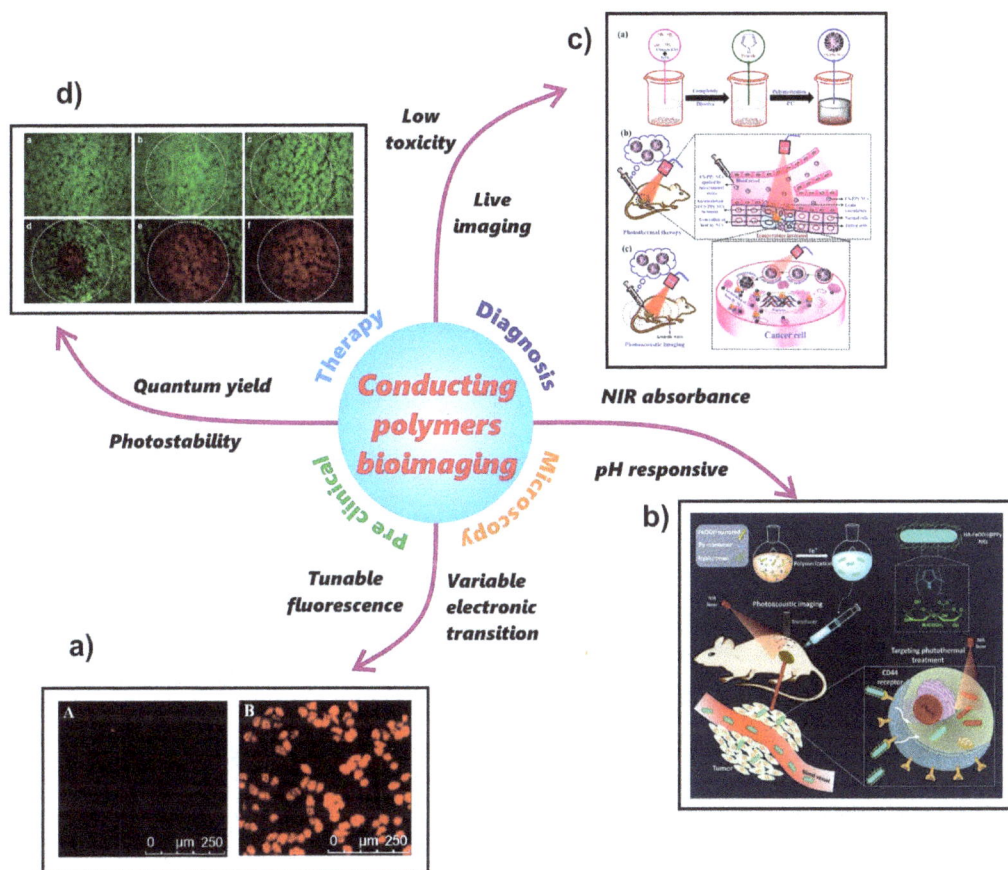

FIGURE 15.2 Illustration of multidimensional aspects of conducting polymers for bioimaging (a) Fluorescent images of (A) HeLa cells cultured with CdSeTe QDs and (B) HeLa cells cultured with CdSeTe@ PABA QDs; (b) Schematic representation for the synthesis and the applications of HA-FeOOH@PPy NRs; (c) (a) Schematic for the preparation of chitosan-polypyrrole nanocomposites (CS-PPy NCs). A scheme showing the possible mechanism of CS-PPy NCs for photothermal therapy (b) and photoacoustic imaging (c) of cancer; (d) Photothermal destruction of HeLa cells treated by various combinations of TaOx@PPy NPs and NIR laser irradiation (808 nm, 6 W/cm²): (a) no irradiation and no nanoparticles; (b) 10 min irradiation, no agent; (c) 5 min irradiation, nanoparticles of 40 μg/mL; (d) 5 min irradiation, nanoparticles of 80 μg/mL; (e) 5 min irradiation, nanoparticles of 160 μg/mL; (f) 5 min irradiation, nanoparticles of 200 μg/mL. White circle indicates the laser spot; live/dead stain for viability shows dead cells as red while viable cells as green. ((a) Reprinted with permission from Ref. [86]. Copyright (2015) American Chemical Society. (b) Reprinted with permission from Ref. [87]. Copyright (2018) Springer Nature. (c) Reprinted with permission from Ref. [88]. Copyright (2017) Springer Nature. (d) Reprinted from Ref. [91], Copyright (2014), with permission from Elsevier.)

the NPs, which increases affinity toward biotin and promotes cellular and subcellular tracking of HeLa and MCF-7 cells. P dots are useful for tracking the biotinylated ligands in the presence of streptavidin [48].

Grafting is the covalent attachment of polymer chains to the CP backbone. It is a useful tool to alter the chemistry, stability, solubility, and mechanical properties in addition to the inclusion of functional groups for further functionalization. It will be an exhilarating new tool for engineering functional electro-active bio-interfaces.

Grafting to is the formation of polymer brushes over the CP upon post-synthesis [49]. It gives the control to direct the polymer functionalization with marking attachment sites and used azide-alkyne

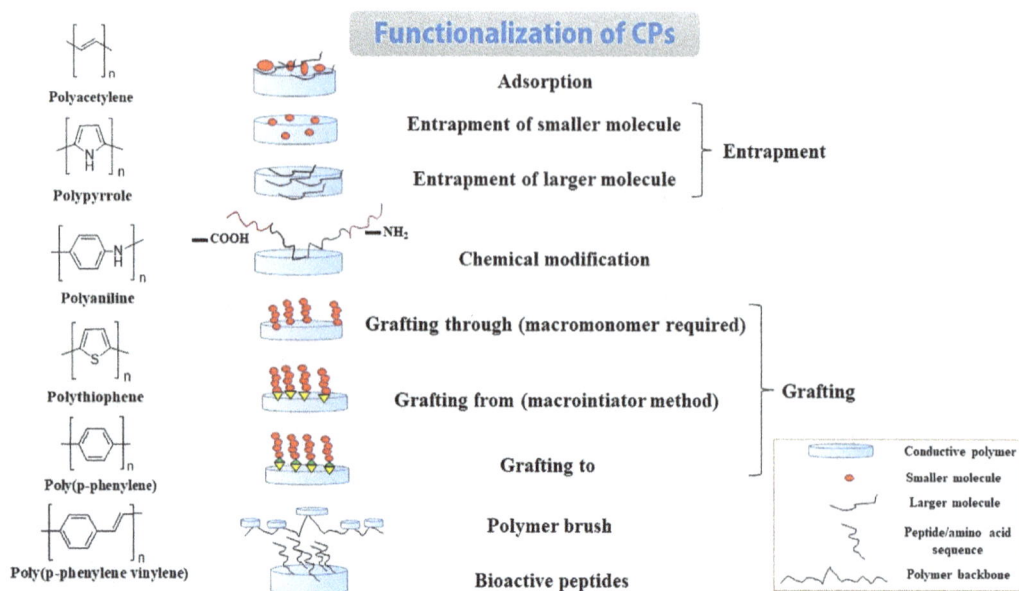

FIGURE 15.3 Overview of various functionalization techniques used for conducting polymers in the perspective of imaging.

to improve the QY of conjugated polyelectrolyte (CPE) with amino-polyethylene glycol (PEG) overcoming its limitation for advanced imaging [50].

Grafting from is used most frequently to provide stable, longer chain polymer brushes. It can be achieved easily by controlled/living radical polymerization (CRP) [51]. One of the important objectives of this strategy is to overcome the insolubility issue faced by CP. Conjugation of polyelectrolyte brush (PB3) over the poly [9,9′-bis(6-N, N, N-trimethylammoniumhexyl) fluorene] via an atomic transfer radical polymerization (ATRP) method have improved the quantum efficiency by 52% and water solubility. Thus, it can be used as a water-soluble fluorescent biomarker due to the sustenance of fluorescence by the PB3 [52].

Grafting through is one of the earliest methods reported by Alkan et al., about the first report of polymer brush made off with PT capped over the poly(methyl-methacrylate) (PMMA) [53]. It showed 3T3 fibroblast adhesion with minimal cytotoxicity, which indicates the composite imbibe the potency to serve for biomedical applications [63]. In further reports in this line, PT backbone was grafted with PEG chains for specific adsorption of protein [54]. Maoine et al. investigated the effect of PEG density over the thiophene; in this, PEG conjugated at the third position of thiophene are referred to as (PTh3-g-PEG) compared with existing PEG conjugated at the fifth position of thiophene (PTh5-g-PEG). PTh3-g-PEG showed the excellent electrical property and electro stability, showed lower optical π-π* transition energy, better cellular adhesion compared to PTh5-g-PEG [55].

15.3.2 Polymer Brushes

15.3.2.1 Addition of Polymer Brushes

CP brushes can be grafted by the addition of CP over non-conducting natural or synthetic polymers. Such types of functionalization have started to pop up in imaging applications. PANi was grafted over the magnetic NPs having polydopamine intermediate layer by this polymer brush approach, which offers advantages like it is free from changing molecular structure [56]. This polymerization has increased the conductivity 2000× times, thereby increases the opportunity in developing a commendable material for imaging and sensing.

15.3.2.2 Grafting Polypeptide Brushes

Polypeptide brushes offer ease to form an α helix and β sheet, they can switch the properties accord-ing to amino acids composition with excellent biodegradability hence influences cell-binding due to the presence of RGD motif's. In general, the covalent attachment of polypeptide helps to overcome the leaching caused due to the physisorption, thus used in the development of electrodes for the sensing. Polypeptide brushes interact with the enzymes and target analytes and cause electrochemi-cal changes in the electrode which can be easily detected and recorded. CP NPs prepared by nano-precipitation conjugated with collagen mimetic peptide were used to detect collagen without any preactivation and emit fluorescence signal [57]. Maeda et al. reported that helix to coil transition of optical active polymer poly (phenylacetylene) poly (γ-benzyl-L-glutamate) poly[PBGAm] or poly (L-glutamic acids) (PGA) as a pendant [58].

15.4 APPLICATIONS OF CONDUCTING POLYMERS IN BIOIMAGING

Biomedical imaging is a powerful technique to perform in vivo analysis of the biological process in molecular diagnostics and drug delivery approaches. CPs, owing to their biocompatibility and intrinsic conductivity, are emerging as promising materials in the field of biomedical engineering. Additionally, CPs can be doped with biological molecules, enzymes, and antibodies to alter their physical, chemical, and electrical properties for specific applications [59]. The use of fluorescence microscopy is a breakthrough in the study of living cells and tissue owing to its high sensitivity and low cost. The probes that emit in the far-red and NIR spectral ranges are highly beneficial for biomedical imaging because they penetrate better in tissues and eliminate autofluorescence [60]. Numerous contrast agents have been employed in imaging applications including but not limited to iodinated, magnetic, QDs-based contrast agents [61]. NIR contrast agents have been used for deeper penetration and high contrast imaging [62].

15.4.1 FLUORESCENCE IMAGING

As compared to X-ray and MRI imaging techniques, fluorescence imaging is noninvasive, tunable with stable optical properties, minimally toxic, and sensitive with easier detection in living bodies. Upon coupling with photosensitizing agents, CPs dually function for diagnostics and treatment to combat bacterial infection and cancer [63]. CPs coupled to QDs show fluorescence enhancement. PPy and GQDs exhibited three times fluorescence enhancement than pristine graphene and detected the picomolar dopamine concentration in human serum and urine samples [64]. Gadolinium-doped bovine serum albumin (BSA)-chlorin e6 stabilized PPy NPs acted as a strong contrast agent in fluo-rescence imaging of breast tumor cells [65]. Peptide SP94 and indocyanine green conjugated BSA stabilized polypyrrole NPs target hepatocellular carcinoma, fluoresce twice more in tumors than in liver cells [66]. Folic acid-functionalized conjugated polymer NPs accumulated more in MCF-7 cancer cells than their non-functionalized counterparts. pH-responsive doxorubicin (DOX)-loaded hyaluronic acid (HA)–PPy NPs function for macrophage imaging and therapy in atherosclerosis [67] and breast cancer [68] utilizing the quenching potential of PPy. (Poly(9,9′-dioctylfluorene-2,7-ylene-vinylene-co-alt-1,4-phenylene)(PFV), poly(styrene-co-maleic anhydride) (PMSA), and dopamine (DA) NPs exhibit intracellular pH imaging in HeLa cells. The DA oxidation potential is reduced when the buffer pH is raised from acidic to alkaline, which favors DA oxidation and, as a result, increases the number of quinones and, thus, fluorescence quenching in the medium. In acidic solution, DA oxidation was substantially decreased, and intense fluorescence emission of PFV/PSMA-DA NPs solution was detected (Figure 15.4a). Hence, PFV/PSMA-DA NPs showed a sensi-tive fluorescence response to pH variations. The two-photon imaging showed that the fluorescence intensity of PFV-Dopamine NPs increases gradually in a linear fashion as the pH decreases from 9.0 to 5.0 (Figure 15.4b) [69]. Two-photon excitation fluorescence microscopy is used for high reso-lution, sensitive, and deep-tissue imaging at the IR region. Polybenzothiadiazole and polythiophene

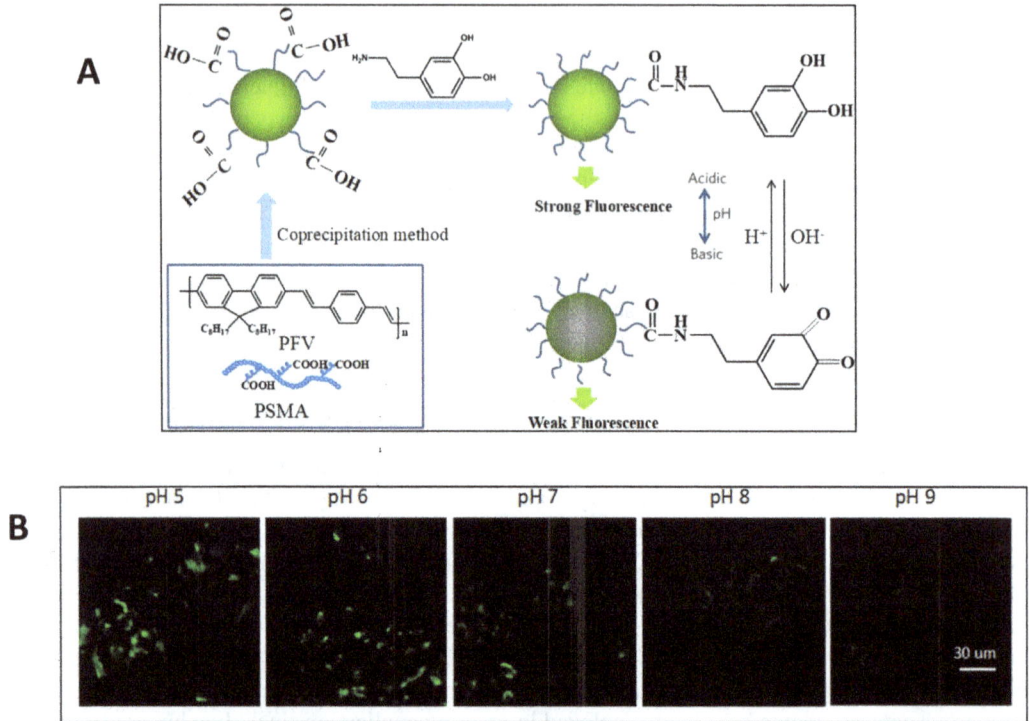

FIGURE 15.4 pH sensing-based two-photon imaging. (a) Schematic representation of pH sensing based on PFV/PSMA-DA NPs, (b) two-photon fluorescence images of PFV/PSMA-DA NPs in Hela cells at pH values of 5.0, 6.0, 7.0, 8.0, and 9.0, respectively. (Reprinted from Ref. [139], Copyright (2019) with permission from Elsevier.)

polymer dots generated 3D images of blood vessels from mouse brain using 1200-nm laser excitation through intact skull and could perform imaging up to 400-μm depth [37]. So NIR-II light excitation and NIR-I fluorescence emission are advantageous for deep and high contrast imaging. NIR-II fluorescent polymeric NPs (poly(benzo[1,2-b:3,4-b′]difuran-alt-fluorothieno-[3,4-b]thiophene) have been used to image in vivo mouse blood vessel and could even track the blood flow offering a straightforward diagnostic tool for observing the tissue's metabolic differences [70].

15.4.2 LUMINESCENCE IMAGING

Luminescent agents have afterglow or persistent luminescence after stopping the light excitation and have been explored in tumor imaging [71] and cell tracking [72]. Semiconducting polymer NPs are important in imaging applications as they can undergo variable electronic transitions resulting in fluorescence [24], luminescence [73], PAI [74], and cancer therapy [75]. PPV functionalized with poly[2-methoxy-5-(2-ethylhexyloxy)-1,4-phenylenevinylene] (MEH) and NIR dye 755 showed persistence luminescence up to an hour after single irradiation and re-excitation could be performed by exposing mice under white light [76]. Afterglow luminescence is due to the formation of PPV-dioxetane by $^1O_2$ under light irradiation. A smart PPV-thiol-based probe analyzed the drug-induced hepatotoxicity in mice using real-time imaging and showed a 25-fold higher signal to background ratio as compared to NIR fluorescence imaging [73]. PPV was functionalized with PEG and another luminescent agent silicon 2,3-naphthalocyanine bis(trihexylsilyloxide) (NCBS) to amplify and shift the signal in the NIR range to form NCBS-doped PPV-PEGL (SPPVN), which showed the smaller size of SVPPN leading to their better distribution and deeper penetration in

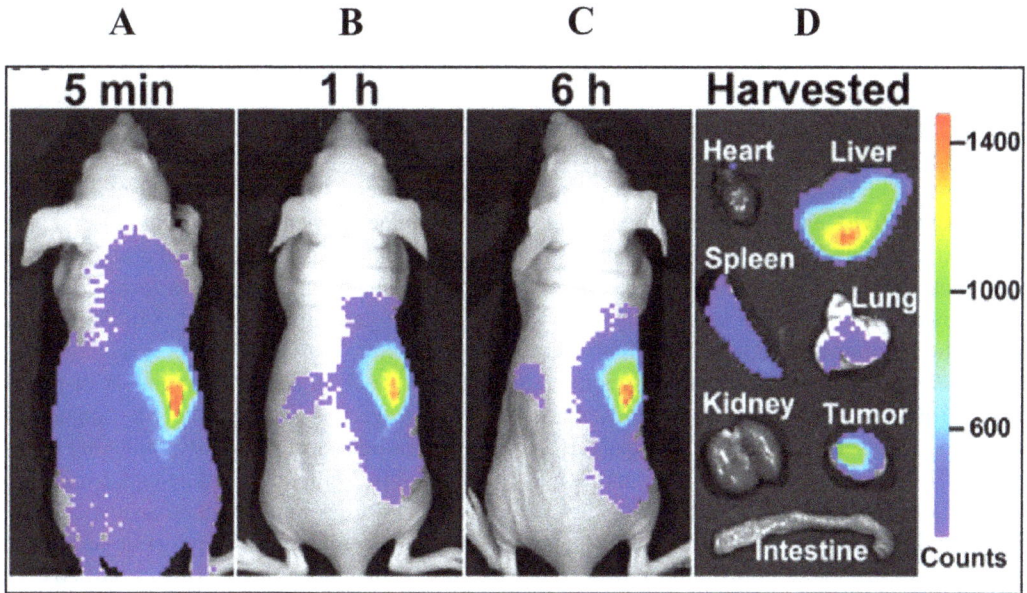

FIGURE 15.5 In vivo PL imaging: (a–c) In vivo PL images of a mouse after intravenous injection with the LPLNP@SPP for 5 min, 1 h, and 6 h. (d) Ex vivo PL images of major organs of the nude mouse and tumor collected from LPLNP@SPP pre-injected mouse for 6 h. (LPLNP@SPP refers to long persistent luminescent nanoparticles prepared with silicon dioxide, polypyrrole, and polyethylene glycol). (Reprinted from Ref. [151], Copyright (2021) with permission from Elsevier.)

metastatic tumor-bearing mice models. Furthermore, the afterglow intensity of the tumor was also significantly higher than the background control. SVPPN could detect even a small tumor of 1 mm^3 within 40 min post-injection [77]. Wu et al. fabricated PPy nanocomposite with persistent luminescence in the NIR region for dual-modal imaging to detect and treat mammary cancer in mice. Intravenous injection of NPs showed accumulation in tumor 1 h post-injection through EPR effect (Figure 15.5b, c); ex vivo PL imaging of major organs and tumor tissue 6 h after injection revealed significant signals in the liver and spleen, with a noticeable fraction of signals in the tumor as shown in Figure 15.5d [78]. Weak intrinsic luminescence of PANi can be improved by amalgamating with other polymers, e.g., PMMA leads to concentration-dependent increase in photoluminescence property of PANi [79]. Positively charged PLNP@PANi-GCS (NIR-emitting $Zn_{1.2}Ga_{1.6}$ $Ge_{0.2}O_4$:Cr^{3+} (PLNP) – glycol chitosan (GCS) has been used for bacterial imaging where it attaches to negatively charged cell wall and performs photothermal therapy in acidic bacterial infected cells without harming neighboring normal cells [80].

15.4.3 PHOTOACOUSTIC IMAGING

PAI, involves the deep penetration of ultrasound waves and high optical contrast and utilizes the optical wavelength covering the visible and NIR region to irradiate the tissue. Absorption by chromophores produces heat resulting in a temperature rise, which leads to an increase in pressure. Relaxation of pressure leads to the emission of acoustic waves onto the surface for detection. The image is constructed by measuring the time and speed of sound waves [81]. The intrinsic contrast agents hemoglobin and lipids absorb light in the visible region, whereas biological tissue shows light scattering to ultimately compromise the resolution of an image. Hence, exogenous contrast agents that can absorb in the NIR region are required for PAI, to minimize light scattering to enhance the sensitivity and resolution of the image. A good contrast agent should possess properties like good photostability, strong extinction coefficient, narrow absorption coefficient, biodegradability, and biocompatibility. PPyNPs with strong absorption in NIR showed high contrast PA images of mouse brain vasculature in comparison to intrinsic contrast agents, e.g., hemoglobin. PPy NPs could be detected in stacked chicken breast muscles to a depth of 4.3 cm, indicating deeper penetration [82]. PPy-based NPs have emerged as a multifunctional theranostic platform for NIR-induced PAI and photothermal therapy in cancer diseased models [45, 46, 83]. The ultra-small iron oxide-PPy-PEG NPs showed photoacoustic signal 24 h post-injection in cancer tissues [84]. Under acidic conditions, conductivity and NIR absorption of PANi are maintained which can be utilized to prepare PA contrast agents for functional imaging of the stomach in case of gastrointestinal disorders. The functionalized PANi is used as a pH-sensitive probe (as a contrast agent) for assessment of gastric acid secretion and real-time monitoring, and quantification of gastrointestinal pH [85], PANi-BSA probe showed four-fold high intensity in the tumor as compared to muscles 5 min post-injection and reached a maximum 60 min post-injection suggesting the role of an acidic environment of tumor in PANi-BSA induced PAI [86]. A series of tunable contrast agents based on indigo π-conjugated semiconducting polymer NPs with strong NIR absorbance has been developed for bioimaging applications [13]. PD conjugated semiconducting NPs (poly[(4,4′-bis(2-ethylhexyl)dithieno[3,2-b:2′,3′-d]silole)-2,6-diyl-alt-(2,1,3-benzothiadiazole)-4,7-diyl] (PSBTBT)) showed the photoacoustic signal in tumor 4 h post-injection in mice [87]. Conjugated polymers combined with inorganic molecules have been utilized as dual imaging contrast agents for cancer theranostics, e.g., BSA-modified conjugated PBTP-DPP/gold NPs showed a concentration-dependent increase in PA signal intensities [88]. The poly(cyclopentadithiophene-alt-benzothiadiazole (PCPDTBT)-based NPs showed 4.0 and 5.8-fold more PA signals than carbon tubes and gold nanorods [74]. Furthermore, intra-tumoral injection of lipid micelle-coated gadolinium functionalized PCPDTBT nanodots displayed a stronger PAI signal in the tumor region in HepG2 tumor-bearing nude mice [89]. The PA contrast agents are passively accumulated at the target location due to increased permeability and tumor cell retention effect. The functionalization with molecules such as folic acid or HA and SP94 peptide to target breast cancer and hepatocellular carcinoma, respectively, have shown strong PA signal as early as 1 h post-injection and many folds increase in PA signals at the site, whereas nontargeting counterparts were imaged around the tumor center [44, 66, 90]. The noninvasive PA imaging capabilities of folate conjugated polymer dots were assessed in mice bearing folate receptor-positive MCF-7 breast cancer xenografts. The images from the transverse and coronal abdomen section of tumor-bearing mice post folate-CP dots injection (Figure 15.6b, c) showed visible enhancement in PA signals as compared to bare CP dots (Figure 15.6d, e). According to ROI analysis, there was four-fold enhancement in PA signal in the tumor of folate-CP dots-treated mice than in the tumor of CP dots-treated mice (Figure 15.6f) [90]. The majority of PAI is accomplished using NIR I contrast agents; Zhang et al. designed the polymer with two acceptors, i.e., diketopyrrolopyrrole (DPP)-benzobisthiadiazole with NIR-II absorption and observed strong PA signals immediately upon injection near the tumor skin and 8 h post injection from deeper tumor tissues [91]. Similarly, cyclo(Arg-Gly-Asp-D-Phe-Lys(mpa)) functionalized thiophene-based NPs (P1RGD) with NIR-II absorption targeted $\alpha V\beta 3$ integrin receptors of endothelial cells to image brain tumor and showed better penetration as compared to NIR-I PA agents [92]. Owing to the presence of targeting ligand, peptide-based NPs showed a higher signal-to-noise ratio (SNR) than the undecorated NPs group. Undoubtedly, for deep tissue, e.g., imaging of brain through scalp and skull the use of NIR-II contrast agents are beneficial.

15.5   CONCLUSION AND FUTURE PERSPECTIVES

To date, a lot of reviews focused on the use of CPs for imaging. Herein, we focused our attention on the compilation of the list of remarkable CPs and their capabilities, with a focus on their applicability. More importantly, various functionalization strategies were discussed for improving the throughput in bioimaging applications. CPNPs show eminent advantages in fluorescence, luminescence, PAI

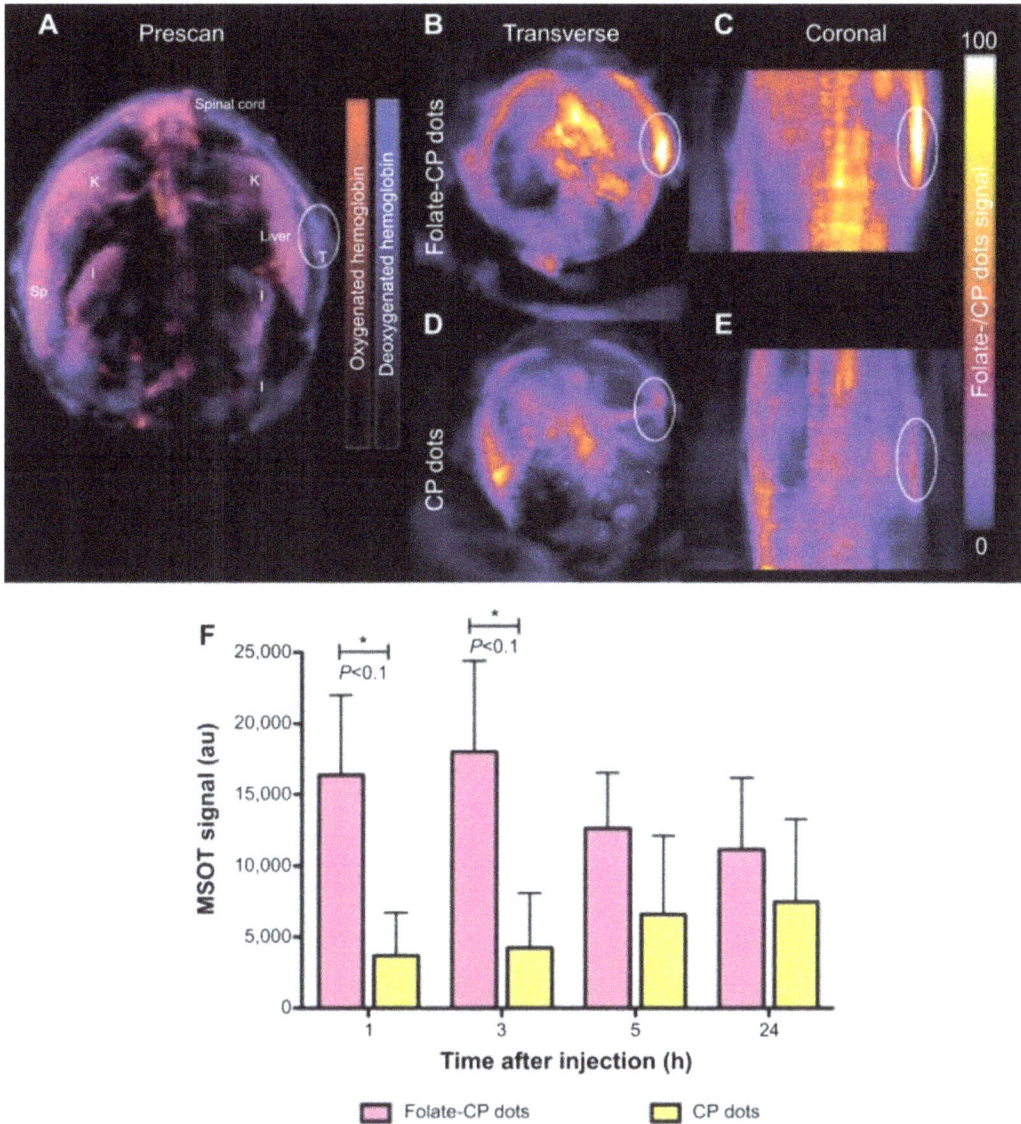

FIGURE 15.6 Molecular photoacoustic imaging of breast cancer. MSOT MIP images of cross-sectional abdomen before systemic administration of the probe – (a) at 3 h after systemic administration of folate-CP dots (b, transverse view and c, coronal view) and CP dots (d, transverse view and e, coronal view). Quantification of PA signal by ROI analysis at the tumor site after systemic administration of folate-CP-CP dots in FR+ve MCF-7 breast cancer tumor model (n=3) (f). Data are represented as mean ± SEM after statistical analysis using unpaired t-test; P<0.1; *P<0.1. (*Note*: White ellipse indicates the location of the tumor.) **Abbreviations**: au, arbitrary unit; CP, conjugated polymer; FR+ve, folate receptor-positive; h, hour; MIP, maximum intensity projection; MSOT, multispectral optoacoustic tomography; PA, photoacoustic; ROI, region of interest; SEM, standard error of the mean; Sp, spleen; I, intestine; K, kidneys; T, tumor. (Reprinted with permission from Ref. [163]. Copyright (2015) Dove Medical Press, UK.)

over other agents. Recent progress is turning the attention of CPs tunable features for expanding emission wavelength toward NIR-I and NIR-II regimes by reducing the bandgap. Even though the long-term reliability and compatibility are moderate, the technical advancements of CPs are viable for biomedical applications. Although the quantum efficiency needs further improvement, altering polymer backbone and side chains may help in boosting the quantum yield. Efforts have been made to improve the functional properties by the introduction of various functional moieties like heparin and chondroitin sulfate, so that biocompatibility and recognition capability together with the conducting capability will be improved and multifunctional benefit can be achieved. In a nutshell, with respect to current advancements in CP-based fabrication technologies, more CPs with outstanding characteristics will emerge and may be commercialized soon.

ACKNOWLEDGMENTS

The authors wish to thank the School of Medical Sciences and Technology, IIT Kharagpur for providing infrastructural support. Mr. Gaurav Kulkarni, Mrs. Krishna Dixit would like to thank the Ministry of Human Resource and Development (MHRD), India for their fellowship. Dr. Hema Bora is thankful to IIT Kharagpur for Institute postdoctoral fellowship. Dr. Jhansi Lakshmi Parimi and Dr. Baisakhee Saha kindly acknowledge the Department of Science and Technology (DST), India for the Woman Scientist Fellowship, WOS-B and WOS-A, respectively.

REFERENCES

1. L. Fass, Imaging and Cancer: A Review, Mol. Oncol. 2 (2008) 115–152. https://doi.org/10.1016/j.molonc.2008.04.001.

2. J. Wallyn, N. Anton, S. Akram, T.F. Vandamme, Biomedical Imaging: Principles, Technologies, Clinical Aspects, Contrast Agents, Limitations and Future Trends in Nanomedicines, Pharm. Res. 36 (2019) 78. https://doi.org/10.1007/s11095-019-2608-5.

3. M. Nishita, S.-Y. Park, T. Nishio, K. Kamizaki, Z. Wang, K. Tamada, T. Takumi, R. Hashimoto, H. Otani, G.J. Pazour, V.W. Hsu, Y. Minami, Ror2 Signaling Regulates Golgi Structure and Transport through IFT20 for Tumor Invasiveness, Sci. Rep. 7 (2017) 1. https://doi.org/10.1038/s41598-016-0028-x.

4. A.B. Chinen, C.M. Guan, J.R. Ferrer, S.N. Barnaby, T.J. Merkel, C.A. Mirkin, Nanoparticle Probes for the Detection of Cancer Biomarkers, Cells, and Tissues by Fluorescence, Chem. Rev. 115 (2015) 10530–10574. https://doi.org/10.1021/acs.chemrev.5b00321.

5. C.M. Pitsillides, J.M. Runnels, J.A. Spencer, L. Zhi, M.X. Wu, C.P. Lin, Cell Labeling Approaches for Fluorescence-based In Vivo Flow Cytometry, Cytometry A. 79 (2011) 758–765. https://doi.org/10.1002/cyto.a.21125.

6. J.A. Gleave, J.P. Lerch, R.M. Henkelman, B.J. Nieman, A Method for 3D Immunostaining and Optical Imaging of the Mouse Brain Demonstrated in Neural Progenitor Cells, PLoS One. 8 (2013) e72039. https://doi.org/10.1371/journal.pone.0072039.

7. K. McKayed, J. Simpson, Actin in Action: Imaging Approaches to Study Cytoskeleton Structure and Function, Cells. 2 (2013) 715–731. https://doi.org/10.3390/cells2040715.

8. L. Fass, Imaging and Cancer: A Review, Mol. Oncol. 2 (2008) 115–152. https://doi.org/10.1016/j.molonc.2008.04.001.

9. A. Carovac, F. Smajlovic, D. Junuzovic, Application of Ultrasound in Medicine, Acta Inform. Med. 19 (2011) 168. https://doi.org/10.5455/aim.2011.19.168-171.

10. S. Iranmakani, T. Mortezazadeh, F. Sajadian, M.F. Ghaziani, A. Ghafari, D. Khezerloo, A.E. Musa, A Review of Various Modalities in Breast Imaging: Technical Aspects and Clinical Outcomes, Egypt. J. Radiol. Nucl. Med. 51 (2020) 57. https://doi.org/10.1186/s43055-020-00175-5.

11. Z. Guo, S. Park, J. Yoon, I. Shin, Recent Progress in the Development of Near-Infrared Fluorescent Probes for Bioimaging Applications, Chem. Soc. Rev. 43 (2014) 16–29. https://doi.org/10.1039/c3cs60271k.

12. M. Monici, Cell and Tissue Autofluorescence Research and Diagnostic Applications, Biotechnol. Annu. Rev. 11 (2005) 227–256. https://doi.org/10.1016/S1387-2656(05)11007-2.

13. T. Stahl, R. Bofinger, I. Lam, K.J. Fallon, P. Johnson, O. Ogunlade, V. Vassileva, R.B. Pedley, P.C. Beard, H.C. Hailes, H. Bronstein, A.B. Tabor, Tunable Semiconducting Polymer Nanoparticles with INDT-Based Conjugated Polymers for Photoacoustic Molecular Imaging, Bioconjug. Chem. 28 (2017) 1734–1740. https://doi.org/10.1021/acs.bioconjchem.7b00185.

14. J. Christensen, D. Vonwil, V.S. Prasad, Non-invasive In Vivo Imaging and Quantification of Tumor Growth and Metastasis in Rats Using Cells Expressing Far-Red Fluorescence Protein, PLoS One. 10 (2015) 1–14. https://doi.org/10.1371/journal.pone.0132725.

15. Y. Li, Z. Li, X. Wang, F. Liu, Y. Cheng, B. Zhang, D. Shi, In Vivo Cancer Targeting and Imaging-Guided Surgery with Near Infrared-Emitting Quantum Dot Bioconjugates, Theranostics. 2 (2012) 769–776. https://doi.org/10.7150/thno.4690.

16. T.D. Yang, K. Park, H.-J. Kim, N.-R. Im, B. Kim, T. Kim, S. Seo, J.-S. Lee, B.-M. Kim, Y. Choi, S.-K. Baek, In Vivo Photothermal Treatment with Real-Time Monitoring by Optical Fiber-Needle Array, Biomed. Opt. Express. 8 (2017) 3482. https://doi.org/10.1364/BOE.8.003482.

17. R. Wang, L. Zhou, W. Wang, X. Li, F. Zhang, In Vivo Gastrointestinal Drug-Release Monitoring through Second Near-Infrared Window Fluorescent Bioimaging with Orally Delivered Microcarriers, Nat. Commun. 8 (2017) 14702. https://doi.org/10.1038/ncomms14702.

18. L. Bachmann, D.M. Zezell, A. da Costa Ribeiro, L. Gomes, A.S. Ito, Fluorescence Spectroscopy of Biological Tissues—A Review, Appl. Spectrosc. Rev. 41 (2006) 575–590. https://doi.org/10.1080/05704920600929498.

19. K.S. Valluru, J.K. Willmann, Clinical Photoacoustic Imaging of Cancer, Ultrasonography. 35 (2016) 267–280. https://doi.org/10.14366/usg.16035.

20. C. Wu, B. Bull, C. Szymanski, K. Christensen, J. McNeill, Multicolor Conjugated Polymer Dots for Biological Fluorescence Imaging, ACS Nano. 2 (2008) 2415–2423. https://doi.org/10.1021/nn800590n.

21. D. Tuncel, H.V. Demir, Conjugated Polymer Nanoparticles, Nanoscale. 2 (2010) 484–494. https://doi.org/10.1039/b9nr00374f.

22. X. Guo, A. Facchetti, The Journey of Conducting Polymers from Discovery to Application, Nat. Mater. 19 (2020) 922–928. https://doi.org/10.1038/s41563-020-0778-5.

23. Y. Park, J. Jung, M. Chang, Research Progress on Conducting Polymer-Based Biomedical Applications, Appl. Sci. 9 (2019) 1070. https://doi.org/10.3390/app9061070.

24. C. Wu, D.T. Chiu, Highly Fluorescent Semiconducting Polymer Dots for Biology and Medicine, Angew. Chem. Int. Ed. 52 (2013) 3086–3109. https://doi.org/10.1002/anie.201205133.

25. W.-K. Tsai, Y.-H. Chan, Semiconducting Polymer Dots as Near-Infrared Fluorescent Probes for Bioimaging and Sensing, J. Chinese Chem. Soc. 66 (2019) 9–20. https://doi.org/10.1002/jccs.201800322.

26. T. Klingstedt, K.P.R. Nilsson, Conjugated Polymers for Enhanced Bioimaging, Biochim. Biophys. Acta, Gen. Subj. 1810 (2011) 286–296. https://doi.org/10.1016/j.bbagen.2010.05.003.

27. C. García-Aljaro, M.A. Bangar, E. Baldrich, F.J. Muñoz, A. Mulchandani, Conducting Polymer Nanowire-Based Chemiresistive Biosensor for the Detection of Bacterial Spores, Biosens. Bioelectron. 25 (2010) 2309–2312. https://doi.org/10.1016/j.bios.2010.03.021.

28. V. Van Tran, N.H.T. Tran, H.S. Hwang, M. Chang, Development Strategies of Conducting Polymer-based Electrochemical Biosensors for Virus Biomarkers: Potential for Rapid COVID-19 Detection, Biosens. Bioelectron. 182 (2021) 113192. https://doi.org/10.1016/j.bios.2021.113192.

29. M. Naseri, L. Fotouhi, A. Ehsani, Recent Progress in the Development of Conducting Polymer-Based Nanocomposites for Electrochemical Biosensors Applications: A Mini-Review, Chem. Rec. 18 (2018) 599–618. https://doi.org/10.1002/tcr.201700101.

30. S. Jadoun, U. Riaz, V. Budhiraja, Biodegradable Conducting Polymeric Materials for Biomedical Applications: A Review, Med. Devices Sens. 4 (2021) 1–13. https://doi.org/10.1002/mds3.10141.

31. Y. Luo, Y. Han, X. Hu, M. Yin, C. Wu, Q. Li, N. Chen, Y. Zhao, Live-Cell Imaging of Octaarginine-Modified Polymer Dots via Single Particle Tracking, Cell Prolif. 52 (2019) e12556. https://doi.org/10.1111/cpr.12556.

32. R. Balint, N.J. Cassidy, S.H. Cartmell, Conductive Polymers: Towards a Smart Biomaterial for Tissue Engineering, Acta Biomater. 10 (2014) 2341–2353. https://doi.org/10.1016/j.actbio.2014.02.015.

33. T.F. Abelha, C.A. Dreiss, M.A. Green, L.A. Dailey, Conjugated Polymers as Nanoparticle Probes for Fluorescence and Photoacoustic Imaging, J. Mater. Chem. B. 8 (2020) 592–606. https://doi.org/10.1039/C9TB02582K.

34. C.G. Qian, Y.L. Chen, P.J. Feng, X.Z. Xiao, M. Dong, J.C. Yu, Q.Y. Hu, Q.D. Shen, Z. Gu, Conjugated Polymer Nanomaterials for Theranostics, Acta Pharmacol. Sin. 38 (2017) 764–781. https://doi.org/10.1038/aps.2017.42.

35. L.R. MacFarlane, H. Shaikh, J.D. Garcia-Hernandez, M. Vespa, T. Fukui, I. Manners, Functional Nanoparticles through π-Conjugated Polymer Self-Assembly, Nat. Rev. Mater. 6 (2021) 7–26. https://doi.org/10.1038/s41578-020-00233-4.

36. Y. Braeken, S. Cheruku, A. Ethirajan, W. Maes, Conjugated Polymer Nanoparticles for Bioimaging, Materials (Basel). 10 (2017) 1420. https://doi.org/10.3390/ma10121420.

37. S. Wang, J. Liu, G. Feng, L.G. Ng, B. Liu, NIR-II Excitable Conjugated Polymer Dots with Bright NIR-I Emission for Deep In Vivo Two-Photon Brain Imaging Through Intact Skull, Adv. Funct. Mater. 29 (2019) 1808365. https://doi.org/10.1002/adfm.201808365.

38. M. Lan, S. Zhao, Y. Xie, J. Zhao, L. Guo, G. Niu, Y. Li, H. Sun, H. Zhang, W. Liu, J. Zhang, P. Wang, W. Zhang, Water-Soluble Polythiophene for Two-Photon Excitation Fluorescence Imaging and Photodynamic Therapy of Cancer, ACS Appl. Mater. Interfaces. 9 (2017) 14590–14595. https://doi.org/10.1021/acsami.6b15537.

39. A. Mostafaei, A. Zolriasatein, Synthesis and Characterization of Conducting Polyaniline Nanocomposites Containing ZnO Nanorods, Prog. Nat. Sci. Mater. Int. 22 (2012) 273–280. https://doi.org/10.1016/j.pnsc.2012.07.002.

40. G. Zotti, B. Vercelli, A. Berlin, Gold Nanoparticle Linking to Polypyrrole and Polythiophene: Monolayers and Multilayers, Chem. Mater. 20 (2008) 6509–6516. https://doi.org/10.1021/cm801836j.

41. I. Cesarino, F.C. Moraes, S.A.S. Machado, A Biosensor Based on Polyaniline-Carbon Nanotube Core-Shell for Electrochemical Detection of Pesticides, Electroanalysis. 23 (2011) 2586–2593. https://doi.org/10.1002/elan.201100161.

42. N. Shoaie, M. Daneshpour, M. Azimzadeh, S. Mahshid, S.M. Khoshfetrat, F. Jahanpeyma, A. Gholaminejad, K. Omidfar, M. Foruzandeh, Electrochemical Sensors and Biosensors based on the Use of Polyaniline and Its Nanocomposites: A Review on Recent Advances, Microchim. Acta. 186 (2019) 465. https://doi.org/10.1007/s00604-019-3588-1.

43. J. Xue, X. Chen, S. Liu, F. Zheng, L. He, L. Li, J.-J. Zhu, Highly Enhanced Fluorescence of CdSeTe Quantum Dots Coated with Polyanilines via In-Situ Polymerization and Cell Imaging Application, ACS Appl. Mater. Interfaces. 7 (2015) 19126–19133. https://doi.org/10.1021/acsami.5b04766.

44. T.T.V. Phan, N.Q. Bui, S.-W. Cho, S. Bharathiraja, P. Manivasagan, M.S. Moorthy, S. Mondal, C.-S. Kim, J. Oh, Photoacoustic Imaging-Guided Photothermal Therapy with Tumor-Targeting HA-FeOOH@PPy Nanorods, Sci. Rep. 8 (2018) 8809. https://doi.org/10.1038/s41598-018-27204-8.

45. P. Manivasagan, N. Quang Bui, S. Bharathiraja, M. Santha Moorthy, Y.-O. Oh, K. Song, H. Seo, M. Yoon, J. Oh, Multifunctional Biocompatible Chitosan-Polypyrrole Nanocomposites as Novel Agents for Photoacoustic Imaging-Guided Photothermal Ablation of Cancer, Sci. Rep. 7 (2017) 43593. https://doi.org/10.1038/srep43593.

46. Y. Jin, Y. Li, X. Ma, Z. Zha, L. Shi, J. Tian, Z. Dai, Encapsulating Tantalum Oxide into Polypyrrole Nanoparticles for X-Ray CT/Photoacoustic Bimodal Imaging-Guided Photothermal Ablation of Cancer, Biomaterials. 35 (2014) 5795–5804. https://doi.org/10.1016/j.biomaterials.2014.03.086.

47. A.B. Sanghvi, K.P.-H. Miller, A.M. Belcher, C.E. Schmidt, Biomaterials Functionalization Using a Novel Peptide that Selectively Binds to a Conducting Polymer, Nat. Mater. 4 (2005) 496–502. https://doi.org/10.1038/nmat1397.

48. X. Liu, Z. Yue, M.J. Higgins, G.G. Wallace, Conducting Polymers with Immobilised Fibrillar Collagen for Enhanced Neural Interfacing, Biomaterials. 32 (2011) 7309–7317. https://doi.org/10.1016/j.biomaterials.2011.06.047.

49. B. Zdyrko, I. Luzinov, Polymer Brushes by the "Grafting to" Method, Macromol. Rapid Commun. 32 (2011) 859–869. https://doi.org/10.1002/MARC.201100162.

50. K.Y. Pu, K. Li, B. Liu, A Molecular Brush Approach to Enhance Quantum Yield and Suppress Nonspecific Interactions of Conjugated Polyelectrolyte for Targeted Far-Red/Near-Infrared Fluorescence Cell Imaging, Adv. Funct. Mater. 20 (2010) 2770–2777. https://doi.org/10.1002/adfm.201000495.

51. R. Barbey, L. Lavanant, D. Paripovic, N. Schüwer, C. Sugnaux, S. Tugulu, H.-A. Klok, Polymer Brushes via Surface-Initiated Controlled Radical Polymerization: Synthesis, Characterization, Properties, and Applications, Chem. Rev. 109 (2009) 5437–5527. https://doi.org/10.1021/cr900045a.

52. Z. Zhang, X. Lu, Q. Fan, W. Hu, W. Huang, Conjugated Polyelectrolyte Brushes with Extremely High Charge Density for Improved Energy Transfer and Fluorescence Quenching Applications, Polym. Chem. 2 (2011) 2369–2377. https://doi.org/10.1039/c1py00213a.

53. S. Alkan, L. Toppare, Y. Hepuzer, Y. Yagci, Block Copolymers of Thiophene-Capped Poly(Methyl Methacrylate) with Pyrrole, J. Polym. Sci., A: Polym. Chem. 37 (1999) 4218–4225. https://doi.org/10.1002/(SICI)1099-0518(19991115)37:22<4218::AID-POLA22>3.0.CO;2-Z.

54. A.-D. Bendrea, G. Fabregat, L. Cianga, F. Estrany, L.J. del Valle, I. Cianga, C. Alemán, Hybrid Materials Consisting of an All-Conjugated Polythiophene Backbone and Grafted Hydrophilic Poly(Ethylene Glycol) Chains, Polym. Chem. 4 (2013) 2709. https://doi.org/10.1039/c3py00029j.

55. S. Maione, G. Fabregat, L.J. del Valle, A.-D. Bendrea, L. Cianga, I. Cianga, F. Estrany, C. Alemán, Effect of the Graft Ratio on the Properties of Polythiophene-g-Poly(Ethylene glycol), J. Polym. Sci., B: Polym. Phys. 53 (2015) 239–252. https://doi.org/10.1002/polb.23617.

56. J. Li, S.J. Yoon, B.-Y. Hsieh, W. Tai, M. O'Donnell, X. Gao, Stably Doped Conducting Polymer Nanoshells by Surface Initiated Polymerization, Nano Lett. 15 (2015) 8217–8222. https://doi.org/10.1021/acs.nanolett.5b03728.

57. M. Doshi, M. Krienke, S. Khederzadeh, H. Sanchez, A. Copik, J. Oyer, A.J. Gesquiere, Conducting Polymer Nanoparticles for Targeted Cancer Therapy, RSC Adv. 5 (2015) 37943–37956. https://doi.org/10.1039/C5RA05125H.

58. K. Maeda, N. Kamiya, E. Yashima, Poly(phenylacetylene)s Bearing a Peptide Pendant: Helical Conformational Changes of the Polymer Backbone Stimulated by the Pendant Conformational Change, Chemistry. 10 (2004) 4000–4010. https://doi.org/10.1002/chem.200400315.

59. D.-H. Kim, S.M. Richardson-Burns, J.L. Hendricks, C. Sequera, D.C. Martin, Effect of Immobilized Nerve Growth Factor on Conductive Polymers: Electrical Properties and Cellular Response, Adv. Funct. Mater. 17 (2007) 79–86. https://doi.org/10.1002/adfm.200500594.

60. J. Frangioni, In Vivo Near-Infrared Fluorescence Imaging, Curr. Opin. Chem. Biol. 7 (2003) 626–634. https://doi.org/10.1016/j.cbpa.2003.08.007.

61. W.K. Chong, V. Papadopoulou, P.A. Dayton, Imaging with Ultrasound Contrast Agents: Current Status and Future, Abdom. Radiol. 43 (2018) 762–772. https://doi.org/10.1007/s00261-018-1516-1.

62. J.S. Jung, D. Jo, G. Jo, H. Hyun, Near-Infrared Contrast Agents for Bone-Targeted Imaging, Tissue Eng. Regen. Med. 16 (2019) 443–450. https://doi.org/10.1007/s13770-019-00208-9.

63. C. Zhu, L. Liu, Q. Yang, F. Lv, S. Wang, Water-Soluble Conjugated Polymers for Imaging, Diagnosis, and Therapy, Chem. Rev. 112 (2012) 4687–4735. https://doi.org/10.1021/cr200263w.

64. X. Zhou, P. Ma, A. Wang, C. Yu, T. Qian, S. Wu, J. Shen, Dopamine Fluorescent Sensors based on Polypyrrole/Graphene Quantum Dots Core/Shell Hybrids, Biosens. Bioelectron. 64 (2015) 404–410. https://doi.org/10.1016/j.bios.2014.09.038.

65. X. Song, C. Liang, H. Gong, Q. Chen, C. Wang, Z. Liu, Photosensitizer-Conjugated Albumin–Polypyrrole Nanoparticles for Imaging-Guided In Vivo Photodynamic/Photothermal Therapy, Small. 11 (2015) 3932–3941. https://doi.org/10.1002/smll.201500550.

66. Y. Jin, X. Yang, J. Tian, Targeted Polypyrrole Nanoparticles for the Identification and Treatment of Hepatocellular Carcinoma, Nanoscale. 10 (2018) 9594–9601. https://doi.org/10.1039/C8NR02036A.

67. D. Park, Y. Cho, S.-H. Goh, Y. Choi, Hyaluronic Acid–Polypyrrole Nanoparticles as pH-Responsive Theranostics, Chem. Commun. 50 (2014) 15014–15017. https://doi.org/10.1039/C4CC06349J.

68. D. Park, K.-O. Ahn, K.-C. Jeong, Y. Choi, Polypyrrole-based Nanotheranostics for Activatable Fluorescence Imaging and Chemo/Photothermal Dual Therapy of Triple-Negative Breast Cancer, Nanotechnology. 27 (2016) 185102. https://doi.org/10.1088/0957-4484/27/18/185102.

69. B. Bao, Z. Yang, Y. Liu, Y. Xu, B. Gu, J. Chen, P. Su, L. Tong, L. Wang, Two-Photon Semiconducting Polymer Nanoparticles as a New Platform for Imaging of Intracellular pH Variation, Biosens. Bioelectron. 126 (2019) 129–135. https://doi.org/10.1016/j.bios.2018.10.027.

70. G. Hong, Y. Zou, A.L. Antaris, S. Diao, D. Wu, K. Cheng, X. Zhang, C. Chen, B. Liu, Y. He, J.Z. Wu, J. Yuan, B. Zhang, Z. Tao, C. Fukunaga, H. Dai, Ultrafast Fluorescence Imaging In Vivo with Conjugated Polymer Fluorophores in the Second Near-Infrared Window, Nat. Commun. 5 (2014) 4206. https://doi.org/10.1038/ncomms5206.

71. N. Liu, J. Shi, Q. Wang, J. Guo, Z. Hou, X. Su, H. Zhang, X. Sun, In Vivo Repeatedly Activated Persistent Luminescence Nanoparticles by Radiopharmaceuticals for Long-Lasting Tumor Optical Imaging, Small. 16 (2020) 2001494. https://doi.org/10.1002/smll.202001494.

72. E. Teston, T. Maldiney, I. Marangon, J. Volatron, Y. Lalatonne, L. Motte, C. Boisson-Vidal, G. Autret, O. Clément, D. Scherman, F. Gazeau, C. Richard, Nanohybrids with Magnetic and Persistent Luminescence Properties for Cell Labeling, Tracking, In Vivo Real-Time Imaging, and Magnetic Vectorization, Small. 14 (2018) 1800020. https://doi.org/10.1002/smll.201800020.

73. Q. Miao, C. Xie, X. Zhen, Y. Lyu, H. Duan, X. Liu, J. V Jokerst, K. Pu, Molecular Afterglow Imaging with Bright, Biodegradable Polymer Nanoparticles, Nat. Biotechnol. 35 (2017) 1102–1110. https://doi.org/10.1038/nbt.3987.

74. A.J. Shuhendler, K. Pu, L. Cui, J.P. Uetrecht, J. Rao, Real-Time Imaging of Oxidative and Nitrosative Stress in the Liver of Live Animals for Drug-Toxicity Testing, Nat. Biotechnol. 32 (2014) 373–380. https://doi.org/10.1038/nbt.2838.

75. J. Li, J. Rao, K. Pu, Recent Progress on Semiconducting Polymer Nanoparticles for Molecular Imaging and Cancer Phototherapy, Biomaterials. 155 (2018) 217–235. https://doi.org/10.1016/j.biomaterials.2017.11.025.

76. M. Palner, K. Pu, S. Shao, J. Rao, Semiconducting Polymer Nanoparticles with Persistent Near-Infrared Luminescence for In Vivo Optical Imaging, Angew. Chem. 127 (2015) 11639–11642. https://doi.org/10.1002/ange.201502736.

77. C. Xie, X. Zhen, Q. Miao, Y. Lyu, K. Pu, Self-Assembled Semiconducting Polymer Nanoparticles for Ultrasensitive Near-Infrared Afterglow Imaging of Metastatic Tumors, Adv. Mater. 30 (2018) 1801331. https://doi.org/10.1002/adma.201801331.

78. S. Wu, Y. Li, R. Zhang, K. Fan, W. Ding, L. Xu, L. Zhang, Persistent Luminescence-Polypyrrole Nanocomposite for Dual-Modal Imaging and Photothermal Therapy of Mammary Cancer, Talanta. 221 (2021) 121435. https://doi.org/10.1016/j.talanta.2020.121435.

79. M. Amrithesh, S. Aravind, S. Jayalekshmi, R.S. Jayasree, Enhanced Luminescence Observed in Polyaniline–Polymethylmethacrylate Composites, J. Alloys Compd. 449 (2008) 176–179. https://doi.org/10.1016/j.jallcom.2006.02.096.

80. L. Yan, L. Chen, X. Zhao, X. Yan, pH Switchable Nanoplatform for In Vivo Persistent Luminescence Imaging and Precise Photothermal Therapy of Bacterial Infection, Adv. Funct. Mater. 30 (2020) 1909042. https://doi.org/10.1002/adfm.201909042.

81. P. Beard, Biomedical Photoacoustic Imaging, Interface Focus. 1 (2011) 602–631. https://doi.org/10.1098/rsfs.2011.0028.

82. Z. Zha, Z. Deng, Y. Li, C. Li, J. Wang, S. Wang, E. Qu, Z. Dai, Biocompatible Polypyrrole Nanoparticles as a Novel Organic Photoacoustic Contrast Agent for Deep Tissue Imaging, Nanoscale. 5 (2013) 4462. https://doi.org/10.1039/c3nr00627a.

83. X. Liang, Y. Li, X. Li, L. Jing, Z. Deng, X. Yue, C. Li, Z. Dai, PEGylated Polypyrrole Nanoparticles Conjugating Gadolinium Chelates for Dual-Modal MRI/Photoacoustic Imaging Guided Photothermal Therapy of Cancer, Adv. Funct. Mater. 25 (2015) 1451–1462. https://doi.org/10.1002/adfm.201402338.

84. X. Song, H. Gong, S. Yin, L. Cheng, C. Wang, Z. Li, Y. Li, X. Wang, G. Liu, Z. Liu, Ultra-Small Iron Oxide Doped Polypyrrole Nanoparticles for In Vivo Multimodal Imaging Guided Photothermal Therapy, Adv. Funct. Mater. 24 (2014) 1194–1201. https://doi.org/10.1002/adfm.201302463.

85. W. Huang, R. Chen, Y. Peng, F. Duan, Y. Huang, W. Guo, X. Chen, L. Nie, In Vivo Quantitative Photoacoustic Diagnosis of Gastric and Intestinal Dysfunctions with a Broad pH-Responsive Sensor, ACS Nano. 13 (2019) 9561–9570. https://doi.org/10.1021/acsnano.9b04541.

86. Q. Tian, Y. Li, S. Jiang, L. An, J. Lin, H. Wu, P. Huang, S. Yang, Tumor pH-Responsive Albumin/Polyaniline Assemblies for Amplified Photoacoustic Imaging and Augmented Photothermal Therapy, Small. 15 (2019) 1902926. https://doi.org/10.1002/smll.201902926.

87. B. Bao, L. Tong, Y. Xu, J. Zhang, X. Zhai, P. Su, L. Weng, L. Wang, Mussel-Inspired Functionalization of Semiconducting Polymer Nanoparticles for Amplified Photoacoustic Imaging and Photothermal Therapy, Nanoscale. 11 (2019) 14727–14733. https://doi.org/10.1039/C9NR03490K.

88. D. Gao, P. Zhang, Y. Liu, Z. Sheng, H. Chen, Z. Yuan, Protein-Modified Conjugated Polymer Nanoparticles with Strong Near-Infrared Absorption: A Novel Nanoplatform to Design Multifunctional Nanoprobes for Dual-Modal Photoacoustic and Fluorescence Imaging, Nanoscale. 10 (2018) 19742–19748. https://doi.org/10.1039/C8NR06197A.

89. D. Zhang, M. Wu, Y. Zeng, N. Liao, Z. Cai, G. Liu, X. Liu, J. Liu, Lipid Micelles Packaged with Semiconducting Polymer Dots as Simultaneous MRI/Photoacoustic Imaging and Photodynamic/Photothermal Dual-Modal Therapeutic Agents for Liver Cancer, J. Mater. Chem. B. 4 (2016) 589–599. https://doi.org/10.1039/C5TB01827G.

90. G. Balasundaram, C.J.H. Ho, K. Li, W. Driessen, U. Dinish, C.L. Wong, V. Ntziachristos, B. Liu, M. Olivo, Molecular Photoacoustic Imaging of Breast Cancer Using an Actively Targeted Conjugated Polymer, Int. J. Nanomed. 10 (2015) 387. https://doi.org/10.2147/IJN.S73558.

91. W. Zhang, X. Sun, T. Huang, X. Pan, P. Sun, J. Li, H. Zhang, X. Lu, Q. Fan, W. Huang, 1300 nm Absorption Two-Acceptor Semiconducting Polymer Nanoparticles for NIR-II Photoacoustic Imaging System Guided NIR-II Photothermal Therapy, Chem. Commun. 55 (2019) 9487–9490. https://doi.org/10.1039/C9CC04196F.

92. Z. Sheng, B. Guo, D. Hu, S. Xu, W. Wu, W.H. Liew, K. Yao, J. Jiang, C. Liu, H. Zheng, B. Liu, Bright Aggregation-Induced-Emission Dots for Targeted Synergetic NIR-II Fluorescence and NIR-I Photoacoustic Imaging of Orthotopic Brain Tumors, Adv. Mater. 30 (2018) 1800766. https://doi.org/10.1002/adma.201800766.

16 Conducting Polymer-Based Micro-Containers for Biomedical Applications

Selcan Karakuş,[1] *Cemal Özeroğlu,*[1] *and*
Mizan İbrahim Kahyaoğlu[2]

[1]Faculty of Engineering, Department of Chemistry,
Istanbul University-Cerrahpasa, Istanbul, Turkey

[2]Department of Chemistry, Faculty of Science and Arts,
Ondokuz Mayis University, Kurupelit, Samsun, Turkey

CONTENT

16.1 MORPHOLOGIES OF MICRO-CONTAINERS

In recent years, stimuli-responsive micro-containers have been among the interesting topics of multidisciplinary research in biomedical applications such as drug delivery systems, imaging, monitoring, and sensors [1]. In particular, it has been shown that the physicochemical properties of nano-/micro-containers such as surface area, shape, size distribution, surface topography, surface morphology, solubility, and structure have an important role in biomedical applications [2]. In the literature, the micro-containers are classified based on their shapes as cubic, spherical/hemispherical, elliptical, cylindrical, conical, and pyramidal geometries in the micrometric size range. As known, conducting polymer (CP)-based nanocomposites have the superior advantageous properties of synergetic effect of CPs and organic/inorganic nanoparticles in biomedical applications. Also, the combination of conductive nano-/micro-fillers such as organic and inorganic nanoparticles in a conductive polymer matrix improves the electrochemical biosensors compared to traditional sensors. Among conductive polymers, poly(aniline) (PANI), poly(3,4-ethylenedioxythiophene) (PEDOT), poly(pyrrole) (PPy), poly(acetylene), poly(thiophene), and poly(p-phenylenevinylene) (PPV) were widely used in supercapacitor, nano-coating, catalysis, biosensor, and biomedical applications. Modarres-Gheisari et al. prepared cubic ultrasonic micro-containers using the sonication method at different excitation frequencies [3]. Mazur et al. developed polypyrrole micro-containers using the emulsion method in the change shapes from spherical to hemispherical. The spherical shaped polymeric micro-containers were characterized using scanning electron microscopy (SEM), atomic force microscopy (AFM), conducting AFM, near-field scanning optical microscopy (NSOM), infrared spectroscopy, UV-Vis spectrophotometer, and cyclic voltammetry. They observed that the droplets of the prepared polymeric micro-containers played the major role of templates that induced the growth of the conductive polypyrrole film in the form of hemispherical micro-containers [4].

To fabricate nanocomposites with different morphologies, Lalegül-Ülker et al. reported the fabrication of dual stimuli-responsive nanocomposites based on silica-coated iron oxide/polyaniline (Si-MNPs/PANI) using the chemical oxidative polymerization method. The prepared Si-MNPs/PANI nanocomposites were characterized to determine chemical, physical, and biological properties using SEM, transmission electron microscopy (TEM), Fourier transform infrared (FTIR) spectrometer, X-ray powder diffraction (XRD), thermogravimetric analysis (TGA), room temperature vibrating sample magnetometer (VSM), and electrical resistivity techniques. According to results, they observed that the prepared Si-MNPs/PANI nanocomposites had a spherical morphology and a diameter range of 250–400 nm and promising potential scaffolds for tissue engineering applications [5]. In literature, many studies based on the fabrication of stimulus-responsive nanocarriers such as drug delivery systems, smart drug release platforms, and imaging systems have been reported. PPy has superior properties such as electrical conductivity, high environmental stability, easy to prepare, and biocompatibility for biomedical applications. For this purpose, Wang et al. developed smart montmorillonite-polypyrrole scaffolds for electro-responsive aspirin release systems. The aspirin-loaded montmorillonite-polypyrrole was characterized by field emission SEM (FESEM), FTIR, electrochemical impedance spectroscopy (EIS), and XRD. According to FESEM results, it was observed that the sample had a cauliflower-shaped morphology with an aggregated and layered structure [6].

To use CP-based micro-/nano-containers for the release of very commonly used drugs such as aspirin, Kong et al. designed a clay polymer nanocomposite (CPN) based on aspirin-loaded palygorskite modified PPy for electrically tunable drug delivery. Palygorskite is a conductive polymer for the development of the nanostructure-based drug delivery system. According to TEM results, the developing nanocomposite exhibited a fibrillar single crystal with a diameter ranging from 20 to 30 nm. The release results showed the Higuchi kinetic model indicating a good fitting to experimental data together with the highest correlation constant and the prepared novel aspirin delivery system could be a promising implantable device for electrically tunable drug delivery systems with the patient's requirement [7]. The morphology (size and shape) of micro-containers is a special key in controlling gap closure in the wound healing biological process. In this regard, smart polymer-based therapeutics as carriers have been prepared with variously shaped micro-containers (Figure 16.1).

Bao et al. reported the elliptical morphology of micro-container is an effective impact on gap closure rate in wound healing studies [8]. Furthermore, they highlighted that biomaterial is commonly preferred to obtain high surface porosity and the small size in tissue engineering strategies. Nielsen et al. emphasized that polymeric micro-containers can be prepared cylindrical shapes in the micrometric size range [9]. Moreover, they showed that the cubosome loaded with ovalbumin in a spray-dried powder form is a promising carrier for oral vaccine delivery in biomedical applications. Huang et al. developed a nano-container of conducting PPy with a conical shape and a uniform inner diameter using a stepwise electropolymerization process. Furthermore, they showed their potential applications under electrochemical voltages in biomedical devices and controlled delivery systems [10]. Lee et al. prepared tetrode tips coated with Au nanoparticles (Au NPs) and dextran (Dex)-doped PEDOT. They aimed to investigate the developed recording performance of modified tetrodes with high electrochemical properties due to their two-dimensional surfaces area using a simple electroplating method. This chapter demonstrated the effect of PEDOT and Au NPs on the surface of tetrodes, which improved the electrochemical properties and neural recording quality. They proved the effect of conductive PEDOT and Au NPs on the well-plated shaped surface of modified tetrodes, which increased the electrochemical properties and cathodic storage capacity and also improved neural recording quality. Results showed that conductive PEDOT-coated tetrode recordings can be used in specific biomedical applications for multiple neuronal recordings to increase the understanding of in vivo mechanism of the complex brain signal activities [11].

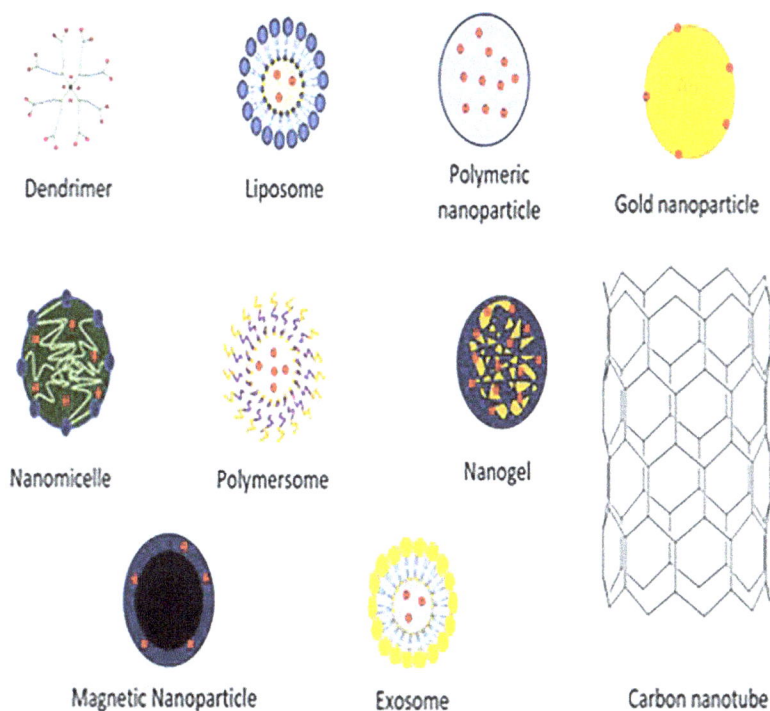

FIGURE 16.1   Smart polymer-based therapeutics.

Chen et al. designed a novel drug nano-carrier based on conductive tumor-targeting folate-poly(ethylene glycol) polymer-acylhydrazone-β-cyclodextrin-polypyrrole (FA-PEG-Ad-β-CD-PPy) nanotubes using a template method to investigate the controlled doxorubicin release system. The results of TEM measurements showed that the morphology of the PPy-Cl was a nanotube shape with a diameter of 200 nm and a length of about 600 nm. The morphology played a major role in the controlled DOX (doxorubicin) release mechanism of FA-PEG-Ad-β-CD-PPy-based nano-carrier systems. According to results, they found that the DOX-loaded FA-PEG-Ad-β-CD-PPy was sensitive and stable under physiological conditions [12]. Recent studies have demonstrated the fabrication of conductive polymer-based micro-containers with excellent electrical and magnetic properties for magnetic resonance imaging systems. For this purpose, Gimi et al. prepared cubic and pyramidal micro-containers for cell encapsulation with high surface area-to-volume ratios and showed noninvasive detection of micro-containers using positive-contrast MRI [13]. Wang et al. prepared the conducting PEDOT/polypeptide (PEP) nanocomposite with a 3D microporous network structure, large surface area, and excellent antifouling ability using the electropolymerization method. The prepared PEDOT/PEP nanocomposite DNA biosensor exhibited high selectivity and sensitivity with a detection limit of 0.0034 pM for breast cancer marker (BRCA1) complementary oligonucleotides detection [14].

Alongside many other biomedical applications, conductive polythiophene-based nanocomposites have been extensively reported on the determination of the performance of conductive polymer-based nano-formulations in drug release systems, gene therapy, and antimicrobial/antifungal studies against antibiotics-resistant pathogenic organisms. Noreen et al. synthesized graphene nanoplatelets/polythiophene nanocomposites by in situ chemical oxidative polymerization method. The obtained results showed that the synthesized irregular small sheet-shaped graphene nanoplatelets/polythiophene nanostructures had excellent properties such as photocatalytic degradation of bromophenol blue with degradation of 94% and antibacterial activities against Escherichia coli and Staphylococcus aureus, a zone of inhibition of 18 mm [15].

16.2   EVALUATION OF THE STRUCTURES OF MICRO-CONTAINERS

Depending on the structure, micro-containers can be divided into six categories, polymeric-based micro-containers, clay-based micro-containers, metal/metal oxide-based micro-containers, ceramic-based micro-containers, cellulose-based micro-containers, and carbon nanotubes. Micro/nano-containers consisted of conductive polymers showed good reproducibility, mechanical properties, biocompatibility, flexibility, and electrical conductivity to tissue in biomedical applications. Therefore, researchers have widely studied conductive polymers and their derivatives as the matrix of choice due to their superior advantages.

16.3   EVALUATION OF THE EFFECT OF CHEMICAL AND MORPHOLOGICAL ADVANTAGES OF MICRO-CONTAINERS

As reported previously, the conductive polymeric micro-containers have been extensively investigated to improve biodegradability, biocompatibility, flexibility, and low cost for pharmaceutical studies due to their primary chemical, physical, biological, and morphological advantages. Natural and synthetic polymers have been preferred for applications in biosensors, diagnosis, and drug delivery systems due to their stimuli responsiveness such as ionic strength, pH, temperature, light, mechanical, magnetic, electric, ultrasound, solvent, electrochemical, receptor, and enzyme. For instance, Massoumi et al. fabricated a conductive nanofibrous scaffold based on polyaniline-*co*-(polydopamine-*g*-polylactide) (PANI-*co*-(PDA-*g*-PLA) as a scaffolding biomaterial for tissue engineering applications using the electrospinning technique and one-step chemical oxidization methods. The surface morphologies of the prepared PANI-*co*-PDA and PANI-*co*-(PDA-*g*-PLA) were showed that the structure had a spherical shape in the size range of 70 ± 20 nm. They examined physicochemical and biological characteristics of the prepared electrically conductive nanofibrous to show their excellent biocompatibility, biodegradability, cells adhesion, and proliferation properties [16].

In efforts to improve the sensitivity of the sensor, Jasim et al. developed an electrically conductive bacterial cellulose/polyaniline/single-walled carbon nanotubes membrane. The prepared nanostructures were characterized by field emission SEM (FE-SEM), FTIR, and XRD techniques. According to obtained results, they observed that the conductive nanotube had a uniform distribution and promising material for biosensors, biofuel cells, and bioelectronic devices with an improved electrical conductivity [17]. Kooti et al. developed a conductive $CoFe_2O_4$/PANI/Ag nanocomposite via a facile chemical reduction method. According to the results, the silver nanoparticles were immobilized successfully on the surface of the prepared $CoFe_2O_4$/PANI nanocomposite with an irregular microcrystal shape. The antibacterial activities of the prepared nanostructures were investigated by determining the diameter of inhibition zone (DIZ) of the antibacterial substances and results proved the antibacterial activities of the prepared nanostructures had high performances against Gram-positive and Gram-negative bacteria [18].

Parkinson's disease is defined as the most common neurodegenerative disease. Nano-medicine has a promising future in the drug delivery system in Parkinson's disease. The controlled release systems of Parkinson's drug-loaded nano-carriers based on conductive polymers are possible, and it is aimed to increase their bioavailability in the treatment of the disease. With this approach, Xue et al. developed an injectable conductive gelatin-PANI hydrogel system that serves as a promising therapeutic agent to deliver bone marrow stromal cells (BMSCs) for Parkinson's disease treatment. According to results, they showed that BMSCs-loaded injectable Gelatin-PANI hydrogels were an effective agent for neuroprotective MPTP-induced PD models [19]. Cancer can develop in the tissues of the body with its characteristics and can exhibit lethal features. Although early diagnosis, rapid and target detection save lives, and effective results have begun to be obtained with tumor microenvironment targeted nano therapies with reduced side effects. Zahed et al. developed a highly efficient electrochemical sensor for the analysis of anticancer drug 5-fluorouracil (5-FU) using silver nanoparticles-polyaniline nanocomposite (AgNPs@PANINTs) in the 5-FU concentration range of 1.0–300.0 μM with a low detection limit of 0.06 μM [20]. Lu et al. developed a nanozyme sensor based on the electro-synthesized molecularly imprinted conducting PEDOT nanocomposite with graphene-like two-dimensional layered molybdenum disulfide (MoS$_2$) for electrochemical detection of luteolin. According to electrochemical results, they showed that the prepared conductive polymer/MoS$_2$-based sensor had a high electrochemical performance to detect luteolin in the linear concentration range from 3×10^{-7} to 3×10^{-5} M with a limit of detection of 0.04 μM under the optimal conditions [21].

Recently, conductive polymer-based core/shell hybrid nanoparticles have become a progressively major topic for advanced materials chemistry and biomedical applications. For instance, Singh et al. developed Fe$_3$O$_4$/polythiophene (Fe$_3$O$_4$/PT) hybrid nanocomposite sensing platform for hydrogen peroxide in electroanalytical application with a limit of detection of 5 μM. The prepared nanostructure was characterized to determine the morphology, structure thermal, and functional groups of the hybrid nanocomposite by TEM, FTIR, TGA, XRD, and superconducting quantum interference device (SQUID) magnetometry. According to morphology results, the prepared nanocomposite showed a uniform spherical shape with a particle size of 11 ± 3 nm. And also, conductive polymer shells tightly encapsulated the Fe$_3$O$_4$ core in the well-dispersed nanostructure. Electroanalytical results of Fe$_3$O$_4$/PT nanocomposite showed the electrocatalytic effects of prepared Fe$_3$O$_4$/PT nanocomposites on hydrogen peroxide reduction with a cathodic response of 19.5 μA [22].

Hatamzadeh et al. prepared nanofibrous electrically conductive scaffolds based on poly(ethylene glycol)-modified polythiophene and poly(ε-caprolactone) using electrospinning technique. The experimental results of the prepared nanostructures were confirmed biocompatibilities, adhesion, viability, and proliferation of human osteoblast MG-63 cells for tissue engineering applications. Also, high electrical conductivity, hydrophilicity, and mechanical properties of the nanostructure were observed with a three-dimensional (3D) interconnected pore morphology with an average diameter in the size range of 100 ± 20 nm. According to electrical conductivity results of the prepared samples showed lower electrical conductivities than that of the pure polythiophene and were found suitable for tissue engineering applications [23]. Wei et al. fabricated a sensitive zirconium-based porphyrinic metal-organic framework decorated on conductive PEDOT films-based sensor for detecting antibiotic chloramphenicol with an ultralow detection limit, good selectivity, and reproducibility. They observed the synergistic effect between the zirconium-based porphyrinic metal-organic framework and the conductive PEDOT films. Also, they optimized a high sensing performance of the fabricated PCN-222-CHIT/PEDOT/ITO sensor. Results showed that the fabricated PCN-222-CHIT/PEDOT/ITO sensor had an ultralow detection limit of 0.0018 μM for antibiotic chloramphenicol [24].

16.4   THE DEVELOPMENT OF DRUG DELIVERY SYSTEMS AND SENSORS

In recent years, great progress on cancer therapy using sensitive target cancer nano-drugs, controlled drug delivery systems, and sensitive electrochemical anticancer drugs sensors has increased. With this approach, Fouladgar et al. fabricated a sensitive electrochemical anticancer drugs sensor for simultaneous determination of raloxifene and tamoxifen drugs using graphene-CuO-conductive polypyrrole nanocomposites. They investigated the electrochemical behaviors of the combination of graphene and CuO nanoparticles on the fabricated electrochemical anticancer drugs sensor. They showed the adsorption mechanism of raloxifene drug based on van der Waals effects on GO-CuO surface due to the work function, charge transfer, energetic, structural, and electronic features. According to results of the fabricated graphene-CuO-conductive polypyrrole nanocomposites sensor, they observed that the limits of detection of graphene-CuO-conductive polypyrrole nanocomposites were obtained 3.0 and 10.5 nmol L^{-1} for the measurement of raloxifene and tamoxifen, respectively [25].

Many studies have been focused on photothermal cancer therapy with a nanotechnological approach based on nanostructures a remarkably increasing interest, including various metal/metal oxide nanostructures, carbon-based nanostructures, organic-based nanostructures, and conductive/

nonconductive polymeric matrices-based nanoparticles. With this viewpoint, Xia et al. prepared photothermal and biodegradable polyaniline/porous silicon hybrid nanocomposites (PANi-PSiNPs) as drug carriers. The hybrid nanocomposites were prepared using chemo-photothermal therapy, known as combination therapy, due to the synergistic anticancer effect. The prepared polyaniline/ PSiNPs hybrid nanocomposites were characterized using advanced techniques such as X-ray photoelectron spectra (XPS), transmission FTIR, SEM, TEM, and thermal imaging to determine the chemical property, size, shape, and photothermal effect. Anticancer drugs were loaded on the prepared polyaniline/PSiNPs hybrid nanocomposites and dual pH/NIR light-triggered release of drug-loaded nanocomposites was examined in different biological environments. They investigated the release profile of the pH-responsive doxorubicin hydrochloride (DOX)-loaded PANi-PSiNPs under near-infrared (NIR) laser irradiation. According to experimental results, they found that the hybrid nanocomposites had excellent biodegradability, biocompatibility, and capacities on chemo-photothermal therapy [26].

As an example of nano-carrier studies, Işıklan et al. fabricated a novel gelatin-decorated magnetic graphene oxide (MGO@GEL) nanoplatform to exhibit their paclitaxel (PAC) as a chemotherapy drug delivery performance at pH 5.5 than at pH 7.4. Also, cytotoxicity analysis results showed that MGO@GEL nanosheets were biocompatible. Consequently, the prepared MGO@GEL-based multifunctional nanostructure is a promising material for chemo-photothermal therapy and drug delivery systems [27]. Khadir et al. reported the study of adsorption parameters, kinetic models, and isotherm equilibrium of a non-steroidal anti-inflammatory drug (ibuprofen) from aquatic media by lignocellulosic biomaterial (Luffa cylindrica) reinforced with conductive polypyrrole. They proved that Luffa/polypyrrole had a significantly better adsorption performance and efficiency than raw Luffa with a maximum monolayer adsorption capacity of 19.157 mg g^{-1}. The adsorption process of Luffa/polypyrrole was exothermic, feasible, and spontaneous with the highest removal percentage of 97% and at an equilibrium time of 90 min. According to adsorption results, they proved that the Luffa/polypyrrole composite is a low-cost and efficient material for the adsorption of ibuprofen from aqueous solutions [28].

To develop a high-sensitivity drug sensor, Karimi-Maleh et al. developed a novel pencil graphite electrode modified with conductive PPy and reduced graphene oxide (rGO) (PGE/PPy/rGO). According to results, they observed that a highly sensitive guanine/adenine DNA-based electrochemical PGE/PPy/rGO biosensor for nanomolar determination of didanosine anticancer drug had a low limit of detection (LOD = 8.0 nM) in the dynamic range of didanosine identified in the range of 0.02–50.0 µM using electrochemical methods for didanosine. Furthermore, F-test and T-test were calculated for confirming performances of guanine/adenine DNA-loaded PGE/PPy/rGO for analysis of didanosine in real samples [29]. As known, polypyrrole is a potential conductive polymer for hydrophobic drug delivery systems. For this reason, Attia et al. fabricated a one-step synthesis of hybrid iron oxide/polypyrrole multifunctional nanoparticles encapsulating hydrophobic drug (ketoprofen) and decorated with polyethylene glycol. According to ketoprofen release results, they calculated the high encapsulation efficiency of iron oxide/polypyrrole multifunctional nanoparticles was 98%, and also the characterization results showed that the average particle size of iron oxide/polypyrrole multifunctional nanoparticles was below 50 nm. Consequently, magnetic relaxometry studies of prepared iron oxide/polypyrrole multifunctional nanoparticles confirmed that the nanostructure is a promising contrast agent for the field of magnetic resonance imaging (MRI) systems [30].

Smart and conductive polymer-based anticancer drug delivery systems have been investigated in recent years. With this nanotechnological approach, Guo et al. reported near-infrared (NIR) light-responsive DOX-loaded polypyrrole nanoparticles to exhibit combined tumor photothermal-chemotherapy properties with high colloidal stability and good biocompatibility. The drug-loading efficiency of prepared samples was calculated in the range of 13.5%–26.4%. The difference in drug-loading efficiency of the prepared polypyrrole nanoparticles was related to the polypyrrole and DOX-loading efficiency [31]. Studies have highlighted that the cumulative amount of a loaded drug changed with different factors, such as the structure of the polymer matrix, the size, and solubility of the drug, and additives in drug delivery systems for biomedical applications. Krukiewicz et al. reported the PEDOT/naproxen release studies and proved the effects of CPs on the release characteristics due to their versatility and tunable properties in biomedical applications. The kinetic mechanisms were analyzed with different models to describe the release kinetics as a function of time. Results given in this study showed that the spontaneous release of drugs from PEDOT was well fitted with the first-order kinetic model with the highest parameters and the mechanism of release was a complex transport. Also, according to results, they showed that drug-loading capacity was based on the electrostatic interactions between charged polymer matrix and drug ions [32].

Dexamethasone phosphate decreases the immune system's response to different diseases to reduce the swelling and allergic-type reactions. Intending to develop these features, it is an anionic corticosteroid drug of choice in controlled release studies. Yasin et al. designed a three-dimensionally ordered macroporous PEDOT thin films using a vapor phase polymerization method with a face-centered cubic structure of 280–290 nm spherical macropores in a PEDOT conductive polymer matrix. The dexamethasone phosphate drug was loaded into three-dimensionally ordered macroporous PEDOT for triggered drug delivery systems and the dexamethasone phosphate release profile was investigated with an alternating pulse stimulation. According to dexamethasone phosphate drug release results, they calculated that the amount of dexamethasone phosphate was released from three-dimensionally ordered macroporous films with a higher amount of 38 ± 25 µg compared to 26.5 ± 5 µg of dexP$^-$ released from non-templated PEDOT films. Also, the amount of dexamethasone phosphate release was 118 ± 9 µg from the three-dimensionally ordered macroporous films compared to 109 ± 11 µg from the non-templated films [33].

5-Fluorouracil is an anticancer agent and nucleobase analogue and it inhibits DNA synthesis and slows tumor growth. Hsiao et al. fabricated the electrically conductive polypyrrole-tungsten disulfide nanocomposite with higher electrical conductivity and biocompatibility properties for in vivo 5-fluorouracil (5-FU) anticancer drug release in mice skin. The fabricated polypyrrole-tungsten disulfide nanocomposites were characterized by different techniques such as UV-Vis, XRD, Raman, and SEM techniques. The prepared polypyrrole-tungsten disulfide nanocomposites had an aggregated flake-like morphology. 5-FU-loaded nanocomposite had a high release (%) of 90% drug under electrical stimulation. The encapsulation efficiency of the synthesized polypyrrole-tungsten disulfide nanocomposites was found to be in the range of 36.58–56.91%. Also, the electrochemical results of nanomaterials showed that the synthesized conductive polypyrrole-tungsten disulfide nanocomposites had a high electroactive property and can be used as an effective and conductive nanocargo for drug release studies [34]. Chahm et al. prepared a novel electrochemical platform to detect oligonucleotides using polypyrroles due to conducting surfaces. Their results showed that the mechanism was based on the excellent organic conducting surfaces of polypyrroles [35].

Currently, the world has dramatic problems than ever before due to the increase of population growth, heterogeneous population distribution, and mobility with the lack of financial protection of basic health-care services in global approach at extraordinary speed. These issues cause a global spread of pathogenic viruses. Using a sensitive, rapid, sustainable, cost-effective sensor for viral detections of pathogenic viruses is a critical issue for control of the epidemic and pandemic potential and prevention of infection. For this purpose, studies focused on the facile and one-step strategy to prepare organic/inorganic nanostructures that detect different viruses such as coronavirus 2 (SARS-CoV-2), hepatitis B virus (HBV), human immunodeficiency virus (HIV), Zika, and dengue in a real sample with the lowest concentrations (picomolar-nanomolar range). Wankar et al. prepared a sensitive fluorescence sensor using conductive polythiophene nanofilms with an approximate thickness of 100–500 nm for the detection of tobacco necrosis virus (TNV) in drinking water. The electrochemical results showed that the prepared nanofilm had an excellent performance in the range of 0.1–10 ng L^{-1} (0.15–15 pg) with the lower calculated detection limit of 2.29 ng L^{-1} (3.4 pg) [36]. Shamsipur et al. developed a sensitive and rapid sensor based on a composite platform reinforced by a conductive polymer sandwiched between two nanostructured layers for sub-femtomolar detection of HIV-1 gene using DNA immobilized with an exclusive performance. The conductive layer consisted of two nanostructured layers as p-aminobenzoic acid (PABA) sandwiched between the electrochemically reduced graphene oxide (ERGO) (the sub-layer) and the gold nanoparticles (AuNPs). Analytical results showed that the prepared biosensor could be used to detect the DNA in a concentration range from 0.1 fM to 10 nM with a detection limit of 37 aM due to the presence of nanosized islands of the conductive polymer matrix [37]. Mutharani et al. developed a temperature-reversible switched antineoplastic drug 5-fluorouracil electrochemical sensor based on adaptable thermo-sensitive microgel based on poly(*N*-isopropylacrylamide) (PNIPAM)-conductive PEDOT-modified glassy carbon electrode (PNIPAM-PEDOT/GCE) in a good sensitivity with a low detection limit of 15 nM at 40°C compared to 25°C (0.37 µM) [38]. Selvi et al. designed a highly soluble polythiophene-based strontium-doped NiO nanocomposite (Sr-NiO). The prepared nanocomposite was used for effective electrochemical detection of catechol in contaminated water. They proved that the sensor had an excellent electrocatalytic performance due to synergistic effects of Sr-NiO nanoparticles and water-soluble polythiophene (P3ThA) with a linear range of 0.009–14.1 and 14.1–404 µM and a detection limit of 6.5 nM (S/N = 3) [39]. Hashemi et al. prepared an ultrasensitive non-enzymatic electrochemical glucose biosensor using the semi-spherical polythiophene silver bromide nanostructure within real samples with high accuracy. The morphology of the prepared nanostructure was well-ordered hybrid cubic/nanorode. Results confirmed the successful doping of silver bromide nanoparticles with conductive polymer polythiophene. The electrochemical results showed that the limit of detection (LOD)/sensitivity was 370 nM/756 µA mM^{-1} cm^{-2} and 310 nM/10.4 µA mM^{-1} cm^{-2} [40]. Faisal et al. developed polythiophene-doped ZnO (PTh/ZnO) nanostructures with a randomly distributed micron size flake-like structure. The prepared nanostructure was used for the photocatalytic degradation of methylene blue dye (MB) and gemifloxacin mesylate (GFM) antibiotics. Results were confirmed that the sample had the highest degradation efficiency (95%) for the methylene blue and the degradation of gemifloxacin mesylate [41]. Kiilerich-Pedersen et al. prepared a conductive poly(3,4-ethylenedioxythiophene) doped with tosylate (PEDOT:TsO) microelectrode-based biosensor for rapid electrochemical detection of virus infection of human cells within 3 h. The PEDOT:TsO layer was prepared using a standard photolithographic processing technique. Results were established that the rapid, label-free, and real-time electrochemical biosensor could be a promising material for real-time and label-free electrochemical detection of the effect of human cytomegalovirus (CMV) on cells in biomedical applications [42].

As known, hepatitis B is a viral infection caused by hepatitis B virus (HBV) and it is liver damage and liver cancer risk factor. In recent decades, studies were reported to evaluate this problem using several biosensors for hepatitis B virus detection. Hu et al. prepared an electrochemical immunosensor based on a 3D carbon nanotube-CP (CNT-CP) network for detection of hepatitis B surface antigen (HBsAg) in human serum with a detection limit of 0.01 ng mL^{-1} [43].

16.5 KINETIC MODELS OF NANOSTRUCTURES

The success of controlled drug delivery micro/nanosystems is to make the drug release as long as possible by accumulating at the target sites and control the drug release rate. Slow drug delivery prevents the rapid release of the drug in the body, causing frequent drug use over long periods and reducing unwanted side effects. Consequently, in drug delivery systems, micro-/nano-carriers are preferred because they prevent rapid drug release and reduce side effects and high-dose drug use. Since 1998, the conductive polymeric carriers have been used widely in the preparation of controlled drug delivery micro/nanosystems. For this purpose, to achieve effective and successful therapy, and to understand the mechanism of the drug release at the target area, the parameters of mathematical kinetic models are calculated for conductive polymeric micro-/nano-carriers [44].

The kinetic models of drug release micro/nanosystems can be calculated by applying different kinetic models, namely, zero-order kinetic, first-order kinetic, second-order kinetic, Higuchi model, Hixson-Crowell model, Korsmeyer-Peppas model, Bhaskar model, parabolic diffusion model, Elovich model, Baker model, Weibull model, and Peppas-Sahlin model, respectively (Eqs. 16.1–16.12) [45–48].

$$M_t - M_\infty = K_0 t \tag{16.1}$$

$$M_t / M_\infty = K_1 t \tag{16.2}$$

$$\frac{1}{Q^2} = k \times t + \frac{1}{Q_0^2} \tag{16.3}$$

$$M_t / M_\infty = K_H t^{1/2} \tag{16.4}$$

$$\sqrt[3]{M_\infty} - \sqrt[3]{M_t} = K_{HC} t \tag{16.5}$$

$$M_t / M_\infty = K_{KP} t^n \tag{16.6}$$

$$M_t / M_\infty = K_B t^{0.65} \tag{16.7}$$

$$1 - M_t / M_\infty = K_{Pd} t^{-0.5} + a \tag{16.8}$$

$$1 - M_t / M_\infty = a \ln t + b \tag{16.9}$$

$$f_t = \frac{3}{2}\left[1 - \left(1 - \frac{M_t}{M_\infty}\right)^{2/3}\right] - \left(\frac{M_t}{M_\infty}\right) = K \times t \tag{16.10}$$

$$m = 1 - \exp^{\left(\frac{-(t-T_i)^b}{a}\right)} \ or \ m = 1 - \exp^{\left(-(t-T_i)^{b/a}\right)} \tag{16.11}$$

$$\frac{M_t}{M_\infty} = k_d \times t^n + k_r \times t^{2n} \tag{16.12}$$

Where M_t is the amount of cumulative drug released at time t, M_∞ is the total amount of loaded drug into the nanostructures. K_0 is the zero-order rate constant, K_1 is the first-order rate constant, K_H is the Higuchi model rate constant, K_B is the Bhaskar rate constant, K_{HC} is the Hixson-Crowell rate constant, K_{KP} is the Korsmeyer-Peppas rate constant, n is the release exponent, K_{Pd} is the parabolic diffusion rate constant, K is the Baker release constant, corresponding to the slope, m is the drug fraction accumulated, t is the time, Ti is the latency time a, and b: Weibull parameters, Kd is the diffusion release rate constant, and Kr is the relaxation release rate constant. Q is the amount (%) of drug substance released at the time (t) and Q_0 is the start value of Q.

As known, the release profile of the drug for the various value of n is as follows: (i) $0.45 \leq n$ diffusion-controlled drug release model, based on a Fickian diffusion mechanism or case I relaxation, (ii) $0.45 < n < 0.89$ diffusions and case II relaxation (swelling), based on non-Fickian transport,

(iii) $n = 0.89$ case II relaxation, and (iv) $n > 0.89$ dissolution or super case II relaxation. Samanta and co-workers prepared pH-sensitive polypyrrole nanoparticles (PPy NPs) loaded with fluorescein sodium salt (FL), a negatively charged model drug, and rhodamine 6G (R6G), a positively charged model drug. Furthermore, the release mechanism and release behavior of drug-loaded PPy NPs were investigated. According to the release results, they showed that drug-loaded PPy NPs had a slow-release behavior and controlled release mechanism [49]. In another study, Jalal et al. developed a graphene oxide nanoribbons/polypyrrole nanocomposite film for the controlled release of leucovorin as a model anticancer drug by electrical stimulation. Various release kinetic parameters were calculated to determine the drug release mechanism by applying such as zero-order, Korsmeyer-Peppas, Higuchi, Weibull, and Peppas-Sahlin models. According to release results, they showed that the release profiles of Leucovorin were best fitted with the Peppas-Sahlin kinetic model with the highest correlation coefficient (R^2) and the mechanism of the release model was based on Fickian diffusion and case II relaxation [48].

The drug release mechanism of drug-loaded conductive polymer-based micro-/nano-containers is involved in the reduction of the conductive polymer under an electric voltage. For this reason, Oktay et al. developed a poly (3,4 ethylenedioxythiophene)/poly(styrenesulfonate)-gelatin methacryloyl hydrogel as an electrically sensitive drug carrier for target-specific 5-fluorouracil (5-FU) delivery in skin cancer treatment. They showed that the amount of released 5-FU from nontoxic poly(3,4-ethylenedioxythiophene)/poly(styrenesulfonate)-gelatin methacryloyl hydrogel was adequate for skin cancer treatment [50].

16.6 CONCLUSION AND FUTURE PERSPECTIVE

The objective of this chapter is to review the novel structures developed for biomedical applications using conductive polymer-based micro-/nano-containers. The major advantages of these micro-/nano-containers are (i) electrical and thermal conductive properties; (ii) the excellent morphology with different shapes, small sizes, and special chemical composition; and (iii) high chemical stability and biocompatibility for novel pharmaceutical products and sensors. A large spectrum of biomedical applications describing sensor and drug delivery systems from conductive polymer-based micro-/nano-containers-based products has been developed. It has been seen that the advanced properties of the conductive micro/nanostructure for a specific purpose depend on various aspects. In this context, it should be expanded to include different mathematical models in drug release studies to be useful in elucidating the mechanism of drug-loaded nanosystems and designing optimization strategies on conductive polymer-based micro-/nano-containers.

REFERENCES

1. S. Ahadian, J.A. Finbloom, M. Mofidfar, S.E. Diltemiz, F. Nasrollahi, E. Davoodi, V. Hosseini, I. Mylonaki, S. Sangabathuni, H. Montazerian, K. Fetah, R. Nasiri, M.R. Dokmeci, M.M. Stevens, T.A. Desai, A. Khademhosseini, Micro and nanoscale technologies in oral drug delivery, Adv. Drug Deliv. Rev. 157 (2020) 37–62.
2. E.K. Efthimiadou, C. Tapeinos, L.A. Tziveleka, N. Boukos, G. Kordas, pH- and thermo-responsive microcontainers as potential drug delivery systems: Morphological characteristic, release and cytotoxicity studies, Mater. Sci. Eng. C. 37 (2014) 271–277.
3. S.M.M. Modarres-Gheisari, M. Mohammadpour, R. Gavagsaz-Ghoachani, P. Safarpour, M. Zandi, Edge fillet radius effect on acoustic energy in an ultrasonic microcontainer for preparing nanoemulsion, Acta Acust. United Acust. 105 (2019) 1243–1250.
4. M. Mazur, A. Krywko-Cendrowska, P. Krysiński, J. Rogalski, Encapsulation of laccase in a conducting polymer matrix: A simple route towards polypyrrole microcontainers, Synth. Met. 159 (2009) 1731–1738.
5. Ö. Lalegül-Ülker, Y.M. Elçin, Magnetic and electrically conductive silica-coated iron oxide/polyaniline nanocomposites for biomedical applications, Mater. Sci. Eng. C. 119 (2021) 111600.
6. R. Wang, Y. Peng, M. Zhou, D. Shou, Smart montmorillonite-polypyrrole scaffolds for electro-responsive drug release, Appl. Clay Sci. 134 (2016) 50–54.

7. Y. Kong, H. Ge, J. Xiong, S. Zuo, Y. Wei, C. Yao, L. Deng, Palygorskite polypyrrole nanocomposite: A new platform for electrically tunable drug delivery, Appl. Clay Sci. 99 (2014) 119–124.

8. M. Bao, J. Xie, A. Piruska, X. Hu, W.T.S. Huck, Microfabricated gaps reveal the effect of geometrical control in wound healing, Adv. Healthc. Mater. 10 (2021) 2000630.

9. L.H. Nielsen, T. Rades, B. Boyd, A. Boisen, Microcontainers as an oral delivery system for spray dried cubosomes containing ovalbumin, Eur. J. Pharm. Biopharm. 118 (2017) 13–20.

10. J. Huang, B. Quan, M. Liu, Z. Wei, L. Jiang, Conducting polypyrrole conical nanocontainers: Formation mechanism and voltage switchable property, Macromol. Rapid Commun. 29 (2008) 1335–1340.

11. D. Lee, H. Moon, B. Tran, D. Kwon, Y. Kim, S. Jung, J. Joo, Y. Park, Characterization of tetrodes coated with Au nanoparticles (AuNPs) and PEDOT and their application to thalamic neural signal detection in vivo, Exp. Neurobiol. 27 (2018) 593–604.

12. J. Chen, X. Li, J. Li, J. Li, L. Huang, T. Ren, X. Yang, S. Zhong, Assembling of stimuli-responsive tumor targeting polypyrrole nanotubes drug carrier system for controlled release, Mater. Sci. Eng. C. 89 (2018) 316–327.

13. B. Gimi, D. Artemov, T. Leong, D.H. Gracias, Z.M. Bhujwalla, MRI of regular-shaped cell-encapsulating polyhedral microcontainers, Magn. Reson. Med. 58 (2007) 1283–1287.

14. J. Wang, D. Wang, N. Hui, A low fouling electrochemical biosensor based on the zwitterionic polypeptide doped conducting polymer PEDOT for breast cancer marker BRCA1 detection, Bioelectrochemistry. 136 (2020) 107595.

15. H. Noreen, J. Iqbal, W. Hassan, G. Rahman, M. Yaseen, A.U. Rahman, Synthesis of graphene nanoplatelets/polythiophene nanocomposites with enhanced photocatalytic degradation of bromophenol blue and antibacterial properties, Mater. Res. Bull. 142 (2021) 111435.

16. B. Massoumi, M. Abbasian, R. Jahanban-Esfahlan, R. Mohammad-Rezaei, B. Khalilzadeh, H. Samadian, A. Rezaei, H. Derakhshankhah, M. Jaymand, A novel bio-inspired conductive, biocompatible, and adhesive terpolymer based on polyaniline, polydopamine, and polylactide as scaffolding biomaterial for tissue engineering application, Int. J. Biol. Macromol. 147 (2020) 1174–1184.

17. A. Jasim, M.W. Ullah, Z. Shi, X. Lin, G. Yang, Fabrication of bacterial cellulose/polyaniline/single-walled carbon nanotubes membrane for potential application as biosensor, Carbohydr. Polym. 163 (2017) 62–69.

18. M. Kooti, P. Kharazi, H. Motamedi, Preparation, characterization, and antibacterial activity of CoFe2O4/polyaniline/Ag nanocomposite, J. Taiwan Inst. Chem. Eng. 45 (2014) 2698–2704.

19. J. Xue, Y. Liu, M.A. Darabi, G. Tu, L. Huang, L. Ying, B. Xiao, Y. Wu, M. Xing, L. Zhang, L. Zhang, An injectable conductive Gelatin-PANI hydrogel system serves as a promising carrier to deliver BMSCs for Parkinson's disease treatment, Mater. Sci. Eng. C. 100 (2019) 584–597

20. F.M. Zahed, B. Hatamluyi, F. Lorestani, Z. Es'haghi, Silver nanoparticles decorated polyaniline nanocomposite based electrochemical sensor for the determination of anticancer drug 5-fluorouracil, J. Pharm. Biomed. Anal. 161 (2018) 12–19.

21. X. Lu, Y. Li, X. Duan, Y. Zhu, T. Xue, L. Rao, Y. Wen, Q. Tian, Y. Cai, Q. Xu, J. Xu, A novel nanozyme comprised of electro-synthesized molecularly imprinted conducting PEDOT nanocomposite with graphene-like MoS2 for electrochemical sensing of luteolin, Microchem. J. 168 (2021) 106418.

22. B. Singh, R.A. Doong, D.S. Chauhan, A.K. Dubey, Anshumali, Synthesis and characterization of Fe3O4/Polythiophene hybrid nanocomposites for electroanalytical application, Mater. Chem. Phys. 205 (2018) 462–469.

23. M. Hatamzadeh, P. Najafi-Moghadam, A. Baradar-Khoshfetrat, M. Jaymand, B. Massoumi, Novel nanofibrous electrically conductive scaffolds based on poly(ethylene glycol)s-modified polythiophene and poly(ε-caprolactone) for tissue engineering applications, Polymer (Guildf). 107 (2016) 177–190.

24. C. Wei, H. Zhou, Q. Liu, PCN-222 MOF decorated conductive PEDOT films for sensitive electrochemical determination of chloramphenicol, Mater. Chem. Phys. 270 (2021) 124831.

25. M. Fouladgar, H. Karimi-Maleh, F. Opoku, P.P. Govender, Electrochemical anticancer drug sensor for determination of raloxifene in the presence of tamoxifen using graphene-CuO-polypyrrole nanocomposite structure modified pencil graphite electrode: Theoretical and experimental investigation, J. Mol. Liq. 311 (2020) 113314.

26. B. Xia, B. Wang, J. Shi, Y. Zhang, Q. Zhang, Z. Chen, J. Li, Photothermal and biodegradable polyaniline/porous silicon hybrid nanocomposites as drug carriers for combined chemo-photothermal therapy of cancer, Acta Biomater. 51 (2017) 197–208.

27. N. Işıklan, N.A. Hussien, M. Türk, Synthesis and drug delivery performance of gelatin-decorated magnetic graphene oxide nanoplatform, Colloids Surfaces A: Physicochem. Eng. Asp. 616 (2021) 126256.

28. A. Khadir, M. Negarestani, A. Mollahosseini, Sequestration of a non-steroidal anti-inflammatory drug from aquatic media by lignocellulosic material (Luffa cylindrica) reinforced with polypyrrole: Study of parameters, kinetics, and equilibrium, J. Environ. Chem. Eng. 8 (2020) 103734.

29. H. Karimi-Maleh, A. Bananezhad, M.R. Ganjali, P. Norouzi, A. Sadrnia, Surface amplification of pencil graphite electrode with polypyrrole and reduced graphene oxide for fabrication of a guanine/adenine DNA based electrochemical biosensors for determination of didanosine anticancer drug, Appl. Surf. Sci. 441 (2018) 55–60.

30. M. Attia, N. Anton, I. Khan, C. Serra, N. Messaddeq, A. Jakhmola, R. Vecchione, T. Vandamme, One-step synthesis of iron oxide polypyrrole nanoparticles encapsulating ketoprofen as model of hydrophobic drug, Int. J. Pharm. 508 (2016) 61–70.

31. B. Guo, J. Zhao, C. Wu, Y. Zheng, C. Ye, M. Huang, S. Wang, One-pot synthesis of polypyrrole nanoparticles with tunable photothermal conversion and drug loading capacity, Colloids Surfaces B: Biointerfaces. 177 (2019) 346–355.

32. K. Krukiewicz, A. Kruk, R. Turczyn, Evaluation of drug loading capacity and release characteristics of PEDOT/naproxen system: Effect of doping ions, Electrochim. Acta 289 (2018) 218–227.

33. M.N. Yasin, R.K. Brooke, S. Rudd, A. Chan, W.T. Chen, G.I.N. Waterhouse, D. Evans, I.D. Rupenthal, D. Svirskis, 3-Dimensionally ordered macroporous PEDOT ion-exchange resins prepared by vapor phase polymerization for triggered drug delivery: Fabrication and characterization, Electrochim. Acta 269 (2018) 560–570.

34. P.F. Hsiao, R. Anbazhagan, H.C. Tsai, Rajakumari krishnamoorthi, S.J. Lin, S.Y. Lin, K.Y. Lee, C.Y. Kao, R.S. Chen, J.Y. Lai, Fabrication of electroactive polypyrrole-tungsten disulfide nanocomposite for enhanced in vivo drug release in mice skin, Mater. Sci. Eng. C 107 (2020) 110330

35. M. Chahma, C. Carruthers, Electrochemical detection of oligonucleotides using polypyrroles, Sensors Actuat. Rep. 3 (2021) 100039.

36. S. Wankar, N. Turner, R. Krupadam, Polythiophene nanofilms for sensitive fluorescence detection of viruses in drinking water, Biosens. Bioelectron. 82 (2016) 20–25.

37. M. Shamsipur, L. Samandari, A. (Arman) Taherpour, A. Pashabadi, Sub-femtomolar detection of HIV-1 gene using DNA immobilized on composite platform reinforced by a conductive polymer sandwiched between two nanostructured layers: A solid signal-amplification strategy, Anal. Chim. Acta 1055 (2019) 7–16.

38. B. Mutharani, P. Ranganathan, S.M. Chen, Temperature-reversible switched antineoplastic drug 5-fluorouracil electrochemical sensor based on adaptable thermo-sensitive microgel encapsulated PEDOT, Sensors Actuat. B: Chem. 304 (2020) 127361.

39. S.V. Selvi, V. Lincy, S.-M. Chen, P.-D. Hong, A. Prasannan, Highly soluble polythiophene-based strontium-doped NiO nanocomposite for effective electrochemical detection of catechol in contaminated water, J. Mol. Liq. 334 (2021) 116490.

40. S.A. Hashemi, S.M. Mousavi, S. Bahrani, S. Ramakrishna, Polythiophene silver bromide nanostructure as ultra-sensitive non-enzymatic electrochemical glucose biosensor, Eur. Polym. J. 138 (2020) 109959.

41. M. Faisal, F.A. Harraz, M. Jalalah, M. Alsaiari, S.A. Al-Sayari, M.S. Al-Assiri, Polythiophene doped ZnO nanostructures synthesized by modified sol-gel and oxidative polymerization for efficient photodegradation of methylene blue and gemifloxacin antibiotic, Mater. Today Commun. 24 (2020) 101048.

42. K. Kiilerich-Pedersen, C. Poulsen, T. Jain, N. Rozlosnik, Polymer based biosensor for rapid electrochemical detection of virus infection of human cells, Biosens. Bioelectron. 28 (2011) 386–392.

43. Y. Hu, Z. Zhao, Q. Wan, Facile preparation of carbon nanotube-conducting polymer network for sensitive electrochemical immunoassay of Hepatitis B surface antigen in serum, Bioelectrochemistry. 81 (2011) 59–64.

44. H.M. Kashani, T. Madrakian, A. Afkhami, Development of modified polymer dot as stimuli-sensitive and 67Ga radio-carrier, for investigation of in vitro drug delivery, in vivo imaging and drug release kinetic, J. Pharm. Biomed. Anal. 203 (2021) 114217. https://doi.org/10.1016/J.JPBA.2021.114217.

45. D. Hirai, Y. Iwao, S.I. Kimura, S. Noguchi, S. Itai, Mathematical model to analyze the dissolution behavior of metastable crystals or amorphous drug accompanied with a solid-liquid interface reaction, Int. J. Pharm. 522 (2017) 58–65.

46. M. Joy, S.J. Iyengar, J. Chakraborty, S. Ghosh, Layered double hydroxide using hydrothermal treatment: morphology evolution, intercalation and release kinetics of diclofenac sodium, Front. Mater. Sci. 2017 114. 11 (2017) 395–408.

47. A.M. dos Santos-Silva, L.B. de Caland, E.G. do Nascimento, A.L.C. de S.L. Oliveira, R.F. de Araújo-Júnior, A.M. Cornélio, M.F. Fernandes-Pedrosa, A.A. da Silva-Júnior, Self-assembled benznidazole-loaded cationic nanoparticles containing cholesterol/sialic acid: Physicochemical properties, in vitro drug release and in vitro anticancer efficacy, Int. J. Mol. Sci. 20 (2019) 2350.

48. N.R. Jalal, T. Madrakian, A. Afkhami, A. Ghoorchian, Graphene oxide nanoribbons/polypyrrole nano-composite film: Controlled release of leucovorin by electrical stimulation, Electrochim. Acta 370 (2021) 137806.
49. D. Samanta, J.L. Meiser, R.N. Zare, Polypyrrole nanoparticles for tunable, pH-sensitive and sustained drug release, Nanoscale 7 (2015) 9497–9504.
50. S. Oktay, N. Alemdar, Electrically controlled release of 5-fluorouracil from conductive gelatin methacryloyl-based hydrogels, J. Appl. Polym. Sci. 136 (2019) 46914.

17 Antimicrobial Activities of Conducting Polymers and Their Derivatives

Palanichamy Nandhini and Mariappan Rajan
Biomaterials in Medicinal Chemistry Laboratory,
Department of Natural Products Chemistry, School of Chemistry,
Madurai Kamaraj University, Madurai, Tamil Nadu, India

CONTENT

17.1 INTRODUCTION

Polymers are known as macromolecules, which consist of many small molecules called monomers. Natural, synthetic, and semisynthetic polymers are the main classifications of polymers. Furthermore, the polymers have been categorized based on the nature of the polymer (Table 17.1). These polymers are widely used in various applications because of their unique properties like chemical, mechanical, and transport properties [1]. Owing to their high resistivity, low conductivity, and dielectric strength, they have become viable alternatives to traditional inorganic materials. These polymeric materials also possess different roles such as binders, drug carriers, thickeners, disintegrants, emulsifiers, suspending agents, and bioadhesives [2]. In 1811, Henri Braconnot has done pioneering works with cellulose derivatives, and in the late 1830s, natural rubber was successfully invented by Charles Goodyear through the process known as "vulcanization". Bakelite is the

TABLE 17.1

Categorization of Polymers

Basis of Categorization	Polymer Category
Source	Natural, semisynthetic, man-made
Line structure	Straight, cross-related, splitted
Mode of construction	Accumulation, abridgment
Application and physical properties	Rubber, fibers
Thermal response	Thermoplastic, thermosetting
Crystallinity	Non-crystalline (amorphous), crystalline
Tacticity	Isotactic, atactic
Chain	Hetro-chain, homo-chain

first synthetic polymer synthesized in 1907 [3]. The Nobel laureates Giulio Natta and Karl Ziegler developed the Ziegler-Natta catalyst, which is an important contribution to synthetic polymer science. In 1974, Paul Flory was awarded the Nobel Prize for his contribution to the field and for recognizing the importance of polymers.

In 1977, a new class of polymers was discovered with conducting effect known as conducting polymers (CPs) (Figure 17.1) [4]. Alan J. Heeger, Hideki Shirakawa, and Alan G. MacDiarmid received the Nobel Prize in 2000, for the invention and improvement of CPs [5–8]. Generally, polymers are heat sensitive [9], adaptable, and electrically insulating shapeless materials [10] due to the existence of covalent bonds in saturated carbon compounds, but the invention of CPs has given another aspect to the present epoch of polymer applications [11]. Electronic properties of

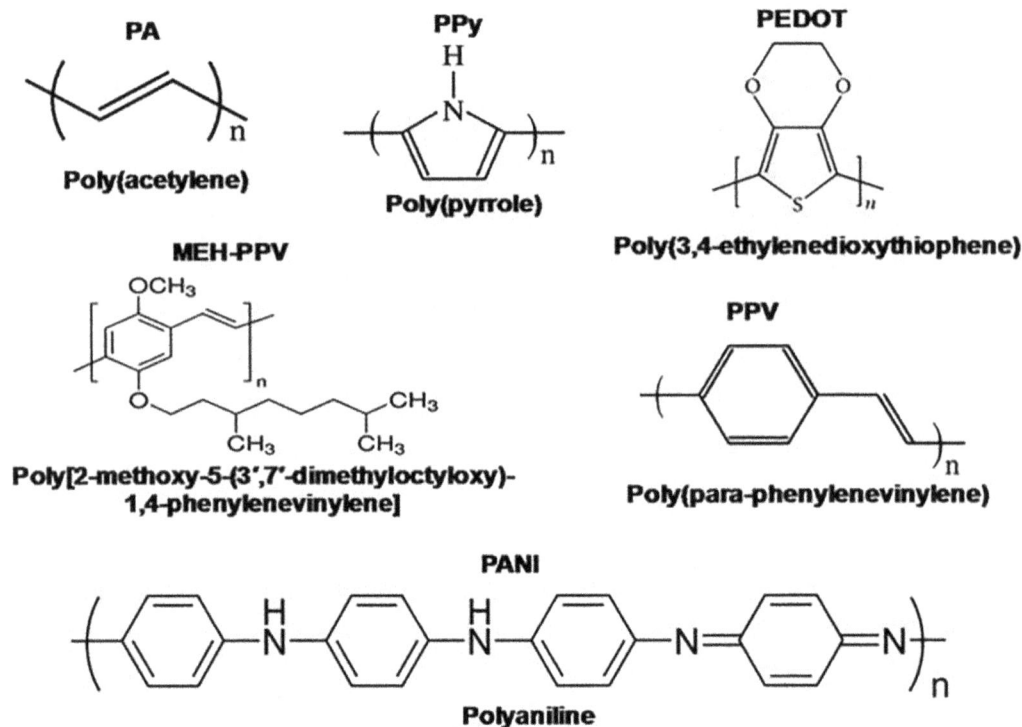

FIGURE 17.1 Conducting polymers. (Adapted with permission from Ref. [4]. Copyright (2016) RSC.)

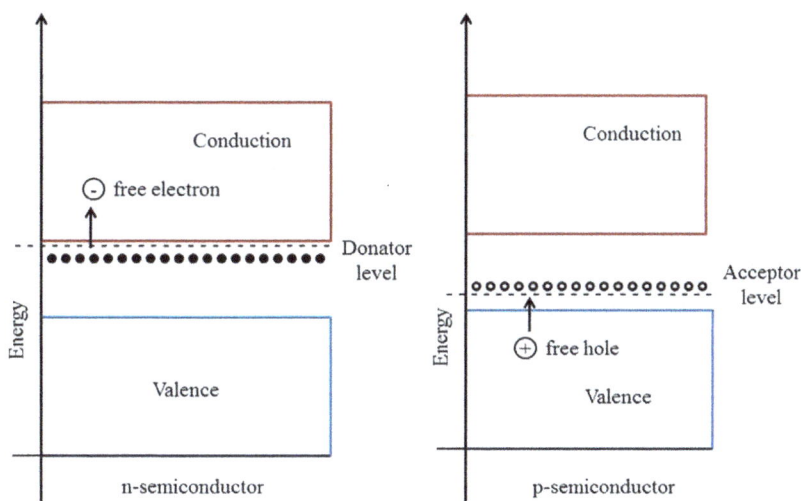

FIGURE 17.2 Schematic delineation of the n- and p-doping processes.

CPs also diverge from the well-known inorganic crystalline semiconductors like silicon in two areas such as long-range order and their molecular nature [12]. These CPs possess an extensive π-conjugation by the side of the polymer backbone while the traditional polymers, such as polypropylene and polyethylene, are made up of σ-bonds. The existence of an extended π-conjugation in CPs gives the necessary versatility to charges and makes them electrically conducting [13].

CPs have the characters of high electron affiliation or low oxidation potential due to the reason CPs can be either diminished and doped with electron donors (n-type) or can be oxidized and doped with electron acceptors (p-type) (Figure 17.2). The conductivity of the doped polymers is as high as the non-doped polymers. Redox p- and n-doping, electrochemical p- and n-doping, and photo-induced doping are the main doping methods to get a highly conductive polymer (Figure 17.3).

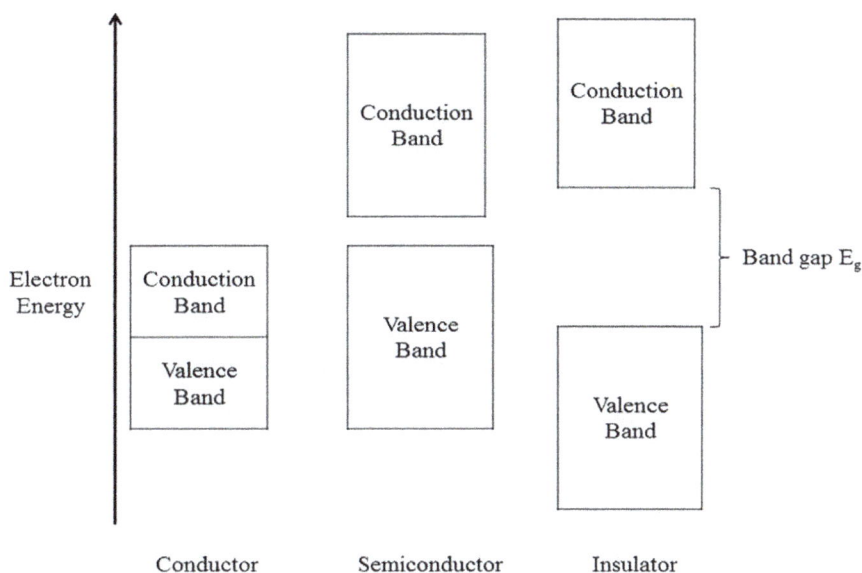

FIGURE 17.3 Schematic representation of band model.

Poly(acetylene)

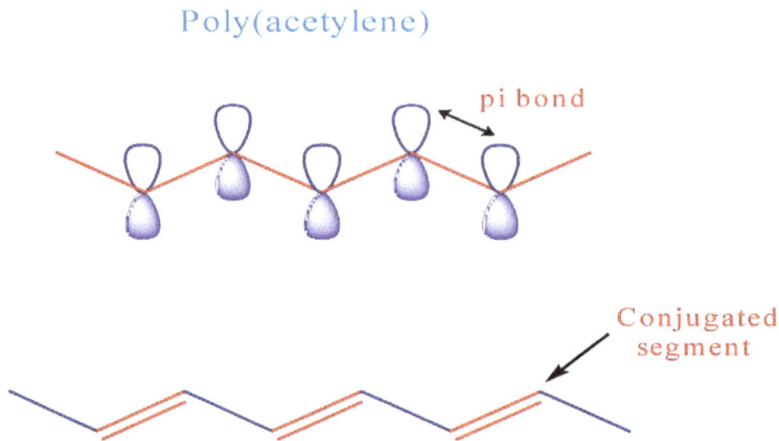

pi bond

Conjugated
segment

FIGURE 17.4 The π-conjugated system of conducting polymers.

Mechanical alignment of polymer chains is one more strategy to amplify the conductivity of polymers. The conductivity of these CPs may vary due to the structure of the polymer or by the type of dopant used and its concentration (Figure 17.4) [14, 15]. Electrical Conductivity of some conducting polymers are given in Table 17.2.

17.2 PROPERTIES OF CONDUCTING POLYMERS

CPs also have some other astonishing properties like electrical and electronic, magnetic, mechanical, optical, wetting, and microwave-absorbing properties (Figure 17.5). These properties are very significantly involved in various applications like energy, environmental, and medicinal applications.

17.2.1 ELECTRICAL AND ELECTRONIC PROPERTIES

The electrical properties of an object can be explained by its electronic band structures. The energy distinction between the conduction band and the valence band helps to know whether the material is an insulator or conductor [16]. Generally, conducting materials contain a diminished bandgap, and the conduction and valence band overlie. Conjugated bonds present in the CPs are accountable for the movement of electrons [17]. The dopant concentration and the pH value also plays a vital role to enhance the conductivity of these polymers. For paradigm, polyaniline (PANI) shows good electrical conductivity if the pH is kept between 0 to 3. The presence of unit cells in the backbone

TABLE 17.2
Electrical Conductivity of Some Conducting Polymers

Conducting Polymer	Formula	Electrical Conductivity (S/cm)
Polyaniline	$[C_6H_4NH]_n$	$10^{-2}-10^0$
Polythiophene	$[C_4H_4S]_n$	10^0-10^3
Polyacetylene	$[C_2H_2]_n$	10^5
Polypyrrole	$[C_4H_2NH]_n$	$2-100$
Poly(p-phenylene)	$[C_6H_4]_n$	$10^{-3}-10^2$

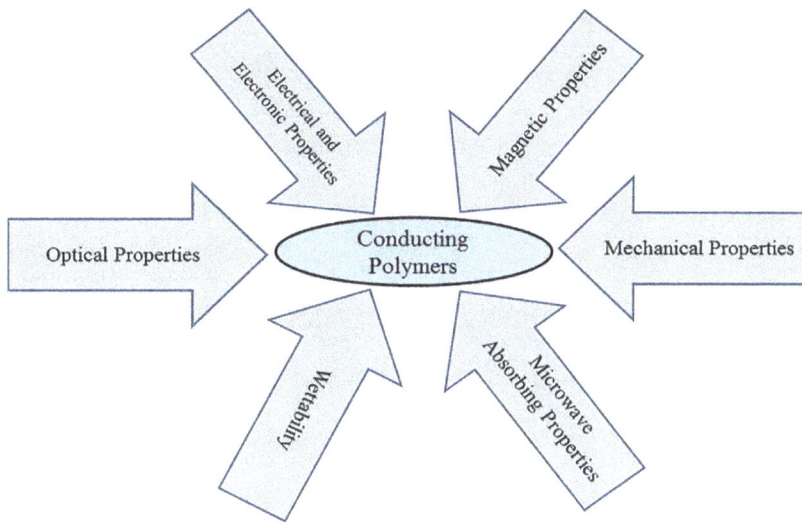

FIGURE 17.5   Properties of conducting polymers.

of the CPs can interact with the adjoining unit and create a valence band and a conduction band. CPs can get lofty energy levels whilst ionizing, and the band structures will be transferred after the ionization of the polymer. Some factors can affect the electrical conductivity of the CPs. They are

- Presence of doping materials
- Temperature
- The density of charge carriers
- Their mobility

17.2.2   MAGNETIC PROPERTIES

CPs was used in many applications due to their interesting magnetic properties and the capacity to manage their electrical conductivity over an extensive range. The magnetic properties of CPs reveal noteworthy information on the charge-carrying group and unpaired spins.

The magnetic properties of CPs can be examined by electron paramagnetic resonance (EPR) and magnetization measurements. EPR can be used to explore low energy changes in the magnetic properties of the prepared polymers, which are associated with unpaired electrons. The magnetization measurements can also be observed by the magnetic moments in the samples. The magnetic and transport properties of conducting PANI/nickel oxide were investigated by Nandapure et al. in 2012 [18]. In this study, researchers proposed that magnetization can be enhanced with the augment in a weight extent of nickel oxide in PANI. Yan et al. prepared the CPs/ferromagnet film using the anodic-oxidation technique [19]. A chemical method using p-dodecyl benzenesulfonic acid sodium salt as a surfactant and dopant produced a polypyrrole (Ppy) composite with ferromagnetic behavior. This CP shows the highest conductivity of 10 S/cm. The evolution of next-generation devices is profoundly correlated to the improvement of novel materials. CPs are the advanced materials ideally used in mobile phones, high-resolution displays, and portable devices due to their magnetic properties and low density.

17.2.3   MECHANICAL PROPERTIES

The microscopic changes in molecular mobility are the reason for the mechanical property of the CPs. Generally, the mechanical properties of CPs rely on the array of the monomers and crystallinity.

While comparing crystalline and semi-crystalline polymers, crystalline polymers have superior mechanical properties than amorphous semi-crystalline polymers. The mechanical properties of some CP composites were studied by Sulong et al. [20]. They have calculated the filler concentrations and chemical functionalization using the mechanical properties of the CP composites. The functionalized CP composites also have the property of better strength, hardness, and elongation. It has the maximum flexural and tensile strength of 80 and 35 MPa, respectively. The resonance frequency estimation and the resilient tensile modulus of Ppy nanotube were studied by Cuenot et al. [21]. This study proves that the flexible modulus increases while the thickness or the outer diameter of the Ppy nanotube decreases. Some other single nanofibers and inorganic nanowires, like silicon and silver nanowires, also exhibit size-dependent mechanical behavior [22–29]. Sgreccia et al. studied the mechanical properties of hybrid proton-CP blends based on sulfonated polyether ketones [30].

17.2.4 OPTICAL PROPERTIES

CPs were widely used in nanophotonic devices due to their unique optical properties [31]. The 1D nano-structured semiconductors can be used to fabricate photodetectors, photonic wire lasers, and photochemical sensors [32, 33]. The optical properties of cadmium sulfide/PANI-fused nano cables were studied by Xi et al. [34]. They proposed that the photoluminescence spectrum of the nano cables has comparable features to cadmium sulfide nanowires. Because of the photo-generated carriers moving from the PANI layer into cadmium sulfide nanowires, the enhancement of signal intensities was taken place. The novel optical properties of CPs doped with molecular dopants have been observed by Yoshino et al. [35]. Fujii et al. investigated the optical properties of disubstituted acetylene polymers, which exhibit enormous photoluminescence quantum competence in distinguish with unsubstituted or monosubstituted polyacetylene [36].

17.2.5 WETTABILITY

Surface wettability is an important property that has an imperative role in the performance of materials and various applications, including aeronautics, self-cleaning windows, cookware coatings, waterproof textiles, mobile phones, liquid transportation, microfluidics, controlled drug liberation, bio-detachment, cell and antibacterial adhesion (Figure 17.6) [37, 38]. Generally, CPs have hydrophilic nature [39]. CPs with super-hydrophobic characteristics can be manufactured by using hydrophobic acids [40]. A reversibly switchable super-hydrophobic or super-hydrophilic surface can be fabricated by controlling the chemical composition of CPs. The wettability of electrochemically deposited CPs depends on the deposition conditions, the dopant, and the roughness of the working electrode. Xu et al. studied the preparation and surface wettability of TiO_2 nanorod films, which are customized with triethoxyoctylsilane [41]. Lin et al. developed the poly(3-alkylthiophene) thin films with tunable wettability characteristics [42]. Teh et al. studied the impact of a redox-induced overhaul of Ppy and its wettability [43].

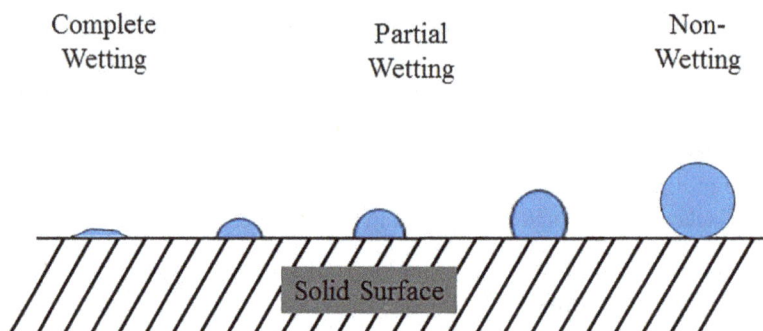

FIGURE 17.6 Surface analysis of polymer films for wettability.

17.2.6 MICROWAVE ABSORBING PROPERTIES

CPs have been used as novel microwave absorbing materials owing to their lower thickness, simple processability, and low cost. Microwave absorbing materials have the capability to employ in the military fields to limit the electromagnetic reflection from the metal plate as shells of ships, tanks, and aircraft [44–46]. Conductivity, complex permittivity, and permeability are the properties required to design microwave absorbing materials [47].

CPs as new semiconducting materials have been widely focused on due to their potential applications in sensors, catalysis, electromagnetic shielding, secondary batteries, and electrochemical displays [48]. Ting et al. studied the microwave absorbing properties using composite permittivity, permeability, and reflection loss in the microwave frequency range [49]. They have also demonstrated that PANI addition was important for attaining a huge absorption over a wide frequency range. Wai et al. investigated the microwave absorbing capacity of PANI nanocomposites, together with TiO_2 nanoparticles [50]. Liu et al. proposed that the doped PANI with a fiber-like morphology has a preferable electromagnetic wave-absorbing property than PANI with particle-like morphology [51]. These studies showed that nanotubes of CPs can be used as high absorption, wide frequency, and lightweight microwave absorbents.

17.3 PREPARATION OF CONDUCTING POLYMERS

Redox and electronically conducting polymers can be synthesized via chemical or electrochemical polymerization techniques. During the polymerization process, electrochemically active groups are integrated into the polymer structure either inside the chain or incorporated as a pendant group added to the polymer phase or fixed by a further step followed by the coating process in the case of polymer film electrodes. Enzymatic-catalyzed polymerization is also an important eco-friendly technique to synthesize CPs.

17.3.1 CHEMICAL POLYMERIZATION

CPs can be prepared by performing the chemical polymerization technique using a monomer, dopant, and an oxidant dissolved solution kept in a definite temperature. Even though there are some disagreements regarding the steps associated with chain growth, the cation-radical mechanism has been adopted by many researchers [52, 53]. The monomer is initially oxidized into a radical cation afterward the coupling of two radical cations results in a dimer. The resulting dimer is capable of oxidized into a radical dimer cation, and the continuous reaction produces oligomers followed by polymers until the extinction of the chain. The polymerization process can end up from minutes to few days, which depends on the reaction condition. Then the mixture is separated, washed, and dried to get unadulterated CPs.

17.3.2 ELECTROPOLYMERIZATION

Electrochemical or electropolymerization is also similar to chemical polymerization. This method depends on the deposition of the polymer on the surface of the solid electrode material. This method is widely used to prepare the CPs from the respective monomer compound (Figure 17.7). In this method, the potential is applied transversely to an electrolyte solution containing a monomer and a dopant. The electrochemical polymerization technique can be detached into two significant methods as anodic and cathodic electrochemical polymerization. The anodic electrochemical polymerization is the appropriate technique of oxidizing monomer species for the preparation of CPs. This process starts with the oxidation of the monomer, which brings about the formation of a cation radical, and the chain proliferation takes place through dimerization of the radicals formed and the disposal of hydrogen ions. The coupling reaction happens with a monomeric cation radical followed by the oxidation of dimer. The radical cation-radical cation coupling (RR coupling) occurs owing to the sturdy coulombic repulsion

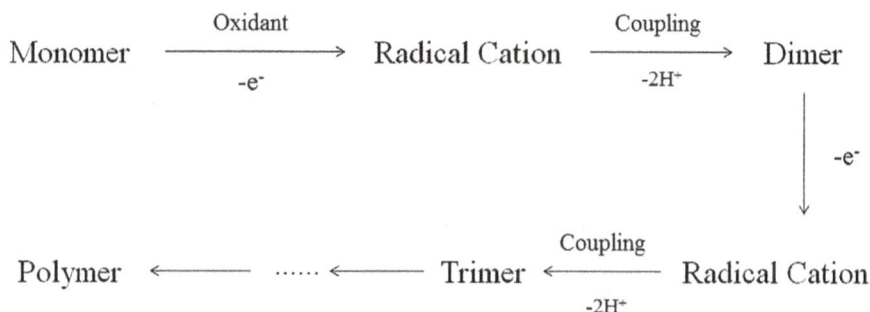

FIGURE 17.7 Polymerization scheme of a conducting polymer.

to form a polymer. In the condensed phase, the RR coupling is able to continue at a substantial rate. The nature of the rate-determining process relies on the oligomer's solubility and the deprotonation rate. The constancy of the cation radical depends on the nature of the solvent, electrolyte, and temperature [54, 55]. Polymers can be produced either by the reaction of radical cations with neighboring radical cations or with unbiased monomers. The formation of polymer film needs two electrons for each molecule of the monomer, which is 2 F/mol. The abundance charge is used for oxidation of the oligomeric compounds that can be observed from the cyclic voltammogram.

The cathodic electrochemical polymerization can also be used for the production of CPs. In this method, a small amount of CPs can only be obtained as yield due to involving the reductive electrosynthesis process. The poly(p-phenylenevinylene) CPs, used in the organic light-emitting devices, can only be synthesized by the cathodic polymerization process [56].

Electrochemical polymerization is identified as a very efficient method for the production of electroactive conductive polymers (Figure 17.8). The pace of polymer development, morphology, and characteristics of the polymer films deposited on the electrode surface is sturdily

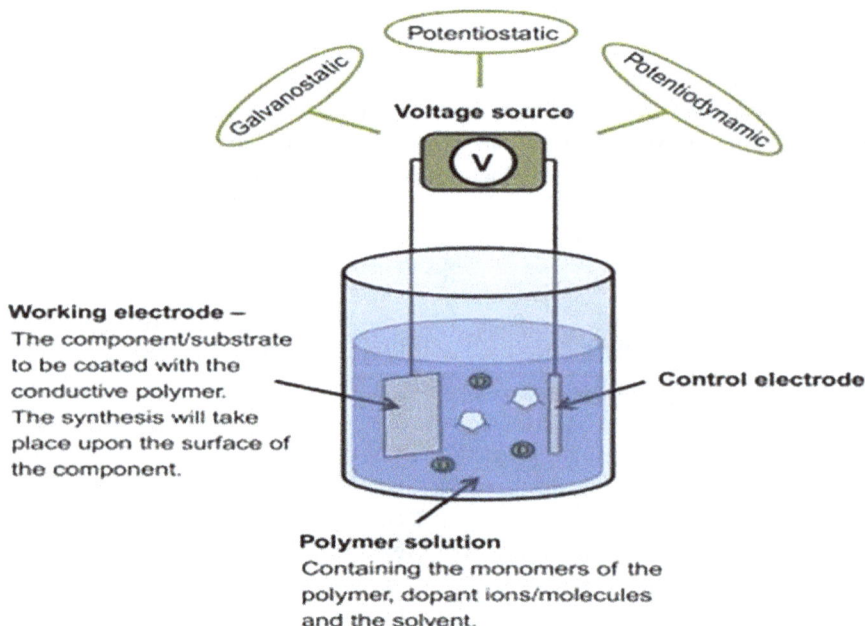

FIGURE 17.8 Schematic illustration of the synthesis of conducting polymers using electropolymerization technique. (Adapted with permission from Ref. [65]. Copyright (2014) Elsevier.)

affected by the nature of the supporting electrolyte [57]. Hence, the supporting electrolyte is an essential parameter that should be considered and enhanced when designing the electrochemical polymerization procedure. Aqueous or organic solutions can be used as supporting electrolytes in the electrochemical polymerization method. Many studies have recognized aqueous media is perfect for the electrosynthesis of CPs due to their low pollution and appropriateness for a wide range of applications. Microemulsions have also been introduced as supporting electrolytes for electropolymerization to resolve the difficulty of the insolubility of certain monomers in aqueous media. The microemulsion is a composite mixture of water, hydrophobic organic material, and surfactant that forms a micellar solution, which can be stabilized thermodynamically by the amphiphilic surfactant [58]. Microemulsions are extensively used for the electrochemical polymerization of aniline derivatives, pyrrole, benzene, and thiophenes, which are typically insoluble in water [59–62]. In addition, the electrochemical cell also contains the solvent in which the monomer is dissolved. The electrochemical reaction rate can also be affected by the nature of the solvent in which the monomer was dissolved [63]. The selection of the solvent relies upon the nucleophilicity and polarity of the solvent [64]. Then the monomer should be unadulterated and oxygen-free to stay away from the reaction of oxygen with the radical intermediates to form hydroxides on the electrode surface [65]. These properties help to decide the interaction between the solvent and the aromatic radical cation intermediates formed through the polymerization process (Figure 17.9). The nature of the monomer, pH of the medium, monomer concentration, temperature, and doping agent are the factors that can affect the electrochemical polymerization process.

17.3.3 ENZYME-CATALYZED POLYMERIZATION

The chemical and electrochemical polymerization techniques are moreover environmentally harmful or produce water-insoluble products; hence, additional modifications are required with surfactants [66]. These drawbacks can be conquered by the enzyme-catalyzed polymerization technique [67]. Li et al. investigated the environmentally non-hazardous technique (enzyme-catalyzed polymerization) to fabricate water-soluble PANI nanoparticles using eco-friendly oxidant hydrogen peroxide [68]. The synthesized PANI nanoparticles have been confirmed to have significant tumor cells killing effect.

17.4 RECENT DEVELOPMENTS AND APPLICATIONS OF CONDUCTING POLYMERS

CPs have admirable conductivity, stability, magnetic, optical, and electrical properties. Due to this reason, CPs have been widely used as a material for the improvement of solar cells, electrochromic devices, sensors, and electronic devices [69–75]. CPs have also been demonstrated to be an appropriate material for improving the functionality of electrochemical capacitors, owing to the amalgamation of high charge density and thermal stability [76]. CPs have arisen as potential materials for developing compact and portable probes and sensors due to their biocompatibility and ease of derivatization [77]. The CPs, including PANI, polyacetylene, polythiophene (PT), polyindole, and Ppy have been used as electrode materials in rechargeable batteries [78, 79]. The applications of the CPs have also been found in biotechnology, food manufacturing, biosensors, organic solar energy production, and health care products (Figure 17.10).

CPs have the ability to perform as a phenomenal material for the immobilization of biomolecules and swift electron shift for the production of competent biosensors [80]. The CP nanocomposites can be used as biosensors to detect triglycerides after being encapsulated with lipase [81]. Malhotra et al. reported the uses of biosensors in food analysis, environmental management, clinical detection and farming industries [82]. Biosensors can also be used for the detection of H_2O_2 in enzymatic reactions.

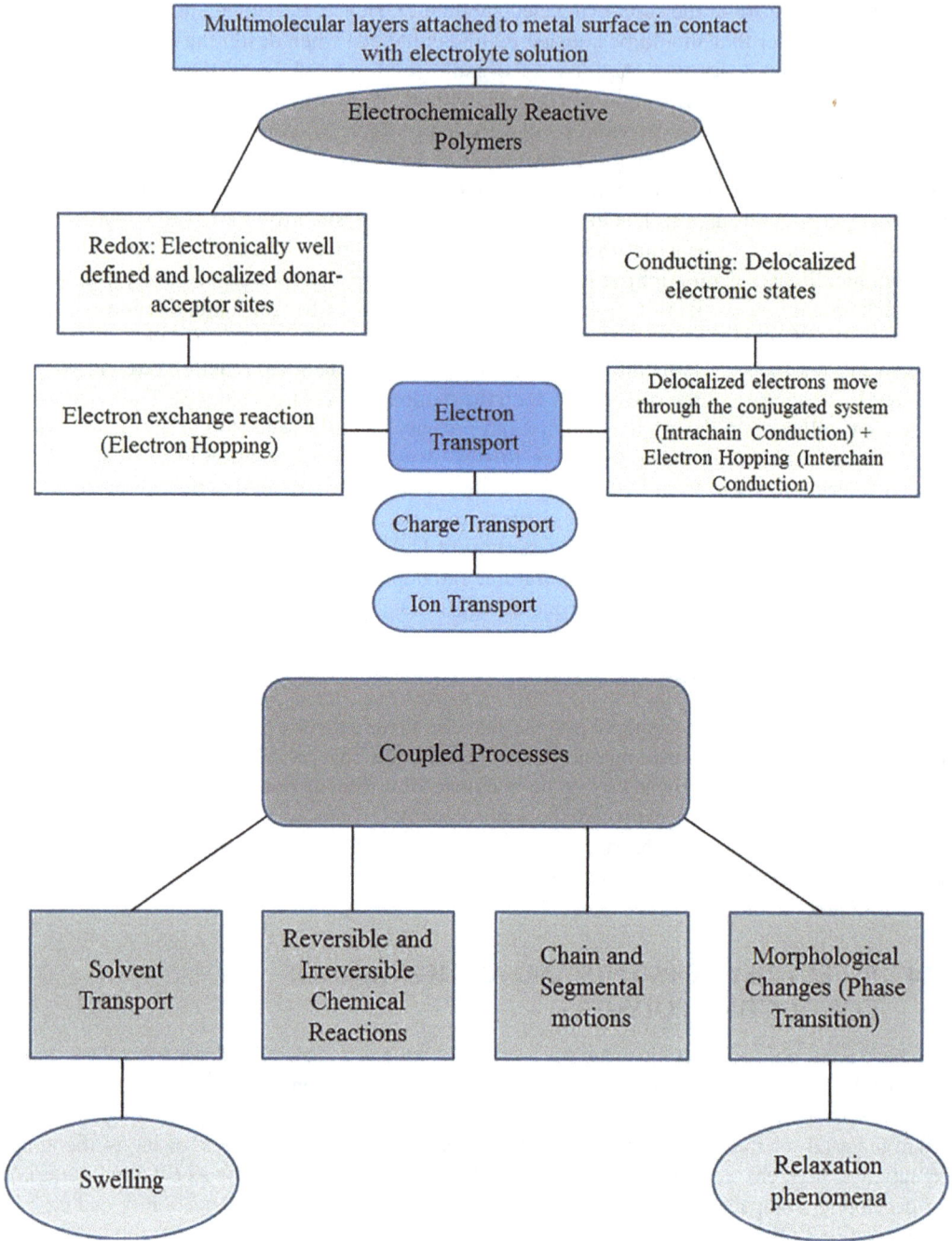

FIGURE 17.9   The charge transport process in the case of polymer film electrodes.

The CPs can be used in tissue engineering to modify cellular behavior. After all, the majority of the biological cells are sensitive to electrical impulses. Many drug liberation system devices have emerged to develop the drug-targeting specificity and reduce the drug toxicity for the treatment of various types of diseases. The polymeric nanofibers, polymeric micelles, and polymeric microspheres are the different types of drug delivery systems. The CPs like Ppy film, polyacrylic

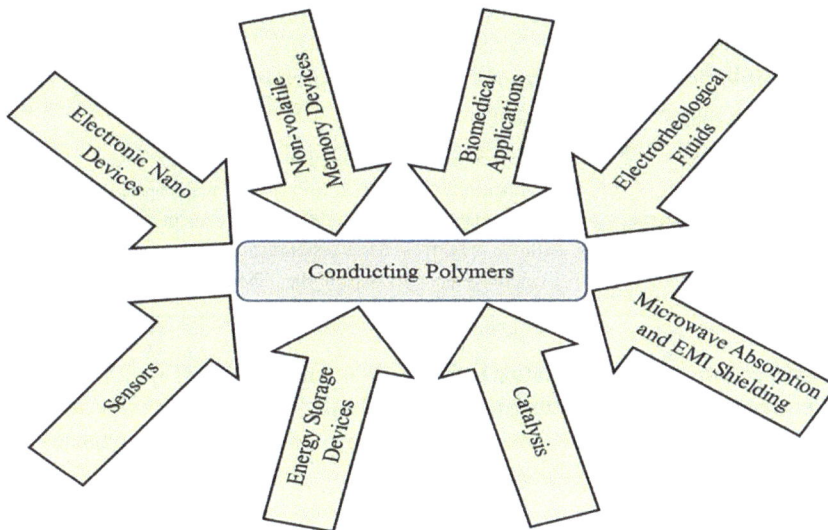

FIGURE 17.10   Applications of conducting polymers.

acid microspheres, and poly(3,4-ethylenedioxythiophene) nanotubes have been used for this purpose [83, 84].

The exclusive characteristics of CPs make them suitable candidates for tissue engineering treatment. The polymer nanofibers should be produced with high porosity, high surface area, biocompatibility, and biodegradability for tissue engineering treatments. The natural protein gelatin and polygalacturonic acid have also been used as a scaffold for tissue engineering treatment [85]. The nanostructures and bulk films of CPs have potential applications as actuators or artificial muscles [86–88]. Actuators can be operated by the energy source, usually by the electric current, hydraulic fluid pressure, and changes of energy into action.

17.5   ANTIMICROBIAL ACTIVITY OF CONDUCTING POLYMERS AND THEIR DERIVATIVES

CPs have given another dimension to the present epoch of polymer applications. Generally, polymers are heat sensitive, flexible, and electrically insulating materials, but when incorporating additives, they will transfer as conducting materials with special characteristics [89]. CPs have an extensive range of applications as antimicrobial agents owing to their non-volatile and chemically firm properties. They do not easily penetrate through the skin of a human or animal but can improve the efficiency of few existing antimicrobial agents by prolonging their lifetime [90]. CPs have various applications in several fields, including drug liberation systems and as antimicrobial polymers (Table 17.3).

The microorganisms, for example, bacteria and fungi pose a serious threat to human beings. The efficiency of many antibiotics is gradually diminishing due to the resistance mechanism of pathogens. Hence, novel antibacterial drugs are important to meet clinical or civilian needs. Many researchers have investigated CPs due to their excellent antibacterial effect. The United States of America spends $21–34 billion as a yearly expense responding to infections caused by the drug-resistant bacteria [91].

Polymeric materials, for example, cationic polymers are harmless antibacterial agents which may destroy bacteria by harming the cell membrane. Antimicrobial polymers have the capability of exhibiting their antibacterial activity through their intrinsic chemical structure; for example, quaternary nitrogen groups, halamines, and polylysine, or it can serve as a backbone to progress the activity of existing antibiotics [92]. Hence, antibacterial polymers can act as a promising alternative

---

TABLE 17.3

Biological Applications of Conducting Polymers

Applications	Depiction of Applications
Biosensors	Devices containing biomolecules as sensing components, incorporated with an electrical transducer
Tissue engineering	Biodegradable, biocompatible materials containing stimuli to augment tissue restoration
Drug liberation system	Devices for storage and controlled liberation of drugs

---

to antibiotics. In contrast, auto-degradation or biodegradation of polymers is an important property for picking precise antimicrobial materials [93]. The tuning of monomer composition and amine functionality can lead to the controlled degradation pace of the polymers and enhance exact control of the existence of antimicrobial activity [94].

17.5.1 MECHANISM OF ANTIBACTERIAL EFFECT

The biological structure of bacteria is primarily made out of cell walls, cell membrane, and intracellular components. In 1884, Hans Christian Gram invented a method to classify bacteria as Gram-positive and Gram-negative bacteria. Both Gram-positive and Gram-negative bacteria have cytoplasmic membranes, which are enormously diverse from mammalian cell membranes (Figure 17.11). Due to this reason, many scientists have been working to discover novel antibacterial medicines which can particularly exterminate bacteria without destroying the mammalian cells.

17.5.1.1 Targeting the Bacterial Membranes

Numerous cationic polymers conflict with bacteria by electrostatic interaction to the cell membrane followed by the hydrophobic inclusion into lipid tails that results in membrane lysis. Generally, cationic polymers contain positively charged functional groups, whilst bacterial membranes are negatively charged due to the reason electrostatic interaction occurs amid the positive and negative

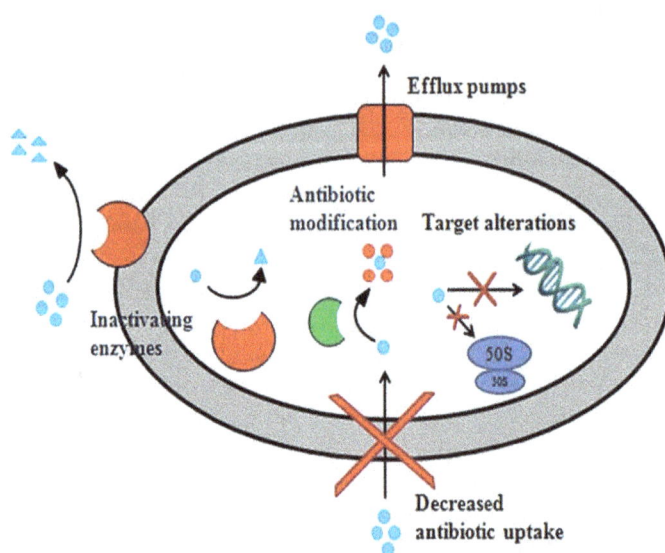

FIGURE 17.11 Schematic representation of antibacterial mechanism.

charges. Hence, the cationic materials penetrate the bacterial cytoplasm, which may lead to membrane breakage. The amphiphilic polymers, which consist of cationic and hydrophobic residues, can kill the bacteria through the membrane lysis method [95].

17.5.1.2 Inhibition of Cell Wall Synthesis

The peptidoglycans present in the bacterial cell wall provide mechanical support to maintain the morphology. The cationic polymer targets the lipid II, which is an indispensable pioneer for the synthesis of peptidoglycan layers to avert cell wall fabrication. It was found that daptomycin can block the cell wall fabrication by the elimination of lipid II synthesis proteins [96]. Lipid II is synthesized in the bacterial cytosol and transported by the lipid screens, undecaprenyl phosphate, and flippase. Hence, the molecules that can focus on the lipid II screens may as well restrain cell wall fabrication.

17.5.1.3 Targeting the Intracellular Molecules

The cationic polymers can cross the external membrane and cytoplasmic membrane barriers; hence, they disrupt the intracellular pieces of machinery. The cationic polymer, polyhexamethylene biguanide (PHMB), is able to hamper cell division by the direct interaction with DNA [97]. The PHMB has the ability to reduce the DNA to form nanoparticles by the electrostatic force amid the cationic PHMB and anionic DNA molecules [98]. Zhou et al. revealed that cationic poly(2-oxazoline) can specifically kill *Staphylococcus aureus* by engaging the sturdy interaction with DNA at a low concentration and that leads to the damage of the membrane and kills the bacteria [99].

17.5.2 Conducting Polymers and Their Derivatives against Microbes

The preliminary electrostatic interaction amid the CP and bacteria results in the linkage of microorganisms to the polymer surface. The physical interaction step takes place after the dissemination of nanoparticles and active counterions particles in the track of the cytoplasmic membrane. This progression can be established by the penetration of the species into the cell, which triggers the demise of bacteria. The zeta potential measurement of the resulting CP gives significant information regarding the surface charge signal and also the overall probability to use the CP-based composite as an antibacterial agent. The interaction of components present in the complex gives the possibility for the synergistic interaction towards more effective antibacterial activity.

The linkage of CP-based composites with cells interrupts the negatively charged cytoplasmic membrane of bacteria and generates the outflow of internal components, which leads to the demise of bacteria. Though, the presence of Gram-negative bacteria in an extracellular membrane can diminish the total negative charge per cell. Due to this reason, these microorganisms have less probability to be electrostatically absorbed on positive charges of the composite surface. Concentration-dependent techniques and the synergistic interaction with nanosized-scale structures correspond to the vital strategies to avoid this drawback against Gram-negative species.

These antibacterial hurdles can be classified as

- An amalgamation of diffusive fillers;
- Interaction with physical processes;
- Inclusion of fillers for ROS fabrication;
- Enhancement of porous supports that minimize the aggregation stage of antibacterial components.

PANI is a phenylene-based polymer that contains a chemically flexible –NH group on either side of the polymer chain. The emeraldine salt forms of copolymers have more antimicrobial activity than the emeraldine base forms of the same copolymer because of the existence of an acidic group in the polymer chain. The acidic dopants on the molecular chains of the copolymers react with the bacteria, which may lead to microbial death. It's because of the electrostatic interaction between the copolymer molecules and the bacteria [100]. The functionalized-PANI (f-PANI) can strongly inhibit the growth of *Pseudomonas aeruginosa, Escherichia coli, S. aureus,* and several antibiotic-resistant pathogens [101]. Pathania et al. synthesized a PANI zirconium (IV) silicophosphate (PANI-ZSP) nanocomposite ion-exchanger, using a sol-gel technique by incorporation of PANI gel into the inorganic precipitates of zirconium (IV) silicophosphate. The PANI-ZSP nanocomposite can remove methylene blue dye from water and can show the antibacterial activity against *E. coli* [102].

The conjugated polymers that are also known as intrinsically CPs (ICPs) have been engaged in the textile field because of their electrical characteristics [103]. The ICPs can be produced by the chemical oxidative polymerization method, which may produce positive charges by the side of the backbone chain of the polymer. The positive charges present in the CPs are responsible for their antibacterial activity. The existence of –NH group and aromatic ring in Ppy increase the partial cationic character and antibacterial activity [104]. Ppy shows higher antibacterial activity against *Streptococcus pneumoniae, S. aureus,* and *Enterococcus faecalis.* The antibacterial activity of Ppy-g-chitosan copolymer has stronger proceedings against bacteria than penicillin, trimethoprim, and rifampicin. Ppy-chitosan nanocomposite can sturdily restrain the augmentation of *E. coli* and shows an imperative enrichment in antibacterial activity than Ppy and chitosan.

PTs have good electrical conductivity, chemical stability, and low redox potential. Due to these reasons, these CPs have an extensive range of applications in bioelectronics, water purification, and hydrogen storage devices. These organic polymers are relatively soft, biocompatible, and have low cytotoxicity. The cationic PTs have competent antimicrobial activity because of their capability to absorb visible light and sensitize the fabrication of reactive oxygen species [105]. The Ag NP-polythiophene (Ag NP-PTh) nanocomposite shows good antibacterial activity against many pathogens due to the electrostatic interaction [106]. The increase in the concentration of Ag NP-PTh nanocomposite and interaction time may lead to the result of the decline in the feasibility of objective pathogens. The smaller particle-sized nanocomposites can effectively interact with the bacterial cells due to an enormous surface region. Huang et al. investigated an imidazolium-functionalized poly(hexylthiophene) that can exhibit lofty biocidal competence to both Gram-positive and Gram-negative bacteria [107].

Polyacetylenic lipids and their derivatives can be secluded from various microbial species and have a broad range of applications in the pharmaceutical industry. Polyacetylene and its derivatives have excellent antimicrobial activity, especially against *P. aeruginosa* [108]. Polyacetylene glycosides are the derivatives of polyacetylenes, which have various biological effects, such as antiallergic, anti-inflammatory, antifungal, and antimicrobial activities [109]. Pollo et al. investigated the antimicrobial activity of polyacetylene against *Mycobacterium tuberculosis* [110]. The synthesized polyacetylene compounds showed superior antimicrobial activity against *M. tuberculosis* with the lowest minimum inhibitory concentration (MIC) value of 17.88 μg/mL. The presence of three conjugated unsaturated moieties in the polyacetylene compounds are responsible for its antimicrobial activity.

17.6 CONCLUSION

CPs have many applications in several fields, including biomedical applications, like antimicrobial polymers and drug liberation systems. The functional CPs can reveal potential benefits over smaller analogous molecules because of the existence of functional groups and polymeric nature. PANI has been widely used in biomedical applications owing to its insolubility in common solvents; hence, it is complicated to degrade. f-PANI-based polymers have the potential to be used as biocompatible scaffolds for tissue engineering and as antimicrobial wound dressings, with the benefit of being able to exterminate microbes without any antiseptics. The presence of positive charges in the CPs enhances its antimicrobial activity. The discovery of CPs helps to develop novel antimicrobial agents because of having unique properties than other materials. CPs can give promising results in biomedical applications, drug delivery systems, and as antimicrobial agents.

REFERENCES

1. Sen D, Mohite B, Kayande N (2019) Review on Polymer. Int. J. Pharm. Sci. Med. 4:1–15
2. Tomar S, Singh L, Sharma V (2016) Miraculous Adjuvants: The Pharmaceutical Polymers. Int. Res. J. Pharm. 7:10–18
3. Cook P, Slessor C (1998) An Illustrated Guide To Bakelite Collectables. London: Quantum. ISBN 9781861602121
4. Ghosh S, Maiyalagan T, Basu RN (2016) Nanostructured Conducting Polymers for Energy Applications: Towards a Sustainable Platform. Nanoscale 8:6921–6947
5. Shirakawa H, Louis EJ, MacDiarmid AG, Chiang CK, Heeger AJ (1977) Synthesis of Electrically Conducting Organic Polymers: Halogen Derivatives of Polyacetylene, $(CH)_x$. J. Chem. Soc. Chem. Commun. 16:578–580
6. Chiang CK, Fincher CR, Park YW, Heeger AJ, Shirakawa H, Louis EJ, Gau SC, MacDiarmid AG (1977) Electrical Conductivity in Doped Polyacetylene. Phys. Rev. Lett. 39:1098–1101
7. Chiang CK, Druy MA, Gau SC, Heeger AJ, Louis EJ, MacDiarmid AG, Park YW, Shirakawa H (1978) Synthesis of Highly Conducting Films of Derivatives of Polyacetylene. J. Am. Chem. Soc. 100:1013–1015
8. Shirakawa H (2001) The Discovery of Polyacetylene Film: The Dawning of an Era of Conducting Polymers. Angew. Chem. Int. Ed. 40:2574–2580
9. Billmeyer FW (1984) Textbook of Polymer Science, Interscience. New York, NY: Oxford University Press
10. Flory PJ (1952) Principles of Polymer Chemistry, Imprint. New York, NY: Cornell University Press
11. Chundawat NS, Chauhan NPS (2019) Conducting Polymers with Antimicrobial Activity. Biocidal Polymers. Berlin, Boston: De Gruyter. 147–170
12. Kausar A, Siddiq M (2016) Conducting Polymer/Graphene Filler-based Hybrids: Energy and Electronic Applications, Polymer Science: Research Advances, Practical Applications and Educational Aspects. Badajoz: Formatex Research Center
13. Ramakrishnan S (2011) Conducting Polymers. Resonance 2:48–58
14. Heeger AJ (2001) Semiconducting and Metallic Polymers: The Fourth Generation of Polymeric Materials. J. Phys. Chem. B 105:8475–8491
15. Wang XX, Yu GF, Zhang J, Yu M, Ramakrishna S, Long YZ (2021) Conductive Polymer Ultrafine Fibers via Electrospinning: Preparation, Physical Properties and Applications. Prog. Mater. Sci. 115:100704
16. Lee MC, Simkovich G (1987) Electrical Conduction Behaviour of Cementite, Fe_3C. Metall. Trans. A 18:485–486
17. Le TH, Kim Y, Yoon H (2017) Electrical and Electrochemical Properties of Conducting Polymers. Polymers 9:150
18. Nandapure BI, Kondawar SB, Salunkhe MY, Nandapure AI (2013) Magnetic and Transport Properties of Conducting Polyaniline/Nickel Oxide Nanocomposites. Adv. Mater. Lett. 4:134–140
19. Yan F, Xue G, Chen J, Lu Y (2001) Preparation of a Conducting Polymer/Ferromagnet Composite Film by Anodic-Oxidation Method. Synth. Met. 123:17–20
20. Sulong AB, Ramli MI, Hau SL, Sahari J, Muhamad N, Suherman H (2013) Rheological and Mechanical Properties of Carbon Nanotube/Graphite/SS316L/Polypropylene Nanocomposite for a Conductive Polymer Composite. Compos. 50:54–61
21. Cuenot S, Demoustier-Champagne S, Nysten B (2000) Elastic Modulus of Polypyrrole Nanotubes. Phys. Rev. Lett. 85:1690–1693
22. Park JG, Lee SH, Kim B, Park YW (2002) Electrical Resistivity of Polypyrrole Nanotube Measured by Conductive Scanning Probe Microscope: The Role of Contact Force. Appl. Phys. Lett. 81:4625–4627
23. Cuenot S, Frétigny C, Demoustier-Champagne S, Nysten B (2003) Measurement of Elastic Modulus of Nanotubes by Resonant Contact Atomic Force Microscopy. J. Appl. Phys. 93:5650–5655
24. Cuenot S, Frétigny C, Demoustier-Champagne S, Nysten B (2004) Surface Tension Effect on the Mechanical Properties of Nanomaterials Measured by Atomic Force Microscopy. Phys. Rev. B 69:165410/1–5

25. Sun L, Han RPS, Wang J, Lim CT (2008) Modeling the Size-Dependent Elastic Properties of Polymeric Nanofibers. Nanotechnology 19:455706/1–8
26. Arinstein A, Burman M, Gendelman O, Zussman E (2007) Effect of Supramolecular Structure on Polymer Nanofibre Elasticity. Nat. Nanotechnol. 2:59–62
27. Burman M, Arinstein A, Zussman E (2008) Free Flight of an Oscillated String Pendulum as a Tool for the Mechanical Characterization of an Individual Polymer Nanofiber. Appl. Phys. Lett. 93:193118/1–3
28. Gordon MJ, Baron T, Dhalluin F, Gentile P, Ferret P (2009) Size Effects in Mechanical Deformation and Fracture of Cantilevered Silicon Nanowires. Nano Lett. 9:525–529
29. Guo JG, Zhao YP (2007) The Size-Dependent Bending Elastic Properties of Nanobeams with Surface Effects. Nanotechnology 18:295701/1–6
30. Sgreccia E, Khadhraoui M, Bonis C, Licoccia S, Vona MLD, Knauth P (2008) Mechanical Properties of Hybrid Proton Conducting Polymer Blends Based on Sulfonated Polyetheretherketones. J. Power Sources 178:667–670
31. Das TK, Prusty S (2012) Review on Conducting Polymers and Their Applications. Polym. Plast. Tech. Eng. 51:1487–1500
32. Evans P, Grigg R, Monteith M (1999) Metathesis of Aniline and 1,2-Dihydroquinoline Derivatives. Tetrahedron Lett. 40:5247–5250
33. Masuda T, Karim SM, Nomura R (2000) Synthesis of Acetylene-Based Widely Conjugated Polymers by Metathesis Polymerization and Polymer Properties. J. Mol. Catal. A Chem. 160:125–131
34. Xi Y, Zhou J, Guo H, Cai C, Lin Z (2005) Enhanced Photoluminescence in Core-Sheath CdS-PANI Coaxial Nanocables: A Charge Transfer Mechanism. Chem. Phys. Lett. 412:60–64
35. Yoshino K, Tada K, Yoshimoto K, Yoshida M, Kawai T, Araki H, Hamaguchi M, Zakhidov A (1996) Electrical and Optical Properties of Molecularly Doped Conducting Polymers. Synth. Met. 78:301–312
36. Fujii A, Hidayat R, Sonoda T, Fujisawa T, Ozaki M, Vardeny ZV, Teraguchi M, Masuda T, Yoshino K (2001) Optical Properties of Disubstituted Polyacetylene Thin Films. Synth. Met. 116:95–99
37. Feng L, Li S, Li Y, Li H, Zhang L, Zhai J, Song Y, Liu B, Jiang L, Zhu D (2002) Super-Hydrophobic Surfaces: From Natural to Artificial. Adv. Mater. 14:1857–1860
38. Zhang X, Shi F, Niu J, Jiang Y, Wang Z (2008) Superhydrophobic Surfaces: From Structural Control to Functional Application. J. Mater. Chem. 18:621–633
39. Zhong W, Chen X, Liu S, Wang Y, Yang W (2006) Synthesis of Highly Hydrophilic Polyaniline Nanowires and Sub-Micro/Nano-Structured Dendrites on Poly(Propylene) Film Surfaces. Macromol. Rapid Commun. 27:563–569
40. Zhu Y, Hu D, Wan M, Jiang L, Wei Y (2007) Conducting and Super-Hydrophobic Rambutan-Like Hollow Spheres of Polyaniline. Adv. Mater. 19:2092–2096
41. Xu C, Fang L, Huang Q, Yin B, Ruan H, Li D (2013) Preparation and Surface Wettability of TiO_2 Nanorod Films Modified with Triethoxyoctylsilane. Thin Solid Films 531:255–260
42. Lin P, Yan F, Chan HLW (2009) Improvement of the Tunable Wettability Property of Poly(3-Alkylthiophene) Films. Langmuir 25:7465–7470
43. Teh KS, Takahashi Y, Yao Z, Lu YW (2009) Influence of Redox-Induced Restructuring of Polypyrrole on Its Surface Morphology and Wettability. Sens. Act. A 155:113–119
44. Feng YB, Qiu T, Shen CY (2007) Absorbing Properties and Structural Design of Microwave Absorbers Based on Carbonyl Iron and Barium Ferrite. J. Magn. Magn. Mater. 318:8–13
45. Jana PB, Mallick K, De SK (1992) Effects of Sample Thickness and Fiber Aspect Ratio on EMI Shielding Effectiveness of Carbon Fiber Filled Polychloroprene Composites in the X-Band Frequency Range. IEEE Trans. Electromagn. Compat. 34:478–481
46. Liu G, Wang L, Chen G, Hua S, Ge C, Zhang H (2012) Enhanced Electromagnetic Absorption Properties of Carbon Nanotubes and Zinc Oxide Whisker Microwave Absorber. J. Alloys Compd. 514:183–188
47. Chandrasekhar P, Naishadham K (1999) Broadband Microwave Absorption and Shielding Properties of a Poly (Aniline). Synth. Met. 105:115–120
48. Fannin PC, Malaescu I, Marin CN (2005) The Effective Anisotropy Constant of Particles within Magnetic Fluids as Measured by Magnetic Resonance. J. Magn. Magn. Mater. 289:162–164
49. Ting TH, Jau YN, Yu RP (2012) Microwave Absorbing Properties of Polyaniline/Multi-Walled Carbon Nanotube Composites with Various Polyaniline Contents. Appl. Surf. Sci. 258:3184–3190
50. Wai PS, Masato T, Jiro W (2008) Microwave Absorption Behaviors of Polyaniline Nanocomposites Containing TiO_2 Nanoparticles. Curr. Appl. Phys. 8:391–394
51. Liu CY, Jiao YC, Zhang LX, Xue MB, Zhang FS (2007) Electromagnetic Wave Absorbing Property of Polyaniline/Polystyrene Composites. Acta Metall. Sin. (Engl. Lett.) 43:409–412

52. Wei Y, Tang X, Sun Y, Focke WW (1989) A Study of the Mechanism of Aniline Polymerization. J. Polym. Sci. A Polym. Chem. 27:2385–2396

53. Ding Y, Padias AB, Hall Jr KH (1999) Chemical Trapping Experiments Support a Cation-Radical Mechanism for the Oxidative Polymerization of Aniline. J. Polym. Sci. A Polym. Chem. 37:2569–2579

54. Heinze J, Frontana-Uribe BA, Ludwigs S (2010) Electrochemistry of Conducting Polymers-Persistent Models and New Concepts. Chem. Rev. 110:4724–4771

55. Cosnier S, Karyakin A (2010) Electropolymerization. Weinheim: Wiley-VCH

56. Kraft A, Grimsdale AC, Holmes AB (1998) Electroluminescent Conjugated Polymers-Seeing Polymers in a New Light. Angew. Chem. Int. Ed. 37:402–428

57. Duic LJ, Mandic Z, Kovacicek F (1994) The Effect of Supporting Electrolyte on the Electrochemical Synthesis, Morphology, and Conductivity of Polyaniline. J. Polym. Sci. A Polym. Chem. 32:105–111

58. Tsakova V, Winkels S, Schultze JW (2000) Anodic Polymerization of 3,4-Ethylenedioxythiophene from Aqueous Microemulsions. Electrochim. Acta 46:759–768

59. Lagrost C, Jouini M, Tanguy J, Aeiyach S, Lacroix JC, Chane-Ching KI, Lacaze PC (2001) Bithiophene Electropolymerization in Aqueous Media: A Specific Effect of SDS and β-Cyclodextrin. Electrochim. Acta 46:3985–3992

60. Fall M, Dieng MM, Aaron JJ, Aeiyach S, Lacaze PC (2001) Role of Surfactants in the Electrosynthesis and the Electrochemical and Spectroscopic Characteristics of Poly (3-Methoxythiophene) Films in Aqueous Micellar Media. Synth. Met. 118:149–155

61. Barr GE, Sayre CN, Connor DM, Collard DM (1996) Polymerization of Hydrophobic 3-Alkylpyrroles from Aqueous Solutions of Sodium Dodecyl Sulfate. Langmuir 12:1395–1398

62. Mani A, Phani KLN (2001) Spherulitic Morphology of Electrochemically-Deposited Polyparaphenylene (PPP) Films. J. Electroanal. Chem. 513:126–132

63. Matyjaszewski K, Davis TP (2002) Handbook of Radical Polymerization. Hoboken, NJ: John Wiley & Sons, Inc.

64. Karon K, Lapkowski M (2015) Carbazole Electrochemistry: A Short Review. J. Solid State Electrochem. 19:2601–2610

65. Balint R, Cassidy NJ, Cartmell SH (2014) Conductive Polymers: Towards a Smart Biomaterial for Tissue Engineering. Acta Biomater. 10:2341–2353

66. Yang J, Choi J, Bang D, Kim E, Lim EK, Park H, Suh JS, Lee K, Yoo KH, Kim EK, et al. (2011) Convertible Organic Nanoparticles for Near-Infrared Photothermal Ablation of Cancer Cells. Angew. Chem. Int. Ed. 50:441–444

67. German N, Ramanaviciene A, Ramanavicius A (2019) Formation of Polyaniline and Polypyrrole Nanocomposites with Embedded Glucose Oxidase and Gold Nanoparticles. Polymers 11:377

68. Li LL, Liang KX, Hua ZT, Zou M, Chen KZ, Wang W (2015) A Green Route to Water-Soluble Polyaniline for Photothermal Therapy Catalyzed by Iron Phosphates Peroxidase Mimic. Polym. Chem. 6:2290–2296

69. Nambiar S, Yeow JTW (2011) Conductive Polymer-Based Sensor for Biomedical Application. Biosens. Bioelectron. 26:1825–1832

70. Bartlett PN, Cooper JM (1993) A Review of the Immobilization of Enzyme in Electropolymerized Films. J. Electroanal. Chem. 362:1–12

71. Cosnier S (2003) Biosensors Based on Electropolymerized Films: New Trends. Anal. Bioanal. Chem. 377:507–520

72. Cosnier S (2007) Recent Advances in Biological Sensors Based on Electrogenerated Polymers: A Review. Anal. Lett. 40:1260–1279

73. Gerard M, Chaubey A, Malhotra BD (2002) Application of Conducting Polymers to Biosensors. Biosens. Bioelectron. 17:345–359

74. Rahman MA, Kumar P, Park DS, Shim YB (2008) Electrochemical Sensor Based on Organic Conjugated Polymers. Sensors 8:118–141

75. Santhanan KSV (1998) Conducting Polymers for Biosensors: Rational Based Models. Pure Appl. Chem. 70:1259–1262

76. Rudge A, Raistrick I, Gottesfeld S, Ferraris JP (1994) A Study of the Electrochemical Properties of Conducting Polymers for Application in Electrochemical Capacitors. Electrochim. Acta 39:273–287

77. McQuade D, Tyler AEP, Swager TM (2000) Conjugated Polymer-Based Chemical Sensors. Chem. Rev. 100:2537–2574

78. Saraswathi R, Gerard M, Malhotra BD (1999) Characteristics of Aqueous Polycarbazole Batteries. J. Appl. Polym. Sci. 74:145–150

79. Kawai T, Kuwabara T, Wang S, Yoshino K (1990) Secondary Battery Characteristics of Poly (3-Alkylthiophene). Jap. J. Appl. Phys. 29:602–605

80. Nambiar S, Yeow JT (2011) Conductive Polymer-Based Sensors for Biomedical Applications. Biosens. Bioelectron. 26:1825–1832

81. Dhand C, Solanki PR, Sood KN, Datta M, Malhotra BD (2009) Polyaniline Nanotubes for Impedimetric Triglyceride Detection. Electrochem. Commun. 11:1482–1486

82. Malhotra BD, Singhal R, Chaubey A, Sharma SK, Kumar A (2005) Recent Trends in Biosensors. Curr. Appl. Phys. 5:92–97

83. Li Y, Neoh KG, Kang ET (2005) Controlled Release of Heparin from Polypyrrole-Poly(Vinyl Alcohol) Assembly by Electrical Stimulation. J. Biomed. Mater. Res. A 73A:171–181

84. Li D, Huang J, Kaner RB (2009) Polyaniline Nanofibers: A Unique Polymer Nanostructure for Versatile Applications. Acc. Chem. Res. 42:135–145

85. Li M, Guo Y, Wei Y, MacDiarmid AG, Lelkes PI (2006) Electrospinning Polyaniline-Contained Gelatin Nanofibers for Tissue Engineering Applications. Biomaterials 27:2705–2715

86. Baker CO, Shedd B, Innis PC, Whitten PG, Spinks GM, Wallace GG, Kaner RB (2008) Monolithic Actuators from Flash-Welded Polyaniline Nanofibers. Adv. Mater. 20:155–158

87. Jager EWH, Smela E, Inganäs O (2000) Microfabricating Conjugated Polymer Actuators. Science 290:1540–1545

88. Okamoto T, Kato Y, Tada K, Onoda M (2001) Actuator Based on Doping/Undoping-Induced Volume Change in Anisotropic Polypyrrole Film. Thin Solid Films 393:383–387

89. Hoppe H, Sariciftci NS (2004) Organic Solar Cells: An Overview. J. Mater. Res. 19:1924–1945

90. Chen CZ, Cooper SL (2002) Interactions between Dendrimer Biocides and Bacterial Membranes. Biomaterials 23:3359–3368

91. Spellberg B, Blaser M, Guidos RJ, Boucher HW, Gilbert DN (2011) Combating Antimicrobial Resistance: Policy Recommendations to Save Lives. Clin. Infect. Dis. 52:S397–S428

92. Jain A, Duvvuri LS, Farah S, Beyth N, Domb AJ, Khan W (2014) Antimicrobial Polymers. Adv. Healthc. Mater. 3:1969–1985

93. Foster LL, Mizutani M, Oda Y, Palermo EF, Kuroda K (2017) Polymers for Biomedicine. Hoboken, NJ: John Wiley & Sons

94. Holden MTG, Hauser H, Sanders M, Ngo TH, Cherevach I, Cronin A, Goodhead I, Mungall K, Quail M, Price C (2009) Rapid Evolution of Virulence and Drug Resistance in the Emerging Zoonotic Pathogen *Streptococcus suis*. PLoS One 4:e6072

95. Chin W, Yang C, Ng VWL, Huang Y, Cheng J, Tong YW (2013) Biodegradable Broad-Spectrum Antimicrobial Polycarbonates: Investigating the Role of Chemical Structure on Activity and Selectivity. Macromolecules 46:8797–8807

96. Strahl H, Hamoen LW (2010) Membrane Potential is Important for Bacterial Cell Division. Proc. Nat. Aca. Sci. USA. 107:12281–12286

97. Chindera K, Mahato M, Sharma AK, Horsley H, Kloc-Muniak K, Kamaruzzaman NF (2016) The Antimicrobial Polymer PHMB Enters Cells and Selectively Condenses Bacterial Chromosomes. Sci. Rep. 6:23121

98. Chin W, Zhong G, Pu Q, Yang C, Lou W, Sessions PFD (2018) A Macromolecular Approach to Eradicate Multidrug Resistant Bacterial Infections While Mitigating Drug Resistance Onset. Nat. Commun. 9:917

99. Zhou M, Qian Y, Xie J, Zhang W, Jiang W, Xiao X, Liu L (2020) Poly (2-Oxazoline)-Based Functional Peptides Mimics: Eradicating MRSA Infections and Persisters While Alleviating Antimicrobial Resistance. Angew. Chem. Int. Ed. 59:6412–6419

100. Allan JE, Marjia GN, Srdjan S (2009) Bioactive Aniline Copolymers. WO2009041837 A1, Patent

101. Marija R, Gizdavic-Nikolaidis JR, Bennett SS, Allan J (2011) Broad Spectrum Antimicrobial Activity of Functionalized Polyanilines. Acta Biomater. 7:4204–4209

102. Pathania D, Sharam G, Kumar A, Kothiyal NC (2014) Fabrication of Nanocomposite Polyaniline Zirconium (IV) Silicophosphate for Photocatalytic and Antimicrobial Activity. J. Alloys Compd. 588:668–675

103. Malinauskas A (2001) Chemical Deposition of Conducting Polymers. Polymer 42:3957–3972

104. Kong M, Chen XG, Xing K, Park HJ (2010) Antimicrobial Properties of Chitosan and Mode of Action: A State of the Art Review. Int. J. Food Microbiol. 144:51–63

105. Brown DM, Yang J, Strach EW, Khalil MI, Whitten DG (2018) Size and Substitution Effect on Antimicrobial Activity of Polythiophene Polyelectrolyte Derivatives under Photolysis and Dark Conditions. Photochem. Photobiol. 94:1116–1123

106. García-Lara J, Masalha M, Foster SJ (2005) *Staphylococcus aureus*: The Search for Novel Targets. Drug Discov. Today 10:643–651

107. Huang Y, Pappas HC, Zhang L, Wang S, Cai R, Tan W, Wang S, Whitten DG, Schanze KS (2017) Selective Imaging and Inactivation of Bacteria Over Mammalian Cells by Imidazolium-Substituted Polythiophene. Chem. Mater. 29:6389–6395

108. Kyi S, Wongkattiya N, Warden AC, OShea MS, Deighton M, Macreadie I, Grachen FHM (2010) Synthesis and Activity of Polyacetylene Substituted 2-Hydroxy Acids, Esters, and Amides against Microbes of Clinical Importance. Bioorg. Med. Chem. Lett. 20:4555–4557

109. Pellati F, Calò S, Benvenuti S, Adinolfi B, Nieri P, Melegari M (2006) Isolation and Structure Elucidation of Cytotoxic Polyacetylenes and Polyenes from *Echinacea pallida*. Phytochemistry 67:1359–1364

110. Pollo LAE, Martin EF, Machado VR, Cantillon D, Wildner LM, Bazzo ML, Waddell SJ, Biavatti MW, Sandjo LP (2021) Search for Antimicrobial Activity among Fifty-Two Natural and Synthetic Compounds Identifies Anthraquinone and Polyacetylene Classes That Inhibit *Mycobacterium tuberculosis*. Front. Microbiol. 11:622629

18 Biodegradable Electronic Devices

Bavatharani Chokkiah,[1] *Muthusankar Eswaran,*[2] *and Ragupathy Dhanusuraman*[1]

[1]Nano Electrochemistry Lab (NEL), Department of Chemistry, National Institute of Technology Puducherry, Karaikal, India

[2]Department of Biology and Biological Engineering, Chalmers University of Technology, Gothenburg, Sweden

CONTENTS

18.1 INTRODUCTION

The potential of a generation to guarantee its desires for the upcoming generations to encounter their specific requirements is called sustainability. Though it is hard to deliver a thorough description for consumption, "Consumption is the human transformation of energy and materials (alongside the production-consumption chain) that creates the transformed energy or materials less accessible for forthcoming use, or undesirably impact biophysical systems in such a way to threaten human health, welfare, or other things people value". In the modern world, waste and plastic consumptions are the two main concerns. In recent years, our needs in the unparalleled rate of technological and scientific progression and a substantial elevation on multifunctional electronic devices are inevitable. Besides, there has been a visible limitation of the electronics lifetime. These elements underwrite the issue of electronic waste (e-waste) [1]. The present circumstances see more than 50 million tons of e-waste being made every year. To solve these issues, progress has been held up by the various nature of e-waste, which is deliberated as important intricate surplus streams. In these wide varieties of existing goods where there is deviance in the product models, materials, and components, compatibility issues, size changes, and additional specifications create the retrieval course of e-waste as a

DOI: 10.1201/9781003205418-18

perplexing one [2]. The difficultly is again built up by the manifestation of harmful and toxic materials like lead and cadmium in certain electrical usages. Such elements afford conservative discarding course not as much than ideal since their harmful impression on the surroundings.

The traditional landfilling waste meting out method can end in the percolating of lead hooked on the groundwater, while the other technique termed ignition outcomes in the toxic fumes emission into the atmosphere [3]. To evade the cost of reusing these trashes, developed nations choose to direct the produced waste to emerging nations in the designation of constricting down digital inequality. Managing waste is mainly controlled by Far Eastern nations, which access bulky waste and thus help as main discarding grounds. Hitherto, such places are hardly well resourced to grip the harmful materials. In India, the disposal methods brought about human health and pollution concerns. The improper processing of e-waste causes the toxic substances (brominated flame retardants, cadmium, chromium, lead, and polychlorinated biphenyls [PCBs]) to stick out into the atmosphere. Youngsters are more defenseless to such contaminants owing to their bodies' functional arrangements and once their acquaintance with such noxious elements hinders further progress and roots irreparable destruction. The harshness of these problems aggregate consideration on the hunt for less hazardous and more sustainable materials in the forms of electronics [4]. Many recent reports ensure various reliable applications based on green or biodegradable electronics were progressively growing up in the sensor, energy storage, drug delivery, health care, biomedicine, point-of-care detections, and so on (Figure 18.1).

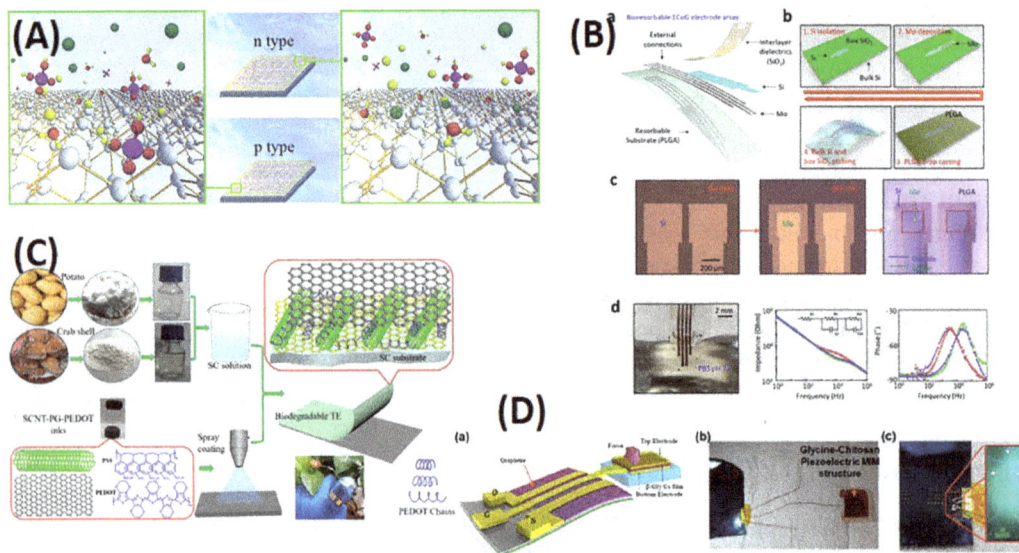

FIGURE 18.1 Graphical abstract for multifunctional degradable green electronic devices applications: (a) Mechanisms of silicon nanomembranes for biodegradable electronics. Adapted with permission from [5]. Copyright (2019) ACS Appl. Mater. Interfaces. (b) Thin, flexible, and neural electrode arrays with fully bioresorbable construction by use of doped Si NMs. (A) Schematic exploded-view illustration of a passive array for electrocorticography. (B) Fabrication process. (C) Optical images of the device. (D) Electrochemical impedance spectrum of four electrodes in PBS solution. Adapted with permission from [6]. Copyright (2017) ACS Nano. (c) Schematic illustration of the fabrication process for the biodegradable and flexible 3D interconnected SCNT-PG-PEDOT-based TE. Adapted with permission from Ref. [7]. Copyright (2018) ACS Appl. Mater. Interfaces. (d) Pressure sensor comprising a GFET with a glycine–chitosan piezoelectric transducer layer: (A) scheme of GFET coupled with glycine/chitosan in an extended gate configuration; (B, C) glycine–chitosan MIM structure connected in an extended gate configuration to the GFET. The inset in (C) shows the contact pads to the GFET. Adapted with permission from [8]. Copyright (2020) ACS Appl. Mater. Interfaces.

18.1.1 What Is Biodegradable Electronics?

The embryonic form of "green" electronics is called biodegradable electronics that reveals a transient nature and before its physical disappearance through breaking into non-harmful constituents, being proficient in helping their purpose over the recommended time. More specifically, the quality of biodegradability has been well explained by the EN13432 standard as the possessions there at least 90% of the substances are transformed into water, biomass, and carbon dioxide in particular oxygen conditions, humidity, and temperature within 6 months, in the incidence of microorganisms/fungi. This special distinctive thus permits for the ensuing incorporation of such schemes with the body/atmosphere with negligible harmful properties [4].

18.1.2 The Necessity for Biodegradable/Compatible ("Green") Electronics?

E-waste managing advises for the growth of novel biodegradable/compatible electronics that are ecologically friendly, disposable, and economical. Apart from backing the ongoing worldwide issue of e-waste, time-invariant electronics, conventionally excellent electronic novelty standards have also been seeing confines in their submissions. Thus, the rise of biodegradable/compatible electronics can also carry around enormous developments to today's technologies [4, 9]. In biomedical fields, electronic equipment and devices have played a historical role in vivo uses because of the intrinsic hazards of enduring implantations, even with their vast potential in solving various medical concerns, like by helping as entrenched sensors and the substitute requisite for surgical elimination persist as main obstacles. Thus, the necessity for raising consideration should be positioned on emerging biodegradable (green) electronics to evade the above-mentioned drawbacks [4, 10].

18.1.3 Conventional and Biocompatible or Degradable Electronics

Conventional or traditional electronics are made of inorganic materials like ceramics and metals, whereas biodegradable electronics use organic materials together with some metals. The degradation of organic materials through biological (e.g. bacteria and enzymes) and chemical (e.g. radical oxidation and hydrolysis) means connecting the damage of major bonds to produce lower molecular weight yields that can be metabolized or dissolved. Hence, the footmark of the by-products' degradation is lesser and usually measured benign to the atmosphere [4]. The ever-raising sustainable bioelectronics has brought about numerous improvements in various fields. In this chapter, we are attempted to recap the highlights of new progress in the biodegradable electronics field, classes of material and its cover insulators and substrates, conductors and semiconductors, and its degradability along with their final applications (in terms of wearable, flexible, and biodegradable highly clinical diagnostics sensors and allied biomedical healthcare applications as shown in Figure 18.2). Also, we concluded with the prospect outlook of biodegradable (green) electronics.

18.2 MATERIALS IN BIOCOMPATIBLE OR DEGRADABLE ELECTRONICS

Scientific developments in materials exploration and dealing out have allowed industries to improve from smaller to feature-rich device electronics and semiconductors for modern clients. The biocompatible or degradable materials include numerous individuals and combinations; some types of biopolymers and their application components are shown in Table 18.1 and elaborated below.

18.2.1 Substrates

The substrate is a solid material on which many active functional covers are deposited (electrode, dielectric, semiconducting, and capsulation layer). The substrate is producing more electronic waste

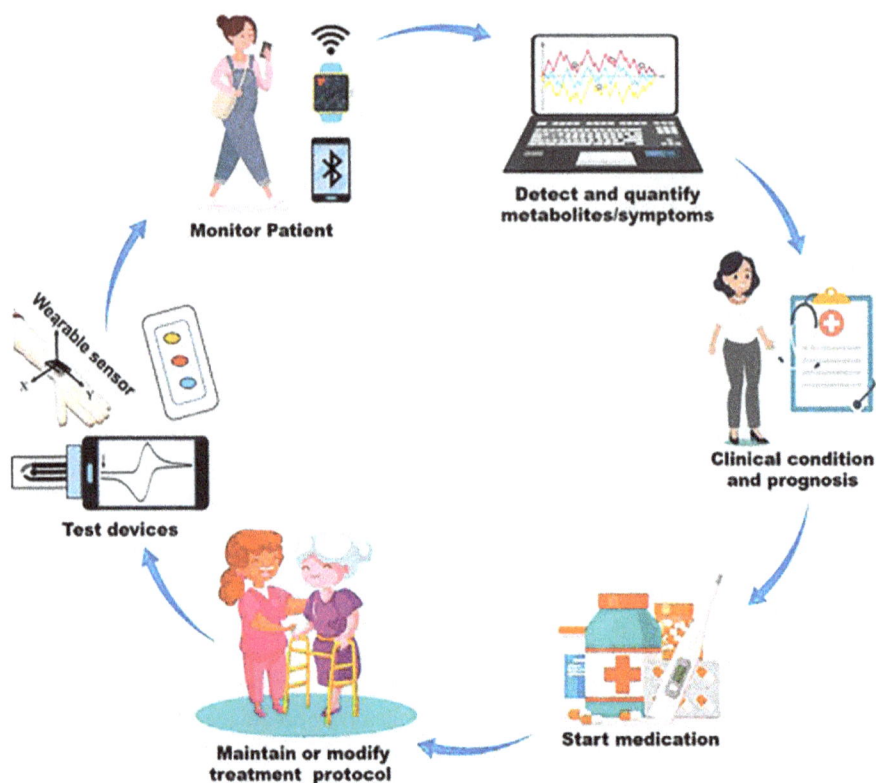

FIGURE 18.2 Schematic representation of potential interactive use of portable devices (wearable and biodegradable sensors) to detect and continuously monitor the clinical condition of a patient submitted to a medical treatment protocol. (Adapted with permission from Ref. [11]. Copyright (2021) ACS Appl. Electron. Mater.)

because it's larger and thicker than any other functional layer. Henceforward, substituting existing with biodegradable substrates will decrease e-waste problems. For the synthesis of organic electronics, numerous bio-origin materials have been used as the right substrates. Such materials show non-toxicity, biodegradability, biocompatibility, low cost, and bioresorbable functionalities for biomedical uses [1, 4].

TABLE 18.1
Types of Biopolymers and Their Application

Type	Polymer Material	Electrical Behavior	Applications	Ref
	Paper	Insulator	Substrate and dielectric	[12]
	Silk	Insulator	Substrate and dielectric	[17]
Natural polymers	Chicken egg white (albumen)	Insulator	Dielectric	[20]
	Gelatin	Insulator	Substrate and dielectric	
	Shellac	Insulator	Substrate and dielectric	[22]
	Polydimethylsiloxane (PDMS)	Insulator	Substrate and dielectric	
	Poly(vinyl alcohol) (PVA)	Insulator	Substrate and dielectric	[4]
	Polypyrrole (PPy)	Conductor (doped)	Conductor	[22]
Synthetic polymer	Polyaniline (PANI)	Conductor (doped)	Conductor	[28]
	Poly(3,4-ethylenedioxythiophene) (PEDOT)	(PEDOT) Conductor (doped)	Conductor	[29]

Paper is one of the most familiar, oldest, and by far cheapest biodegradable "substrate" materials of natural origin from plant-derived cellulose. The science of processing paper with required surface properties and mechanical assets is advanced. OFETs and OFET circuit's arrays have been made on paper, representing flexible devices, which shows on a par with more conventional substrates [12]. Low-voltage active circuits have been recognized on a dollar bill for anti-counterfeiting appliances. OFETs working at less than 1 V with 0.2 cm^2/Vs mobility could be made up consistently [13]. Currently, paper-based electronic circuits design has been the topic of a wide review. Paper substratum has been utilized for thermochromic show screens, as well as flexible electrowetting displays, for flushable consumer products [14]. For paper-based organic photovoltaics, promising performance has been demonstrated. Roll-to-roll solution printing by means of flexographic and gravure printing methods was employed to produce photovoltaics printed on paper. These devices are inverted design, with a transparent top electrode made of conducting polymer (CP) and a printed ZnO/Zn back electrode. Such a flexible final product solar cells associate economical elements with great throughput low-temperature roll-to-roll printing. One more accountable paper-printed photovoltaic is prepared via low-temperature chemical vapor deposition with features like organic active layer, thin semi-transparent paper as a substrate, reflective back electrode, and a CP transparent electrode [15]. These cells can be doubled over repetitively with no array performance degradation.

Silk (polypeptide polymer) is another type of natural material. It is composed of two types: fibroin and sericin as the main protein. Silk can be implanted without harm into the body owing to its complete bioresorbability and causes no immune reply. One of the studies presented that the electronic sensor ultrathin array can be prepared on silk and positioned in vivo onto uncovered brain tissue [16]. The silk securely resorbs and dissolves, ensuing in the conformal covering of doubled-over brain tissue with the sensor array. Bioresorbable electronics of silicon-based combinations to be synthesized onto silk and employed as a transporter (in vivo) to acquaint with the electronic component [17]. In OFETs, silk can perform as an active solution-processed gate insulator assisting superior mobilities of ~23 cm^2/Vs in pentacene united with a low-voltage procedure [18]. And also, in passive RF-ID circuits, silk has been used as a substrate that is incorporated straight onto foodstuff, i.e., eggs and apples, as food quality sensors. The transistors that are silicon-based on freestanding silk fibroin covers were prepared via PDMS stamp transfer printing. After implantation, the completely biodegradable therapeutic device over silk facades possibly will be wirelessly started to deliver the essential thermal therapy or activate drug delivery. On account of the biocompatibility, programmable biodegradability and non-toxicity silk facade could recognize the environmental and implantable electronic therapeutic devices (a research investigation on advanced engineering materials and electronic devices for being bioresorbable, implantable toward biomedicine as depicted in Figure 18.3) [1].

For oral drug ingestion, gelatin, a protein-based material, is generally utilized for capsules. It is completely biodegradable and biocompatible. Electronics based on hard gelatin might effortlessly be consumed for particular biomedical uses aiming at short examination time. OFET devices are fabricated straight onto hard gelatin cases and have been established. The protein albumin from chicken egg whites has been exposed as a high-enactment cross-link solution treated for OFET device materials [20]. The features of the albumen dielectrics such as smoothness, hydrophobicity, and electrical breakdown were powerfully connected to baking sequences and thermal treatment conditions.

Polysaccharides can be employed as biodegradable substrate materials, besides protein-based polymers. Likewise, deoxyribonucleic acid (DNA-building block) is one of the attractive molecules, which has been stimulated by numerous scientists to spread on in real-world submissions in organic electronics, photonics, etc. Many reports have shown DNA materials used in photonic arrays, nonlinear optoelectronic modulators, and organic light-emitting diodes (OLEDs). For low operating voltage OFETs, cross-linked and solution-processed DNA has effectively been employed as a gate dielectric stratum. The specific nucleobases (adenine, guanine, cytosine, and thymine)

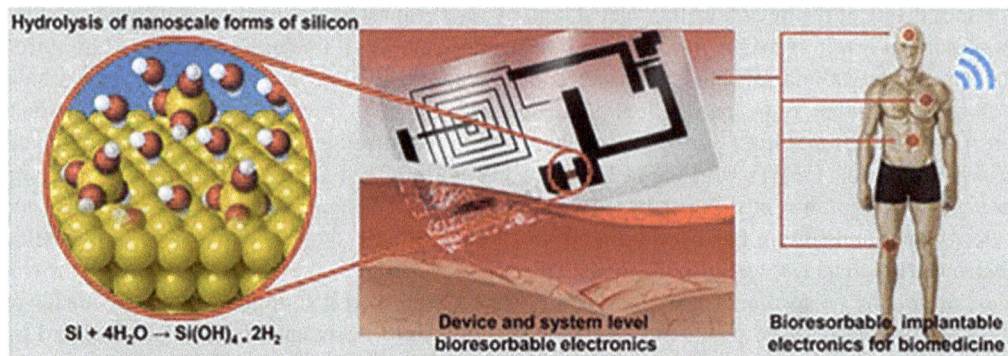

FIGURE 18.3 Graphical abstract for Advanced Materials and Devices for Bioresorbable, implantable Electronics for biomedicine.

Adapted with permission from Ref. [19]. Copyright (2018) Advanced Materials and Devices for Bioresorbable Electronics, Acc. Chem. Res.

are taken out for cosmetic and medical-based commercial claims and have been instigated as gate dielectrics for OFETs. The combination of adenine thin film with electrochemically attained aluminum oxide dielectric and fullerene C60 possibly pays for OFETs through a low working voltage (~0.5 V) along with high semiconductor agility (~5.5 cm²/Vs) [1, 21].

An elastic polymer, transparent PDMS shows brilliant biocompatibility. It has been accepted for the authentication of both in vivo as well as in vitro assessments in the biomaterial's estimation. PVA is one more illustration of artificial biodegradable polymer that has been employed as a base material for biocompatible/biodegradable electronics. The united devices on the exterior part of a PDMS thin foil over a water-soluble PVA material calculate the electrical indications formed by the skeletal muscles, brain, and heart. The collection of multifunctional sensors holds by an integrated circuit scheme mounted on the skin; the water-soluble PVA material can be splashed out and the device is able to be simply unwrapped out [4].

These shortcomings of dielectrics and natural substrates expose that nature provides privileged circumstances of materials with high quality that could be joined to several organic electronic devices, contributing substitutes for biodegradable/biocompatible, and even bioresorbable and bio-implantable, claims.

18.2.2 Dielectric Materials

A dielectric material is known to be an electrical insulator (drive apart by an external electric arena). The positive charges are moved in the way of the field and negative charges change in a reverse way, producing an in-house electric pitch that decreases the whole field inside the dielectric.

18.2.2.1 Natural Polymers

As discussed earlier, chicken egg white (albumen) is a protein-based dielectric material. The dielectric layer for organic field-effect transistors (OFET) based on pentacene and C60 via albumen thermal treatments and spin-coating engaged straight from a fresh egg devoid of no further synthesis or extraction. The albumen dielectric features include hydrophobicity, smoothness, and electrical breakdown were intensely connected to baking sequences and thermal conduct process.

18.2.2.2 Synthetic Materials

In an organic complementary inverter, a dielectric material based on cellulose is utilized as a gate dielectric by an extremely great small signal gain to 1600, a large noise margin of up to 92.5%, and a low working voltage of 4 V only outperforming other organic inverter systems. FET devices with gelatin or shellac gate dielectrics also display superior mobility and low working voltage, demonstrating their higher dielectric assets [22].

18.2.3 CONDUCTORS AND SEMICONDUCTORS

In recent years, biocompatible and degradable conducting materials' exploration has been an exciting zone. Various CPs are exclusively apt as interface materials for bioelectronics, energy sectors, and healthcare diagnostics [23–27] since they possess excellent electroactivity with a tuneable stable matrix that can transport both electronic and ionic currents. For traditional metallic conductors, this is not available. A demo of a thin-film transistor device made on proton-conducting polysaccharide managed by the electronic field effect of a gate is efficient recognition of the protonic/electronic communication. This device develops the chitosan polymer, acquired from the deacetylation of chitin, the organizational polymer constituting the crustacean's exoskeletons. Chitosan can also have derived from shrimp commercially. A transistor by solution-processed chitosan proton conductors might be prepared on paper as substrates materials [1].

The molecular basis electrical conductivity containing electrons' delocalization along the conjugated backbones via the π-orbitals overlaps along with π-π stacking amid the polymer repetitive units in the conjugated polymers (CoPs). Many highly conjugated polymers include polythiophene, poly(3,4-ethylenedioxythiophene) (PEDOT), polypyrrole (PPy), and polyaniline (PANI), which have been established and effectively combined into numerous optoelectronic devices [22]. In the initial period of the 1980s, PANI has fascinated by rigorous consideration due to the reawakening of their high electrical conductivity. PANI has been briefly explored for numerous probable applications, such as transparent electronic conductors, biological sensors, energy sectors, and chemical and electrochromic coatings [23, 26]. The biocompatibility/degradability of PANI to tissues and cells has been exposed in vitro and in vivo studies. PANI won't aggravate inflammatory replies in the dermal tissues during the complete implantation time of 2 years. In comparison to the tissue-culture-treated polystyrene, the cardiac muscle cells showed a comparable proliferation rate and little lower adhesion on PANI films [22, 28].

The structure poly(3,4-ethylenedioxythiophene) added together to the polyanion poly (styrenesulfonate) (PEDOT:PSS) is one of the renowned electron CPs in organic electronics and it has been used in various bio-sensing applications. In vivo electrocorticography, PEDOT:PSS applied as conformal polymer electrodes on living brain tissue and screening a greater signal/noise ratio (SNR) than conventional assessments. For neural recording, the interconnected conducting PEDOT nanotubes have also stayed effective. PEDOT can also be electrochemically polymerized in situ in an active brain, achieving a therapeutic outcome. PEDOT has been exposed to be an active anions conductor as well, while PSS can perform as a conducting space for cations like Ca^{2+}, K^+, Na^+, and acetylcholine (the neurotransmitter) [1, 29].

In electronic devices, the chief electrical action occurs at the layer of active materials. The vital principle of most electronic devices is typically semiconducting to understand some degree of governable conductivity. Organic semiconductors can be branched into two major classes: the first one is small molecules and the second one is conjugated polymers (CoPs). The important benefits of CoPs than inorganic semiconducting materials are potential in mechanical flexibility and processing, allowing the preparation of flexible electronic devices economically. Nature is filled in pi-conjugated molecules that can be utilized mostly as semiconductors. Many safe and nontoxic organic conjugated dyes have been produced by the synthetic dye industry and are determinately used as textiles or colorants and inks. Among the various survey of biodegradable materials includes substrates and insulators; the investigation into biodegradable semiconductors persists scarcely.

The carotenoids (β-carotene) are linear π-conjugated molecules that behave as hole-transporting semiconductors. Natural gate insulators and β-carotene devices like glucose prepared on biocompatible or degradable plastic substrate devices are truly the demonstrations of "natural" OFETs. Solution-processed β-carotene is utilized in solar cells and displayed only new acts. The primary investigations of completely "green" OFET devices contained biodegradable/compatible natural dielectrics and substrates and active nontoxic organic synthetic textile dyes like perylene bisimides and anthraquinones with 10^{-2} to 10^{-1} mobility array. The complete OFETs biomaterial-based ambipolar charge transport movements in the array of 10^{-2} to 0.4 were confirmed with indigo and its results [1, 9, 21].

In worldwide, indigo is one of the utmost ready-made dyes and is largely engaged in the coloring of blue jeans. However, nowadays, it is manufactured unnaturally from numerous plants species and has been taken out and employed as a dye from olden periods and valued as supplies as treasured as gold. In nature, several indigos are derived from animals and plants and indigo itself has been stated to be nontoxic and biodegradable. Owing to inter- and intramolecular hydrogen bonding amid carbonyl groups and amine hydrogen, indigo and its derived products are photochemically and thermally stable [30]. Conventional molecular organic semiconductors, as it has marginal intramolecular conjugation, with amine and carbonyl groups, realized to interfere conjugation in the resonance model; however, indigo "breaks this rule". Though, the outstanding charge transport nature of indigoid dyes is ascribed to the robust intermolecular π-stacking connections incorporated by hydrogen attachment. Owing to the π-stacking (normally along the crystallographic b-axis for most of the indigoids) directionality charge transportation is extremely anisotropic. To attain noble OFET enactment, molecules require an essential "standing-up" structure, with π-stacking analogous to the gate dielectric. Aliphatic dielectric materials are utilized to reach this and include natural oligoethylene tetratetracontane and polyethylene. Tyrian purple and indigo show small bandgaps (1.7–1.8 eV) and reversible two-electron reduction and oxidation processes and hence are appropriate for ambipolar OFETs and voltage inverters [1]. The electronic device displays well-balanced electron and hole transport channels with 0.3–0.4 cm²/Vs mobility [31]. This research work displays complementary like inverters prepared with a Tyrian purple channel and Au source and drain electrodes. The gain of ~250–290 is amongst the greatest stated for a sole semiconductor with a single-type interaction electrode. These outcomes describe that natural-originated nontoxic and cheap materials can strive for the greatest synthetic organic semiconductors. Investigation into the biocompatibility and biodegradation of organic semiconductors rests inadequate.

18.2.4 ELECTRODES

The electrodes are accountable for carrying charged carriers from the device to the peripheral circuits. For example, conducting materials like metals and greatly doped metal oxides and organic polymers are usually appropriate for electrodes for multifunctional applications [24, 25, 27]. Noble metals like gold (Au) and silver (Ag), because of their corrosion-resistant and nonreactive assets, have extensively been related to medical usages like dental stuffing. Titanium (Ti) and Ti-based alloys have also yielded applications in bone and dental implants. These metals have greater electrical conductivity (σ = 3–60 × 10^6 S/m). Though these metals are typically expensive owing to their rarity and resistance to breaking down, they might reason for considerable construction and ultimate obstruction in the body if engaged in bulk. Hence, care has been located to walk around physiologically pleasant metals like molybdenum (Mo), magnesium (Mg), iron (Fe), manganese (Mn), zinc (Zn), or tungsten (W). These metals are metabolized in physiological circumstances and the subsequent metabolites can either be passed out or safely absorbed by the body.

Some polymer electrodes like melanin and PEDOT are also established for usage in distinctive requests in organic electronics with certain limitations. The effectiveness of electrodes based on carbon in physiological situations is unsubstantiated. So, metals are tranquil for the foremost selection of electrode materials because of their facile processing of successive device layers and higher intrinsic carrier mobility [4].

18.3   BIODEGRADABLE ELECTRONIC APPLICATIONS

Even though, to emerging electronic devices with a definite "use-by-date or shelf-life", the recent-day user ways have finished a circumstance where the e-waste management produced has developed a problem, and substitute solutions are needed to stop an approaching pollution pandemic. Though, it will yield constant education and time for the consistent customer can admit a novel idea (biocompatibility) as associated with conventional nondegradable electronics. Eco-friendly electronics can presently achieve place applications, mainly in the medical area, where transient device implantation is anticipated to circumvent a subordinate surgical process for deletion and degraded device by-products can be safely excreted or absorbed from the body and others [4].

18.3.1   Sensing and Diagnostics

In recent years, sensing and diagnostic electronic devices for biomedical and healthcare claims have grabbed much improvement. One specific area that biodegradable electronics have a huge possibility in is that of profound tissue and slightly invasive continuing observing. The biodegradability of such sensors excludes the necessity for a further operation to recuperate the device, thus offering a notable quantity of accessibility and efficacy to trauma patients and the aged [4, 32].

For detection of visible light, a transparent, flexible, and nontoxic phototransistor was prepared using biodegradable CNF substrates [33]. Along with the phototransistors' transparency, flexibility, and biocompatibility, this study specifies the substantial probable of NCs as environmentally friendly and inexpensive sensors.

A multifunctional and healable E-tattoo constructed on a graphene/silk fibroin (SF)/Ca^{2+} blend was reported by Wang et al. [34] The stretchy E-tattoos are made by writing or printing by an SF/graphene/Ca^{2+} suspension. The graphene nanosheets are evenly spread in the medium and develop an electrically conducting route, which can delicately reply to the variations of the closest location, with temperature, humidity, and strain. This asset supports the E-tattoo to be employed as a sensor, observing this variable quantity with higher sensitivity, good stability, and rapid response. Also, the E-tattoo shows excellent healable properties.

To develop novel conductive nanocomposites sensors, Ling et al. [35] utilized graphene. Using SF as a base material and fabricated graphene/SF nanocomposites over a homogeneously distributed and extremely stable graphene/SF suspension structure. The synthesized graphene/SF nanocomposites conserve not only the graphene's electronic merits but also the mechanical assets of SF. Their electrical resistance owing to the changes in the chemical environment, humidity, body movement, and sensitivity to deformation display a capable future for real claims as human-machine interfaces, wearable sensors, healthcare/biomedical devices, and intelligent skins.

Likewise, Scaffaro et al. [36] fabricated a piezoresistive sensor by manipulating amphiphilic graphene oxide (GO) to give the polylactide (PLA) and poly (ethylene glycol) (PEG) combinations with electrical assets sensitive to vary the strain and pressure. Han et al. [37] stated a natural rubber (NR) matrix along with CNF-PANI nanostructured complexes to prepare conductive hybrid elastomers, in which PANI offers the electrical conductivity, while CNFs support the material. The prepared bio-based elastomers with evenly arranged assemblies exhibited intrinsic stretchability, decent flexibility, enhanced mechanical assets, desired electric, and low density. Then, the strain sensor with excellent sensitivity and repeatability was prepared by utilizing an elastomer, which could display the human body motion in instantaneous time.

18.3.2   Drug Delivery Applications

Nowadays, a multifunctional biomedical/healthcare device (integrating both drug delivery functions with bio-sensing capabilities) can be established for point-of-care automated claims. These innovative devices united with electrically conductive and electroactive combinations with biocompatible polymers would deliver harmless and further effective drug delivery on-site and with a real-time on-off regulation system.

For modulated dexamethasone (DEX) release, the drug-wrapped chitosan/CNT composite films were prepared. It stayed observed that electrical stimulation of the electrically conductive films enhanced the DEX release; by applied voltage via electromigration, the proclamation rate of the negatively charged drug would be pretentious. Electrical stimulus led to an improved average release rate of drug and a whole proclamation of DEX within 12h, whereas only drug release of 50% for the inactive sample. In addition, the drug release rate possibly will be modulated by changing the polarity and the amplitude of the voltage applied. Programmable release of chlorpromazine via a biocompatible conductive heparin-doped PPy film was prepared as a high-capacity cation exchanger. Ampicillin-wrapped silk film was united into the device for wirelessly activated release of drug and ampicillin proclamation outlines by thermal triggering were noted. The antibiotic-loaded silk device was noted to be biocompatible and biodegradable in vivo, and antibiotics released were able to obstruct the bacteria growth, including Escherichia coli and S. aureus for contagion decline in implantation [4, 38].

Conductive materials, like PPy, PANI, and GO, have been integrated into dissimilar hydrogels as electrically stimulated drug release delivery systems. To inspect the in vitro electro-stimulated release of the drug, hydrocortisone was used as a model drug. The regulation of the release profile was observed owing to the change in the duration time and strength of the applied electrical potential. A controllable release of drug method by release of 7% per switch was observed at a working potential of 3 V for 1 min. AFM outcomes displayed that electrical stimulus might dynamically alter the hydrogel's pore size and porosity, additionally stimulus the drug release. This analysis proved that the electroactive films and hydrogels could be utilized as capable drug delivery devices by current/voltage stimulation. The biocompatible polymers in the arrangements permit the electronics to be personalized to contest physiological progressions for the release of drugs in a controllable manner [4, 39].

18.3.3 TISSUE ENGINEERING APPLICATIONS

Recently, in tissue engineering (TE), the quick advancement in the electrode's fabrication on biocompatible substrates has unlocked a novel investigation zone in the implantable devices field. The myocardium and natural nervous system have distinctive conductive paths for the propagation of electrical signals to attain their distinct electrophysiological roles. Traumas or diseases, like myocardium infarction, lead to slippage of the cell, formation of scar tissue, and electrical conductivity loss. It is the main task of scientists in the TE field to reinstate the native nerve and cardiac tissue's electrophysiological functions. Biocompatible electronics are called "smart" scaffolds for the next generation in TE claims [4]. Over the years, both synthetic and natural polymers have been extensively engaged to prepare TE scaffolds. Nevertheless, to induce a complete tissue function recovery, a huge number of these biomaterials do not have adequate bioactivity. Organic CPs, like PANi, PPy, PEDOT, and PTh have been mostly estimated as biomaterial scaffolds for TE claim because of their excellent biocompatibility and conductivity [40].

In cardiac TE applications, conductive biodegradable scaffolds have been engaged. CPs, like PEDOT, PPy, and PANI, have stayed integrated into biocompatible scaffolds to stimulus the cardiac cell's function and growth due to their good biocompatibility property. An electrically active scaffold made of united electro-spun nanofibers (NFs) comprising PLGA and PANI were prepared for directing the cardiomyocytes' synchronous beatings [4]. In mammalian skin, a natural CP generally found in is melanin. Kai et al. combined with biodegradable CP to poly (L-lactide- co-ε-caprolactone)/gelatin NFs to offer electrophysiological signals for cardiac cells and imitate the native myocardial atmosphere [41]. PPy has been usually stated to progress neurite extension, neural activities, and axonal outgrowth. In one study, PPy and poly(D,L-lactide-*co*-epsilon-caprolactone) (PDLLA/CL) combination of a biodegradable/compatible electrically conducting polymeric compound was recognized.

The initial polymer scaffolds employed for tissue regeneration and cell transplantation purpose were felt-like materials or mesh based on PLGA and PGA. Over a year, many tissue engineers have worked on mesh-like or nonwoven scaffolds made of polymer fibers of PGA/PLLA, and PGA/PDLA, and PGA. Foam-like scaffolds can be intended and prepared to encounter the exact targets for TE claims, e.g. balancing pores of high interconnectivity (3D internal geometry) with complete organizational integrity (mechanical characteristics) [42]. A conducting, recyclable electroactive shape memory polymer assembly has been gathered for the myogenic differentiation of myoblasts acceleration for regeneration of skeletal muscle tissue by using the electroactive aniline tetramer. Also, analysis has informed the substantial enhancement of the neurotrophin secretion of Schwann cells and myelin gene expression, through the utilization of conducting biocompatible copolymer films, which were after used to improve the growth of peripheral nerve and neuronal cells revival [43].

For TE applications, carbon nanotubes (CNT) have also continually fused into biocompatible polymers. CNTs were combined into assembled poly (glycerol sebacate)/gelatine NFs for cardiac engineering constructs. The CNTs introduction not only improved electrical conductivity and fiber alignment but also retained the alignment, viability, and cardiomyocytes' contractile activities planted on the scaffolds. Shin et al. established CNT-incorporated gelatin-based hydrogels and planted neonatal rat cardiomyocytes on them for bio-actuators and cardiac engineering constructs [44]. The CNT-decorated gelatin hydrogels exposed higher anisotropic electrical conductivity and mechanical properties. Likewise, GO was also introduced into gelatin-based hydrogels to progress electroactive scaffolds for revival requests in myocardial systems [45].

By polymerization in the existence of an Et2Zn catalyst, poly(dioxanone) (PDS), a p-dioxanone homopolymer, has been synthesized. PDS is a monofilament suture material, which causes a low tissue response; it can be observed as a PGA-modified version. It preserves a resorbable property owing to the ester bond occurrence, and its stretchability is enhanced because of the change in the ester link to an ether linkage. Also, PDS is used in other biomedical device-based applications like braided mesh, as foil, fixation pins, and staples, for the orbita wall reconstruction [42]. By integrating gold nanowires (NWs) within alginate scaffolds, 3D electroactive cardiac patches were prepared. The electrically conductive NWs linked the electrically resistant alginate pore walls and allowed electrical interface amid in-line cardiomyocytes. The cells in the matrix exhibited contraction synchronously with decent alignment under electrical stimulation. Additionally, by casting collagen hydrogel with freestanding NWs, the same assembly designed and prepared nanoelectronic 3D macroporous scaffolds [46]. Such media revealed strong electronic characteristics and possibly can be employed as extracellular scaffolds for 3D TE of cardiomyocytes, smooth muscle cells, and neurons. Planted with neonatal rat cardiomycytes, the 3D cardiac arrangement displayed striations feature of cardiac tissue and the native cardiomyocytes electrical output signals, signifying that the 3D hybrid scaffolds could be engaged as electroactive cardiac patches for both real-time monitoring and cardiac regeneration of the local electrical action. These electrically conductive substrates can be employed as the biosensor's boundary or cardiac TE electrodes. Nearby, these innovative 3D electroactive biocompatible devices, when combined and entrenched collected with coronary stents or cardiac pacemakers, encourage and lead the differentiation and growth of the cardiac cell by sensing the resident bio-surroundings. Using distant switch machinery, coupled with software and hardware incorporation, cardiologists will be able to display patients' physiological indexes and even supervisor conduct distantly [4].

Even though TE has set courage in the direction of emerging a trustworthy technique for replacing injured organs and body parts, a different vital stage in the understanding of incorporating such superficially developed fragments with the present body is neural stimulation and interfacing. Notoriously, neural cells are known as those that are hard to develop and endure in a preferred manner. By itself, not only do electroactive biocompatible devices act as provisional neural transduction channels for early incorporation, but they also encourage the new neural cells' growth that will ultimately proceed as the enduring feature though the device degrades. Such progress can stimulate the development of bionic or artificial limbs that are impartially linked to

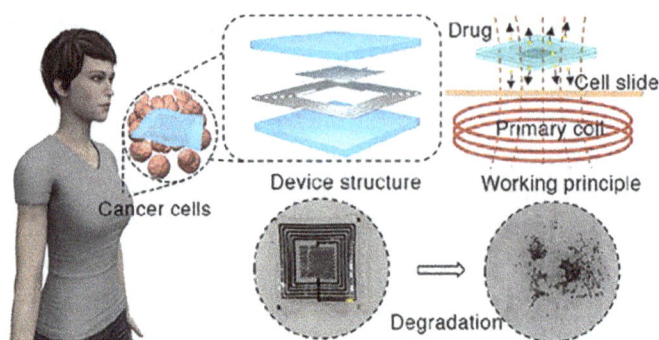

FIGURE 18.4 Graphical abstract for Biodegradable flexible electronic device for controllable drug release for cancer treatment. (Adapted with permission from Ref. [50]. Copyright (2021) Biodegradable Flexible Electronic Device with Controlled Drug Release for Cancer Treatment, ACS Appl. Mater. Interfaces.)

the host nervous scheme, relatively than depending on outwardly power-driven machines, giving courage to paraplegic publics of some independence of movement. A biocompatible and biodegradable polymer built on polyethylene glycol and desaminotyrosyl-tyrosine was prepared for neural interface toward the upcoming generation [47]. The polymer film was engaged as a biocompatible/degradable, multielectrode, and multi-luminal path for neural recording. Outcomes exhibited that the channel with several operative electrodes conserved recording spot impedance in the one-digit kΩ value and possibly will also note neural signals for higher than 4 weeks' post-implantation in a rabbit template [48]. In a different motivating analysis, wireless electronic schemes were made on biocompatible substrates of PCL for creating alternating electric field stimulus to direct the sensory neurons' alignment [49]. The investigation directed that the innovative circuit over a flexible biodegradable medium may be an efficient electrical stimulation device to repair the peripheral nervous scheme. All these devices displayed exceptional device performance and worthy biodegradability both in vivo and in vitro. The development of a real-world biocompatible or degradable neural probe will greatly hang on the perfect considerate of the development of novel biomaterial and neural biology to incorporate nerve/probe communication as well as to set of scales in an in vivo device running span and biodegradation time frame [4]. Also, other interdisciplinary applications such as low-scale biocompatible electronic devices can identify the determination in cancer therapy research as well, where it can be used for the distant pointing and removal of cancerous cells. Also, it can be focused on reducing the treatment's side effects on the patient (Figure 18.4).

18.4 CONCLUSION AND FUTURE OUTLOOK

E-waste management necessitates commencing now beforehand it makes worse into an irretrievable effluence issue. Moreover, raising awareness and active recycling amid customers, alternative sourcing's to conventional electronics could be a trustworthy answer in the prospect. Producing a biocompatible or degradable lodging with numerous natural and bio-derived materials reveals possible electronic assets as semiconductors, electrode materials, dielectrics, and substrates. To prepare stretchable and flexible compatible devices, synthetic materials afford superior regulation over mechanical assets and are hence used. Improved consideration of the surface chemistries of different physiological and natural surroundings might expose the entrance for biocompatible surface chemistries hitherto to be revealed with semiconducting, dielectric, substrate, and CP schemes. Thus, these green or biodegradable electronics systems have superior probability and are encouraged and composed to provide a constructive effect in the upcoming generation. In summary, "green" biodegradable organic electronics have already demonstrated their science entrance; researchers need to envision the promises.

REFERENCES

1. Irimia-Vladu, Mihai, Eric D. Głowacki, Gundula Voss, Siegfried Bauer, and Niyazi Serdar Sariciftci. "Green and biodegradable electronics." Materials Today 15, no. 7–8 (2012): 340–346.
2. Babu, Balakrishnan Ramesh, Anand Kuber Parande, and Chiya Ahmed Basha. "Electrical and electronic waste: a global environmental problem." Waste Management & Research 25, no. 4 (2007): 307–318.
3. Kurian, Jospeh. "Electronic waste management in India–issues and strategies." Eleventh International Waste Management and Landfill Symposium, 5, no. 1, (2007): 74–82.
4. Tan, Mein Jin, Cally Owh, Pei Lin Chee, Aung Ko Ko Kyaw, Dan Kai, and Xian Jun Loh. "Biodegradable electronics: cornerstone for sustainable electronics and transient applications." Journal of Materials Chemistry C, 4, no. 24 (2016): 5531–5558.
5. Wang, Liu, Yuan Gao, Fanqi Dai, Deying Kong, Huachun Wang, Pengcheng Sun, Zhao Shi, Xing Sheng, Baoxing Xu, and Lan Yin. "Geometrical and chemical-dependent hydrolysis mechanisms of silicon nanomembranes for biodegradable electronics." ACS Applied Materials & Interfaces, 11, no. 19 (2019): 18013–18023.
6. Lee, Yoon Kyeung, Ki Jun Yu, Enming Song, Amir Barati Farimani, Flavia Vitale, Zhaoqian Xie, Younghee Yoon, Yerim Kim, Andrew Richardson, Haiwen Luan, Yixin Wu, Xu Xie, Timothy H. Lucas, Kaitlyn Crawford, Yongfeng Mei, Xue Feng, Yonggang Huang, Brian Litt, Narayana R. Aluru, Lan Yin, and John A. Rogers. "Dissolution of monocrystalline silicon nanomembranes and their use as encapsulation layers and electrical interfaces in water-soluble electronics." ACS Nano, 11, no. 12 (2017): 12562–12572.
7. Miao, Jinlei, Haihui Liu, Yongbing Li, and Xingxiang Zhang. "Biodegradable transparent substrate based on edible starch–chitosan embedded with nature-inspired three-dimensionally interconnected conductive nanocomposites for wearable green electronics." ACS Applied Materials & Interfaces, 10, no. 27 (2018): 23037–23047.
8. Yogeswaran, Nivasan, Ensieh S. Hosseini, and Ravinder Dahiya. "Graphene based low voltage field effect transistor coupled with biodegradable piezoelectric material based dynamic pressure sensor." ACS Applied Materials & Interfaces 12, no. 48 (2020): 54035–54040.
9. Irimia-Vladu, Mihai, Pavel A. Troshin, Melanie Reisinger, Lyuba Shmygleva, Yasin Kanbur, Günther Schwabegger, Marius Bodea, Reinhard Schwödiauer, Alexander Mumyatov, Jeffrey W. Fergus, Vladimir F. Razumov, Helmut Sitter, Niyazi Serdar Sariciftci, Siegfried Bauer. "Biocompatible and biodegradable materials for organic field-effect transistors." Advanced Functional Materials 20, no. 23 (2010): 4069–4076.
10. Yin, Lan, Huanyu Cheng, Shimin Mao, Richard Haasch, Yuhao Liu, Xu Xie, Suk-Won Hwang, Harshvardhan Jain, Seung-Kyun Kang, Yewang Su, Rui Li, Yonggang Huang, John A. Rogers. "Dissolvable metals for transient electronics." Advanced Functional Materials 24, no. 5 (2014): 645–658.
11. Baldo, Thaisa A., Lucas Felipe de Lima, Leticia F. Mendes, William R. de Araujo, Thiago RLC Paixao, and Wendell KT Coltro. "Wearable and Biodegradable Sensors for Clinical and Environmental Applications." ACS Applied Electronic Materials 3, no. 1 (2020): 68–100.
12. Bollström, Roger, Anni Määttänen, Daniel Tobjörk, Petri Ihalainen, Nikolai Kaihovirta, Ronald Österbacka, Jouko Peltonen, and Martti Toivakka. "A multilayer coated fiber-based substrate suitable for printed functionality." Organic Electronics 10, no. 5 (2009): 1020–1023.
13. Tobjörk, Daniel, and Ronald Österbacka. "Paper electronics." Advanced Materials 23, no. 17 (2011): 1935–1961.
14. Siegel, Adam C., Scott T. Phillips, Benjamin J. Wiley, and George M. Whitesides. "Thin, lightweight, foldable thermochromic displays on paper." Lab on a Chip 9, no. 19 (2009): 2775–2781.
15. Barr, Miles C., Jill A. Rowehl, Richard R. Lunt, Jingjing Xu, Annie Wang, Christopher M. Boyce, Sung Gap Im, Vladimir Bulović, and Karen K. Gleason. "Direct monolithic integration of organic photovoltaic circuits on unmodified paper." Advanced Materials 23, no. 31 (2011): 3500–3505.
16. Kim, Dae-Hyeong, Jonathan Viventi, Jason J. Amsden, Jianliang Xiao, Leif Vigeland, Yun-Soung Kim, Justin A. Blanco, Bruce Panilaitis, Eric S. Frechette, Diego Contreras, David L. Kaplan, Fiorenzo G. Omenetto, Yonggang Huang, Keh-Chih Hwang, Mitchell R. Zakin, Brian Litt and John A. Rogers. "Dissolvable films of silk fibroin for ultrathin conformal bio-integrated electronics." Nature Materials 9, no. 6 (2010): 511–517.
17. Kim, Dae-Hyeong, Yun-Soung Kim, Jason Amsden, Bruce Panilaitis, David L. Kaplan, Fiorenzo G. Omenetto, Mitchell R. Zakin, and John A. Rogers. "Silicon electronics on silk as a path to bioresorbable, implantable devices." Applied Physics Letters 95, no. 13 (2009): 133701.

18. Wang, Chung-Hwa, Chao-Ying Hsieh, and Jenn-Chang Hwang. "Flexible organic thin-film transistors with silk fibroin as the gate dielectric." Advanced Materials 23, no. 14 (2011): 1630–1634.

19. Kang, Seung-Kyun, Jahyun Koo, Yoon Kyeung Lee, and John A. Rogers. "Advanced materials and devices for bioresorbable electronics." Accounts of Chemical Research 51, no. 5 (2018): 988–998.

20. Chang, Jer-Wei, Cheng-Guang Wang, Chong-Yu Huang, Tzung-Da Tsai, Tzung-Fang Guo, and Ten-Chin Wen. "Chicken albumen dielectrics in organic field-effect transistors." Advanced Materials 23, no. 35 (2011): 4077–4081.

21. Irimia-Vladu, Mihai, Pavel A. Troshin, Melanie Reisinger, Guenther Schwabegger, Mujeeb Ullah, Reinhard Schwoediauer, Alexander Mumyatov, Marius Bodea, Jeffrey W. Fergus, Vladimir F. Razumov, Helmut Sitter, Siegfried Bauer, Niyazi Serdar Sariciftci. "Environmentally sustainable organic field effect transistors." Organic Electronics 11, no. 12 (2010): 1974–1990.

22. Cao, Yue, and Kathryn E. Uhrich. "Biodegradable and biocompatible polymers for electronic applications: A review." Journal of Bioactive and Compatible Polymers 34, no. 1 (2019): 3–15.

23. Muthusankar, E., Saikh Mohammad Wabaidur, Zeid Abdullah Alothman, Mohd Rafie Johan, Vinoth Kumar Ponnusamy, and D. Ragupathy. "Fabrication of amperometric sensor for glucose detection based on phosphotungstic acid–assisted PDPA/ZnO nanohybrid composite." Ionics 26, no. 12 (2020): 6341–6349.

24. Chokkiah, Bavatharani, Muthusankar Eswaran, Saikh Mohammad Wabaidur, Zeid Abdullah Alothman, Pei-Chien Tsai, Vinoth Kumar Ponnusamy, and Ragupathy Dhanusuraman. "Novel PDPS-SiO2 nanospherical network decorated graphene nanosheets composite coated FTO electrode for efficient electro-oxidation of methanol." Fuel 279 (2020): 118439.

25. Muthusankar, E., Soo Chool Lee, and D. Ragupathy. "Enhanced electron transfer characteristics of surfactant wrapped SnO2 nanorods impregnated poly (diphenylamine) matrix." Sensor Letters 16, no. 12 (2018): 911–917.

26. Bavatharani, C., E. Muthusankar, Saikh Mohammad Wabaidur, Zeid Abdullah Alothman, Khalid M. Alsheetan, Murefah mana AL-Anazy, and D. Ragupathy. "Electrospinning technique for production of polyaniline nanocomposites/nanofibres for multi-functional applications: A review." Synthetic Metals 271 (2020): 116609.

27. Bavatharani, C., E. Muthusankar, Zeid Abdullah Alothman, Saikh Mohammad Wabaidur, Vinoth Kumar Ponnusamy, and D. Ragupathy. "Ultra-high sensitive, selective, non-enzymatic dopamine sensor based on electrochemically active graphene decorated Polydiphenylamine-SiO2 nanohybrid composite." Ceramics International 46, no. 14 (2020): 23276–23281.

28. Bidez, Paul R., Shuxi Li, Alan G. MacDiarmid, Everaldo C. Venancio, Yen Wei, and Peter I. Lelkes. "Polyaniline, an electroactive polymer, supports adhesion and proliferation of cardiac myoblasts." Journal of Biomaterials Science, Polymer Edition 17, no. 1–2 (2006): 199–212.

29. Tybrandt, Klas, Karin C. Larsson, Sindhulakshmi Kurup, Daniel T. Simon, Peter Kjäll, Joakim Isaksson, Mats Sandberg, Edwin WH Jager, Agneta Richter-Dahlfors, and Magnus Berggren. "Translating electronic currents to precise acetylcholine–induced neuronal signaling using an organic electrophoretic delivery device." Advanced Materials 21, no. 44 (2009): 4442–4446.

30. Zollinger, Heinrich. Color chemistry: syntheses, properties, and applications of organic dyes and pigments. John Wiley & Sons, (2003), Germany.

31. Kanbur, Yasin, Mihai Irimia-Vladu, Eric D. Głowacki, Gundula Voss, Melanie Baumgartner, Günther Schwabegger, Lucia Leonat, Mujeeb Ullah, Hizir Sarica, Sule Erten-Ela, Reinhard Schwödiauer, Helmut Sitter, Zuhal Küçükyavuz, Siegfried Bauer, Niyazi Serdar Sariciftci. "Vacuum-processed polyethylene as a dielectric for low operating voltage organic field effect transistors." Organic Electronics 13, no. 5 (2012): 919–924.

32. Owens, Róisín M., and George G. Malliaras. "Organic electronics at the interface with biology." MRS Bulletin 35, no. 6 (2010): 449–456.

33. Park, Junsu, Jung-Hun Seo, Seung-Won Yeom, Chunhua Yao, Vina W. Yang, Zhiyong Cai, Young Min Jhon, and Byeong-Kwon Ju. "Flexible and transparent organic phototransistors on biodegradable cellulose nanofibrillated fiber substrates." Advanced Optical Materials 6, no. 9 (2018): 1701140.

34. Wang, Qi, Shengjie Ling, Xiaoping Liang, Huimin Wang, Haojie Lu, and Yingying Zhang. "Self-healable multifunctional electronic tattoos based on silk and graphene." Advanced Functional Materials 29, no. 16 (2019): 1808695.

35. Ling, Shengjie, Qi Wang, Dong Zhang, Yingying Zhang, Xuan Mu, David L. Kaplan, and Markus J. Buehler. "Integration of stiff graphene and tough silk for the design and fabrication of versatile electronic materials." Advanced Functional Materials 28, no. 9 (2018): 1705291.

36. Scaffaro, Roberto, Andrea Maio, Giada Lo Re, Antonino Parisi, and Alessandro Busacca. "Advanced piezoresistive sensor achieved by amphiphilic nanointerfaces of graphene oxide and biodegradable polymer blends." Composites Science and Technology 156 (2018): 166–176.

37. Han, Jingquan, Kaiyue Lu, Yiying Yue, Changtong Mei, Chaobo Huang, Qinglin Wu, and Xinwu Xu. "Nanocellulose-templated assembly of polyaniline in natural rubber-based hybrid elastomers toward flexible electronic conductors." Industrial Crops and Products 128 (2019): 94–107.

38. Naficy, Sina, Joselito M. Razal, Geoffrey M. Spinks, and Gordon G. Wallace. "Modulated release of dexamethasone from chitosan–carbon nanotube films." Sensors and Actuators A: Physical 155, no. 1 (2009): 120–124.

39. Tsai, Tong-Sheng, Viness Pillay, Yahya E. Choonara, Lisa C. Du Toit, Girish Modi, Dinesh Naidoo, and Pradeep Kumar. "A polyvinyl alcohol-polyaniline based electro-conductive hydrogel for controlled stimuli-actuable release of indomethacin." Polymers 3, no. 1 (2011): 150–172.

40. Kenry, and Bin Liu. "Recent advances in biodegradable conducting polymers and their biomedical applications." Biomacromolecules 19, no. 6 (2018): 1783–1803.

41. Kai, Dan, Molamma P. Prabhakaran, Guorui Jin, and Seeram Ramakrishna. "Biocompatibility evaluation of electrically conductive nanofibrous scaffolds for cardiac tissue engineering." Journal of Materials Chemistry B 1, no. 17 (2013): 2305–2314.

42. Hutmacher, D. W., J. C. H. Goh, and S. H. Teoh. "An introduction to biodegradable materials for tissue engineering applications." Annals-Academy of Medicine Singapore 30, no. 2 (2001): 183–191.

43. Wu, Yaobin, Ling Wang, Baolin Guo, Yongpin Shao, and Peter X. Ma. "Electroactive biodegradable polyurethane significantly enhanced Schwann cells myelin gene expression and neurotrophin secretion for peripheral nerve tissue engineering." Biomaterials 87 (2016): 18–31.

44. Shin, Su Ryon, Sung Mi Jung, Momen Zalabany, Keekyoung Kim, Pinar Zorlutuna, Sang bok Kim, Mehdi Nikkhah, Masoud Khabiry, Mohamed Azize, Jing Kong, Kai-tak Wan, Tomas Palacios, Mehmet R. Dokmeci, Hojae Bae, Xiaowu (Shirley) Tang, and Ali Khademhosseini. "Carbon-nanotube-embedded hydrogel sheets for engineering cardiac constructs and bioactuators." ACS Nano 7, no. 3 (2013): 2369–2380.

45. Shin, Su Ryon, Behnaz Aghaei-Ghareh-Bolagh, Tram T. Dang, Seda Nur Topkaya, Xiguang Gao, Seung Yun Yang, Sung Mi Jung, Jong Hyun Oh, Mehmet R. Dokmeci, Xiaowu (Shirley) Tang, Ali Khademhosseini. "Cell-laden microengineered and mechanically tunable hybrid hydrogels of gelatin and graphene oxide." Advanced Materials 25, no. 44 (2013): 6385–6391.

46. Tian, Bozhi, Jia Liu, Tal Dvir, Lihua Jin, Jonathan H. Tsui, Quan Qing, Zhigang Suo, Robert Langer, Daniel S. Kohane, and Charles M. Lieber. "Macroporous nanowire nanoelectronic scaffolds for synthetic tissues." Nature Materials 11, no. 11 (2012): 986–994.

47. Lewitus, Dan, Karen L. Smith, William Shain, and Joachim Kohn. "Ultrafast resorbing polymers for use as carriers for cortical neural probes." Acta Biomaterialia 7, no. 6 (2011): 2483–2491.

48. Lewitus, Dan, R. Jacob Vogelstein, Gehua Zhen, Young-Seok Choi, Joachim Kohn, Stuart Harshbarger, and Xiaofeng Jia. "Designing tyrosine-derived polycarbonate polymers for biodegradable regenerative type neural interface capable of neural recording." IEEE Transactions on Neural Systems and Rehabilitation Engineering 19, no. 2 (2010): 204–212.

49. Martin, Christopher, Théophile Dejardin, Andrew Hart, Mathis O. Riehle, and David RS Cumming. "Directed nerve regeneration enabled by wirelessly powered electrodes printed on a biodegradable polymer." Advanced Healthcare Materials 3, no. 7 (2014): 1001–1006.

50. Li, Hangfei, Fei Gao, Peng Wang, Lan Yin, Nan Ji, Liwei Zhang, Lingyun Zhao, Guohui Hou, Bingwei Lu, Ying Chen, Yinji Ma, and Xue Feng. "Biodegradable flexible electronic device with controlled drug release for cancer treatment." ACS Applied Materials & Interfaces 13, no. 18 (2021): 21067–21075.

19 Microfluidic Devices with Integrated Conductive Polymeric Electrodes for Biosensing Applications

Khairunnisa Amreen and Sanket Goel
MEMS, Microfluidics and Nanoelectronics Lab,
Department of Electrical and Electronics Engineering,
Birla Institute of Technology and Science, Hyderabad, India

CONTENTS

19.1 INTRODUCTION

Conducting polymers (CPs) are considered significant materials owing to their supreme chemical, physical, and structurally tunable properties. CPs can be further functionalized with various chemical functional groups and nanoparticles to improve their electrical conductivity. This observation was first reported by Shirakawaet et al. in 1977 wherein they reported the preparation of halogen derivatives of polyacetylene as conductive organic polymers [1]. CPs acquired much attention as a key matrix for developing biosensors after the groundbreaking discovery by Diaz et al., for electrodeposition of polypyrrole thin films [2]. Since then, CPs have been explored substantially for immobilization of biomolecules, like enzymes, proteins, antibodies-antigen, etc., as they offer feasibility of electron transfer mechanism. CPs improve sensitivity and selectivity toward bioanalytics giving a remarkable limit of detections. Generally, polymers with structures comprising conjugated, π-electrons, alternate sigma, and pie bonds show good conductivity due to the delocalization of electrons. Based on these structural attributes, a few of the common CPs used in the electrochemical biosensing are polyaniline (PANI), polypyrrole (PPy), poly(3,4-ethylenedioxythiophene) polyacetylene (PEDOT), polyfluorene, polyphenylene, polycarbazole, etc.

Several factors like chemical structure, temperature, pH, applied potential, and chain length govern the conductivity and polymerization. Hence, the synthesis of CPs can be varied and manipulated by changing the parameters based on the desired type of application. There are several methods to prepare such CPs like chemical vapor deposition [3], polycondensation[4], and oxidative polymerization [5] with the use of oxidants like $KMnO_4$, $K_2Cr_2O_7$, etc., or non-oxidative approaches by using chemicals like Grignard reagent[6]. Despite these traditional methods for CPs as biosensors, the electrochemical polymerization approach is preferred as it is easier to control the polymerization and deposition over the electrode.

19.1.1 ELECTRODEPOSITION OF POLYMER

The CPs can be electrochemically prepared via the in situ electrodeposition method. Herein, a conventional three-electrode system is employed comprising a working (where the CP film is formed), counter, and reference electrodes where the potential is applied. As shown in Figure 19.1, the experimental setup has the above-mentioned electrodes kept in an electrolytic cell consisting of the monomer (to be polymerized) dissolved in the respective solvent and a supporting buffer salt solution of the desired pH depending upon the monomer. There are three basic types of electropolymerization techniques that can be employed in this setup. (a) A potentiodynamic technique like cyclic voltammetry (CV) wherein a current response is generated due to oxidation/reduction of the monomer with linear cycling of potential sweep between a set potential window. (b) Potentiostatic techniques like amperometry wherein a fixed optimized potential is applied for instigating the polymerization. (c) Galvanostatic technique wherein a fixed optimal current is applied for starting the polymerization. Each of the methods has its advantage like potentiodynamic technique gives uniform deposition of CP films that are firmly adsorbed over electrodes. The potentiostatic approach allows control over the thickness of deposited films, while the galvanostatic approach helps to monitor deposition based on time; hence, conductivity is controlled. Nevertheless, the deposition method is dependent on the ambient parameters like the chemical composition of the monomer, pH, and temperature. Furthermore, this approach can also be slightly modified, where the monomer can be dropped and cast over the working

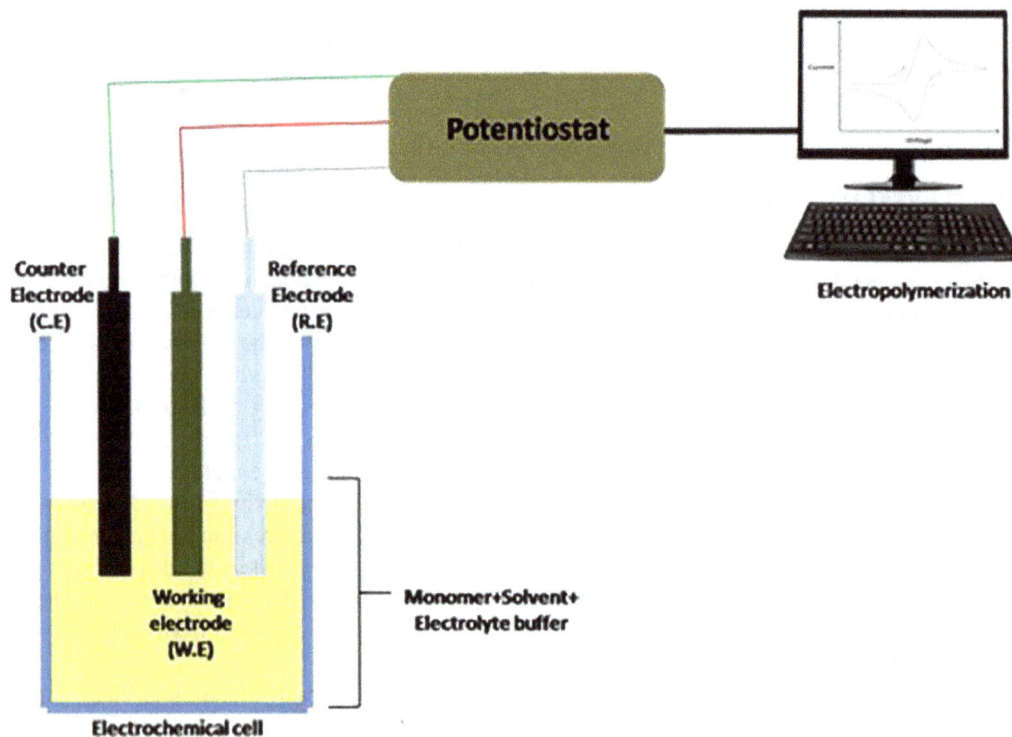

FIGURE 19.1 Schematic representation of experimental setup for electrodeposition.

electrode, air-dried, followed by potential cycling in desired pH buffer solution. The monomer starts to polymerize over the electrode. However, controlling the polymerization is difficult in this method. Either of these approaches can be used for the preparation of biosensors as the interaction of functionalized polymers and biomolecules can take place with ease post CPs' film formation.

19.1.2 SIGNIFICANCE OF CPs IN BIOSENSORS

CPs have proven to be an extremely useful matrix for entrapping biocatalysts. The procedures adopted for immobilizing these onto the CPs via electrodeposition give an enhanced, sensitive, and selective detection of analytes and help in multi-analyte detection with interference mitigation. CPs can be chemically modified as needed to improve the affinity toward the binding; therefore, they can be used for protein binding. Moreover, during the electrodeposition, it is possible to entrap the biocatalyst into the matrix while depositing it over the electrode. In addition, the special arrangement, film thickness of the biocatalyst can be modulated and the activity of the biological entity can also be altered with varying the state of the CPs. The interaction and facile electron flow between CPs and the electroactive site of the biocatalysts decide the fate of the biosensor. For this, the electrical wiring approach has been regarded as the best fit [7]. In further detail, the CPs are known to be stable in neutral pH; henceforth, the biological components can retain their activity. Another crucial feature of CPs is that they can be doped or undoped easily leading to manipulation of their electrical properties. The biggest advantage of CPs is that they can effectively transfer the electrical charge generated via biochemical reaction to the electrical circuit; hence, a quantifiable readout is possible. Since CPs get electrodeposited over a specified geometrical surface active area of the electrode, they have size-exclusion properties. Henceforth, they have been extensively utilized for the development of electrochemical biosensors.

19.1.3 BIOSENSORS

An electroanalytical device that utilizes biological entities, such as living cells, enzymes, antibodies, etc., to investigate the presence of chemicals, biochemicals, bioanalytes, etc., is regarded as a biosensor. These devices aim to produce electrical signals that are directly proportional to the concentration of the analyte. Biosensors have three components. (a) Bioreceptors: the biological element of choice like DNA, enzyme, living cell, etc. (b) Transducer: converts the quantifiable biological signal into a measurable electrical signal. (c) Reader output device: that converts the electrical signals received from the transducer to readable analog display. In a biosensor, the biological entity is either entrapped within the matrix or integrated with the transducer. Based on the integration level, the biosensors can be classified into three types. (a) First generation: herein, the biological component is entrapped or attached to a membrane that is bound to the transducer. (b) Second generation: covalent bonding of biological entity with the transducer. (c) Third generation: herein, the direct bonding of biological moiety with the electronic device that traduces and then amplified the signal. CPs based biosensors are often regarded as third-generation biosensors. Literature has several biosensors reported wherein the bulk volume of sample has been used. A few of the remarkable recent advances are discussed here.

For instance, Virutkar et al. reported a graphite electrode-based biosensor. Herein, carbon nanotubes/polyaniline/polypyrrole-based nanocomposite film was deposited electrochemically, via CV. This matrix was used to entrap the enzyme acetylthiocholine chloride, which detects the pesticide acephate. The technique they adapted was chronoamperometry and achieved a remarkable limit of detection as 0.007 ppm [8]. To summarize various CPs based biosensors, several review articles have been reported. For example, in 1994, Contractor et al. reviewed CPs based biosensors [8]. Lai et al., in 2016, reported a detailed review about polyaniline-based glucose biosensors [9]. Moon et al. in 2018, published a detailed review about neurotransmitter biosensors based on CP where they summarized biosensors for various kinds of neurotransmitters[10]. Likewise, Aydemir et al. reported a brief review about various techniques in CPs based biosensors [11]. Park et al. have also published a detailed review about advances in CP-based nanobiosensors [12]. Naseri et al. briefly reviewed recent trends in the development of CPs based nanocomposites for electrochemical biosensors [13]. Park et al. reported an overview of the progress in research for biomedical applications using CPs [14].

19.1.4 BIOCATALYSTS OR BIOCOMPONENTS

Substances like enzymes and live cells like bacteria, fungi, tissues, antibodies-antigens, cell organelles, liposomes, aptamers, etc., which are electroactive can be used as biocatalysts. Even though these biocomponents exhibit high selectivity, their activity and stability are dependent on parameters like pH and temperature. Often these biocatalysts are unstable and lose their activity at elevated temperatures and pH. Thus, their incorporation into compatible matrix increases their shelf life. Several underlying matrices like gels, silica, membranes, nanomaterials, carbons, and polymers have been reported. However, particularly CPs have gained much weight due to their unique characteristics. In addition, the geometrical surface area, hydrophilicity, porosity, matrix composition, and most importantly the immobilization technique were chosen to trap them, determining the activity of the biocomponents. In this context, varied approaches have been studied.

For instance, **(a) entrapment**: Gambhiret et al. reported co-immobilization of two different enzymes; urease and glutamate dehydrogenase through electrochemical entrapment in the polypyrrole-polyvinyl sulphonate conducting films deposited over ITO glass as the base electrode. Herein, a measured amount of enzymes were added in the mixed solution of pyrrole and polyvinyl sulphonate. Furthermore, ITO as working electrode, platinum as counter, and saturated calomel as reference electrode were used. CV in the potential window range of −0.8 to 1.4 V was carried out at 50 mV/s. The polymer films got deposited over ITO glass along with entrapped enzymes[15]. **(b) Physical adsorption**: Gokoglanet al. reported a simple method of direct immobilization of glucose oxidase enzyme over poly(4,7-bis(thieno[3,2-b]thiophen-2-yl)benzo[c] [1, 2, 5] selenadiazole) polymer via physical adsorption. The polymer was electrodeposited over gold electrode through electrodeposition. Pyranose oxidase enzyme solution was dropped and cast over the polymer-modified electrode followed by the addition of glutaraldehyde. The modified electrode was air-dried and used for the detection of glucose in beverages through voltammetric and electrical impedance techniques [16]. **(c) Cross-linkage**: Chaubeyet al. reported lactate biosensors with polypyrrole-polyvinylsulphonate composite CP, immobilized with lactate dehydrogenase enzyme through cross-linking with glutaraldehyde [17]. (iv) Covalent bonding: Ramanathan et al. developed a glucose biosensor. Herein, glucose oxidase was immobilized covalently to poly (o-amino benzoic acid). The carboxyl groups of polymers bonded covalently with the enzyme and the amperometric technique were used here for biosensing [18]. Any of these methods can be used for immobilizing biocatalysts to the polymer. Nevertheless, the type of electrodes used also influences the stability of the biocomponent. There is thus a great need to design electrodes that are compatible with the biological component that can lead to rapid electron transfer at the electrode surface. In this context, CPs are attractive as possible materials for such applications. The major research gap observed in all these reported protocols is that they are bulk volume systems wherein large reagent, sample volume, bulky hardware potentiostat are used; hence, they are confined to laboratory spaces and cannot be used for real-time applications. Therefore, no matter how significant these biosensors are, they lack practicality. To overcome this, the incorporation of CP-based electrodes into microfluidic devices has enabled a gateway for real-time applications.

19.1.5 MICROFLUIDIC/MINIATURIZED DEVICES

Microfluidic and miniaturized devices are compact, integrated devices that allow the analysis or detections of various analytes with a minimal sample, reagent volume. Furthermore, these devices can be operated without any skilled training and hence can be used as point-of-care testing (POCT). From biomedical applications to energy devices, clinical and forensic investigations, biomarker and pathogen detection for diagnostic applications, thermal to physical sensors, electrochemical biosensors, chemical reactors to nanoparticles synthesis, and microfluidic devices are being explored. This chapter discusses microfluidic and miniaturized devices fabricated by integration of CP-based electrodes for biosensing applications. As mentioned before, CPs are an excellent matrix for biocomponent immobilization; these biosensors can be used as POCT devices. While designing a microfluidic POCT biosensor, generally, the following salient features are considered:

- **Portability:** The device should be portable and compact like a handheld device. It should not include any bulkier hardware or lab-based equipment.
- **Rapid results:** The results should be obtained with minimal waiting time preferably instant as POCT is crucial to decide the course of action for treatment in critical emergency cases.
- **Ease of operation:** There should be a simplistic, user-friendly procedure to handle and perform the analysis using the device as even a layman without skilled training should be able to operate.
- **Minimal sample/reagent consumption:** A microfluidic device should operate with minimalistic sample and reagent volume with precision.
- **No sample preparation:** A device should give results with the whole sample.
- **Sensitivity:** POCT should be able to detect the analyte of interest with high selectivity, sensitivity, and negligible interference.
- **Robust:** The external and environmental parameters should not disrupt the performance of the device.
- **Cost-effective:** The microfluidic device should be affordable.
- **Integrated calibration:** An inbuilt calibration integrated within the device to mitigate the error.

Based on the above-mentioned attributes and the type of use, the microfluidic POCT biosensing devices can be broadly classified into two categories: single use and multiple uses. Figure 19.2 is the schematic representation of the general classification of these devices. Dipsticks, available commercially for blood and urine test, are examples of single-use quantitative POCT devices. Various strip-based devices with digital output readout are single-use quantitative POCT devices. Multiple-use cassette devices and benchtop devices, like pH meters, dissolved oxygen sensors, etc., are examples of qualitative and quantitative multiple usage POCT devices. All these devices are governed by microfluidics principles and used for

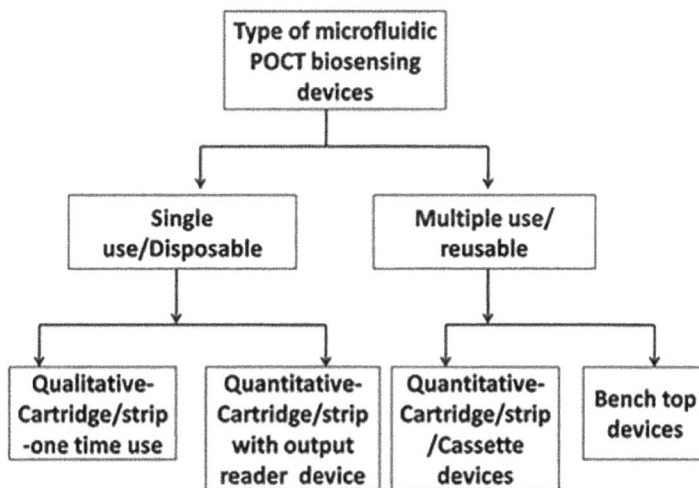

FIGURE 19.2 Schematic representation of the general classification of POCT biosensing devices.

the analysis of biomarkers and disease pathogens in real-time biological samples at clinical laboratories, diagnostic centers, hospitals, etc. wherever primary care is given to the patient.

19.2   FABRICATION OF MICROFLUIDIC DEVICES

There are several methods of fabrication of microfluidic devices. Primarily, the technique adapted depends upon the substrate material used. If the substrate of the micro-device is flexible like paper and polyimide sheet, rigid like glass, then techniques like screen-printing, inkjet printing, laser ablation, laser-cutting, and pencil drawing are used. If the substrate of the micro-device is of conductive filaments like polylactic acid (PLA), wood fiber (cellulose + PLA), polyvinyl alcohol (PVA), acrylonitrile butadiene styrene (ABS), and polyethylene terephthalate (PET), polymers like polydimethylsiloxane (PDMS) and poly(methyl methacrylate) PMMA and techniques like 3D printing, soft lithography, photolithography, lamination, embossing, and molding are used. Figure 19.3 summarizes various techniques used. A brief description of these methods is as follows [19]:

3D printing: Also known as additive manufacturing is a layer-by-layer deposition of desired conductive filament material into a 3D device. The procedure involves designing the device using CAD software, further converting it to a 3D printer compatible format followed by printing of 2D design into a 3D object.

Soft lithography: Herein, liquid polymers like PDMS, polyimides, polyurethanes, etc., are converted into solid elastomers by pouring liquid polymer over a dummy device mold, followed by curing and heating to solidify.

Photolithography: Optical beams like e-beam, X-ray, ion beam, UV beam, etc., are used for cutting and designing the patterns and channels over the substrates.

Molding: This method utilizes high temperatures, heat, and pressure onto various liquid polymers that are poured over dummy master molds made up of materials like silicon followed by solidifying these to form micro-devices.

Laminating: Herein, materials like glass slides, PMMA or acrylic, polycarbonate, etc., are used wherein these are cut independently, stacked, and bonded together as a device using adhesives.

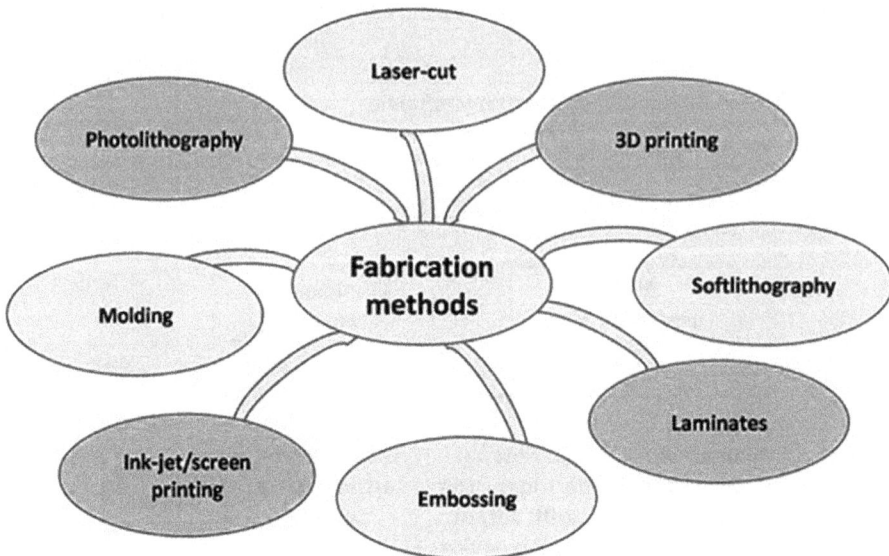

FIGURE 19.3   Schematic representation of various fabrication methods for microfluidic devices.

Screen-printing: Substrates like paper and glass are used here. A conductive ink of choice is used and applied over a mask designed as per the required device or electrode patterns followed by drying, hence leaving behind the desired conductive pattern.

Inkjet printing: A relatively newer technique wherein an inkjet printer with a printing nozzle is employed. Conductive ink with specific viscosity is fed into the printer, and the design of the device or electrodes is made using CAD software and converted to a compatible file. The nozzle sprays the ink and it deposits on the substrate material as per the given design followed by drying in a hot air oven.

Laser-cutting or laser ablation: Numerous substrates, like carbon, silicon, paper, glass, plastic, polyimide, etc., can be exposed to laser ablation for patterning laser-induced graphene as electrodes and microchannels in microfluidic devices. Different types of lasers like CO_2, UV, pulsed, diode, etc., are used to carry out this process.

19.3   CONDUCTING POLYMER-INTEGRATED MICROFLUIDIC BIOSENSORS

Several research groups have developed micro-devices with the integration of these polymer matrix electrodes in microfluidic and miniature devices for biosensor applications. For instance, Huang et al. developed a protein microfluidic sensor using CP-nanowire matrix aptasensor. They carried out IgE protein detection whereby aptamers for IgE were prepared using an integrated DNA approach. A single-step electrochemical method for preparing polypyrrole nanowire incorporated with aptasensors has been reported here. For this, a PMMA-based nanochannel device integrated with gold electrodes was used. A drop of a mixture of pyrrole monomer, sodium chloride, and aptamer as an electrolyte was used between these electrodes. A constant current is supplied to the electrodes that generated the nanowires between the electrodes. These were separated and washed with distilled water and were characterized using fluorescence spectroscopy. Furthermore, these CP-modified aptasensors were integrated into a PDMS-based microfluidic device. PDMS was chosen since it is biocompatible. The micro-device was fabricated using a replica molding process and consisted of six microchannels. Holes were drilled in the PDMS and tubings were connected with the microchannels for supplying the analyte. Change in conductance with respect to change in IgE concentration was measured. The microfluidic aptasensor gave a linear range of IgE detection from 0.01 to 100 nM with an exceptional limit of detection (LOD) as 0.01 nM [20].

Ko et al. reported a polymer-based microfluidic device for the fabrication of biochips for immunosensing application. Herein, PDMS and PMMA polymer-based device was fabricated by laminating the layers. The top layer was made up to PDMS using molding, the bottom was made with PMMA using embossing and then these layers were laminated together. Two inlets with air vent were made in the PDMS layer and two gold electrodes in the bottom layer. Immunosensing of biotin and ferritin was carried out with antiferritin and streptavidin as bioconjugates. The capillary force was used to introduce the sample into the device [21]. Similarly, Park et al. developed a dopamine sensor with a CP nanofiber membrane, using a human dopamine receptor. The membrane had carboxylated poly(3,4-ethylenedioxythiophene) polymer designated as (MCPEDOT) nanofibers. A field-effect transistor (FET) biosensor was realized by first creating a PMMA nanofiber template by electrospinning process, which was collected over a cellulose substrate. Over this template, MCPEDOT nanofibers were developed and human dopamine receptor was covalently attached to this. Glass wafers were exposed to UV for creating the patterns and Cr/Au electrodes were spin-coated. As shown in Figure 19.4, these electrodes post modification with MCPEDOT nanofibers-human dopamine receptor were integrated into a PDMS microchannel[22]. Seo et al. developed a unique microfluidic device that helped in separation as well sensing of neurotransmitters. The device was tested for seven different neurotransmitters, dopamine, epinephrine, 5-hydroxytryptamine, norepinephrine, 3,4-dihydroxy-L-phenylalanine, 5-hydroxyindoleacetic acid, and 5-hydroxytryptophan in blood plasma, showing no interference. A microfluidic channel was developed to separate these analytes via AC field perturbation post which the sensor was attached to the end of the microchannel for detection after separation.

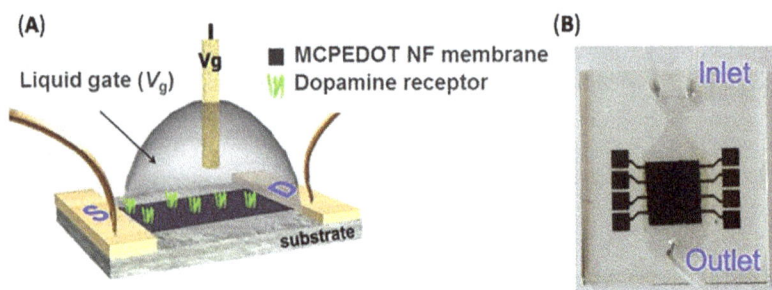

FIGURE 19.4 (A) Schematic representation (B) Real image of the device fabricated by Park et al. [22].

(Adapted with permission from Ref. [22]. Copyright (2016), American chemical society.)

The sample was injected through the inlet and separation of analyte happened in the microchannel of the glass microfluidic device. The separated analytes were exposed to a modified electrode integrated at the outlet of the device. The electrode used here was a carbon screen-printed electrode chemically modified with 2, 2´:5´, 5″-terthiophene-3´-p-benzoic acid CP. This was electropolymerized along with nitrogen and sulfur-doped porous carbon to give a CP composite with the redox mediator. The sensor probe showed remarkable performance toward detection in the linear range 0.05–130 nM with LOD in the range of 0.034–0.044 nM. A human blood plasma sample was used for real sample analysis [23]. Kwon et al. reported a FET sensor using graphene. For fabrication of this sensor, a copper substrate was used over which nitrogen-doped polypyrrole with graphene was polymerized and deposited through the chemical vapor deposition method. Later on, this formed layer was transferred to a flexible substrate. A cancer biomarker, antivascular endothelial growth factor, was the analyte detect using this device. The biocomponent used here was RNA aptamer conjugated to the polymer composite prepared. The sensor had good sensitivity and LOD up to 100 fM. Since the substrate was flexible, it exhibited great durability and bendability [24]. Figure 19.5 is the reprint of their schematic diagram for fabrication and sensing of aptasensor with copyright permission.

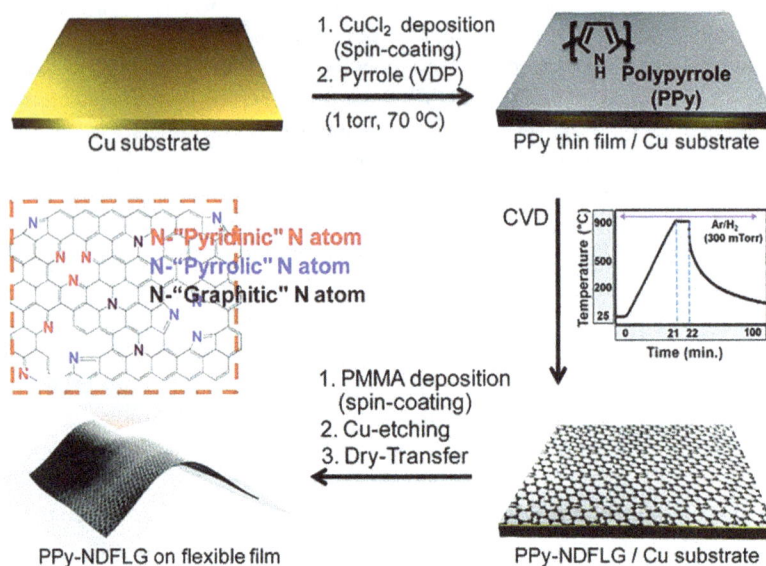

FIGURE 19.5 Schematic representation of the formation of CP on a substrate to give a flexible film.(Adapted with permission from Ref. [24]. Copyright (2012), American chemical society.)

Liao et al. developed a glass microfluidic device as a glucose sensor using glucose oxidase enzyme, graphene, and chitosan polymer. Employing sputtering, Ti and Pt drain and gate electrodes were patterned on the substrate via shadow mask. A 10-nm thick Ti layer served as adhesive for Pt 100-nm thick film. The microchannel was 0.2×6.0 mm in dimension. For sensing, an organic electrochemical transistor modified with chitosan polymer (biocompatible) and graphene and glucose oxidase over the Pt gate electrodes were used. The device gave a linear range of 10nM–1µM with LOD as 10 nM. Furthermore, the device was highly selective for glucose and gave no interference from ascorbic and uric acid [25]. Wang et al. developed a screen-printed electrode modified with a nanocomposite of redox polymer nanobeads and applied for glucose biosensing. CP used here was polyethyleneimine bounded with ferrocene. Furthermore, fabricated CP nanobeads were used to immobilize glucose oxidase enzymes as biocatalysts. To improve the performance of the modified electrode, PEDOT was used to further modify the sensor. They discovered that PEDOT-modified device gave 2.5 times more sensitivity.

Negligible interference from coexisting biochemicals, uric acids, dopamine, and ascorbic acid was observed [26]. Kwon et al. developed HIV immunoassay on a flexible FET microfluidic device, which could be produced on a large scale. CP used here was a carboxyl group functionalized polypyrrole forming a nanohybrid composite with graphene. Chemical vapor deposition was used to grow graphene and it was transferred to the flexible platform. HIV antigen was adhered to the CP-nanohybrid matrix as the biocomponent. These modified nanohybrids were incorporated into the liquid-ion FET system. In further detail, PDMS microchannels were patterned and integrated into the designed FET. This system showed detection of HIV antibody with an appreciable sensitivity, selectivity, and limit of detection as 1 pM. Real sample analysis was carried out with HIV-2 gp36 antibody. Since the nanohybrid film was flexible, it was characterized by examining unbending, bending, and durability [27].

Yang et al designed a unique approach for the preparation of CP-nanofiber-based biosensors for the detection of biomolecules. A platinum microelectrode array was used as the base electrode. Glucose-oxidase-incorporated PEDOT was electrodeposited on the microelectrode array films and nanofibers. Comparative study of both films and nanofiber electrode arrays were performed at the same parameter. Impedance and amperometric studies were carried out in response to the glucose analyte concentration. They discovered that the prepared nanofibers gave more sensitive detection than the fabricated films [28]. Likewise, Mishra et al. reported an ITO-glass plate device modified with ZnS nanocrystals capped with mercaptopropionic acid-polypyrrole nanocomposite film. The ZnS crystals functionalized with carboxyl groups due to mercaptopropionic acid gave an increased surface area for protein antibodies to immobilize. Covalent linkage was formed with the matrix and the biocomponent. The analyte detected here was C-reactive protein by recording impedance measurements with changes in the concentration of the analyte. The synthesized film exhibited high biocompatibility toward protein antibodies. The fabricated immunosensor showed a linear detection range of 10 ng–10 µg/mL [29]. In another report, Zhu et al. developed an organic electrochemical transistor-based glucose biosensor. Herein, functionalized PEDOT was used as the CP matrix embedded with glucose oxidase enzyme. Glucose sensing was carried out in the neutral pH medium [30]. The extension of this work was carried by Macaya et al by exploring the sensor with various concentrations of glucose up to micromolar level. The sensor had platinum gate electrodes with PEDOT:PSS channel. Furthermore, the sensor was tested for real sample analysis of glucose in human saliva [31].

In another remarkable work, Zhu et al. designed gold nanoparticles decorated with polyaniline (PANI) CP and TiO_2 nanotubes for photoelectrochemical sensing on an ITO base electrode. As shown in Figure 19.6, PANI was deposited over TiO_2 nanotube via oxidative polymerization. The CP-modified TiO_2 nanotube was exposed to reducing agent 12-phosphotungstic acid to deposit gold nanoparticles. Furthermore, this matrix was used to entrap lactate dehydrogenase and NAD^+. Lactate was the analyte sensed here; the electrode gave a linear range of 0.5–210 µM with LOD of 0.15 µM [32]. In an impressive work by Tang et al., a PEDOT: PSS-based organic electrochemical transistor

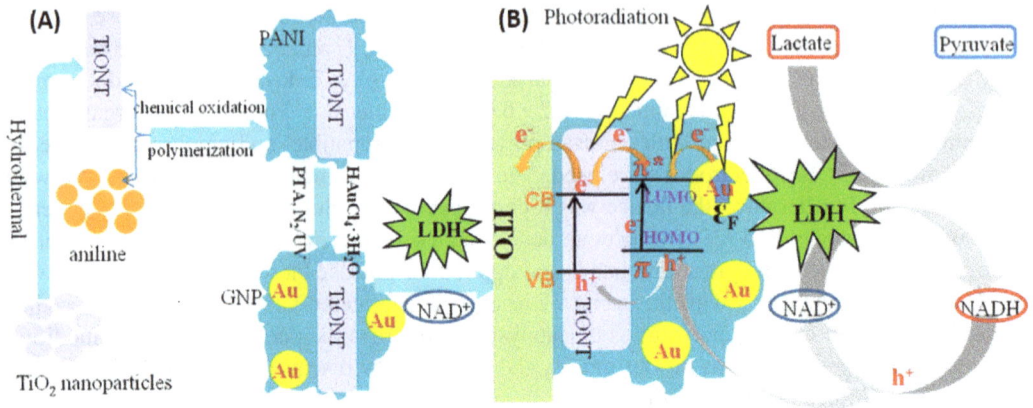

FIGURE 19.6 (A) Schematic representation to prepare a composite. (B) Schematic representation showing the process to detect lactate.(Adapted with permission from Ref. [32]. Copyright (2016), American chemical society.)

sensor for dopamine was designed. Herein, three different types of gate electrodes, gold, platinum, and graphite, were used. The sensitivity of the developed sensor was influenced by the type of electrode and operation voltage. The optimized electrode that gave a better response was platinum at 0.6 V. The LOD obtained was 5nM [33]. In further detail, the same research group improved the sensitivity of this sensor by coating chitosan and Nafion along with graphene over the platinum gate electrode. Post-modification with Nafion, the interference from coexisting biochemicals, i.e., ascorbic acid and uric acid, was negligible. However, the LOD remained the same as 5nM [34].

Similarly, Park et al. reported a FET sensor using polypyrrole nanotubes and reduced graphene oxide composite. Graphene oxide was prepared through the Hummers method from graphite powder. The prepared GO was sonicated with polypyrrole nanotubes and finally treated with hydrazine to form a composite of reduced graphene oxide-polypyrrole nanotubes. Figure 19.7A gives the schematic representation of the composite formation reprinted with copyright permission. The synthesized composite was coated over a microarray FET sensor. Interdigitated 80 gold microelectrodes were designed on a glass substrate using photolithography. As depicted in Figure 19.7B, the dimensions of the electrode were 10 µM wide, 50 nM thick, and 4000 µM in length with interspacing of 10 µM. Liquid-ion gate of FET sensor was developed with phosphate buffer. Their sensor gave selective and sensitive detection of hydrogen peroxide up to 100 pM with no interference from the other biological fluid [35]. Very recently, Tran et al., in 2021, reported CP-based biosensor for virus biomarker detection with potential for application of Covid-19 detection. Herein, various forms of CP-like nanotubes, porous, nanowire, and functionalized, were used to trap antibodies for DNA/RNA, whole virus, antigen, and proteins [36].

19.4 CONCLUSION AND FUTURE OUTLOOK

Microfluidic and miniaturized biosensors have proven to be great potential tools for fast diagnostic and point-of-care device development. They offer extreme selectivity, sensitivity, and portability with rapid analysis. These have been recognized as the most promising approaches in health-care management. Most preferably, the biocatalysts like enzymes, aptamers, antibodies, and protein are used for these biosensors. They are prone to get affected by environmental factors and hence may lose stability and have a short lifetime. Therefore, to protect them, and improve their mechanical, electrical properties, these are incorporated in the polymer matrixes. CPs like polyaniline, polypyrrole, poly(3,4-ethylenedioxythiophene) polyacetylene, polyfluorene, polyphenylene, and polycarbazole have been extensively used so far for the preparation of biosensors. Usually, these polymers are

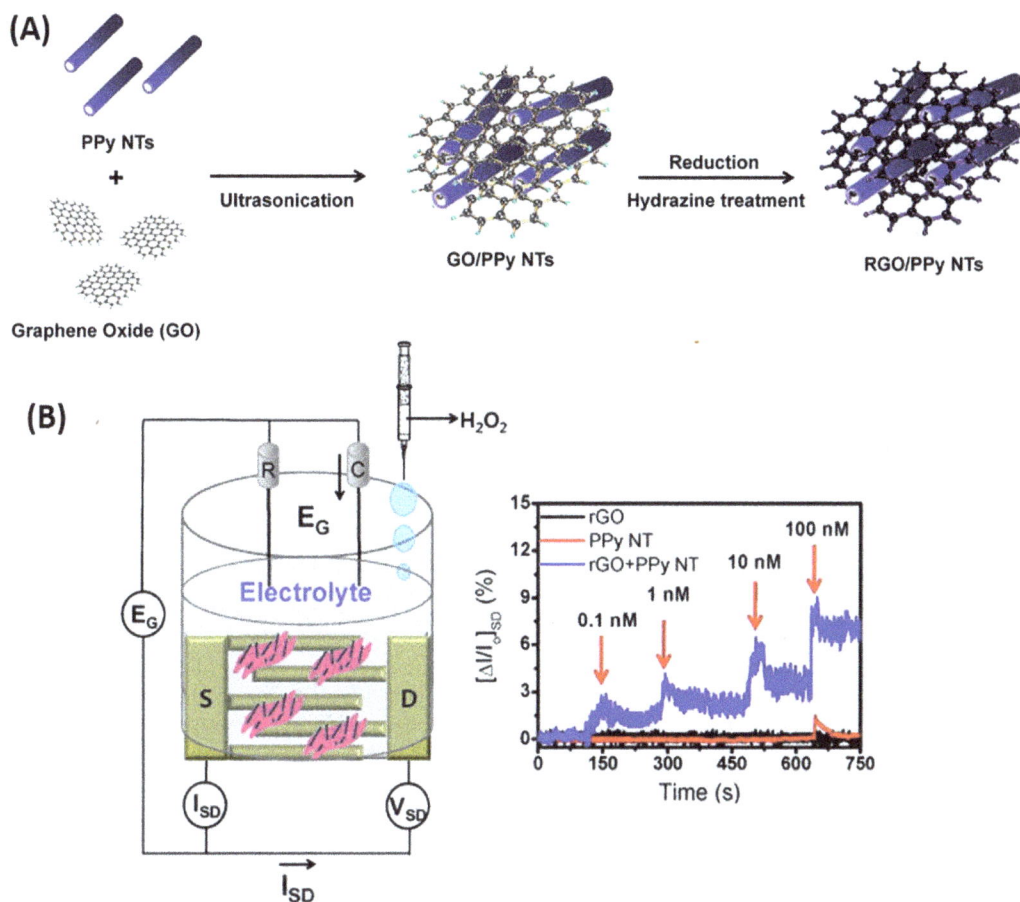

FIGURE 19.7 (A) Schematic representation of the CP-graphene oxide composite formation (B) The experimental setup of the FET sensor and response graph showing a linear increase in the current recorded as a function of increase in concentration in a linear range of 0.1 nM–100 Nm.(Adapted with permission from Ref. [35]. Copyright (2014), American chemical society.)

electrochemically deposited over electrode active surfaces. Different approaches like cross-linking, covalent bonding, entrapment, and physical adsorption are employed for the immobilization of biocomponents over the electrode. Several research groups have reported a significant amount of biosensors in a couple of decades for the detection of various biomarkers, proteins, antigens, biochemicals, etc.; however, there are two major research gaps in terms of practical real-time applications: (a) miniaturization of the bulk systems for point-of-care applications and (b) improvement in the stability of the biosensors. Since the majority of the reported procedures are with bulk volume and lab bound, they are difficult for real-time sensing. Nevertheless, the advent of microfluidics, nanofabrication methods, and nanotechnology has led to extraordinary growth in the fabrication of CP electrodes-based biosensors. The integration of IoT and automation in these systems at present and future has escalated the point-of-care applications. In fact, a normal smartphone can also be converted into a complete POCT electrochemical platform for biosensing.

Despite the significant and remarkable novelties, there is a lot of room for improvement in terms of wireless and wearable sensing for miniaturizing. In further detail, more CPs can be explored for working on the longevity of the biosensor. A combination of CP and nanomaterials can improve the sensitivity, stability, and selectivity of these sensors. Hence, for successful biosensing future applications in microsystems, fabrication of CP, new materials play a key role. Furthermore, the preparation of biocomponents like aptamers can be beneficial for POCT analysis. This area of research is not much explored and is in the early stages. With an emphasis on miniaturization, interdisciplinary approach, more efficient micro-devices for real-time applications can be achieved. Most of the proof-of-concept micro-devices can be developed into real-time commercial products in the future.

REFERENCES

1. H. Shirakawa, E.J. Louis, A.G. MacDiarmid, C.K. Chiang, A.J. Heeger, Synthesis of electrically conducting organic polymers: Halogen derivatives of polyacetylene, (CH)x, Journal of the Chemical Society, Chemical Communications. (1977) 578–580. https://doi.org/10.1039/C39770000578.
2. S. A Chen, Y. C Tsai, Electrochemical polymerization of pyrrole on a fabric, Die Angewandte Makromolekulare Chemie. 169 (1989) 153–157. https://doi.org/10.1002/apmc.1989.051690114.
3. M. Wang, X. Wang, P. Moni, A. Liu, D.H. Kim, W.J. Jo, H. Sojoudi, K.K. Gleason, CVD polymers for devices and device fabrication, Advanced Materials. 29 (2017)1604606. https://doi.org/10.1002/adma.201604606.
4. S. Tamba, K. Fuji, H. Meguro, S. Okamoto, T. Tendo, R. Komobuchi, A. Sugie, T. Nishino, A. Mori, Synthesis of High-molecular-weight head-to-tail-type Poly(3-substituted-thiophene)s by Cross-coupling Polycondensation with [CpNiCl(NHC)] as a catalyst, Chemistry Letters. 42 (2013) 281–283. https://doi.org/10.1246/cl.2013.281.
5. A. Malinauskas, Chemical deposition of conducting polymers, Polymer. 42 (2001) 3957–3972. https://doi.org/10.1016/S0032-3861(00)00800-4.
6. R.S. Loewe, P.C. Ewbank, J. Liu, L. Zhai, R.D. McCullough, Regioregular, head-to-tail coupled poly(3-alkylthiophenes) made easy by the GRIM method: Investigation of the reaction and the origin of regioselectivity, Macromolecules. 34 (2001) 4324–4333. https://doi.org/10.1021/ma001677+.
7. A. Heller, Electrical wiring of redox enzymes, Accounts of Chemical Research. 23 (1990) 128–134. https://doi.org/10.1021/ar00173a002.
8. P.D. Virutkar, A.P. Mahajan, B.H. Meshram, S.B. Kondawar, Conductive polymer nanocomposite enzyme immobilized biosensor for pesticide detection, Journal of Materials NanoScience. 6 (2019) 7–12. http://pubs.thesciencein.org/journal/index.php/jmns/article/view/84.
9. J. Lai, Y. Yi, P. Zhu, J. Shen, K. Wu, L. Zhang, J. Liu, Polyaniline-based glucose biosensor: A review, Journal of Electroanalytical Chemistry. 782 (2016) 138–153. https://doi.org/10.1016/j.jelechem.2016.10.033.
10. J.M. Moon, N. Thapliyal, K.K. Hussain, R.N. Goyal, Y.B. Shim, Conducting polymer-based electrochemical biosensors for neurotransmitters: A review, Biosensors and Bioelectronics. 102 (2018) 540–552. https://doi.org/10.1016/j.bios.2017.11.069.
11. N. Aydemir, J. Malmström, J. Travas-Sejdic, Conducting polymer based electrochemical biosensors, Physical Chemistry Chemical Physics. 18 (2016) 8264–8277. https://doi.org/10.1039/c5cp06830d.
12. C.S. Park, C. Lee, O.S. Kwon, Conducting polymer based nanobiosensors, Polymers. 8 (2016) 1–18. https://doi.org/10.3390/polym8070249.
13. M. Naseri, L. Fotouhi, A. Ehsani, Recent progress in the development of conducting polymer-based nanocomposites for electrochemical biosensors applications: A mini-review, Chemical Record. 18 (2018) 599–618. https://doi.org/10.1002/tcr.201700101.
14. Y. Park, J. Jung, M. Chang, Research progress on conducting polymer-based biomedical applications, Applied Sciences (Switzerland). 9 (2019) 1070. https://doi.org/10.3390/app9061070.
15. A. Gambhir, M. Gerard, A.K. Mulchandani, B.D. Malhotra, Coimmobilization of urease and glutamate dehydrogenase in electrochemically prepared polypyrrole-polyvinyl sulfonate films, Applied Biochemistry and Biotechnology – Part A Enzyme Engineering and Biotechnology. 96 (2001) 249–257. https://doi.org/10.1385/ABAB:96:1-3:249.
16. T.C. Gokoglan, S. Soylemez, M. Kesik, S. Toksabay, L. Toppare, Selenium containing conducting polymer based pyranose oxidase biosensor for glucose detection, Food Chemistry. 172 (2015) 219–224. https://doi.org/10.1016/j.foodchem.2014.09.065.
17. A. Chaubey, M. Gerard, R. Singhal, V.S. Singh, B.D. Malhotra, Immobilization of lactate dehydrogenase on electrochemically prepared polypyrrole-polyvinylsulphonate composite films for application to lactate biosensors, Electrochimica Acta. 46 (2001) 723–729. https://doi.org/10.1016/s0013-4686(00)00658-7.

18. K. Ramanathan, S.S. Pandey, R. Kumar, A. Gulati, A. Surya, N. Murthy, B.D. Malhotra, Covalent immobilization of glucose oxidase to poly(o-amino benzoic acid) for application to glucose biosensor, Journal of Applied Polymer Science. 78 (2000) 662–667. https://doi.org/10.1002/1097-4628(20001017)78:3<662::AID-APP220>3.0.CO;2-T.

19. K. Amreen, S. Goel, Review—Miniaturized and Microfluidic Devices for Automated Nanoparticle Synthesis, ECS Journal of Solid State Science and Technology. 10 (2021) 17002. https://doi.org/10.1149/2162-8777/abdb19.

20. J. Huang, X. Luo, I. Lee, Y. Hu, X.T. Cui, M. Yun, Rapid real-time electrical detection of proteins using single conducting polymer nanowire-based microfluidic aptasensor, Biosensors and Bioelectronics. 30 (2011) 306–309. https://doi.org/10.1016/j.bios.2011.08.016.

21. J.S. Ko, H.C. Yoon, H. Yang, H.B. Pyo, K.H. Chung, S.J. Kim, Y.T. Kim, A polymer-based microfluidic device for immunosensing biochips, Lab on a Chip. 3 (2003) 106–113. https://doi.org/10.1039/b301794j.

22. S.J. Park, S.H. Lee, H. Yang, C.S. Park, C.S. Lee, O.S. Kwon, T.H. Park, J. Jang, Human dopamine receptor-conjugated multidimensional conducting polymer nanofiber membrane for dopamine detection, ACS Applied Materials and Interfaces. 8 (2016) 28897–28903. https://doi.org/10.1021/acsami.6b10437.

23. K.D. Seo, M.M. Hossain, N.G. Gurudatt, C.S. Choi, M.J.A. Shiddiky, D.S. Park, Y.B. Shim, Microfluidic neurotransmitters sensor in blood plasma with mediator-immobilized conducting polymer/N, S-doped porous carbon composite, Sensors and Actuators, B: Chemical. 313 (2020) 128017. https://doi.org/10.1016/j.snb.2020.128017.

24. O.S. Kwon, S.J. Park, J.Y. Hong, A.R. Han, J.S. Lee, J.S. Lee, J.H. Oh, J. Jang, Flexible FET-Type VEGF aptasensor based on nitrogen-doped graphene converted from conducting polymer, ACS Nano. 6 (2012) 1486–1493. https://doi.org/10.1021/nn204395n.

25. C. Liao, M. Zhang, L. Niu, Z. Zheng, F. Yan, Highly selective and sensitive glucose sensors based on organic electrochemical transistors with graphene-modified gate electrodes, Journal of Materials Chemistry B. 1 (2013) 3820–3829. https://doi.org/10.1039/c3tb20451k.

26. J.Y. Wang, L.C. Chen, K.C. Ho, Synthesis of redox polymer nanobeads and nanocomposites for glucose biosensors, ACS Applied Materials and Interfaces. 5 (2013) 7852–7861. https://doi.org/10.1021/am4018219.

27. O.S. Kwon, S.H. Lee, S.J. Park, J.H. An, H.S. Song, T. Kim, J.H. Oh, J. Bae, H. Yoon, T.H. Park, J. Jang, Large-scale graphene micropattern nano-biohybrids: High-performance transducers for FET-type flexible fluidic HIV immunoassays, Advanced Materials. 25 (2013) 4177–4185. https://doi.org/10.1002/adma.201301523.

28. G. Yang, K.L. Kampstra, M.R. Abidian, High-performance conducting polymer nanofiber biosensors for detection of biomolecules, Advanced Materials. 26 (2014) 4954–4960. https://doi.org/10.1002/adma.201400753.

29. S.K. Mishra, R. Pasricha, A.M. Biradar, Rajesh, ZnS-nanocrystals/polypyrrole nanocomposite film based immunosensor, Applied Physics Letters. 100 (2012) 053701. https://doi.org/10.1063/1.3681580.

30. Z. Zhu, J.T. Mabeck, C. Zhu, N.C. Cady, A. Batt, G.G. Malliaras, A simple PEDOT:PSS transistor for glucose sensing at neutral pH, Materials Science. (2004) 1556–1557.

31. D.J. Macaya, M. Nikolou, S. Takamatsu, J.T. Mabeck, R.M. Owens, G.G. Malliaras, Simple glucose sensors with micromolar sensitivity based on organic electrochemical transistors, Sensors and Actuators, B: Chemical. 123 (2007) 374–378. https://doi.org/10.1016/j.snb.2006.08.038.

32. J. Zhu, X. Huo, X. Liu, H. Ju, Gold nanoparticles deposited polyaniline-TiO2 nanotube for surface plasmon resonance enhanced photoelectrochemical biosensing, ACS Applied Materials and Interfaces. 8 (2016) 341–349. https://doi.org/10.1021/acsami.5b08837.

33. H. Tang, P. Lin, H.L.W. Chan, F. Yan, Highly sensitive dopamine biosensors based on organic electrochemical transistors, Biosensors and Bioelectronics. 26 (2011) 4559–4563. https://doi.org/10.1016/j.bios.2011.05.025.

34. C. Liao, M. Zhang, L. Niu, Z. Zheng, F. Yan, Organic electrochemical transistors with graphene-modified gate electrodes for highly sensitive and selective dopamine sensors, Journal of Materials Chemistry B. 2 (2014) 191–200. https://doi.org/10.1039/c3tb21079k.

35. J.W. Park, S.J. Park, O.S. Kwon, C. Lee, J. Jang, Polypyrrole nanotube embedded reduced graphene oxide transducer for field-effect transistor-type H2O2 biosensor, Analytical Chemistry. 86 (2014) 1822–1828. https://doi.org/10.1021/ac403770x.

36. V. V. Tran, N. H. T. Tran, H. S. Hwang, M. Chang, Development strategies of conducting polymer-based electrochemical biosensors for virus biomarkers: Potential for rapid COVID-19 detection, Biosensors and Bioelectronics. 182 (2021)113192. https://doi.org/10.1016/j.bios.2021.

20 Advantages and Challenges of Biodegradable Electronic Devices

Maimoona Ilyas,[1] Muntaha Ilyas,[2,3] Farooq Sher,[4] Umer Liaqat,[2] Eder C. Lima,[5] Ayesha Zafar,[3,6] Jasmina Sulejmanović,[7] and Mika Sillanpää[8]

[1]Sustainable Development Study Centre, Government College University, Lahore, Pakistan

[2]Department of Zoology, Wildlife and Fisheries, University of Agriculture Faisalabad, Faisalabad, Pakistan

[3]International Society of Engineering Science and Technology, Nottingham, United Kingdom

[4]Department of Engineering, School of Science and Technology, Nottingham Trent University, Nottingham, United Kingdom

[5]Institute of Chemistry, Federal University of Rio Grande do Sul (UFRGS), Porto Alegre, RS, Brazil

[6]Institute of Biochemistry and Biotechnology, Faculty of Biosciences, University of Veterinary and Animal Sciences, Lahore, Pakistan

[7]Department of Chemistry, Faculty of Science, University of Sarajevo, Sarajevo, Bosnia and Herzegovina

[8]Chemistry Department, College of Science, King Saud University, Riyadh, Saudi Arabia

CONTENTS

DOI: 10.1201/9781003205418-20

20.1 INTRODUCTION

Nowadays, the ubiquitous things in modern society are electronic devices and have become dominant in every aspect of human affairs. Although electronic gadgets have provided significant ways of convenience and value, there is an insatiable need for newer and more alluring technologies [1]. The most globally accepted technologies nowadays, the vastly rising demand to utilize electronics, can produce waste electronic and electronic equipment (WEEE) or e-waste in unsustainable quantities. The WEEE consists of obsolete equipment (i.e., DVD players, TV cast-off cathode rays, and VHS tapes) applications used for household purposes, and devices are in personal use, i.e., tablets and smartphones. The dangerous elements present in e-wastes comprise halogenated flame retardants, Hg and Pb, these induce environmental issues. Among the various ecological issues rising due to electronic devices is e-waste management [2]. The two-third of the world population comprising 67 countries deals with WEEE with the application of rules and laws [3].

Nowadays, only 20% of WEEEs are reprocessed; the main reason behind this unsatisfactory percentage is due to lack of e-waste collecting facilities and due to security concerns of important data [4]. However, various robots exist that disassemble the smartphone in no minutes, while in different countries, manual dismantling and organizing are still present in inferior working circumstances [5]. Low-cost recycling (informal recycling) has an unpleasant and unfavorable impact on both the environment and workers' health [3, 5, 6]. The basic agreement was signed by 186 nations and different other conventions are also present at the regional level to standardize the trade of electronic wastes, but, the illegal dealing still occurs through trans-boundaries [5]. Approximately, 60 elements of the periodic table are present in e-devices. The e-waste has different intentional substances like indium (In) and various other precious metals like gold (Au) [7, 8]. The continuous demand for these elements to make e-devices can induce a shortage in means of minerals and agreement of conditions in the mining process that highlight the output over protection. For example, it was thought that the gallium and indium (used in GaAs and GaN, and indium tin oxides, compounds) [9] shortage can be an unavoidable threat for the upcoming 20 years [10].

The geographical encounters can occur due to the shortage of these minerals means: the elements used in mobile phones (includes Ta [tantalum], Au [gold], Sn [tin], and W [tungsten]) are already considered as "conflict elements" [2]. The chemical society of Europe made a form of the periodic table to inform the public of the insufficiency of these elements used for the manufacturing of electronic and electrical tools [2, 6]. Thus, a substitute for these conventional devices must be required. Intrinsically, this growing green electronics class is not having the ability to decrease the rising electronic waste problems, but also justifies niche applications that cooperating with the body of humans. To tackle the hazardous impacts induced due to uncontrollable electronic wastes, the currently emerged biodegradable electronic devices must be considered as the viable solution. Therefore, this study provides a solution to the toxic impacts of e-devices in the form of biodegradable electronic devices [11].

20.1.1 NEED FOR A CIRCULAR VISION IN THE ELECTRONICS SECTOR

Considering the problems discussed previous, the current new circular approaches and tactics for the manufacturing, use, and discarding of electrical and electronic tools are wanted, as appropriate to linear ones centered on the paradigm "Take, Make and Dispose", which is effective-able for the plant with limited resources and options of land waste management [4]. Alterations in a vision for the applications of electronics need collaborations between miners, manufacturers, traders, designers, investors, producers of raw materials, policymakers, and consumers [2]. The initiation will be taking to promote this green vision in the sector of electronics is as follows: encouraging closed-loop systems built on organized assemblage; developing products with stability, reparability, and renewal/reuse. Furthermore, basic initiatives to build green electronics must need more investigations to build the most suitable products, possibilities include, such as constructing and designing e-devices that have compostable and biodegradable parts [4].

20.2 IMPORTANCE OF BIODEGRADABLE ELECTRONIC DEVICES

20.2.1 RATE OF BIODEGRADATION

The electronics devices are called biodegradable devices which comprise the materials that can be partially and completely degraded by the actions of microorganisms (such as algae, fungi, and bacteria) in the ecosystem [12–15]. The biological procedure intricate in the biodegradation process is influenced by various symbiotic physiochemical aspects, such as the presence of water and oxygen, suitable temperature, optimum pH, and molecular and supramolecular assemblage of the materials [16]. To elude misperception, it is suggested to use "biodegradable" once the application of organic electronics infers an end-of-life situation that involves microorganisms debasing the e-devices in an atmosphere other than a body [17]. The process of biodegradation starts when microorganisms deal with organic polymers, which includes the changing in arrangements of the polymer chains due to high temperature relatively 58°C and typically 50% level of water in optimal composting situations. These 1st variations assisted afterwards the process of polymers fragmentations and polymers breakdown due to enzymatic actions [17].

The compounds with higher molecular weight and unsalable in water are biodegraded by the action of bacterial catalysts and enzymes present in extracellular regions [16, 17]. The enzymes impart in the polymers cleavage process are water-soluble, reduced the polymers into monomers and oligomers and these enzymes transmit inside the cells of microbes that exist in environ of the material. Then respiration of these subunits of organic molecules happens, i.e., utilized as carbon and energy sources that ultimately lead to CO_2 production. It means less than the phenomenon of respiration happens due to microbes, comprises a variety of approaches that are established by the microorganisms by the process of evolution over billion years. Each approach is precise for the environmental redox potential where these microorganisms are raised. The ultimate alteration process of organic molecules is mineralization, in ecosystem includes landfill (anaerobic process) and compost (aerobic process). These microbial activates resulted in the change of organic molecules into components like CO_2 (compost), CH_4 (methane, landfill), H_2O water, sulfite, sulfide and sulfate, phosphate, nitrates, nitrites, chlorides, fluorides, and phosphides [18, 19].

Landfilling of municipal organic wastes (consisting of yard and food wastes) is occurred or managed underneath the anaerobic situations, which induced negative effect on the environment because it imparts in the emissions of greenhouse gases (GHGs). More accurately in the setting of the landfill, biodegradation occurs anaerobically and yielded CH_4 that harms global warming than that of CO_2 [20–22]. Therefore, the landfills are electronic devices, hence is used in the above describe atmosphere and causes health safety issues. The main misunderstanding about biodegradation is that those polymers consist of bio-elements that can be biologically degraded, while polymers of petroleum cannot. Hence, the petroleum-based polybutylene adipate-*co*-terephthalate and poly-e-caprolactone are polymers that are biologically degraded [23, 24]. Alternatively, the polymers include nylon 11, terephthalate, and polyethylene which are not biologically degraded; therefore, these polymers may be produced from natural bio-based components (bio-components) [25].

20.2.2 WASTE MANAGEMENT THROUGH COMPOSTING

The municipal solid wastes (MSWs) which are also called "landfills" have different numbers of organic and inorganic components. Nowadays, MSW is managed by two designed services, i.e.,

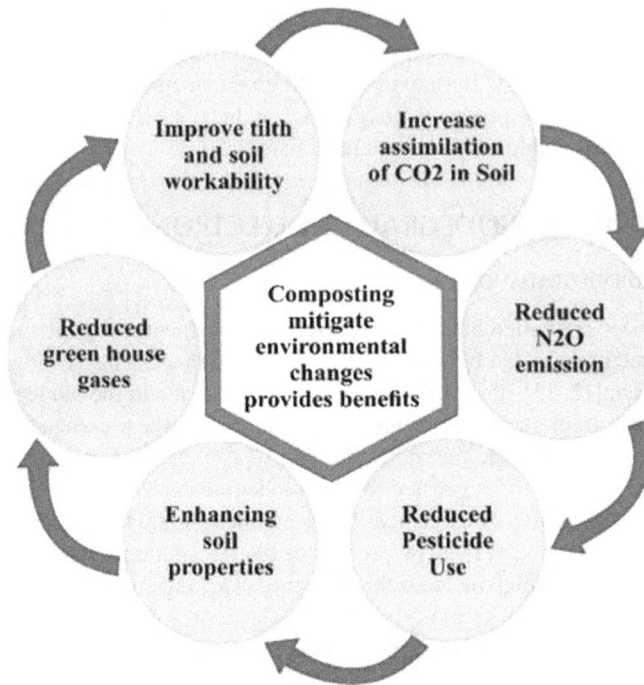

FIGURE 20.1 Composting as the mitigation measure for the environmental impacts.

through industrial composting and anaerobic digestion [20]. These describe the process breakdown of the heterogeneous and complex mixture of municipal organic waste biomass, through the enzymatic chains produced by microorganisms. These enzymes partly convert them into nitrogen, carbon, and other components (like sulfur, potassium, and phosphorous, etc.). Two basic variations are present among the industrial composting and anaerobic digestion, which means the industrial composting occurs in the existences of oxygen, while the anaerobic process occurs without oxygen; in the earlier organic waste changed into methane, which is done by an electrical process [20], while subsequently the organic wastes changed to mature compost, like that of the horizon (top of soil). At the end of the reaction (complete decomposition), the remained material can be used as fertilizer for agriculture or home gardening, then these organic materials return to natural cycles [26, 27]. It is priceless that PLA (polylactic acid), considered as the most significant polymers which are compostable by the industrial method are available in the market, but it is not biologically degraded in marine and soil ecosystems [20]. Composting comprises a significant way to overcome climate variations through its numerous benefits, shown in Figure 20.1:

- Unnatural growth in GHG removal by enhancing the ability of soil to assemble carbon (i.e., soil act as sequestering agent).
- The inclusion of organic nutrients and matter in soil increased its quality (i.e., increased properties of soil, which allowed retention of moisture content and prevent erosion, in shot changing fertilizer chemicals and decreased the release of GHGs associate with their creation).
- Decreased the use of pesticides related to GHG emissions by reducing them.
- Enhanced tillage and viability of soil by the less use of machinery and decreasing emissions.
- Changing the peat use.
- Decreasing N_2O removal by eluding N_2O production from nitrogenous fertilizers.
- Retain the concentration of CO_2 at its basic level and returning the concentrations to their ideal level by improving the soil with compost of rich carbon [28].

The multiple investors demand composting for "Cradle to grave" [29]; production to confirm the making of effective compostable devices, which are disposed of easily by the consumers; and industrial services to compost e-devices. Moreover, incorporating biodegradable organic electronic devices in recyclable consumer items meant for disassembling e-devices in such a way would justify that organic e-waste resources are the management alternatives, including municipalities and governments to assure effective functioning of management plants as well as consumer instructions. The e-products with dissembling ability have been markedly successful in Europe used in vehicles, enhancing reuse of materials by 95% [30].

Composting considers an ascendable solution to tackle organic waste as it can enhance disposal, management, and collection of food and managed them as organic waste by municipalities throughout the world [31, 32]. That's way, we recommended manufacturing the biodegradable organic e-devices having organic components that compost easily (experimentally and effectively certified) can changed electronic life and form landfills and elude the extensive organizing necessary for adequate reusable, diminish toxic chemicals discharge in the ecosystem, and used to handle waste in an eco-friendly beneficial manner, that is ultimately decreased the pressure on the cities that are devoted to the planning of zero waste [2].

20.2.3 TOXICITY ASSESSMENT

About organic e-devices which are biodegradable analyzing the compost for the toxicity related to biodegradation (at the last part of a composting experiment) is a necessary step to find out the ecosystem safety of electronically made biodegraded material [20]. To avoid possible toxic catalyzes of biodegradation to be presented into the soil environment, an ecological test of toxicity having particular biomarkers is required to check the risk related to these catalysts (intermediates) [33, 34]. These analyses check the bioavailability of possibly toxic materials manufactured during the process of electronic substances biodegradations. Moreover, it is convincible, utilizing biomarkers, to evaluate whether these biodegradable substances are gathered by living things. The water toxicity level and dissolved material were tested by the nonlabor and microtox process in the *Vibrio fischeri* a bacterium [35]. In an alternative analysis, the *Lumbriculus variegatus* (earthworm) is used to check the process of bioaccumulation inlining ones in the soil environment [36].

20.3 PREDICTING THE BIODEGRADABILITY OF MOLECULES AND POLYMERS

The biodegradability of minor organic substances was predicted by developing a relationship among reactivity relationships with quantitative structures [37]. The extent and degree of splitting showed by the substances as well as their heteroatom members of substructures (like position, number, and type) influenced the process of biodegradation [38]. Molecular abilities that increased biodegradation process aerobically have been recognized as the class of vulnerable enzymes for degradation, mainly esters, and under unexpected circumstances, they comprise amides also, anomeric chains of liner alkyl (mainly carbon atoms ≥ four), and rings of phenyl compounds. The large series of one or more subunits make the polymers chains. These polymer chains can fold many times, making either shapeless crystalline regions stacked to each other. As a concern, a supreme ability for biodegradation is the availability of the chains of polymers to water tolerated extracellular enzymes that changed the polymers by its breakdown ability into subunits, which are sufficiently small to be gathered by the cell of microorganisms [39].

Mainly, the ability of polymers that affect biodegradability is water-loving in nature (hydrophilic); crystallinity; the backbone of polymer with the linkage of chemical substances; unresolved groups; with their chemical reactivity and position; last-groups; and the reaction ability [16]. Furthermore, chemical bonds that consist of multi-atoms are appropriate for degradation by the use of enzymes. Due to enzymes specificity which depends on arrangements (stereochemical) choice, the spatial arrangement of the monomer subunits along with a chain of polymers influences the rate of biodegradation. The chains of polymers can show collaborative and opposed impact with acclaim to the biodegradation of each part of the chains, with the morphology of the two stages of polymer mixture playing a basis role [20].

20.3.1 Biodegradation and Compost Standards

The principles of present biodegradability are issued by the administrations such as the American Society for Testing Materials (ASTM) International, the International Standardization Organization (ISO), and the Committee of European Standardization (CEN). These organizations give certificates after performing analyses on products to label them as "compostable" and "biodegradable" under particular situations such as freshwater, marine, and terrestrial environment. For persons who used the applicable method to test the process of industrial composting, they set parameters to quantify and studied biodegradation based on respiration activity of microbes (CO_2 and O_2 production), reduction of trial materials, recorded the reacting substances, and variations in the properties of testing substances [16]. In the United States, for example, ASTM-D5338 is the standard that describes "the trial procedure and specifies the results that essentially getting to label the biodegradable substances which biodegraded in composting situations, incorporating thermophilic temperatures." Within 180 days, 90% of at least the carbon amount of the trial substances has to be respired into CO_2 by the microbes through composting [40, 41]. Nonetheless, its value is less than during the process of composting, while some amount of carbon from test substances is combined by the microbes of the compost to form biomass, which changed into humus-like substances after the 180 days of incubation, i.e., complete compost. That's way, the gathered carbon which is qualifying and quantifying neglects standards [15].

The effort of quantifying the bonding of carbon into biomass has been presented in the research literature [41] or, sometimes the problem is not provided [20]. The ability of a material to degrade under the composting situation is considered as a requirement within the three necessities to label the substances/polymers as "compostable". The other process involved within these three is disintegration and safety of soil, as stated in ASTM D-6400 [15]. The process of disintegration was done after 12–24 weeks of test performed for biodegradability in composting situations, the 10% of less than 10 products dry weight left afterword sieving, through 2-mm strainer [15]. The terrestrial safety comprising both the calculations of the amounts of metals that exist in the compost is recorded by the eco-toxicity test using the methods of OECD-208 [42]. These standards were used in phytotoxicity evaluation through the growth and seedling of the plant. In the biodegradability experiment in a composting situation at the endpoint, the lowest of two types of seeds are germinating in a sample of compost to assess the effect of the residual test substances on the development and germination of the plant [10, 42].

20.4 BIODEGRADABLE MATERIALS AND DEVICES

20.4.1 Biodegradable Substrates

In respect to novel technologies known as green electronics, the materials must show the properties of mechanical, physical, and chemical combinations which are suitable for the required application but must be comprising biodegradable elements. Table 20.1 shows different organic substrates and gate dielectrics used in devices.

20.4.2 Biodegradable Paper Substrate

At the last, natural fiber-based paper substrate and different other synthetic polymers PLA, polydimethylsiloxane (PDMS), and polyvinyl alcohol (PVA) are all explored and identified [10]. The natural substrates, the identified substances like paper are taken as a natural substrate used in

TABLE 20.1
Different Organic Substrates, Gate Dielectric [43]

Biodegradable Materials	Functional Layer
PLGA	Substrate
TascPLA	Substrate
CNF	Substrate
Cellulose paper	Substrate
Cellulose	Substrate
Starch paper	Substrate
PPC	Substrate
Shellac resin	Substrate
Shellac resin	Substrate
Silk fibroin	Gate dielectric
Silk fibroin	Gate dielectric
Cellulose	Gate dielectric
Silk fibroin	Gate dielectric
Chicken feather keratin	Gate dielectric
Silk fibroin	Gate dielectric
Human hair keratin	Gate dielectric
WCNs	Gate dielectric
Y_2O_3 nanopartical/chitosan	Gate dielectric
Chicken albumen	Gate dielectric
PMTA	Gate dielectric
PVA	Gate dielectric
Gelatine	Gate dielectric
PVP	Gate dielectric
PDI	Semiconductor
Vat Yellow 1	Semiconductor
β-Carotene	Semiconductor
Vat Orange 3	Semiconductor
Indigo	Semiconductor

printing electronics and showed revolution in printing procedures [44]. The large uneven surface commercial paper which comes from porous formations of fiber network can be beneficial to make devices used for storage of energy, which required greater surface area for the absorption of important electrolytes and combines with nanoparticles. Furthermore, the assimilation of conductive substances (i.e., conductive polymers, graphene, metal oxides, silver nanowire, and carbon nanotube (CNT) among the paper results in efficient electrical conduction for the following hybrid substances, used in supercapacitors as substrate [45].

Undeniably, some devices are manufacture on these altered conducting substrate papers have executed relatively with analogous devices placed on flat directing substrates of the polymer [44, 46], although the high roughness and porosity of the paper can be harmful to devices made of thin films, specifically those using organic materials as dielectric and semiconductor layers. This issue is mainly solved by coating with kaolin, latex, starch, and wax or polymers, including polyurethane, PVA, polypropylene, and polyethylene. Until now, this paper is used in different e-devices like solar cells, organic LEDs, sensors, displays, radio capacitors, transistors, and bio-batteries [47]. To maximize the range of paper-based electrical things and their application [48], a simple procedure is provided to make the translucent paper substrate, directly from a thin slice of wood (400 um/mm) cut in the direction perpendicular to growth.

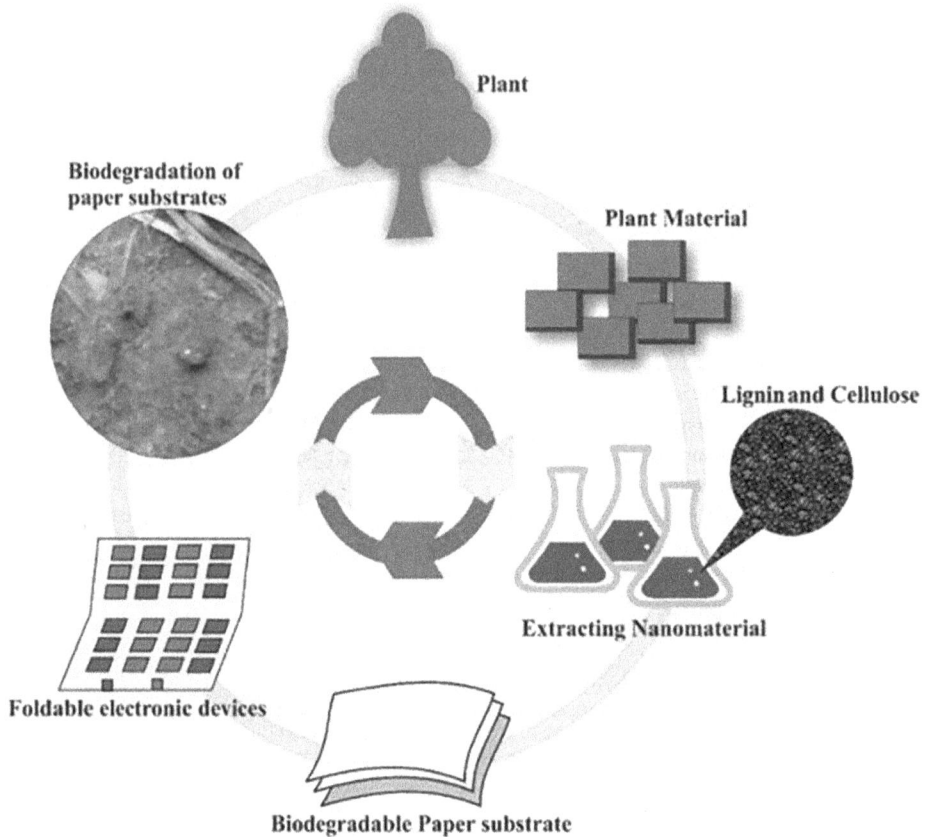

FIGURE 20.2 Mechanism of organic paper substrate electronic devices production and its biodegradation.

Then lignin is removed, the remaining loose collection of the lumen (the nanofibers and micro-fibers made up of tubular structure) which was directed to parallel aligned development was pushed along the axis of the lumen, which causes a breakdown of the lumen in a different direc-tion. The remaining material was consist of an array of isotopic fibers and had a thick layered structure with a low tendency to reflecting light; a transmission of about 90% was recorded in the remaining "isotric paper". A simple device made by combining a flake of graphene with two electrodes of gold, placed through a mask of shadow, presented ohmic contact; however, the stater of electrode gate and electrolyte addition showed a response of bipolar transistor. In soil, this paper is easily degraded. The natural fiber base other substrate is chitin nanofiber (poly-β-(1,4)-*N*-acetyl-*d*-glucosamine) and utilized in organic light emitting diodes (OLED) devices as substrate. The fibers of silk and its films are also considered as appropriate natural materials for the engineering of electronics looks like paper substrate, which showed biodegradation and bio-compatibility rates that can be converted from minutes to days via suitable alteration [49] shown in Figure 20.2.

20.4.3 BIODEGRADABLE SILK SUBSTRATE

Silks are freely accessible from large sources which are costless and processed easily. These features in combined stimulate more exploration of these substances for possible application to e-devices in the biomedical manufacturers (i.e., sensing application of food [50], the perfection of physiological demography and directed drug supply, and implanting) [51, 52]. Similarly, fibroin, a protein present in silk is insoluble, produced by different insects of various genera, provided numerous features that can be utilized in the production of substrates consisting of paper substances that are used to make green electronics; these abilities have outstanding biocompatibility, in vivo, less degeneration, easily changed chemically, and having the ability from either organic and aqueous mixtures [53]. Currently, Zhao et al. [54] proposed that the fibroin films look like the paper used as a material to make memory devices. The phosphate-buffered saline solution can be used to degrade these films within 24 hours at pH 7.4.

The devices made up of fibroin showed outstanding flexibility and converting abilities, and the status of resistance high and low does not change in these fibroin-based devices because bending and mechanical stresses do not affect the described characteristics [11, 55–57]. Various natural substances that are commonly present and offer possible application as biodegradable and biocompatible substances comprise collagen, hard gelatin, chitosan, dextran, alginate and shellac, and many others between them. For example, the female lac beetles secrete "shellac" from the tunnel-like tubes as they feed tree branches and pass through it, the place where this chemical (shellac) was collected [58]. After refining, the films of shellac are mainly smooth and are highly used in e-devices. The similar properties of hard gelatin are that of shellac and having smooth surface with biocompatible films. However, the hard gelatin and shellac used as the substrate for e-devices are still in their present stage [59].

Hence these nature-based materials have the benefits of abundance in availability, low price, low immunoreaction, and most importantly its biodegradable, while some limitations still exist for their wide performance and application of this material. For example, efficient biodegradability showed comparatively low stability in the ecosystem. Therefore, for the fabrication of the device, various processes and conditions are required, such as harsh solvent and high temperature, which may not be suitable as substrates in e-devices. Furthermore, flexible degradation rates of various substrates cannot fulfill the requirements of various long-term devices. The defined drawbacks effectively reduced their use in e-devices and its improvement. Synthetic substrates: the different types of synthetic polymers, used as the biomedical implant can also be used to manufacture film-like substrate for current green technologies [60], for example PDMS is used in national heart, blood, and lung organization after its improvement, as a stretchable and flexible device with an outstanding substrate [61–63].

The easily altered chemical structure of the polymers enables us to better engineering [64]. Therefore, different synthetic polymer-based devices have been manufactured and fixed into tissues of living things to check their functions in therapy and diagnosis of diseases [65]. The different biocompatible and biodegradable artificial polymers comprising poly 3-hydroxyactanoate, poly(*l*-lactide), poly(4-hydroxybutyrate), poly(lactideco-e-caprolactone), poly(3-hydroxybutrate), poly(glycolide-*co*-caprolactonel), poly(glycerol-sebacate), poly(1,8-octanediol-*co*-citrate), poly (glycolic acid) and polycaprolactone (PCL) have been discovered in similar devices. The next sections will go through the specifics of electrical devices made from these biodegradable substrates [11].

20.4.4 BIODEGRADABLE ELECTRODES OR CONTACT MATERIALS

The high conductivity of metals and alloys enable them to be used as electrodes and become the first choice in e-devices [13, 57, 58, 61, 63, 66]. It's a common concept that these substances are tough to degrade and can injure the body of a human when the devices made of these materials are utilized in the field of medicine. Nonetheless, the electrodes of metals used in electronics are made of ultrathin films with the lowest damage ability. Moreover, these metals, including Cu (copper), Pt (platinum), Au (gold), and Ag (silver), cannot be appropriate for the production of green electronics due to their nonbiodegradable nature, even in the condition of highly thin films. The generally utilized metals as contacts and electrodes in green electronics are Mg (magnesium), Fe (iron), and Zn (zinc) or their oxides and alloys [66].

(a)

(b)

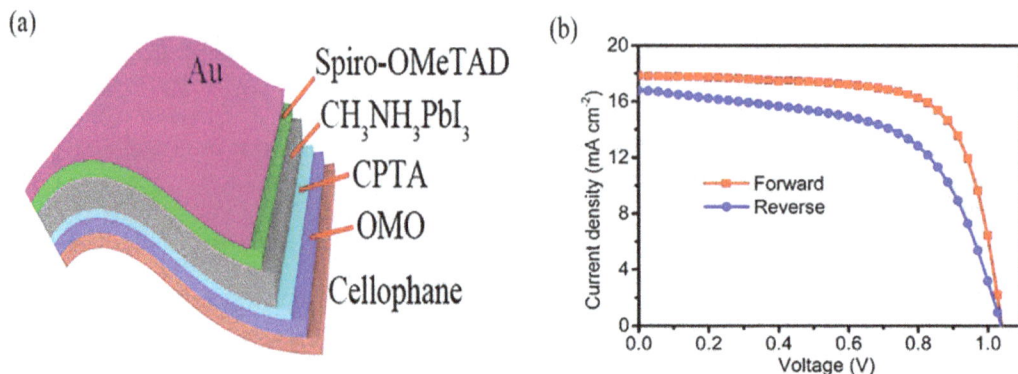

FIGURE 20.3 (a) Main elements in a solar cell and (b) density and voltage effect of a recyclable solar cell. (From Ref. [70]. Copyright (2019) Elsevier.)

Such thin films of metal are compatible with the body of a human and disposed of easily [67]. Rogers et al. evaluated the suspension rates of numerous metallic nano-membranes and evaluated that the different suspension rates are present, the thin film of some metals may be melted in deionized water and stimulated fluid of body (HBSS, hanks solution of salt which is stable). So, these films of metals have been utilized to make the device with a short life, that's why they showed their flexibility to form green electronics as biodegradable substances [68]. However, the termination rates are different, the material of thin films can be biodegraded in deionized water or HBSS (at pH 5–8). Large numbers of metals can be dissolved in solutions of acids at pH < 7. In respect to biodegradable contents or electrodes, the basic problem is in concern is their metals are radioactive and toxic; therefore, these can be injurious to the body of humans [69].

Electrodes of gold can be utilized to form biocompatible electronics while it's not biodegradable. The iron electrode-based devices showed the voltage in a highly negative threshold is related to the devices made of gold. This is due to the disparity of iron function with the greatest occupied elements orbital energy level in semiconductors, and outstanding agility of hole with value $0.12 \ cm^2/Vs$ can be attained. The defined e-devices that are biodegradable have iron electrodes that could be degraded quickly in 1 hour, in an acidic mixture at pH 4.6 [59]. Magnesium (Mg) is also a highly used metal electrode due to its dissolving ability in deionized water, as well as in the mixtures of base and acid. The transient electronics used the magnesium oxides (MgO) and Mg function as an electrode, in connection with other biodegradables things like silicon dioxide and silicon [71].

These e-devices disintegrate in a water within 10 minutes. When magnesium and silicon react with water the hydroxides of both metals are formed ($Mg(OH)_2$ and $Si(OH)_4$) by the process called hydrolysis, while in some cases the MgO and SiO_2 can be generated as middle compounds that again react with water to make hydroxides which are terminated in deionized water and reacted with acid to form salts which are soluble in water [71]. Scientists have made effective attempts to make various organic substances into electrodes. Various basic highly conducting polymers such as polyaniline (PANI), polythiophene (PT), and polypyrrole (PPy), like that poly-(3,4-ethylenedioxythiophene) (PEDOT), have been used effectively as electrode substances [67]. These polymers of similar metals are classically required to be doped in such a way that enhances conductivity (at 101–105 S/cm conduction); thus, the electrodes can be interconnected using this conducting state. After degradation, the remained particles related to these substrates are not harmful and toxic, which is in addition degraded or eliminated from the human body by the immune system and removed from the ecosystem [11].

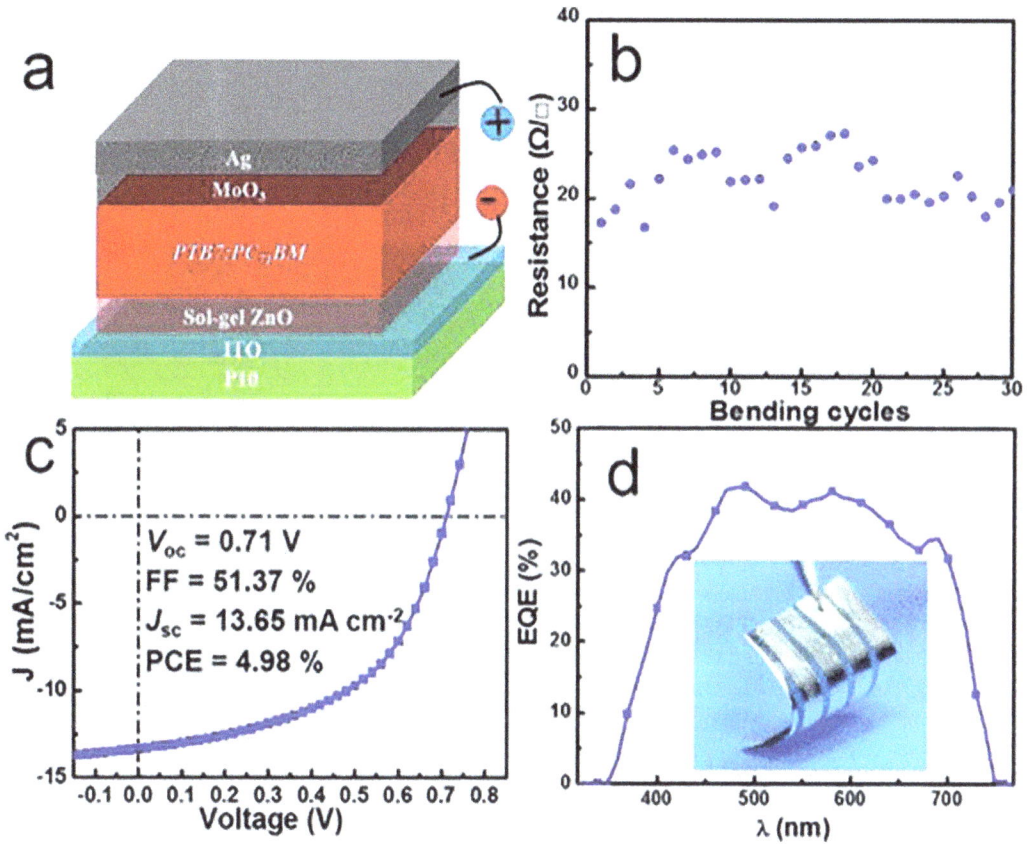

FIGURE 20.4 (a) Organic components of solar cell, (b) resistance per binding cycles, (c) relation of energy storage and voltage, and (d) EQE and wavelength. (From Ref. [73]. Copyright (2018) American Chemical Society.)

20.5 RECENT DEVELOPMENT IN E-DEVICES WITH BIODEGRADABLE MATERIALS

20.5.1 SOLAR CELLS

Considering two types of energy, solar energy belongs to renewable sources of energy, which can be replenished again and again. Humans being superior tend to use this energy in their daily usage by converting this solar energy into electrical energy. Solar cells basically work on the photovoltaic effect that is actually the production of voltage in any material by exposure to light. In this modern era, multiple solar cells technologies are developed which are highly productive but still require work to improve their life span or durability and uses semiconductors as their basic or primary element. The research communities are working with the emphasis on improved performance and service life of cells with the help of different semiconductors layering making multijunctions, more efficient solar cells. The biodegradable and recyclable solar cells are highly desired to overcome energy shortage over the world as well as it also reduces carbon dioxide emission. Each layer is covered by biodegradable materials but to enhance its electrical efficiency, many other nondegradable materials are also used, including metals [72] like cadmium shown in Figure 20.4.

Contemporary reports on the generation of new solar cells (mainly organic solar cells [OSCs] and perovskite solar cells [PSCs]) having multiple biodegradable layers made up of silicon or any other biodegradable material are studied. It is noticeable that thin-film solar cells and silicon solar

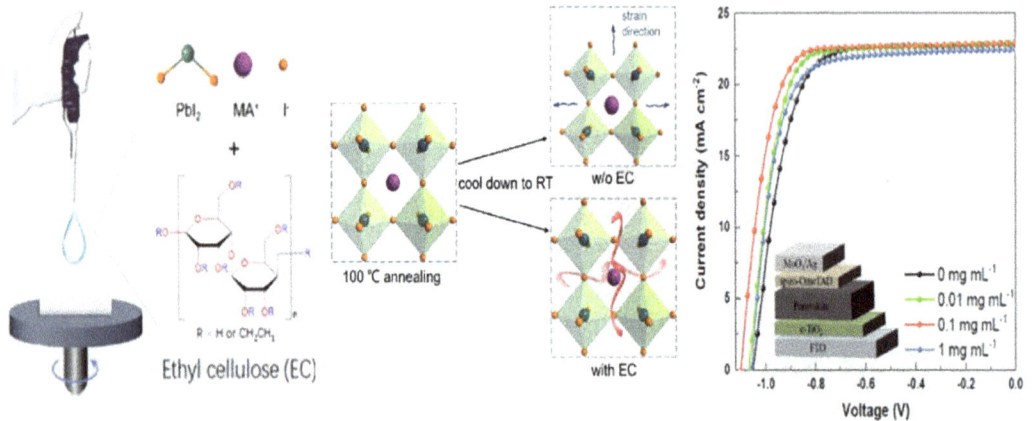

FIGURE 20.5 Schematic illustration of ethyl cellulose (EC) passivation. (From Ref. [77]. Copyright (2019) American Chemical Society.)

cells can retrieve 90% of the cover glass for future usage as well as 95% of semiconductor materials for use in the next solar unit formation. Cellulose can be used as a transparent and elastic surfactant in OSCs due to high optical haze which can spread out the transmitted light and in this way enhance the solar cell efficiency [74]. From this fact, we conclude that we can use easily available biodegradable transparent materials replacing the conventional glass and polyethylene terephthalate substrate like cellulose which is economically effective as well, it is nontoxic and has polymeric properties. Soil bacteria can easily degrade cellulose [75].

Although the efficiency of these biodegradable substances is not in comparison with PSCs but this restructuring denoted the great capabilities in the assembling of biodegradable solar cells. To use solar energy for daily needs, the substances involved in making solar cells should have the properties of transparency and haze so that they can capture sunlight more efficiently. The substrate of extraordinary high-speed transmission and mist is mandatory to take up sunlight more effectively [76]. Intended for the production of best performing eco-friendly OSCs and PSCs substrate might be a helpful decision in this act to be successful. Along with developing decomposable cellulose and chitosan materials as the substrate, possibly they may be operated to inactivate the flaws of PSCs because they work as a recombination center of charge moreover control charge-transmission, otherwise interfacial coating, in OSCs. Precursors of perovskite familiarized with variable intensities of inexpensive and recyclable ethyl cellulose (EC). Hydrogen bonding developed among the ethylcellulose and perovskite excellently causes the inertness of charge defect traps, thus intensifying the power conversion efficiency (PCE) from 17.11 to 19.41% [77]. The effect of EC passivation on perovskite flaws and solar cell performance is depicted schematically in Figure 20.5.

Due to the development of organized layer-by-layer structure along with interfacial and molecular dipoles, the work function of the ITO was fruitfully reduced from 4.6 to 4.3 eV as they assisted the charge transfer. By using chitosan-derived interlayer, the efficacy of the solar cell approximated 10.18% in the same way is presented in Figure 20.6 [75]. Chitosan also behaved as an interfacial layer to transform a ZnO electron transport layer (ETL) designed for an OSC with PTB7-Th: PC71BM as a bulk heterojunction (BHJ) active layer, indicating the proficiency of 7.54% [78] biodegradable substances are mainly consumed as substrates, additives, or interlayers in solar cells as discussed earlier. Conversely, the posterior electrodes in solar cells still contain customary metals, for example, Au, Ag, Cu, Al, or their resembling alloys. These metallic electrodes are usually not affected by corrosion and have great conductance ability; however, they are pricey and uncommon [1]. Mg, Mo, Mn, Fe, W, and Zn are eco-friendly besides this they undergo to break down when

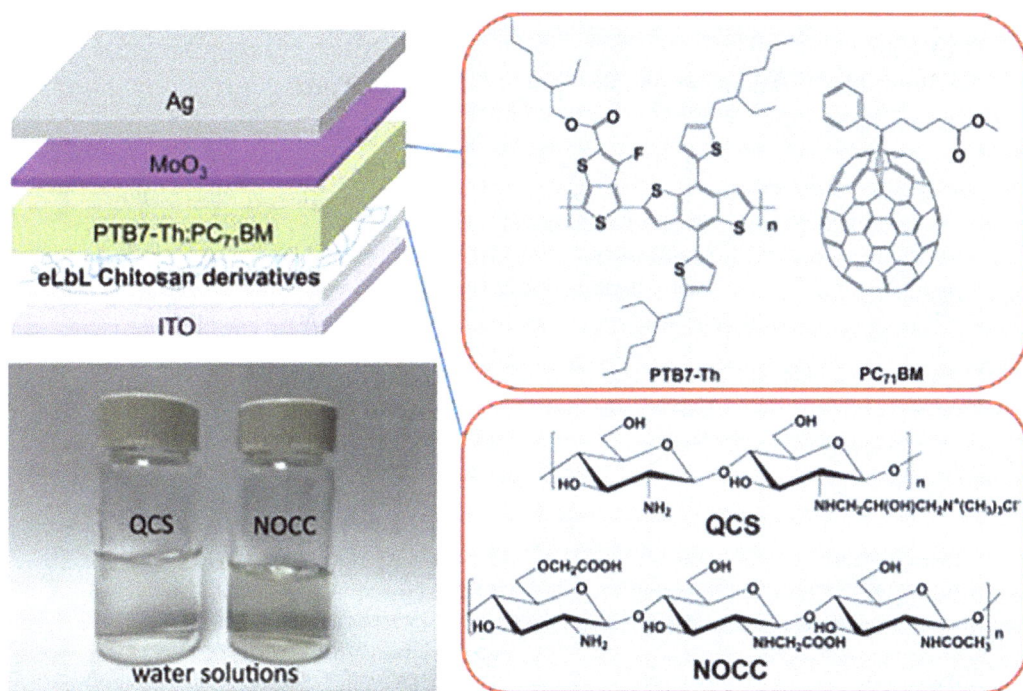

FIGURE 20.6 Use of biodegradable materials in the manufacturing of solar cells and their chemical structure. (From Ref. [75]. Copyright (2017) Elsevier.)

suitable physiological requirements are met. From these metals, Mg is a biodegradable material, shows a work function of 3.7 eV, the conductance of 22.6×106 S/m, and can be metabolized with water [79].

Carbon-containing materials can be a substitute for noble metals, because of their hydrophobic and biodegradable natures. They are also cost-effective and have great stability. The PCE comes close to 15.7% when PSCs have been utilized with the carbon-containing electrode [80]. Techniques operated for carbon electrodes include screen-printing, doctor-blading, rolling-transfer, and hot-pressing deposition [81]. Appreciably, carbon-containing substances are capable of taking the place of the hole-transport layer (HTL), permitting the extensive fabrication of PSCs deprived of the HTL. Despite this advancement, there is yet no report concerning a calculation for the biodegradability of light-absorbing materials or the alteration of materials to accelerate biodegradation. To formulate a greatly biodegradable solar cell, these parts will require to be advanced in the future. Table 20.2 shows different biodegradable materials used in e-devices.

20.5.2 Other Electrical Appliances

The use of biodegradable materials would also be highly beneficial for other devices, e.g., sensors, solar cell implantable products, drug delivery, organic field-effect transistors (OFETs), LED, and batteries [83–86]. A power supply is an essential constituent concerned with e-devices that are biodegradable; conversely, an entirely the power source of biodegradable devices is confirmed to be the main contest in the functioning of such gadgets (for example, therapeutic implants and self-power diagnostic), high-performing, biodegradable battery like an in vivo onboard power source by using a biodegradable Mg anode. This anode owned an extraordinary theoretical energy density and admirable biocompatibility. In the meantime, alginate hydrogel worked as the electrolyte, and molybdenum trioxide (1 g/L in aqueous solution) behaved as the cathode material [87].

TABLE 20.2
Different Biodegradable Materials Used in E-Devices [39, 82]

Biodegradable Materials	Functional Layers	Device Structures
DNA	Hole-blocking layer	ITO/PEDOT:PSS/MEH-PPV/DNA-CTMA/Al
DNA-CTMA	Helo-transporting/ electron-blocking layer	ITO/PEDOT-DNA-CTMA(20 nm)/PFO:MEH. ppv/Cs$_2$CO$_2$/Al
BSA	Submonolayer	ITO/PEDOT:PSS/BSA/NPB/Alq$_3$/Mg/Al
Cytochrome c	Submonolayer	ITO/PEDOT:PSS/cytochrome c/NPB/Alq$_3$/Mg/Al
BSM	Emitting layer	ITO/BSM-RGB with MWNT/Al-Au
Vitamin B12	Emitting layer	ITO/Vitamin B12/Al
RFLT	Emitting layer	ITO/PEDOT:PSS/PYK/RFLT/Al
Cytochrome c	Emitting layer	ITO/cytochrome c/Al
Eumelanin-PEDOT:PSS (Eu-PH)	Electrode	Eu-PH/α-NDP/Alq$_3$/Ca/Al
(ChNF) AgNW/SF	Substrate	AgNW/SF/MoO$_3$/TAPC/CBP:lr(ppy)$_3$(10 wt%)/ TPBi/Liq:Al
Chitin nanofibrils	Substrate	ChNF/ZnS:Ag:MoO$_3$/NPB/Alq$_3$/Liq:Al
CNC	Substrate	CNC on glass/MoO$_3$/Au/α-NDP/LiF/Al
Nanofibrilated cellulose	Substrate	Cellulose/CNT/MoO3/PEDOT:PSS/green polyfluorene/Ca/Al
BC nanofibers	Substrate	BC/SiO$_2$/ITO/Alq$_3$

20.6 CONCLUSION

Electronic devices are the need of life nowadays and make life easy but on the other hand, e-devices have negative impacts on the environment and human beings and become the cause of environmental pollution. Therefore, for tackling these problems the biodegradable electronic devices are an emerging need for sustainable development. The electronics devices are called biodegradable devices which comprise materials that can be partially and completely degraded by the actions of microorganisms in the environment. Biodegradable e-devices can degrade easily by natural factors as well as by the activity of microorganisms and have no impacts on the human and environment. It is noticeable that thin-film solar cells and silicon solar cells can retrieve 90% of the cover glass for future usage as well as 95% of semiconductor materials because of their applications in the formation of biodegradable layers for use in the next recyclable solar unit formation. The efficacy of the solar cell is increased by about 10.18% by using chitosan-derived interlayers. Fortunately, the scientific communities are aware of the importance of green electronics and have made action to attain this goal. The most recent biodegradable material and green process for green electronics such as OFETs, LEDs, solar cells, and other electronic devices are summarized. The current and some previous reports have provided in-depth knowledge about the potential for green electronics fabrication. Natural and synthetic biodegradable materials can be used as substrates, electrodes, and active materials for green electronics. However, the hard gelatin and shellac use as the substrate for e-devices are still in their present stage. Currently, chitosan and cellulose are the most commonly investigated biodegradable materials. The potential of biodegradable electronics is infinite, in the future nonbiodegradable material could be replaced by biodegradable material in the production processes of e-devices.

ACKNOWLEDGMENT

The authors are grateful for the financial supports from the Foundation for Research Support of the State of Rio Grande do Sul – FAPERGS [19/2551-0001865-7] and National Council for Scientific and Technological Development – CNPq [303.622/2017-2].

REFERENCES

1. Tan, M.J., C. Owh, P.L. Chee, A.K.K. Kyaw, D. Kai, and X.J. Loh, Biodegradable electronics: cornerstone for sustainable electronics and transient applications. Journal of Materials Chemistry C, 2016. **4**(24): p. 5531–5558.
2. Zvezdin, A., E.D. Mauro, D. Rho, C. Santato, and M. Khalil, En route toward sustainable organic electronics. MRS Energy & Sustainability, 2020. **7**: p. 1–8.
3. Park, M., Closed for repair: identifying design affordances for product disassembly (paper presentation), in Proceedings of the PLATE product lifetimes and the environment, 3rd PLATE. 2019: Germany: Conference Berlin.
4. Meloni, M., F. Souchet, and D. Sturges, Circular consumer electronics: an initial exploration. 2018: Ellen MacArthur Foundation. p. 1–17.
5. Awasthi, A.K., J. Li, L. Koh, and O.A. Ogunseitan, Circular economy and electronic waste. Nature Electronics, 2019. **2**(3): p. 86–89.
6. Perkins, D.N., M.-N.B. Drisse, T. Nxele, and P.D. Sly, E-waste: a global hazard. Annals of Global Health, 2014. **80**(4): p. 286–295.
7. Odeyingbo, O., I. Nnorom, and O. Deubzer, Odeyingbo, O., I. Nnorom, and O. Deubzer, Person in the Port Project: assessing import of used electrical and electronic equipment into Nigeria. 2017.
8. Lee, D., D. Offenhuber, F. Duarte, A. Biderman, and C. Ratti, Monitour: tracking global routes of electronic waste. Waste Management, 2018. **72**: p. 362–370.
9. Fang, S., T. Tao, H. Cao, M. He, X. Zeng, P. Ning, H. Zhao, M. Wu, Y. Zhang, and Z. Sun, Comprehensive characterization on Ga (In)-bearing dust generated from semiconductor industry for effective recovery of critical metals. Waste Management, 2019. **89**: p. 212–223.
10. Irimia-Vladu, M., "Green" electronics: biodegradable and biocompatible materials and devices for sustainable future. Chemical Society Reviews, 2014. **43**(2): p. 588–610.
11. Li, W., Q. Liu, Y. Zhang, C.A. Li, Z. He, W.C. Choy, P.J. Low, P. Sonar, and A.K.K. Kyaw, Biodegradable materials and green processing for green electronics. Advanced Materials, 2020. **32**(33): p. 2001591.
12. Feron, K., R. Lim, C. Sherwood, A. Keynes, A. Brichta, and P.C. Dastoor, Organic bioelectronics: materials and biocompatibility. International Journal of Molecular Sciences, 2018. **19**(8): p. 2382.
13. Li, R., L. Wang, D. Kong, and L. Yin, Recent progress on biodegradable materials and transient electronics. Bioactive Materials, 2018. **3**(3): p. 322–333.
14. Leja, K. and G. Lewandowicz, Polymer biodegradation and biodegradable polymers – a review. Polish Journal of Environmental Studies, 2010. **19**(2): p. 255–266.
15. ASTM, N., D6400-12: standard specification for labeling of plastics designed to be aerobically composted in municipal or industrial facilities. 2012: West Conshohocken, PA: ASTM International.
16. Platt, D.K., Biodegradable polymers: market report. 2006: United Kingdom: Smithers Rapra Publishing.
17. Feig, V.R., H. Tran, and Z. Bao, Biodegradable polymeric materials in degradable electronic devices. ACS central science, 2018. **4**(3): p. 337–348.
18. Muñoz, I., M.J. Gómez-Ramos, A. Agüera, A.R. Fernández-Alba, J.F. García-Reyes, and A. Molina-Díaz, Chemical evaluation of contaminants in wastewater effluents and the environmental risk of reusing effluents in agriculture. TrAC Trends in Analytical Chemistry, 2009. **28**(6): p. 676–694.
19. Mara, D. and N.J. Horan, Handbook of water and wastewater microbiology. 2003: Elsevier.
20. Narancic, T., S. Verstichel, S. Reddy Chaganti, L. Morales-Gamez, S.T. Kenny, B. De Wilde, R. Babu Padamati, and K.E. O'Connor, Biodegradable plastic blends create new possibilities for end-of-life management of plastics but they are not a panacea for plastic pollution. Environmental Science & Technology, 2018. **52**(18): p. 10441–10452.
21. Malinconico, M., Soil degradable bioplastics for a sustainable modern agriculture. 2017: Berlin, Heidelberg: Springer.
22. Narayan, R., Carbon footprint of bioplastics using biocarbon content analysis and life-cycle assessment. MRS Bulletin, 2011. **36**(9): p. 716–721.
23. Martino, L., L. Basilissi, H. Farina, M.A. Ortenzi, E. Zini, G.D. Silvestro, and M. Scandola, Bio-based polyamide 11: synthesis, rheology and solid-state properties of star structures. European Polymer Journal, 2014. **59**: p. 69–77.
24. Reddy, M.M., S. Vivekanandhan, M. Misra, S.K. Bhatia, and A.K. Mohanty, Biobased plastics and bionanocomposites: current status and future opportunities. Progress in Polymer Science, 2013. **38**(10–11): p. 1653–1689.

25. Lambert, S. and M. Wagner, Environmental performance of bio-based and biodegradable plastics: the road ahead. Chemical Society Reviews, 2017. **46**(22): p. 6855–6871.

26. Stahel, W.R. and E. MacArthur, The circular economy: a user's guide (1st ed.). 2019: London: Routledge. p. 118.

27. Ghodrat, A.G., M. Tabatabaei, M. Aghbashlo, and S.I. Mussatto, Waste management strategies; the state of the art, in Biogas. 2018: Cham: Springer. p. 1–33.

28. Favoino, E. and D. Hogg, The potential role of compost in reducing greenhouse gases. Waste Management & Research, 2008. **26**(1): p. 61–69.

29. Bastianoni, S., M. Porcelli, and F. Pulselli, Energy evaluation of composting municipal solid waste, in WIT transactions on ecology and the environment (vol. 56). 2002. https://www.witpress.com/elibrary/wit-transactions-on-ecology-and-the-environment/56/1057

30. Sakai, S.-I., H. Yoshida, J. Hiratsuka, C. Vandecasteele, R. Kohlmeyer, V.S. Rotter, F. Passarini, A. Santini, M. Peeler, and J. Li, An international comparative study of end-of-life vehicle (ELV) recycling systems. Journal of Material Cycles and Waste Management, 2014. **16**(1): p. 1–20.

31. Schröder, J., Creative food cycles towards urban futures and circular economy, in Food interactions catalogue: collection of best practices. 2019. p. 9–15.

32. Pandyaswargo, A.H. and D.G.J. Premakumara, Financial sustainability of modern composting: the economically optimal scale for municipal waste composting plant in developing Asia. International Journal of Recycling of Organic Waste in Agriculture, 2014. **3**(3): p. 1–14.

33. Love, S.A., M.A. Maurer-Jones, J.W. Thompson, Y.-S. Lin, and C.L. Haynes, Assessing nanoparticle toxicity. Annual Review of Analytical Chemistry, 2012. **5**: p. 181–205.

34. Tarazona, J. and M. Ramos-Peralonso, Ecotoxicology, terrestrial. In: Encyclopedia of Toxicology (3rd edition). 2014: Elsevier.

35. Doherty, F.G., A review of the Microtox® toxicity test system for assessing the toxicity of sediments and soils. Water Quality Research Journal, 2001. **36**(3): p. 475–518.

36. Rillig, M.C., Microplastic in terrestrial ecosystems and the soil?. Environmental Science & Technology, 2012. **46**(12): p. 6453–6454. https://www.degruyter.com/document/doi/10.1515/9781501511967-001/html

37. Mansouri, K., T. Ringsted, D. Ballabio, R. Todeschini, and V. Consonni, Quantitative structure – activity relationship models for ready biodegradability of chemicals. Journal of Chemical Information and Modeling, 2013. **53**(4): p. 867–878.

38. Boethling, R., E. Sommer, and D. DiFiore, Designing small molecules for biodegradability. Chemical Reviews, 2007. **107**(6): p. 2207–2227.

39. Van Der Zee, M., 1. Methods for evaluating the biodegradability of environmentally degradable polymers, in Handbook of biodegradable polymers. 2020, edited by Catia Bastioli. Berlin, Boston: De Gruyter. p. 1–22.

40. Di Mauro, E., D. Rho, and C. Santato, Biodegradation of bio-sourced and synthetic organic electronic materials towards green organic electronics. Nature Communications, 2021. **12**(1): p. 1–10.

41. Sintim, H.Y., A.I. Bary, D.G. Hayes, M.E. English, S.M. Schaeffer, C.A. Miles, A. Zelenyuk, K. Suski, and M. Flury, Release of micro-and nanoparticles from biodegradable plastic during in situ composting. Science of the Total Environment, 2019. **675**: p. 686–693.

42. No, O.T., 208: terrestrial plant test: seedling emergence and seedling growth test, in OECD guidelines for the testing of chemicals, section (vol. 2). 2006. p. 1–21. https://doi.org/10.1787/9789264070066-en

43. Boethling, R., E. Sommer, and D. DiFiore, Designing small molecules for biodegradability. Chemical Reviews, 2007. **107**(6): p. 2207–2227.

44. Tobjörk, D. and R. Österbacka, Paper electronics. Advanced Materials, 2011. **23**(17): p. 1935–1961.

45. Hu, L., J.W. Choi, Y. Yang, S. Jeong, F.L. Mantia, L.-F. Cui, and Y. Cui, Highly conductive paper for energy-storage devices. Proceedings of the National Academy of Sciences, 2009. **106**(51): p. 21490–21494.

46. Lang, A.W., A.M. Österholm, and J.R. Reynolds, Paper-based electrochromic devices enabled by nanocellulose-coated substrates. Advanced Functional Materials, 2019. **29**(39): p. 1903487.

47. Kim, T.S., S.I. Na, S.S. Kim, B.K. Yu, J.S. Yeo, and D.Y. Kim, Solution-processible polymer solar cells fabricated on a papery substrate. Physica Status Solidi (RRL)–Rapid Research Letters, 2012. **6**(1): p. 13–15.

48. Zhu, M., C. Jia, Y. Wang, Z. Fang, J. Dai, L. Xu, D. Huang, J. Wu, Y. Li, and J. Song, Isotropic paper directly from anisotropic wood: top-down green transparent substrate toward biodegradable electronics. ACS Applied Materials & Interfaces, 2018. **10**(34): p. 28566–28571.

49. Jin, J., D. Lee, H.G. Im, Y.C. Han, E.G. Jeong, M. Rolandi, K.C. Choi, and B.S. Bae, Chitin nanofiber transparent paper for flexible green electronics. Advanced Materials, 2016. **28**(26): p. 5169–5175.

50. Tao, H., M.A. Brenckle, M. Yang, J. Zhang, M. Liu, S.M. Siebert, R.D. Averitt, M.S. Mannoor, M.C. McAlpine, and J.A. Rogers, Silk-based conformal, adhesive, edible food sensors. Advanced Materials, 2012. **24**(8): p. 1067–1072.

51. Benfenati, V., S. Toffanin, R. Capelli, L.M. Camassa, S. Ferroni, D.L. Kaplan, F.G. Omenetto, M. Muccini, and R. Zamboni, A silk platform that enables electrophysiology and targeted drug delivery in brain astroglial cells. Biomaterials, 2010. **31**(31): p. 7883–7891.

52. Kim, D.-H., J. Viventi, J.J. Amsden, J. Xiao, L. Vigeland, Y.-S. Kim, J.A. Blanco, B. Panilaitis, E.S. Frechette, and D. Contreras, Dissolvable films of silk fibroin for ultrathin conformal bio-integrated electronics. Nature Materials, 2010. **9**(6): p. 511–517.

53. Rockwood, D.N., R.C. Preda, T. Yücel, X. Wang, M.L. Lovett, and D.L. Kaplan, Materials fabrication from *Bombyx mori* silk fibroin. Nature Protocols, 2011. **6**(10): p. 1612.

54. Ji, X., L. Song, S. Zhong, Y. Jiang, K.G. Lim, C. Wang, and R. Zhao, Biodegradable and flexible resistive memory for transient electronics. The Journal of Physical Chemistry C, 2018. **122**(29): p. 16909–16915.

55. Mahmoudi, S., N. Huda, and M. Behnia, Photovoltaic waste assessment: forecasting and screening of emerging waste in Australia. Resources, Conservation and Recycling, 2019. **146**: p. 192–205.

56. Awasthi, A.K., X. Zeng, and J. Li, Environmental pollution of electronic waste recycling in India: a critical review. Environmental Pollution, 2016. **211**: p. 259–270.

57. Puangprasert, S. and T. Prueksasit, Health risk assessment of airborne Cd, Cu, Ni and Pb for electronic waste dismantling workers in Buriram Province, Thailand. Journal of Environmental Management, 2019. **252**: p. 109601.

58. Irimia-Vladu, M., E.D. Głowacki, P.A. Troshin, G. Schwabegger, L. Leonat, D.K. Susarova, O. Krystal, M. Ullah, Y. Kanbur, and M.A. Bodea, Indigo – a natural pigment for high performance ambipolar organic field effect transistors and circuits. Advanced Materials, 2012. **24**(3): p. 375–380.

59. Irimia-Vladu, M., P.A. Troshin, M. Reisinger, L. Shmygleva, Y. Kanbur, G. Schwabegger, M. Bodea, R. Schwödiauer, A. Mumyatov, and J.W. Fergus, Edible electronics: biocompatible and biodegradable materials for organic field-effect transistors (Adv. Funct. Mater. 23/2010). Advanced Functional Materials, 2010. **20**(23): p. 4017–4017.

60. Serrano, M.C., E.J. Chung, and G.A. Ameer, Advances and applications of biodegradable elastomers in regenerative medicine. Advanced Functional Materials, 2010. **20**(2): p. 192–208.

61. Larmagnac, A., S. Eggenberger, H. Janossy, and J. Vörös, Stretchable electronics based on Ag-PDMS composites. Scientific Reports, 2014. **4**(1): p. 1–7.

62. Jeong, S.H., S. Zhang, K. Hjort, J. Hilborn, and Z. Wu, PDMS-based elastomer tuned soft, stretchable, and sticky for epidermal electronics. Advanced Materials, 2016. **28**(28): p. 5830–5836.

63. Baëtens, T., E. Pallecchi, V. Thomy, and S. Arscott, Cracking effects in squashable and stretchable thin metal films on PDMS for flexible microsystems and electronics. Scientific Reports, 2018. **8**(1): p. 1–17.

64. Kowalewska, A. and W. Stańczyk, Highly thermally resistant UV-curable poly(siloxane)s bearing bulky substituents. Chemistry of Materials, 2003. **15**(15): p. 2991–2997.

65. Yeo, J.C. and C.T. Lim, Emerging flexible and wearable physical sensing platforms for healthcare and biomedical applications. Microsystems & Nanoengineering, 2016. **2**(1): p. 1–19.

66. Lucas, N., C. Bienaime, C. Belloy, M. Queneudec, F. Silvestre, and J.-E. Nava-Saucedo, Polymer biodegradation: mechanisms and estimation techniques – a review. Chemosphere, 2008. **73**(4): p. 429–442.

67. Li, R., L. Wang, and L. Yin, Materials and devices for biodegradable and soft biomedical electronics. Materials, 2018. **11**(11): p. 2108.

68. Yin, L., H. Cheng, S. Mao, R. Haasch, Y. Liu, X. Xie, S.W. Hwang, H. Jain, S.K. Kang, and Y. Su, Dissolvable metals for transient electronics. Advanced Functional Materials, 2014. **24**(5): p. 645–658.

69. Lei, T., M. Guan, J. Liu, H.-C. Lin, R. Pfattner, L. Shaw, A.F. McGuire, T.-C. Huang, L. Shao, and K.-T. Cheng, Biocompatible and totally disintegrable semiconducting polymer for ultrathin and ultralightweight transient electronics. Proceedings of the National Academy of Sciences, 2017. **114**(20): p. 5107–5112.

70. Li, H., X. Li, W. Wang, J. Huang, J. Li, S. Huang, B. Fan, J. Fang, and W. Song, Ultraflexible and biodegradable perovskite solar cells utilizing ultrathin cellophane paper substrates and $TiO_2/Ag/TiO_2$ transparent electrodes. Solar Energy, 2019. **188**: p. 158–163.

71. Hwang, S.-W., H. Tao, D.-H. Kim, H. Cheng, J.-K. Song, E. Rill, M.A. Brenckle, B. Panilaitis, S.M. Won, and Y.-S. Kim, A physically transient form of silicon electronics. Science, 2012. **337**(6102): p. 1640–1644.

72. Tsang, M.P., G.W. Sonnemann, and D.M. Bassani, Life-cycle assessment of cradle-to-grave opportunities and environmental impacts of organic photovoltaic solar panels compared to conventional technologies. Solar Energy Materials and Solar Cells, 2016. **156**: p. 37–48.

73. Cheng, Q., D. Ye, W. Yang, S. Zhang, H. Chen, C. Chang, L. Zhang, Construction of transparent cellulose-based nanocomposite papers and potential application in flexible solar cells. ACS Sustainable Chemistry & Engineering, 2018. **6**(6): p. 8040–8047.

74. Zhu, H., S. Parvinian, C. Preston, O. Vaaland, Z. Ruan, and L. Hu, Transparent nanopaper with tailored optical properties. Nanoscale, 2013. **5**(9): p. 3787–3792.

75. Zhang, K., R. Xu, W. Ge, M. Qi, G. Zhang, Q.-H. Xu, F. Huang, Y. Cao, and X. Wang, Electrostatically self-assembled chitosan derivatives working as efficient cathode interlayers for organic solar cells. Nano Energy, 2017. **34**: p. 164–171.

76. Jia, C., T. Li, C. Chen, J. Dai, I.M. Kierzewski, J. Song, Y. Li, C. Yang, C. Wang, and L. Hu, Scalable, anisotropic transparent paper directly from wood for light management in solar cells. Nano Energy, 2017. **36**: p. 366–373.

77. Yang, J., S. Xiong, T. Qu, Y. Zhang, X. He, X. Guo, Q. Zhao, S. Braun, J. Chen, and J. Xu, Extremely low-cost and green cellulose passivating perovskites for stable and high-performance solar cells. ACS Applied Materials & Interfaces, 2019. **11**(14): p. 13491–13498.

78. Lin, P.-C., Y.-T. Wong, Y.-A. Su, W.-C. Chen, and C.-C. Chueh, Interlayer modification using eco-friendly glucose-based natural polymers in polymer solar cells. ACS Sustainable Chemistry & Engineering, 2018. **6**(11): p. 14621–14630.

79. Hosseini, N.R. and J.S. Lee, Biocompatible and flexible chitosan-based resistive switching memory with magnesium electrodes. Advanced Functional Materials, 2015. **25**(35): p. 5586–5592.

80. Tian, C., A. Mei, S. Zhang, H. Tian, S. Liu, F. Qin, Y. Xiong, Y. Rong, Y. Hu, and Y. Zhou, Oxygen management in carbon electrode for high-performance printable perovskite solar cells. Nano Energy, 2018. **53**: p. 160–167.

81. Meng, F., A. Liu, L. Gao, J. Cao, Y. Yan, N. Wang, M. Fan, G. Wei, and T. Ma, Current progress in interfacial engineering of carbon-based perovskite solar cells. Journal of Materials Chemistry A, 2019. **7**(15): p. 8690–8699.

82. Di Mauro, E., D. Rho, and C. Santato, Biodegradation of bio-sourced and synthetic organic electronic materials towards green organic electronics. Nature Communications, 2021. **12**(1): p. 1–10.

83. Fu, R., W. Luo, R. Nazempour, D. Tan, H. Ding, K. Zhang, L. Yin, J. Guan, and X. Sheng, Implantable and biodegradable poly(L-lactic acid) fibers for optical neural interfaces. Advanced Optical Materials, 2018. **6**(3): p. 1700941.

84. Ding, S., Q. Cai, J. Mao, F. Chen, L. Fu, Y. LV, and S. Zhao, Highly conductive and transient tracks based on silver flakes and a polyvinyl pyrrolidone composite. RSC Advances, 2020. **10**(55): p. 33112–33118.

85. Lee, G., S.K. Kang, S.M. Won, P. Gutruf, Y.R. Jeong, J. Koo, S.S. Lee, J.A. Rogers, and J.S. Ha, Fully biodegradable microsupercapacitor for power storage in transient electronics. Advanced Energy Materials, 2017. **7**(18): p. 1700157.

86. Mahajan, B.K., X. Yu, W. Shou, H. Pan, and X. Huang, Mechanically milled irregular zinc nanoparticles for printable bioresorbable electronics. Small, 2017. **13**(17): p. 1700065.

87. Huang, X., D. Wang, Z. Yuan, W. Xie, Y. Wu, R. Li, Y. Zhao, D. Luo, L. Cen, and B. Chen, A fully biodegradable battery for self-powered transient implants. Small, 2018. **14**(28): p. 1800994.

21 Conducting Polymers
An Efficient Way to Deal with Medical and Industrial Fouling

N. Raghavendra Naveen,[1] Girirajasekhar Dornadula,[2] Pamayyagari Kalpana,[2] and Lakshmi Narasimha Gunturu[3]

[1]Department of Pharmaceutics, Sri Adichunchanagiri College of Pharmacy, Adichunchanagiri University, B.G. Nagar, India

[2]Department of Pharmacy Practice, Annamacharya College of Pharmacy, Rajampet, India

[3]Scientimed Solutions Private Limited, Mumbai, India

CONTENTS

DOI: 10.1201/9781003205418-21

21.1 INTRODUCTION

Fouling is an intricate and undesirable process where reversible/irreversible adhesion of the unwanted stuff on surfaces occurs. This phenomenon alters dynamically with time. It is precipitated by the deposition of various organic, inorganic, biological, and particulate matters. The foulant may differ extensively in composition, deposition type, and physicochemical properties. Based on the nature of the foulants and their occurrence in various sectors, literature shows various definitions of fouling. Industrially, regarding heat exchangers and membranes, fouling may be defined as the accumulation of unwanted material like scale, suspended solids, insoluble salts, and even algae on the internal surfaces. In marine systems, fouling is regarded as the attachment of micro biofilms or macro bio-species to the hull, engine, or submerged parts of the ship. Whereas in the medical sector, fouling is described as the formation of biofilms by microorganisms (bacteria, yeast) on the living tissues and medical devices. Fouling may also rely on the surface characteristics such as morphology, micro texture, wettability, and surface energy. This phenomenon is ubiquitous and imposes significant problems relating to commercial, environmental, and health issues [1, 2].

21.1.1 TYPES OF FOULING

On the basis of the nature of the foulants, fouling can be majorly classified into three types.

21.1.1.1 Organic Fouling

It is the deposition of the carbon-based bulk organic macromolecules. Additionally, some organic colloids and organic biotic forms also contribute to this fouling. Thus, organic fouling may sometimes converge with colloidal fouling and biotic forms of biological fouling [3]. For instance, in water treatment plant applications, organic fouling is a major restraint where the allochthonous and/or autochthonous natural organic matter and wastewater effluent accumulate on the membranes, which affect operational cleaning cycles and productivity.

21.1.1.2 Inorganic Scaling/Fouling

It is the precipitation of inorganic materials that arise from crystallization, erosion, soil, ice, and suspended particles. Moreover, the minerals such as calcium and magnesium in water may also involve scaling, such as (a) corrosion of heat exchanger tube in nuclear power plant, (b) freeze fouling crystal depositions in cold region pipelines [4]. Scaling may result in extensive costs due to functional losses, high energy demands, and uninterrupted maintenance of the fouling systems.

21.1.1.3 Particulate Fouling

It is the accumulation of the suspended solid or colloidal particles.

21.1.1.4 Biological Fouling

It is the accumulation of biotic matter, which leads to biofilm formation [5]. It has a complex and highly regulated architecture, which results in an intercalated network of micron inches. Biofilms may develop on various solid surfaces such as living tissues, medical devices, industrial equipment systems, piping, and innate aquatic systems. Clinically, attachment and colonization of microorganisms on tissues and indwelling medical devices may lead to the risk of local and systemic infections in a patient.

21.1.2 FOULING MECHANISMS

1. Chemical reactions
   Most of the organic and inorganic fouling occurs through chemical reactions. The basic mechanism involves the conversion of the reactants to soluble precursors and later the precursor transport and deposits on the surface in an insoluble form. The common type of reactions includes polymerization, auto-oxidation, precipitation, and decomposition.

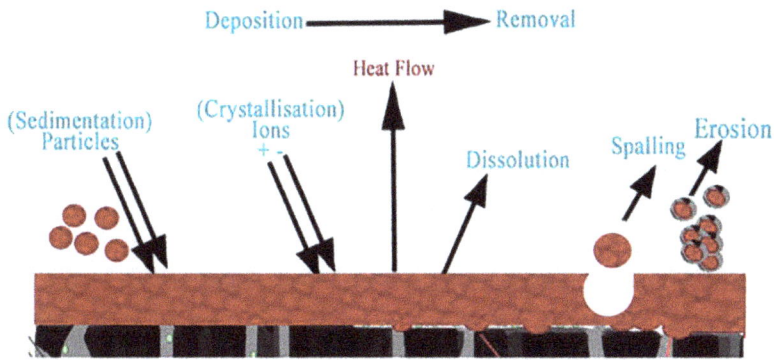

FIGURE 21.1   Different deposition and removal processes in fouling.

Auto-oxidation reaction is found to be a major source of fouling in heat exchangers and fuel lines in the jet. Autocatalytic oxidation of hydrocarbons involves steps, initiation, propagation, and termination, where thermal decomposition of the hydrocarbons takes place and further oxidation of free radical propagates the reaction. Also, various metal ions, nitrogen, sulfur, and unsaturated species may involve in the reaction. Various factors such as temperature, velocity, reactant composition, feedstock, and other impurities affect the rate of reaction and fouling process [6, 7].

2. Crystallization
   Crystallization fouling is most common in surfaces of cold heat transfer, aquatic systems, and cooling systems with waxy hydrocarbons. The hard salts in the aqueous systems, which are employed for cooling purposes, are more vulnerable to forming crystals. The process includes various events such as (a) supersaturation of inverted salts at high temperatures, (b) formation of stable nuclei and crystallites, and (c) deposition of crystals on a surface or suspended in solutions. Crystal shape is determined by the minimal surface energy, adsorption, and diffusion rate. In organic systems, the hydrocarbon waxy molecules get cooled and wax crystals are formed at cloud temperature [8]. Different deposition and removal processes in fouling were shown in Figure 21.1.

3. Corrosion
   Corrosion fouling is most common in heat exchangers, electronic equipment components, commercial water heating systems, and cooling systems of combustion engines. The basic mechanism of corrosion is the electrochemical reactions that occur concurrently at anode and cathode. A corrosion cell is formed when the anode and cathode possess high and low potential energies, respectively. The movement of electrons from anode to cathode and the related transfer of charges from cathode to anode may initiate the corrosion currents. The degree of corrosion is determined by electrode potential and pH of the aqueous solution. The pattern of corrosion may be usually uniform (deposition of foulants occur equally on the surface), pitting type, or localized form [9].

4. Biological fouling
   Biofilm formation involves a complex cyclic process with various steps. Initial attachment of the planktonic/bacterial cell takes place where they adhere to the suitable surface of the substrate. The cell becomes immobilized and binds irreversibly through surface adhesion molecules followed by a monolayer formation. The cells continue to grow and maturation occurs by the release of inducer signals and expression of biofilms-specific genes. Later, dispersion occurs through which further progression of the biofilms continues (Figure 21.2).

Steps Involved In Biofilm Formation			
STEP 1	**STEP 2.**	**STEP 3**	**STEP 4**
Adsorption and irreversible attachment of bacteria	Division and growth of cells[single to multilayer colonies]	Production of EPS and quorum quenching molecules	Maturation and dispersal of biofilms

FIGURE 21.2 Various steps involved in the biofilm formation.

The nature and physical characteristics of the biofilms are insistent on the type of extracellular polymer substances (EPS) that are produced by the microbes. The EPS enfolds the bacterial cells and maintains the biofilms matrix. Clinically, microbial colonization on the surface of biomedical materials may impose serious local and systemic effects. Also, the microbes lodged in biofilms are mostly unaffected by the antimicrobials as they develop resistance and are less receptive to host cells [10].

21.1.3 CONSEQUENCES OF FOULING

Fouling is widespread in various fields such as the health, industrial, and transport sectors. Clinically, biofouling accelerates health risks like augmented infection spread, implant rejections, diminished device effectiveness, device rejections, and impaired biosensor functioning. Despite following the sterility operations, the prosthetic devices and implants are easily prone to contamination. The National Institute of Health (NIH) stated that biofilms are liable for 65% of all microbial infections in humans. In industries, the formation of fouling may have numerous effects like reduction in heat transfer, impair fluid flow, influence safety equipment operations, initiate or aid in corrosion, and result in contamination of fluids. Mostly the effects of fouling are proportional to the amount of deposit formed. In the food processing industry, fouling is considered the potential source of biological contamination and hence removed at regular intervals. In marine systems, fouling may influence the operating system and efficiency of the propulsion components. Fouling is attributed to enormous costs related to the penalties of designing, energy losses, operational and for other fouling mitigation strategies [11]. Various industrial costs ascribed to fouling are as follows (Figure 21.3):

Capital costs: The cost of installation of antifouling equipment and fuel/energy losses.
Operating costs: Cost of cleaning in pace operation, and loss of production time since cleaning-in-place (CIP) time can be substantial.
Maintenance costs: The costs of removing fouling deposits, and the costs of chemicals or other operating costs of antifouling devices.

Table 21.1 summarizes the various issues of fouling that can be seen in various areas.

21.2 PROPERTIES OF ANTIFOULING CONDUCTING POLYMERS

21.2.1 IONIC EFFECT

Many researchers have extensively studied the impact of the ionic effect [24] on the antifouling property, especially for zwitterionic surfaces [25]. In general, on the basis of their propensity to salt-out or salt-in proteins, ions are classified as chaotropes and kosmotropes, which is shortened as the Hofmeister series [26]. Earlier studies have made known that weakly hydrated chaotropes

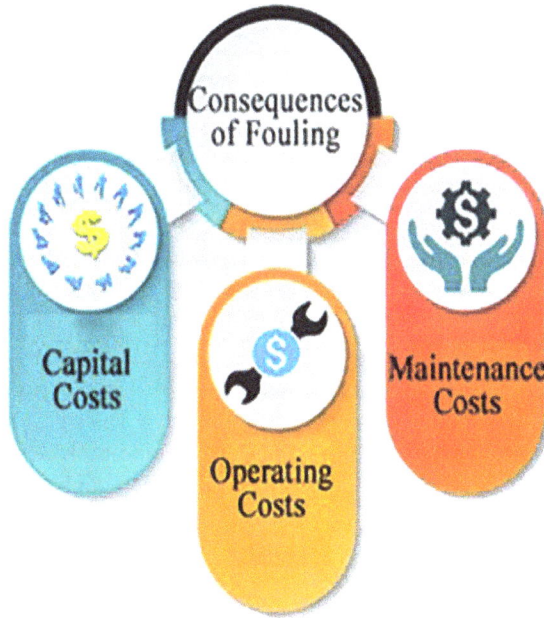

FIGURE 21.3   Types of consequences of fouling in industrial aspect.

TABLE 21.1
A List of General Problems of Fouling in Various Fields and Specific Examples

Type	Problem	Reference
	Medical	
Catheters	Increased biofilms-based catheterized urinary	[12]
	infections/catheter-related bloodstream infections	[13]
	Increased resistance to antibiotics	
Orthopedic implants	Implant associated infections and antimicrobial resistance	[14]
Contact lens	Increased risk of contact lens associated with corneal infection	[15]
Hemodialysis	Increased bacteremia and endotoxemia	[16]
	Peritonitis and technique failure	
Dental implants	Inflammation of peri-implant tissue	[17]
Biosensors	Rejection of devices	
	Industrial	
Heat exchanger	Decreased efficiency	[18]
Membrane	Decreased permeate flux	[19]
Fuel	Contamination of fuels, oil fouling	[20]
Food	Spoilage of food and dairy products	
	Transport	
Marine fouling in ship hulls and engine	Increased fuel consumption, frictional and viscous resistance, influence on the operating system and propulsion components, and extra drag and stress on the engine	[21, 22]
Aviation aircrafts	Chemical contamination, corrosion, and biological fouling in aircrafts	[23]

can effectively form ion pairs with zwitterionic moieties in enhancing the hydration of zwitterionic polymers, whereas, hydrated kosmotropes can degrade the antifouling property by dehydrating the zwitterionic polymers.

Several new investigations have focused specially on the ionic effect of the conducting polymers in antifouling property. Ionic effects of PEDOT derivatives [poly EDOT (4-thylenedioxythiophene)-PC and poly EDOTEG3] were studied by Chen and Luo [27]. The adsorption of positively charged lysozyme and negatively charged bovine serum albumin (BSA) was assessed as a function of ionic concentration in the presence of three different anions (ClO^{4-} chaotropes, SO_4^{2-} kosmotropes, and Cl^- ions). Low oxidation potential makes the backbone of PEDOT positively charged in contrast to other antifouling polymers. Additionally, antifouling property of poly EDOTEG3 was relying on the ion concentration and its species. Highly concentrated SO_4^{2-} ions can cause the dehydration of the ethylene group on polymer films, thus reducing the antifouling property. While the antifouling property also depends on the ionic strength, an increase in the ionic strength can initially promote the nonspecific binding, owing to the adsorbed anions on poly(EDOT-PC) fortified the electrostatic repulsion between polymer and lysozyme. The nonspecific binding increased with increasing the concentration of SO_4^{2-} ions from 100 mM to 500 mM. Even though the antifouling property is related to ionic strength, the surface-immobilized with antifouling moieties usually, present at low nonspecific binding sites of proteins. The antifouling property of many of the polymers can be easily manipulated by modifying the ionic strength of solutions, which can be the probable application of antifouling surfaces.

21.2.2 SURFACE POTENTIAL

Another critical component is to permit electric transmission, which can ensure the antifouling conducting polymers are not the same as other polymeric materials [28]. The movement (approach/leave) of ions or biomolecules from the conducting polymers is based on the surface potential applied to the surface. Certainly, the antifouling properties are affected by the difference in the ionic concentrations near the surface. Conversely, changes in the orientation of surfactant-type dopants can alter the surface potential [29]. If the conducting polymers are positively charged, then the anionic sulfate head groups will move toward conducting polymers and hydrophobic tails face outward. Chen and Luo established that the zwitterionic poly(EDOT-PC) surfaces; the surface potential dominates the nonspecific binding of proteins in deionized water [27]. When the surface potential is applied to induce the electrostatic interaction, then the proteins can be absorbed on antifouling poly(EDOT-PC) [30]. Several studies have demonstrated the importance of implantable biomaterials by integrating antifouling properties of polymers. Consequently, the elemental studies on the surface potential and antifouling property need to be explored.

21.2.3 ZWITTERIONIC PHOSPHORYLCHOLINE (PC)

The zwitterionic phosphorylcholine groups have been utilized to forestall the unnecessary, nonspecific binding of proteins and cells. PC will specifically bind to CRP (C-reactive protein) and elicit a series of immune responses in the presence of calcium ions. Currently, CRP is identified as a biomarker for cardiovascular diseases, tissue damage, and inflammation [31]. Several studies have projected for PC-based polymers as sensing platforms to detect CRP [32]. Goda et al. demonstrated the application of PC-functionalized PEDOT, as a biosensor for detection of CRP and impact of several copolymer thin films of different concentrations. For CRP identification, the presence of calcium ions is mandatory, because each of the protomers has a calcium pocket in the PC-binding domain [33]. The structure and mobility of the bound water are very critical to exhibit antifouling properties [34]. When CRP binds to PC groups, the water near the PC groups will be released simultaneously and rapidly. No further information is clear if the CRP binding to PC groups is the only scenario for the particular affinity with antifouling moieties. Since specific interaction can significantly reduce the antifouling properties, it is very essential to understand the nature of this interaction and identify the molecules that possibly will bind to antifouling moieties.

21.3   SYNTHESIS OF CONDUCTING POLYMERS

Conductive polymers (CPs) are generally synthesized by various methods such as electrochemical polymerization, chemical polymerization/oxidation, hydrothermal, electrospinning, self-assembly, and other photochemical methods, where the first two are the most prominent. The chemical structures of a few CPs are shown in Figure 21.4.

1. Electrochemical polymerization
   In practice, the monomer leads to the formation of an oligomer of low molecular weight, which finally leads to the formation of oligomer (high molecular weight) polymers through oxidation at a lower potential. Several methodologies were utilized for the synthesis of CPs by this method, such as potentiostats (using constant voltage) or galvanostatic (using constant current) or potentiodynamic (flexible current and voltage), etc. [35, 36]. Electrochemical polymerization carried out in a typical three-electrode setup consists of reference, working, and counter electrodes. Through the polymerization process, the polymer gets deposits on the working electrode (made of glassy carbons, indium tin oxide, or platinum). The type of CPs decides the type of electrolyte to be used. For instance, to polymerize aniline, ionic liquids or inorganic salts are required, whereas the same cannot

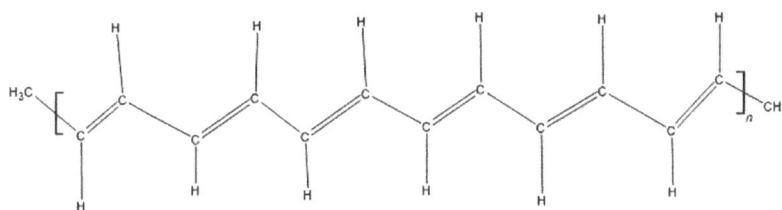

polyacetylene

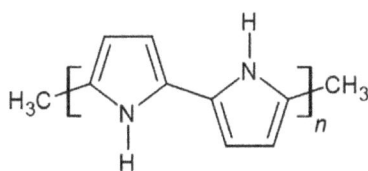

polypyrrole

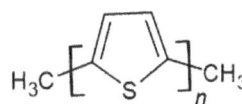

polythiophene

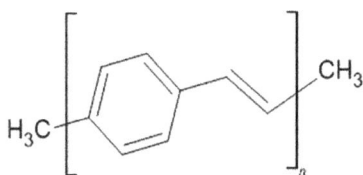

Poly (p- phenylene vinylene) (PPV)

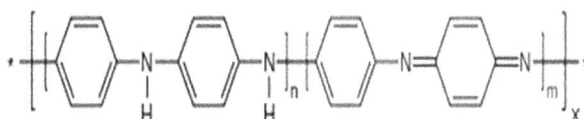

Polyaniline

FIGURE 21.4   Structures of a few CPs.

$$HC\equiv CR \longrightarrow H_3C-\left(CH{=}CR\right)_n-CH_3$$

$$HC\equiv CR \longrightarrow H_3C-\left(CR{=}CR^1\right)_n-CH_3$$

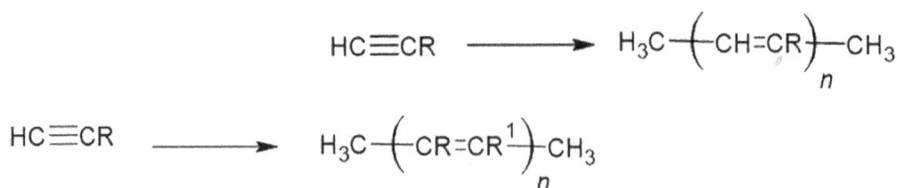

FIGURE 21.5   Structures of mono- or disubstituted polyacetylene.

be used for polypyrrole (PPy) or PEDOT. The protons that are present in the acidic pH range are useful to evade the formation of undesired, branched products and assist in the generation of doped CPs. As such majority of CPs have low electrical conductivity and optical properties in their pristine form, yet these properties can be enhanced with doping of suitable materials.

2. Chemical polymerization/oxidation
   In this process, oxidizing agents like ferric chloride, ammonium persulfate, etc., will initiate the polymerization process and oxidize the monomer. The oxidation process will be carried out in different stages such as oxidation of monomer into radical cation, then into dimers (through radical coupling), and finally, propagation of polymeric chain to form the final product with desired properties. Polyaniline polymerization can be done effectively by using ferric chloride, while the oxidation potential of ferric chloride is much lower than other oxidizing materials. Polymerization of aniline through this process requires excess protons; consequently, the reaction mixture was maintained around pH < 3.

21.3.1   POLYACETYLENE

Polyacetylene and its derivatives possess unique features such as photoconductivity, electrical conductivity, chiral recognition, and liquid crystal properties. The main chain of polyacetylene is composed of a linear polyene chain, which can be added with a few pendant groups to form monosubstituted or disubstituted polyacetylene (Figure 21.5). Polyacetylene, in its intact form, has a low conductivity (10^{-5} S/cm), whereas doping can rise the conductivity to 10^2–10^3 S/cm and also gives modulated optical-mechanical or electrochemical properties. There are many synthetic methods for the synthesis of polyacetylene, as depicted in Table 21.2.

TABLE 21.2
Synthetic Methods to Prepare Polyacetylene [37]

S. No	Method	Characteristic Feature
1.	Catalytic polymerization (Figure 21.6)	• Ziegler-Natta catalysts used for synthesis. Produces highly crystalline films • Luttinger catalysts produce high molecular weight. Polyacetylene produces compounds with low catalytic activity than Ziegler-Natta catalysts formed product
2.	Non-catalytic polymerization	• Electrochemical polymerization technique (galvanostatic, potentiostatic, and cyclic voltammetry)-direct deposition of polymer film on the metal
3.	Nonacetylene synthesis	• Ring-opening reaction of 1,3,5,7-cyclooctatetraene with a metathesis catalyst
4.	Light induced synthesis	• Irradiation of acetylene with UV

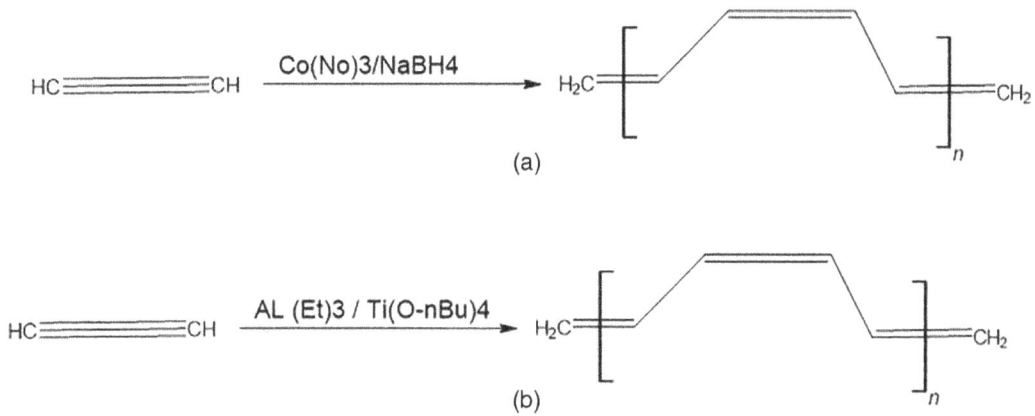

FIGURE 21.6 Synthesis of polyacetylene using Ziegler-Natta catalyst (a) and Luttinger catalyst (b).

21.3.2 POLYANILINE

Polyaniline is the most explored and promising CPs owing to its high processability, optical properties, stability, and tunable properties. Dopant concentration decides the conductivity nature of polyaniline [38]. It can have conductivity near to the metal if the reaction pH was maintained around 3 [39]. Classification of polyaniline can be done on their oxidation states as shown in Table 21.3. The backbone of the polymer consists of both benzoid and quinoid rings in different ratios. This difference in the ratio causes the existence of three different oxidation states.

Polyaniline can be synthesized by using a monomer precursor of the corresponding polymer along with an oxidizing agent. This reaction can be carried out under ambient conditions in the presence of a suitable acid. End reaction is identified by a color change to green color. Oxidizing agents like ammonium peroxydisulfate, potassium bichromate, and ceric nitrate are used. The pH of the acid can be maintained in the range of 1–3 to make the polymer and composite with good conductivity. A few methods for the synthesis of polyaniline are given in Table 21.4.

21.3.3 POLYPYRROLE

For a few years, considerable interest in PPy was increased owing to its enhanced conductivity, high stability, and ease to synthesize the composite and homopolymer from it. For the first time, PPy was prepared by the chemical oxidation of the monomer. The reaction was carried out in the presence

---

TABLE 21.3
Types of Polyaniline, Structures, and Their Oxidation States

S. No	Type of Polyaniline	Structure	Oxidation States
1.	Leucoemeraldine		Fully reduced is in a quinoid state
2.	Emeraldine		Equal ratio of benzoid and quinoid rings
3.	Pernigraniline		Fully oxidized form—benzoid state

TABLE 21.4

Various Synthesis Methods and Their Characteristic Features for Polyaniline

S. No	Method	Characteristic Feature
1.	Interfacial polymerization	Polymerization takes place in between two immiscible liquids
2.	Microemulsion technique	Polymerization takes place with the help of surfactant
3.	Electropolymerization	Without using an oxidant
4.	Electrospinning	Fibrous nano/micro-polymers were prepared using strong electrical field

of hydrogen peroxide. Formed PPy in its undoped form was an insulating material, but when doped with bromine or iodine, the conductivity of 10^{-5} S/m was identified [40]. The exceedingly CP can be prepared by controlling the oxidation potential of a used aqueous solution. The electrochemical synthesis method was used to synthesize highly conductive PPy. Reduced anode size can decrease the yield of the product. But this method can favor in controlling morphology and thickness by regulating the electrochemical parameters in an appropriate way [41]. PPy can also be synthesized using iodine or bromide, in various aqueous/nonaqueous solvents.

21.3.4 POLY(P-PHENYLENEVINYLENE) (PPV)

Poly(p-phenylenevinylene) (PPV) was credited as the first electroluminescent material to fabricate organic light-emitting diodes, owing to its high optical property. PPV behaves as an insulator in pristine form. The conductivity property of pristine PPV depends on reaction conditions and structural behavior. Doping can increase the conductivity from 10^{-13} to 10^3 S/cm [42]. Improved mechanical properties make its application in the field of LED panels. Several methods exist for the synthesis of PPV. Wittig and Suzuki coupling reactions were studied broadly. In Witting reaction, coupling between bisaldehyde and aromatic biphosphonium results in the formation of PPV. Pd-catalyzed reaction between dobromo aromatic compounds and alkyl-substituted aryldiboron acids yields PPV in the Suzuki coupling reactions. Few other methods found in the literature were ring-opening polymerization, chemical vapor deposition, benzoin condensation, metathesis polymerization, etc. [35].

21.3.5 POLYTHIOPHENES

Polythiophene and its derivatives were studied broadly for their admirable optical property, thermal and environmental stability. Polythiophenes have their use in fabricating polymer LEDs, energy storage devices, anticorrosion coatings, photochromic devices, etc. PEDOT [poly(3,4-ethylenedioxythiophene)] is one of the most significant derivatives of polythiophenes, having high electrical and electro-optical properties. The insolubility problem of PEDOT can be overcome by adding a polyelectrolyte-like polysulfonates (PSS) into the matrix. PSS acts as both stabilizer and dopant by a charge balance conductivity [43]. The formed PEDOT:PS complex has good mechanical flexibility, high conductivity, and thermal stability [44]. The conductivity of polythiophene can be enhanced by solvent treatment and changing the concentration of PSS. In and around the 1980s, polythiophene was synthesized [36]. Current advanced techniques include oxidative synthesis, template-assisted synthesis, organometallic coupling reaction, and solvothermal techniques. Especially, for the synthesis of derivatives like PEDOT and PEDOT:PSS, microfluid systems, green synthesis, and electropolymerization techniques were used [45].

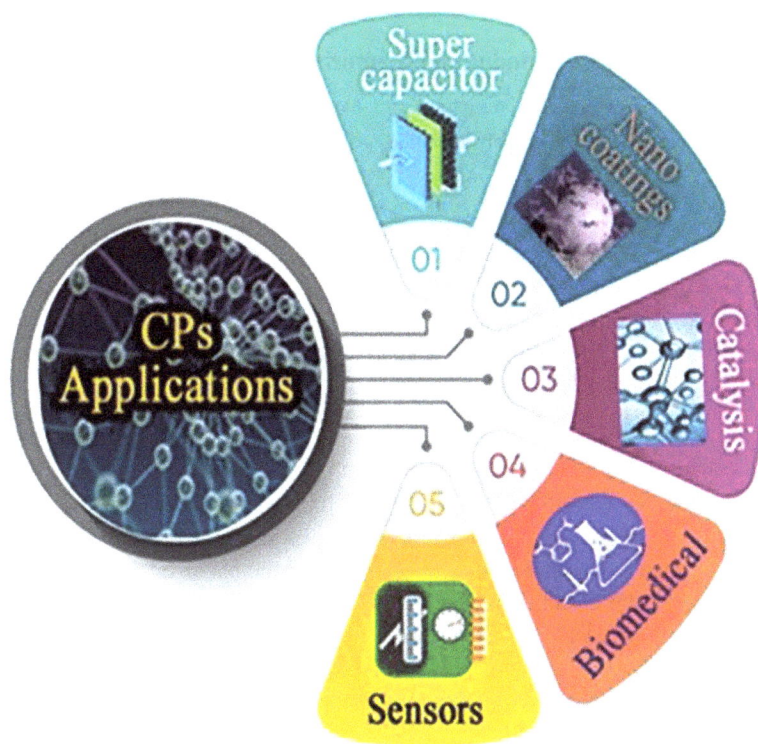

FIGURE 21.7   Applications of antifouling CPs.

21.4   APPLICATIONS OF ANTIFOULING CONDUCTING POLYMERS IN VARIOUS SECTORS

Increased attention to conducting polymers for various industrial and biomedical applications has been raised for years, as CPs have similar optical and electrical properties such as semiconductors and metals. Applications of antifouling CPs in various sectors are shown in Figure 21.7.

21.4.1   ANTIFOULING IN MEDICAL SECTOR: CONTROLLING GROWTH OF BIOFILMS

In the medical field, a daily large number of patients were treated for various diseases with different treatment strategies that include drugs, medical devices, and implants. Hence, in these treatment techniques, it is very important to consider the prevention of fouling in the devices that were implanted internally into the body. For diagnostic devices and medical implants, it is essential to reduce the interactions between the surface of material and biological fluids to get the desired therapeutic outcome. Avoiding the interactions with body fluids, proteins, and tissues is essential because such interactions lead to severe health problems and can also affect the device's safety and efficacy. Conducting polymers are found to modulate various properties like hydrophobicity, surface roughness, redox state, and exposed chemical groups. Studies are being performed to establish the antifouling nature of conducting polymers. Gomez-Carretero et al. demonstrated a study on the prevention of bacterial colonization using electrically conducting polymers. The study was conducted with composites of poly(3,4-ethylenedioxythiophene) (PEDOT) intricated with either chlorine (Cl), heparin (Hep), or dodecylbenzenesulfonate (DBS) against *Salmonella* biofilms. PEDOT served as an electron mediator for the metabolism of the bacteria and showed

TABLE 21.5

Applications of CPs as Conducting Membranes

Membrane Material	Preparation Method	Application	Reference
Polypyrrole/polyacryl onitrile (PPy–PAN)	Chemical polymerization	Removal of Azo dyes from waste water in water treatments and desalination	[48]
CNTs-polyaniline (PANI)	Electro polymerization	Defouling and electro-oxidation of organic compound	[49]
Polypyrrole (PPy)-terylene	Oxidative chemical polymerization in a vapor phase	Reduction of fouling in a membrane bioreactor	[50]

a significant effect on the growth of biofilms. The study also determined the functionality of poly(hydroxymethyl 3,4-ethylenedioxythiophene) with polystyrene sulfonate and silver nanoparticles and its possible use in bacterial sensors for detecting the bacterial colonies [46]. In another study by Ashlyn Young, the effect of p-doped semiconductive poly(3,4-ethylenedioxythiophene) polystyrene sulfonate (PEDOT:PSS) was studied at different oxidative states against *Escherichia coli* biofilms. Reduced PEDOT:PSS was found to have a negative influence on bacterial adhesion and the growth of the biofilms. This observation provides evidence for the use of PEDOT and polystyrene polymers as antifouling agents.

21.4.2 ANTIFOULING IN INDUSTRIAL SECTOR

21.4.2.1 Conducting Membranes

In recent times, electrically active polymers are employed in the preparation of membranes, which are used for multiple purposes. These membranes are prepared by electro or chemical polymerization with an appropriate porous base. The process of averting fouling and cleansing with conductive membranes relies on the characteristics of the conductive surface/foulants and also on electrochemical reactions or electrodynamics interactions at the membrane surface. In an electrochemical setting, the membrane behaves as the electrode at which the foulants may undergo direct or indirect oxidation at the surface of the membrane or the foulants may be eliminated in the form of bubbles [47]. The general composites of CPs used for fabricating conducting membranes include poly(vinylidene fluoride/polyaniline), sulfonated poly(phenylsulfone), and a blend of poly(carbonate)-poly(pyrrole). Examples of conductive membranes and their applications are given in Table 21.5.

21.4.2.2 CP in Corrosion Protection

CPs have been established as anti-corrosive agents for metals and their alloys. They can be employed either as primer alone, can be blended with conventional coatings, or used as an additive in coating formulations. The possible mechanisms of corrosion protection of CPs include (a) forming an anode shield layer of metal oxides on the surface, (b) accumulation on the substrate and undergo reduction to release anion dopant, (c) generation of electric circuits that restricts electron flow to oxidizing species, and (d) coatings may form a heavy, low porous adhering film on the metal surface that limits the approach of oxidants and thus prevents oxidation. Depending on the nature of the metal and corrosive environment, various strategies like copolymerization, multiple layer CPs, different dopants, composites, and nanocomposites are being employed (Table 21.6).

TABLE 21.6
Applications of CPs as a Corrosive Protectant

S. No	Strategy	Description	Example
1.	Copolymers of CP	Moderation by copolymer in the CP structural backbone. This process has an impact on the characteristics like conductance, porosity, and stability of the conducting polymer	Copolymerization of pyrrole with aniline to improve the anti-corrosive properties
2.	Multiple layers of CPs	Coating many layers of conducting polymers on the metal surfaces	Double layer PPy/PANI depositions exhibited higher performance than the plain coatings
3.	Doped CPs	Doping of CPs with suitable charge transfer agents to achieve desired properties	PANI doped with tungstate and diffused in vinyl resin primer for enhancing corrosion resistance
4.	Nanostructure design of CPs	Applying nanotechnology to prepare nanoparticles of conducting polymers. This improves the physical and chemical properties like hydrophobicity, stability, absorption, and magnetic effects of the CPs	Chemically prepared PANI nanostructures for improved corrosion protection

Polyaniline and its derivatives are extensively used in corrosion prevention coatings. It is stable, possesses different redox states, and can be synthesized easily. Various dopants may be used to augment the performance and conductivity of the polymer. For instance, PANI doped with poly(methylmethacrylate-*co*-acrylic acid) and benzoate-doped PANI with vinyl coatings were found to be effective in corrosion protection. PANI along with epoxy primer coating was found to be antagonistic against chlorine strikes and can be employed in the protection of concrete steel bars. Chemically prepared 10–15% concentrated polyaniline powders in 3.5% sodium chloride were observed to possess high corrosion resistance against mild steel. Integrating nanoparticulate PANI with aqueous-based alkyd paints for metals resulted in improved anticorrosion activity. The role of PPy in anticorrosion is also well established. A total of 1% w/w PPy with epoxy polyamide was found to improve protection activity than the control sample. Augmented anti-corrosive properties were observed by coating poly(5-amino-1-naphthol) on PPy films through cyclic voltammetry. Attempts have been made to impart PPy in a copper substrate. The PPy depositions that were formed on the copper surface exhibited good corrosion resistance in 3% NaCl solution. Few reports suggest the anti-corrosive nature of polythiophene and its derivatives against metal surfaces. PTh derivative [poly(3-decylthiophene-2,5-diyl)] when combined with epoxy paints in 2% w/w concentration improved the corrosion resistance [51].

21.4.2.3 Antifouling in Marine Systems: CP-based Antifouling Coatings

The conventional antifouling paints produce toxins to combat fouling in marine systems. The toxicity of the released chemical entities may lead to detrimental effects in marine environments and ecology. The best other possible method for defouling is the use of charged conducting polymers. The electrically active conducting polymers are employed in the manufacturing of paints and protective coatings for combating fouling in ships. Various studies are being conducted to prepare CP-based paints and establish their antifouling properties (Table 21.7).

TABLE 21.7

Application of CPs as an Antifouling Agent in the Marine System

Materials	Application	Reference
Polyaniline-based acrylic paint (0.001 M polyacrylic acid and 0.01 M aniline)	Antifouling and anti-corrosive in marine systems	[52]
Doped polyaniline (PAni-ES and PAni/DBSA) and its sulfonated form (SPAN)	Antifouling paint additives improving the efficiency of antifouling coatings	[53]

21.5 FUTURE PROSPECTS AND CONCLUSION

CPs are novel and promising materials in biomedical fields, yet they have not been investigated completely till today. This further increases the attention toward the development of antifouling CPs owing to their distinct properties. Various strategies are applied to explore and elaborate the application of antifouling CPs especially for implants, controlled and antibacterial coatings. Conversely, there are several challenges to face, while converting these into commercialized products. Durability and stability are the fundamental challenges to overcome especially, with respect to long-term applications like devices used for continuous monitoring. The development of new antifouling polymers must be studied constantly and should not be limited only to the current molecular design. More endeavors are needed to properly infer the impact of properties of CPs on antifouling nature. CPs possess few limitations relating to biocompatibility, the difference in vitro and in vivo studies, and cytotoxicity nature. There has been less research and still many studies are required to promote CPs to answer these queries.

REFERENCES

1. Maan AMC, Hofman AH, de Vos WM, Kamperman M (2020) Recent Developments and Practical Feasibility of Polymer-Based Antifouling Coatings. Adv. Funct. Mater. 30(32)2000936.
2. Halvey AK, Macdonald B, Dhyani A, Tuteja A (2019) Design of Surfaces for Controlling Hard and Soft Fouling. Philos. Trans. R. Soc. A Math. Phys. Eng. Sci. 377(2138)20180266.
3. Amy G (2008) Fundamental Understanding of Organic Matter Fouling of Membranes. Desalination 231(1–3)44–51.
4. Fane T (2016) Inorganic Scaling. Encyclopedia of Membranes; Drioli E, Giorno L, Eds. Berlin, Heidelberg: Springer, 1–2.
5. LoVetri K, Gawande PV, Yakandawala N, Madhyastha S (2010) Biofouling: Types. Impact and Anti-Fouling; Nova Science Pub Inc.: Hauppauge, NY, 105–128.
6. Watkinson AP, Wilson DI (1997) Chemical Reaction Fouling: A Review. Exp. Therm. Fluid Sci. 14:361–374.
7. Watkinson AP (1992) Chemical Reaction Fouling of Organic Fluids. Chem. Eng. Technol. 15(2)82–90.
8. Bott TR (1997) Aspects of Crystallization Fouling. Exp. Therm. Fluid Sci. 14:356–360.
9. Somerscales EFC (1999) Fundamentals of Corrosion Fouling. Br. Corros. J. 34:109–124.
10. Khatoon Z, McTiernan CD, Suuronen EJ, Mah TF, Alarcon EI (2018) Bacterial Biofilm Formation on Implantable Devices and Approaches to Its Treatment and Prevention. Heliyon 4(12)01067.
11. Steinhagen R, Müller-Steinhagen H, Maani K (1993) Problems and Costs Due to Heat Exchanger Fouling in New Zealand Industries. Heat Transf. Eng. 14:19–30.
12. Gahlot R, Nigam C, Kumar V, Yadav G, Anupurba S (2014) Catheter-Related Bloodstream Infections. Int. J. Crit. Illn. Inj. Sci. 4:161.
13. Almalki MA, Varghese R (2020) Prevalence of Catheter Associated Biofilm Producing Bacteria and Their Antibiotic Sensitivity Pattern. J. King Saud. Univ. Sci. 32:1427–1433.
14. Zoubos AB, Galanakos SP, Soucacos PN (2012) Orthopedics and Biofilm – What Do We Know? A Review. Med. Sci. Monit. 18:RA89–RA96.
15. McLaughlin-Borlace L, Stapleton F, Matheson M, Dart JKG (1998) Bacterial Biofilm on Contact Lenses and Lens Storage Cases in Wearers with Microbial Keratitis. J. Appl. Microbiol. 84(5)827–838.

16. Dasgupta MK (2002) Biofilms and Infection in Dialysis Patients. Semin. Dial. 15(5)338–346.
17. Saini R (2011) Oral Biofilm and Dental Implants: A Brief. Natl. J. Maxillofac. Surg. 2(2)228–229.
18. Ogbonnaya SK, Ajayi OO (2017) Fouling Phenomenon and Its Effect on Heat Exchanger: A Review. Front. Heat Mass Transf. 9(31)1–12.
19. Du X, Shi Y, Jegatheesan V, Ul Haq I (2020) A Review on the Mechanism, Impacts and Control Methods of Membrane Fouling in MBR System. Membranes (Basel) 10(2)24.
20. Wilson DI (2018) Fouling During Food Processing – Progress in Tackling This Inconvenient Truth. Curr. Opin. Food Sci. 23:105–112.
21. Farkas A, Degiuli N, Martić I, Dejhalla R (2020) Impact of Hard Fouling on the Ship Performance of Different Ship Forms. J. Mar. Sci. Eng. 8(10)748.
22. Song S, Demirel YK, De Marco Muscat-Fenech C, Tezdogan T, Atlar M (2020) Fouling Effect on the Resistance of Different Ship Types. Ocean Eng. 216:107736:1–17.
23. Civil Aviation Authority (2017) Corrosion and Inspection of General Aviation Aircraft CAP 1570:1–59.
24. Wang T, Wang X, Long Y, Liu G, Zhang G (2013) Ion-Specific Conformational Behavior of Polyzwitterionic Brushes: Exploiting It for Protein Adsorption/Desorption Control. Langmuir 29(22)6588–6596.
25. Fujii S, Kido M, Sato M, Higaki Y, Hirai T, Ohta N, Kojio K, Takahara A (2015) pH-Responsive and Selective Protein Adsorption on an Amino Acid-Based Zwitterionic Polymer Surface. Polym. Chem. 6(39)7053–7059.
26. Okur HI, Hladílková J, Rembert KB, Cho Y, Heyda J, Dzubiella J, Cremer PS, Jungwirth P (2017) Beyond the Hofmeister Series: Ion-Specific Effects on Proteins and Their Biological Functions. J. Phys. Chem. B 121(9)1997–2014.
27. Chen Y, Luo SC (2019) Synergistic Effects of Ions and Surface Potentials on Antifouling Poly(3,4-Ethylenedioxythiophene): Comparison of Oligo(Ethylene Glycol) and Phosphorylcholine. Langmuir 35(5)1199–1210.
28. Zhang H, Molino PJ, Wallace GG, Higgins MJ (2015) Quantifying Molecular-Level Cell Adhesion on Electroactive Conducting Polymers Using Electrochemical-Single Cell Force Spectroscopy. Sci Rep 5:13334.
29. Marzocchi M, Gualandi I, Calienni M, Zironi I, Scavetta E, Castellani G, Fraboni B (2015) Physical and Electrochemical Properties of PEDOT:PSS as a Tool for Controlling Cell Growth. ACS Appl. Mater. Interfaces 7(32)17993–18003.
30. Wu JG, Chen JH, Liu KT, Luo SC (2019) Engineering Antifouling Conducting Polymers for Modern Biomedical Applications. ACS Appl. Mater. Interfaces 11(24)21294–21307.
31. Casas JP, Shah T, Hingorani AD, Danesh J, Pepys MB (2008) C-Reactive Protein and Coronary Heart Disease: A Critical Review. J. Intern. Med. 264(4)295–314.
32. Iwasaki S, Kawasaki H, Iwasaki Y (2019) Label-Free Specific Detection and Collection of C-Reactive Protein Using Zwitterionic Phosphorylcholine-Polymer-Protected Magnetic Nanoparticles. Langmuir 35(5)1749–1755.
33. Goda T, Kjall P, Ishihara K, Richter-Dahlfors A, Miyahara Y (2014) Biomimetic Interfaces Reveal Activation Dynamics of C-Reactive Protein in Local Microenvironments. Adv. Healthc. Mater. 3(11)1733–1738.
34. Ueda T, Murakami D, Tanaka M (2018) Analysis of Interaction between Interfacial Structure and Fibrinogen at Blood-Compatible Polymer/Water Interface. Front. Chem. 6:542.
35. Peres LO, Varela H, Garcia JR, Fernandes MR, Torresi RM, Nart FC, Gruber J (2001) On the Electrochemical Polymerization of Poly(p-Phenylene Vinylene) and Poly(o-Phenylene Vinylene). Synth. Met. 118:65–70.
36. Kiari M, Berenguer R, Montilla F, Morallón E (2020) Preparation and Characterization of Montmorillonite/PEDOT-PSS and Diatomite/PEDOT-PSS Hybrid Materials. Study of Electrochemical Properties in Acid Medium. J. Compos. Sci. 4:51.
37. Park C-S, Kim DH, Shin BJ, Kim DY, Lee H-K, Tae H-S (2016) Conductive Polymer Synthesis with Single-Crystallinity via a Novel Plasma Polymerization Technique for Gas Sensor Applications. Materials (Basel) 9:812.
38. Wang Y, Levon K (2012) Influence of Dopant on Electroactivity of Polyaniline. Macromol. Symp. 317–318:240–247.
39. Yang L, Yang L, Wu S, Wei F, Hu Y, Xu X, Zhang L, Sun D (2020) Three-Dimensional Conductive Organic Sulfonic Acid Co-doped Bacterial Cellulose/Polyaniline Nanocomposite Films for Detection of Ammonia at Room Temperature. Sens. Actuators B: Chem. 323:128689.

40. Kiebooms R, Menon R, Lee K (2001) Synthesis, Electrical, and Optical Properties of Conjugated Polymers. Handbook of Advanced Electronic and Photonic Materials and Devices. Academic Press, Elsevier, 1–102.

41. Rasmussen SC (2020) Conjugated and Conducting Organic Polymers: The First 150 Years. ChemPlusChem 85(7)1412–1429.

42. Namsheer K, Rout CS (2021) Conducting Polymers: A Comprehensive Review on Recent Advances in Synthesis, Properties and Applications. RSC Adv. 11:5659–5697.

43. Faisal M, Harraz FA, Jalalah M, Alsaiari M, Al-Sayari SA, Al-Assiri MS (2020) Polythiophene Doped ZnO Nanostructures Synthesized by Modified Sol-Gel and Oxidative Polymerization for Efficient Photodegradation of Methylene Blue and Gemifloxacin Antibiotic. Mater. Today Commun. 24: 101048.

44. Babudri F, Farinola GM, Naso F (2004) Synthesis of Conjugated Oligomers and Polymers: The Organometallic Way. J. Mater. Chem. 14:11–34.

45. Kim Y, Kim J, Lee H, Park C, Im S, Kim JH (2020) Synthesis of Stretchable, Environmentally Stable, Conducting Polymer PEDOT Using a Modified Acid Template Random Copolymer. Macromol. Chem. Phys. 221:1900465.

46. Gomez-Carretero S, Libberton B, Svennersten K, Persson K, Jager E, Berggren M, Rhen M, Richter-Dahlfors A (2017) Redox-Active Conducting Polymers Modulate *Salmonella* Biofilm Formation by Controlling Availability of Electron Acceptors. NPJ Biofilms Microbiomes 3:19.

47. Ahmed F, Lalia BS, Kochkodan V, Hilal N, Hashaikeh R (2016) Electrically Conductive Polymeric Membranes for Fouling Prevention and Detection: A Review. Desalination 391:1–15.

48. Salehi E, Madaeni SS (2010) Influence of Conductive Surface on Adsorption Behavior of Ultrafiltration Membrane. Appl. Surf. Sci. 256(10)3010–3017.

49. Duan W, Ronen A, Walker S, Jassby D (2016) Polyaniline-Coated Carbon Nanotube Ultrafiltration Membranes: Enhanced Anodic Stability for in Situ Cleaning and Electro-Oxidation Processes. ACS Appl. Mater. Interfaces 8(34)22574–22584.

50. Liu L, Liu J, Gao B, Yang F, Chellam S (2012) Fouling Reductions in a Membrane Bioreactor Using an Intermittent Electric Field and Cathodic Membrane Modified by Vapor Phase Polymerized Pyrrole. J. Membr. Sci. 394–395.

51. Deshpande PP, Jadhav NG, Gelling VJ, Sazou D (2014) Conducting Polymers for Corrosion Protection: A Review. J. Coat. Technol. Res. 11:473–494.

52. Maral M, Dhumal PV, Amar Charkha P, Engineering DOA, Patil DY, Lohegaon SOET (2016) Conducting Polymer Based Antifouling Coating. IJSRD Int. J. Sci. Res. Dev. 4:215–220.

53. Baldissera AF, De Miranda KL, Bressy C, Martin C, Margaillan A, Ferreira CA (2015) Using Conducting Polymers as Active Agents for Marine Antifouling Paints. Mater. Res. 18(6)1129–1139.

22 Antifouling Properties and Biomedical Applications of Conducting Polymers

Trinath Biswal,[1] *Dharmendra K. Jena,*[2] *and Prafulla K. Sahoo*[2,3]

[1]Department of Chemistry, VSS University of Technology, Burla, India

[2]Department of Chemistry, Utkal University, Bhubaneswar, India

[3]School of Applied Sciences, Centurion University of Technology and Management, Jatni, India

CONTENTS

22.1 INTRODUCTION

The conducting polymers (CPs) are treated as a unique class of organic polymeric materials having optical and electrical properties almost similar to metals and semiconductors. Diversified techniques are adopted to modify CPs and to make them suitable for biomedical applications including biomaterials for medical implants, biosensors, biomedicine (regenerative medicine, bioengineering), tissue engineering, drug delivery, and actuator [1]. The surface characteristics of CPs play a vital role in controlling and regulating their interactions with various kinds of proteins present in the human body tissues or cells in biological surroundings. CP proved itself as a promising candidate to promote different cellular mechanisms, which makes it suited for different medical applications. The CPs are highly attractive in medical sectors, because of their brilliant response toward the electric field from different kinds of tissues such as muscle, epithelium, nervous tissue, and various connective tissues. The CPs are used for increasing the sensitivity, stability, and speed of different biomedical devices along with their interfaces associated with the biological tissues. Because of conductive bioelectrodes or biointerfaces, it acts as a bridge between the usual electronics with the tissues and cells [2, 3]. Although CPs associated with organic bioelectronics exhibit a noticeable achievement in different biomedical applications, but many challenges are associated with it especially manufacturing of implant devices application for long-term basis. One of the vital concerns of CP is the wide adsorption of cells and biomolecules, which produces inflammatory activity, which is the reason for electrode malfunction. Hence, antifouling biomaterials or antifouling coating has to be developed to decrease the damages or destructions produced by nonspecific adsorption [4, 5]. The antifouling concept is applied to different inorganic and organic biomaterials for applications in biomedical sectors. A UV-irradiated photografting technique was developed, which uses benzophenone in combination with additional fragmentation chain transfer (RAFT) having living polymerization technique, which is reversible for modification and enhancement of antifouling surface properties of the material poly (dimethylsiloxane) (PDMS). A polymer brush having antifouling properties was grafted on the surface of the polyurethane by the photoinduced anchoring of the initiators and also surface-initiated atom transfer radical polymerization (SI-ATRP). The coating of antifouling on the surface of the polyurethane efficiently decreases the adsorption of bacteria, bovine serum albumin (BSA), and barnacle cyprid settlement [6, 7]. The various technique of grafting of betaine-based polymer of zwitterionic structure on the cellulose membranes stimulates the property of antifouling and also the hemocompatibility property of the cellulose membranes. The antifouling poly(ethylene glycol) ($H-(O-CH_2-CH_2)_n-OH$) on grafting with PVC or PMMA polymer blends forms a new material, which possesses the excellent capability of coagulation prevention. The TiO_2 surface can be immobilized and modified by using a dopamine layer (adhesive) with a moiety of zwitterionic sulfobetaine, which provides the outstanding property of bacterial adsorption. Hence, conducting polymeric materials having antifouling characteristics are expected to be suitable for stimulating the new interesting biomedical applications [8].

22.2   APPLICATIONS OF CONDUCTING POLYMER

The CPs can easily interact with biological tissues, therefore biocompatible. Hence, it was found that CP is used as an excellent candidate material for various medical implants and applications. There is a huge variety of CPs are now developed and the classification of these CPs is normally based upon the nature of the electric charge including the conductive nanomaterials, ions, and delocalized π electrons. The popularly known CPs are polyacetylene, polyphenylene, polythiophene, polyaniline, and polypyrrole. There are different natural biopolymers such as chitosan, cellulose, collagen, pectin, gelatine, and fibroin that can be converted into CPs by using different fabrication techniques [9]. The various applications of CPs are represented in Figure 22.1.

The CPs on conjugation or blending with suitable materials can be used for manufacturing the electrodes of the prosthetic device and neural tissue. There are huge numbers of clinical applications of CPs currently developed, which are highly important for various medical applications.

22.3   APPLICATION OF CPs IN BIOMEDICAL FIELDS

In the last few years, more focus is given to the CPs for biomedical applications. Most of the biological cells are highly sensitive toward electrical impulses, hence can be successfully applied in tissue engineering for modification of cellular activities. The optical and electrical properties of CPs are almost comparable to that of semiconductors and metals and mechanical properties comparable to polymers. The CPs can be easily synthesized and used in different biomedical fields because of their attractive properties such as

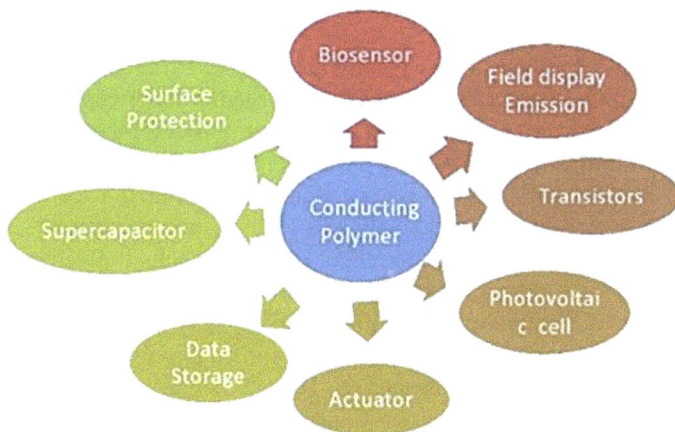

FIGURE 22.1 Various applications of conducting polymer.

- Organic nature of polymeric material
- Highly sensitive to electrical stimuli
- Effective functionalization with living tissues and biomolecules
- Electronic and ionic conductivity [10]

The CPs are highly effective toward drug delivery, biosensors, tissue engineering applications, actuators, and neural prosthetic devices. The conventional CPs are friable and stiff having characteristic elastic moduli; therefore, their use in the biomedical field is limited and not suitable. Hence, to solve this problem new amazing composite or hybrid CP-based materials are designed and developed having comparatively soft, flexible, and higher robust mechanics [11]. The various important biomedical applications of CPs are presented in Figure 22.2.

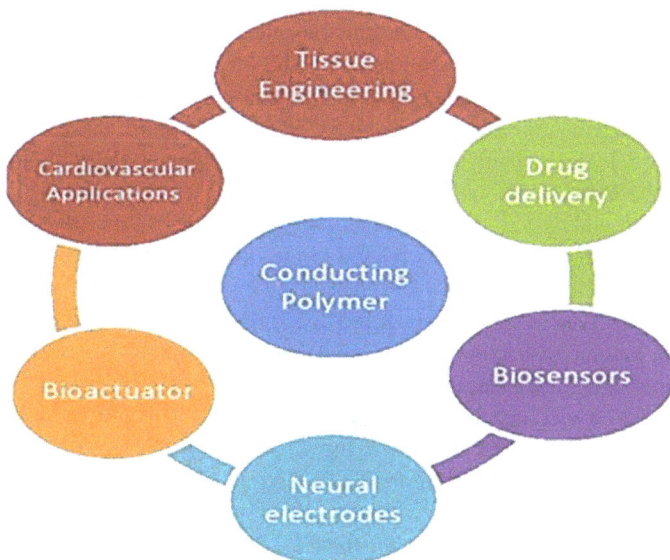

FIGURE 22.2 Different biomedical applications of conducting polymers.

22.3.1 Drug Delivery System

Presently, drug-delivery techniques are used which are effective in the controlled release of drugs to the targeted tissues, but the application of drugs to the targeting cell groups is effective rather than specific cells. The drug delivery technique has a potential impact on the effectiveness of therapy. In the conventional technique of drug-delivery systems, the drug releases immediately and quickly after entering into the human regulatory system. Hence, at the initial stage, the concentration of the drug is much more than a therapeutic range of use, and gradually the concentration of drugs decreases and comes below the therapeutic range. Hence, the actual effective time for the drug activeness is short. The multiple releases of drug release are normally due to differential rates of degradation of the reservoir membrane, which is formed by poly(lactic-co-glycolic acid) (PLGA). The degradation of the membrane was regulated and controlled by monitoring the molecular weight, thickness of PLGA, and chemical composition [12]. The microchip device fabricated by the CPs can effectively release drugs without any external stimulation and offers the triggered drug release system. The CPs can be effectively used primarily biomolecules and release drugs by the application of electrical stimulation. The development of drug-delivery methodologies will provide amazing opportunities, which are previously not applied for conventional oral formulations [13].

The soluble form of polyaniline (PANI) on grafting to lignin, polypyrrole (PPy), and poly (anilinesulfonic acid) serve as efficient scavengers of the stable free radical of 1, 1-diphenyl-2-picrylhyd (DPPH). This characteristic is specifically useful in the living tissues affecting oxidative stress and has the capability of decreasing the excess reactive radical species, which is quite desirable. The reducing capability of the free radical of CPs tallies with the scavenging capability of DPPH radicals having four or two radical scavenged/pyrrole or aniline monomer. Hence, this proves that CPs are found to be a powerful antioxidant in a biological system. The hydrogel layer of composites of CPs is excellently active for application in drug delivery, which can promote biomolecules of even higher concentrations like heparin. The CPs in the redox state possess the capability of promoting the controlled drug release like dexamethasone [14].

The CPs in electrical stimulation have been successfully utilized in a huge number of therapeutic proteins and some drugs such as nerve growth factor (NGF), which included heparin and dexamethasone. The increase in the release of heparin from the hydrogel immobilized onto films of PPy is due to electrical stimulation. Again, it was found that the polymerization of nanofibers of electrospun PLGA produces poly(3,4-ethylenedioxythiophene) nanotubes (PEDOT NTs), which can be used potentially in the release of dexamethasone drug. Hence, dexamethasone was introduced in the nanofibers of PLGA, and after that PEDOT was undergoing polymerization surrounding the dexamethasone-loaded PLGA nanofibers. The nanofibers of PLGA gradually degraded and the molecules of dexamethasone persisted inside PEDOT NTs. Again, these PEDOT NTs promote the controlled drug release by using electrical stimulation. The reason behind this is the change in the volume of the PEDOT NT, which is due to electrical stimulation and is the cause of the removal of anions. In addition to the enormous potential of the CPs in the applications of drug delivery, they also possess some demerits related to the initial burst release of the added drug and the hydrophobic property of the CPs [15]. Despite some demerits, still, CPs are of interest to develop a suitable drug delivery system specially targeted drug release to cancer, cardiovascular and neural applications. There is some restriction in the drug delivery system based on CPs, which depends on the molecular weight of the drug and the nature of the dopant added to it. The used drugs must have less molecular weight for ejecting the drugs out from the film of the CPs. To solve these demerits drug molecules are attached to the surface of the CPs and not incorporated inside the structure of the CPs. They used PPy (as CP) in the form of substrate and biotin as dopant materials during the process of electrochemical polymerization. In the solution form, biotin is usually negatively charged, which is treated as a counter-ion for the CP (PPy). The columbic forces of attraction in between the positively charged PPy and negatively charged biotin facilitate the attachment of biotin molecules on the surface of the PPy film. The attachment of the biotin molecule on the surface of the PPy is highly stable when the activation is almost negligible. The streptavidine possesses four binding sites and has the capability of binding biotin and biotinylated drugs like NGF. When the electric potential of 3 V was applied to the CP mainly PPy, the reduction reaction generally happened and PPy changed electrically neutral [16, 17].

The mesoporous form of silica nanospheres (MSNs) was highly effective because of their excellent property of biocompatibility, nontoxic and chemical stability. The NGF was loaded on the MSNs, which entered in its pores and a new advanced composite is formed, which immobilizes through the electrochemical polymerization process of PPy at the electrode surface. The composite formed served as a dopant material along with poly (styrenesulfonate) (PSS) during the process of electropolymerization of PPy. MSNs have a high aspect ratio, which is the cause of more efficiency in drug loading and preserves the drugs during the process of loading or implantation [18].

22.3.2 USED AS BIOACTUATORS

The CPs are effectively used as actuators in the biomedical field, because of their volume change during the oxidation or reduction reactions. The actuators fabricated by using CPs have been used in the development of biomedical devices and artificial muscles. The piezoelectric polymeric materials, electrostrictive polymers, and shape-memory alloys have been used as significant materials for actuators. During the passage of electric current through the alloy, it is heated and the shape-memory alloy undergoes a change in phase transformation and blending occurs. The piezoelectric and electrostrictive polymeric materials change shape by the application of an external electric field. The CPs possess many benefits over any actuator materials [19]. The CPs possess higher tensile strength and Young's modulus, which fit these materials as appropriate for action of large forces. The electric potential needed for CPs to exert the necessary force is always less than other materials. The CPs can easily be implanted and function in an aqueous environment, such as blood, because of their excellent corrosion resistance. However, CPs have lower efficiency for converting electrical energy into mechanical energy. The CPs are comparatively more impressive than other materials related to an actuator, because of their unique properties associated with biomedical applications [20].

22.3.3 FEATURES OF AN IDEAL CP

The features of an ideal CP are as follows.

- It can be controlled and regulated electrically
- It possesses the property of large strain, which is encouraging for bending or volumetric, linear actuators
- Possess adequately higher strength
- Needed 1 V or less voltage for actuation
- Continuously attains maximum and minimum levels
- It is operating even at room temperature
- It is lightweight and can be easily microfabricated
- It can be easily and perfectly functioning within body fluids [21]

The structure of some important CPs are shown in Figure 22.3.

22.3.4 APPLICATIONS IN TISSUE ENGINEERING

The pure CPs are highly brittle and very tough to synthesize as film or any other desired size. Hence, the CPs are blended with some other degradable polymeric materials, and blended polymeric biomaterials are used effectively for tissue engineering. The natural and synthetic polymeric materials

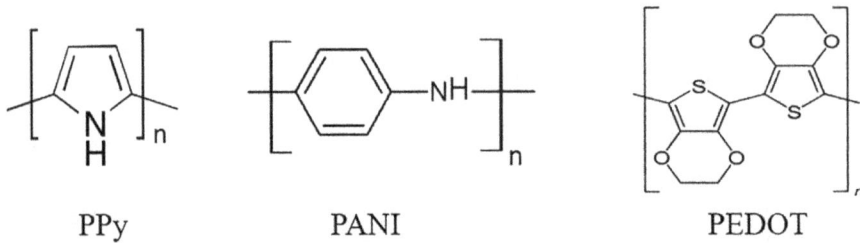

FIGURE 22.3   Structure of some important conducting polymer.

such as PLA, PCL, PLGA, silk fibroin (SF), and chitosan are identified as useful candidates, which on blending with the CPs like PPY and PANI are highly applicable in tissue engineering [22]. The usual properties of CPs desired for the application in tissue engineering include redox stability, hydrophobicity, conductivity, reversible oxidation, surface topography biocompatibility, and three-dimensional (3D) geometry. Because of the capability of cells toward electrical stimulation, CPs are widely and more popularly applied in tissue engineering. But their applications are prevented because of their nondegradable nature [23].

22.3.4.1   Copolymer Films of CP for Tissue Engineering Applications

Although CPs are inherently nonbiodegradable, the pyrrole/aniline-based copolymers possess similar electroactivity like CPs along with the property of biodegradability. Now PEDOT is found to be more suitable in the biomedical field because of its excellent chemical stability and electrical conductivity. For example, the property of biocompatibility of PEDOT and PPy in the NT or film structure is used for the culture of neuronal cells. Although CP provides many applications in the trapping of the protein molecules, the subsequent trapping of protein molecules and sustaining their bioactivity are the major challenge [24].

The different properties of the CP including huge surface area, electrical properties, and electro-activities make it a suitable candidate used in tissue engineering. For living cells, electrical stimulation is the cause of modulating cell-to-cell attachment, cell differentiation, cell proliferation, and cell migration. The polymer nanofibers synthesized by using CP and some other biocompatible natural polymer forms nanofiber composite, which is conductive, possesses high porosity, huge surface area, good biodegradability, and excellent biocompatibility. These are the major properties required for tissue engineering applications. The various kinds of CPs such as PLCL nanofibers, PLA-PPY nanofibers, PLGA nanofibers, PLCL nanofibers, and PCL nanofibers are found to be highly effective in the application of tissue engineering [25].

22.3.4.2   Nanofibers of CP for Application in Tissue Engineering

The biomaterials are the mimic structure of the extracellular matrix (ECM). The various kinds of synthetic and natural biodegradable polymers blending with CPs on further electrospun to the nano-fiber form are highly appreciable in tissue engineering applications. The fibers synthesized by elec-trospinning possess a large surface area, adequate porosity, and the diameter of the fiber is tunable in between several micrometers. The composite of conductive nanofiber associated with PANI and gelatin was synthesized through the process of electrospinning by using N, N-dimethylformamide/H_2O as the solvent, which is then stabilized by the addition of an appropriate quantity of glutaral-dehyde, which serves as cross-linking agent. The scaffolds of conductive nanofibrous synthesized by electrospun using PLA/PANI, and the blend of PLA and PANI/CSA in hexafluoroisopropanol is found to be highly suitable for cardiac tissue engineering. An electroactive nanofiber synthesized by using PPy and PCL in the process of vapor phase electrospinning technique possesses good mechanical strength and dimensional stability. The conductive nanofibers PCL/PANI have been synthesized in the process of electrospinning, which possesses excellent properties that are required for application in tissue engineering. The PPy on coating over poly(L-lactide) (PLLA) forms the biomaterial by the process of electrospinning, which exhibits better cell biocompatibility and mechanical property, therefore highly useful for tissue engineering [26, 27].

22.3.4.3   Conducting Hydrogels for Application in Tissue Engineering

Hydrogel is hydrophilic polymers having a 3D network, which can retain a huge quantity of H_2O and swells because of physical or chemical cross-linking in between the individual polymer molecules. Hence, hydrogel is termed a superabsorbent material. Based on definition, the material which can absorb more than 10% by volume or weight of the material is called hydrogel, which is highly flexible analogous to natural tissue because of high water content. It is rubbery like nature same as soft tissues, possesses the exceptional property of biocompatibility and tunable properties. Different kinds of conducting hydrogels are developed, which are highly applicable for tissue engineering [28].

- An electroactive hydrogel of poly(ethylene dioxythiophene) (PEDOT)/poly(acrylic acid) (PAA) was synthesized through a free radical method of polymerization of acrylic acid with the double bond modification of poly(ethylene glycol) diacrylate and PEDOT. The scaffolds of this CP hydrogel possess highly tailored properties including adequate swelling ratio, electroactivity, and appropriate mechanical properties. Therefore, it is suited for application in tissue engineering [29].
- Conductive scaffold biomaterials based on heparin were synthesized. The degradable hydrogels of PANI and gelatin were synthesized. The copolymer of gelatin-g-PANI (GP) having electroactive was developed via grafting PANI onto the backbone of gelatin, where gelatin was added as a cross-linking agent, which cross-linked the GP at normal body temperature. The conductivity of hydrogel gradually increases with the increase in percent of PANI in copolymer GP.
- A conductive copolymer of chitosan-g-PANI (QCSP) was developed through grafting PANI through *in situ* process onto the backbone of QCS. The oxidized dextran was added as a cross-linking agent to QCSP through Schiff base reaction resulting in the formation of injectable conductive hydrogels [30].
- A conductive hydrogel having self-healing properties was synthesized based on chitosan-g-aniline tetramer and C_6H_5CHO through Schiff base reaction. This hydrogel synthesized exhibits excellent adhesiveness, quick self-healing capability to host tissue, and also excellent antibacterial properties. The conductivity of hydrogel is normally comparable to cardiac tissue. Although the conducting hydrogels are popularly used for the development of different kinds of biomaterials applicable as implant materials, the presence of cross-linking agents and initiators such as isocyanate and glutaraldehyde are exceptionally toxic to cells. Again the hydrogel synthesized through chemically cross-linking is less beneficial than synthesized through physical cross-linking because the physical cross-linking normally experience mild gelation, for which no cross-linking agent is necessary [31].
- The synthesis of degradable electroactive hydrogel poly-(caprolactone)-poly(ethylene glycol)-poly (caprolactone) on grafting with aniline tetramer through thermo-gelling process possesses the excellent property of injectability and electroactivity [32].

22.3.5   3D Scaffolds of Conducting Composite for Application in Tissue Engineering

Apart from the conductive hydrogel, the conductive nanofibers, composites, and blends normally in the form of membrane or film do not show the 3D structures, which is highly beneficial for tissue regeneration. Hence, 3D scaffolds of the conductive composite were designed and synthesized for application in tissue engineering.

- A scaffold of 3D conductive material was developed by using PEDOT poly(4-styrenesulfo-nate) (PEDOT:PSS) into a nanocomposite of bioactive glass and gelatin. The PEDOT:PSS was cross-linked through physical method into bioactive glass and gelatin matrix and cross-linked chemically through N-hydroxysuccinimide and N-(3-(dimethylamino)propyl)-N′-ethylcarbodiimide. The PEDOT:PSS is the cause of an increase in thermal stability, which offers resistance to the mechanical properties and enzymatic degradation of the scaffold of CP.
- An electroactive form of the 3D scaffold was synthesized from conductive poly (aniline-co-N-(4-sulfophenyl)aniline) (PASA) conductive polymer and SF, possesses adequate porosity, mechanical properties, which is necessary for cell culture. Because of excellent processing flexibility and thermal-induced phase separation (TIPS), the nanofi-brous scaffolds of these materials are highly effective for the regeneration of tissue.
- The nanofibrous scaffolds of PLA/PANI nanocomposite were synthesized by in situ pro-cess of polymerization of PLA/THF solution in aniline. The scaffolds designed from it are normally structurally mimicked the native tissues and the 3D cellular structure of it, which is highly appreciable for cardiac muscle and skeletal use [33, 34].

22.3.6 USED AS NEURAL ELECTRODES

The CPs are also be used for stimulation and neural recording of neural prostheses mainly deep brain stimulators and cochlear implants. The electrode designed can able to induce specific responses toward neurons through the application of electrical stimulation, which offers drive or diagnostics external robotics. The implantations of conventional metals show poor performance for neural recording and stimulation, probably because of incongruity between metal electrodes and brain tissue. By application of the CPs at the interface of the electrode-tissue, the fibrous tissue width may be decreased. In addition to that, the CPs are the cause of enhancing the effective sur-face area with reduction of impedance and, hence, improve the performance of signal intensity. The CP nanofibers within a scaffold of hydrogel coating were synthesized, which exhibit high charge density, low impedance, and very good capability of controlling and regulating the drug release. The biodegradable nanofiber synthesized by the electrospun process facilitates for controlling drug release and the layer of hydrogel sustains the drug release. The PEDOT NTs can be used for bio-sensing to identify between chronic and acute responses within the brain tissues. The PEDOT and PPy NTs show better adherence on the electrode surface than their film counterparts. A platinum electrode on coating with PEDOT provides more changes, which permits the transmission to neural tissues without any destruction or breaking the electrode. In addition to that, PEDOT possesses bet-ter electrical stimulation in comparison to platinum. The PEDOT NT, if used for neural electrodes, improves the quality of signals [35, 36]. The properties of some important CPs and their applica-tions are presented in Table 22.1.

22.4 CARDIOVASCULAR APPLICATIONS

Cardiovascular disease is treated as a prominent cause of death worldwide; therefore, the appropri-ate therapeutic interventions are now focused on by scientists and researchers. The surgical and pharmacological process histrionically enhances the lifestyle because of controlling the cardiovas-cular disease of the patients. The biomaterials from natural origin possess significant potential in cardiac regeneration and repair either as a scaffold of ECM substitute or as a carrier for application of drug delivery. The complication of the pathways for electrical conduction in our heart is now given amazing attention in the area of cardiac tissue engineering. The cardiac cells must form appropriate matrix architecture and intercellular connections for permitting the electrical impulses to travel in the proper direction at the usual velocity. Again the cardiac muscle shrinkage is the cause of generating electrical impulsive waves, which is the cause of impacting mechanical stretch

TABLE 22.1
Properties and Applications of Some Important CPs in the Biomedical Field [37]

Name of the CP	Technique of Synthesis	Properties	Application
Polypyrrole (PPy)	Chemical and electrochemical synthesis	High conductivity on doping with iodine, brittle in nature, opaque, and amorphous material	Biosensors, neural prosthetics, drug delivery, bioactuators antioxidants, and cardiovascular applications
Poly(3,4-ethylenedioxythiophene) (PEDOT)	Chemical and electrochemical synthesis	High temperature stability; capability to suppress 'thermal runaway' of the capacitor; transparent conductor, intermediate band gap and less redox potential, good conductivity	Biosensors, neural prosthetics. drug delivery, antioxidants
Polythiophenes (PT)	Chemical and electrochemical synthesis	Good optical property and excellent electrical conductivity	Food industry and biosensors
Polyaniline (PANI)	Chemical and electrochemical synthesis	In the class of semiflexible polymer rod, needed simple doping process exists as in the form of bulk films, high conductivity	Biosensors, bioactuators, food-processing industry, antioxidants, drug delivery, and cardiovascular application

to our heart. Any of the deficiencies in the conducting pathway of the electrical signal is the cause of cardiovascular diseases. Hence, a scaffold based on CP can be potentially beneficial in the treatment of cardiovascular diseases. The PANI is an electroactive CP used effectively for a culture of excitable cells for the application of neuronal and cardiac tissue engineering. The PANI is coupled with oligopeptides by covalent bonds. The nanofiber scaffold of PANI–gelatin synthesized by electrospinning proved as a potential candidate for the application of cardiac tissue engineering. The composite PPy–heparin is found to be highly effective for endothelial cell growth and the composite PPy–hyaluronic acid (HA) stimulates vascularization. The PPy containing erythrocyte possesses the capability of binding more antibodies as compared to unchanged PPy. The PANI–collagen complexes exhibit viability in a cardiac cell, which was well-maintained with an increase in biocompatibility by applying scaffolds of modified CP. The appropriate mechanical properties of CPs exhibit a significant role in the application in the cardiac field. Hence, continuous attempts were made to enhance the mechanical properties of CPs either by blending or composite or doping with appropriate dopant materials to get the mimic of native myocardium [38, 39].

22.5 ANTIFOULING PROPERTIES OF THE CONDUCTING POLYMERS

Any substance that prevents fouling is known as an antifouling substance. Many of the CPs show antifouling properties. The antifouling CPs are categorized into three groups such as polymers are having oligo(ethylene glycol) (OEG) or poly (ethylene glycol) (PEG) having zwitterion structure having some hydrophilic parts like hydrophilic peptides or glycan. For immobilization of the antifouling groups on the surfaces of the CPs, some groups have to start their molecular designing forming monomers of antifouling moieties. The antifouling CPs can be synthesized by oxidation polymerization reaction or may be directly synthesized through electropolymerization. The antifouling moieties create a structure similar to the monolayer on the surface of the CP. The density/unit of the antifouling moieties is found to increase with an increase in roughness of the surface or the creation of nanostructure.

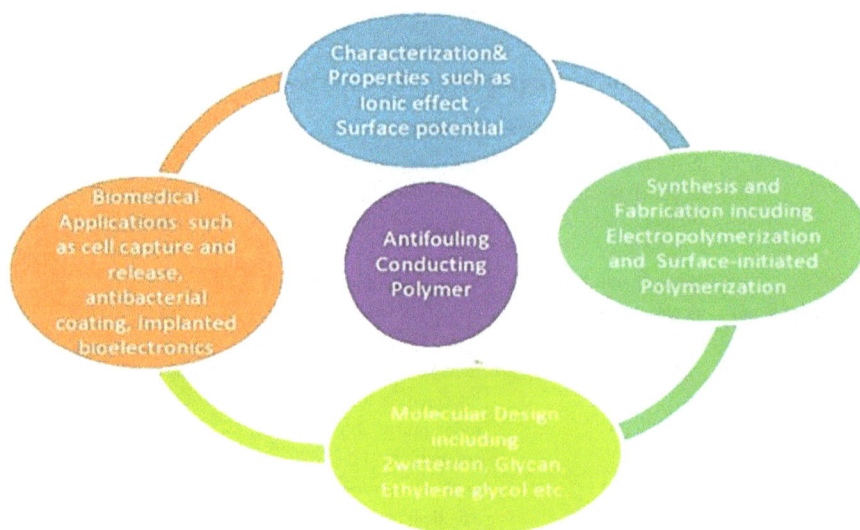

FIGURE 22.4 Various properties and applications of antifouling conducting polymer.

The antifouling layer may be developed by using the process of surface-initiated polymerization mainly RAFT and ATRP or via covalent bonding directly. The three significant applications, where the antifouling CPs, are highly important are sustainable biosensors, controlled cell release and capture, and antibacterial coating. To control the nonspecific adsorption properly, the formation of antifouling sensing interfaces and various electrochemical CP antifouling biosensors have to be designed and developed. The manufacturing of CP-based electrochemical biosensors having antifouling ability requires three different kinds of strategies. The initial strategy is copolymerization. The antifouling materials such as HA or polyethylene glycol were doped on the surface of the CP-like PEDOT in the electrochemical process of polymerization on the surface of the electrode and after those biorecognition elements like CEA or AFP antibody have to be immobilized on the surface of antifouling materials to fabricate the required biosensor. The second strategy is the deposition of PANI or PEDOT on the surface of the electrode and after that immobilization of antifouling substances with bio-recognition of the molecules has to be developed for manufacturing biosensors. The third strategy is the design and development of antifouling monomers of functional CPs including sulfobetaine-EDOT, hyperbranched polyglycerol-EDOT, and phosphocholine-EDOT, and after that, these were electrodeposited on the surface of the electrode for detecting the small molecules like dopamine or glucose or fit it for biosensors by immobilizing the bio-recognition molecules like AFP antibody [40, 41]. The various properties and applications of antifouling CPs were represented in Figure 22.4.

Although the first process is very simple, many of the antifouling substances are entrapped by the CPs, and only those molecules found at the surface show the ability of antifouling. The second process needs the CPs for attaching the functional groups for the CPs, which does not contain any active functional groups like PEDOT, and is required for doping or modification of the material with other substances as per the necessity along with the stability of the CPs. The stability and conductivity of the polymeric materials will be influenced after the immobilization of the antifouling substances. The third process, the antifouling functional monomers is time-consuming and complex and till now not commercialized [42]. Some specific antifouling properties of the CPs are as follows:

22.5.1 Ionic Effect on Antifouling Properties

The ionic effect is observed, because of the antifouling property of the CPs, specifically for surfaces of zwitterion. The ions are normally divided into two types such as kosmotropes and chaotropes according to their capability to salt-out or salt-in proteins. The feebly hydrated chaotropes can effectively produce ion pairs on coupling with the zwitterionic groups, which is the cause of an increase in hydration characteristics of zwitterionic polymers leading to the degradation of antifouling property. The two important antifouling PEDOT derivatives are poly(EDOT-EG3) and poly(EDOTPC). The adsorption of positively charged lysozyme (LYZ) and negatively charged BSA show the direct function of the ionic concentration in the presence of three kinds of anions, such as Cl^-, SO_4^{2-}, and ClO_4^- ions. Like other important antifouling polymeric materials, the backbone of PEDOT is normally positively charged due to low oxidation potential [43]. The poly (EDOT-EG3) and poly(EDOT-PC) can decrease most of the common binding from LYZ and BSA. Due to columbic interaction with the positively charged PEDOT, the binding site of BSA was found to be increased than the binding of LYZ in deionized H_2O. In the case of BSA, the nonspecific binding reduces with an increase in ionic concentrations. For this case, two kinds of mechanisms are suggested. In the first mechanism, the columbic interaction between the poly(EDOT-PC) and negatively charged BSA is the cause of weakening after binding the negatively charged ions on PEDOT. Hence, the negatively charged ion is the cause of decreasing the intra or inter-chain linkage within zwitterion leading to the synthesis of a comparatively compressed layer of hydration on the surface of poly(EDOT-PC). More is the compactness of the hydration layer, greater is the resistance toward BSA binding. Both the LYZ and BSA exhibit the least binding strength particularly in deionized water, because of columbic repulsion between the poly (EDOT PC) and positively charged LYZ. The enhancement of ionic strength at first stimulates nonspecific binding because the anions adsorbed on the surface of poly(EDOTPC) undergoes shielding the columbic repulsion between polymers and LYZ. The BSA binding with an increase in ionic strength against the nonspecific binding against LYZ is prevented because of the production of an improved hydration layer on the surface of poly(EDOT-PC). In the case of both LYZ and BSA, the strength of nonspecific binding enhanced due to an increase in the concentration of SO_4^{2-} ions, which is the cause of greater affinity toward the H_2O molecule from SO_4^{2-} ions, resulting in the effective dehydration of the PC group on the surface of the polymer films with a subsequent decrease in property of antifouling. The concentration of the ions present influences the property of antifouling of poly (EDOT-EG3). On the other hand, the PC group of zwitterion, neutral ethylene glycol moiety unable to form columbic interaction within the protein, ion pairs, and the backbone of PEDOT dominates the property of antifouling of poly (EDOT-EG3). The increase in the concentration of SO_4^{2-} ions are the cause of the effective ethylene glycol group on the surface of polymer films with subsequent reduction of the property of antifouling. Although the property of antifouling is connected with the ionic strength, the immobilization of the surface with antifouling moieties normally exhibits the decrease in nonspecific binding with the molecules proteins. In the case of CPs, the columbic interaction in between the proteins and backbone is specifically required [44, 45].

22.6 APPLICATION OF ANTIFOULING CONDUCTING POLYMERS IN THE BIOMEDICAL FIELD

Antifouling CPs have several applications in the biomedical field. Some important applications of these are as follows.

22.6.1 *In Vivo* Electrochemical Biosensing along with High Sensitivity

The PEGylated PANI nanofibers offer the required properties of antifouling in the samples of human serum. The capture probes on subsequent immobilization on their nanofibers of PEGylated PANI via NHS/EDC coupling, actually a DNA electrochemical biosensor having high sensitivity was created, which was successfully used for breast cancer susceptibility gene (BRCA1). Again, BRCA1 possesses the capability of quantifying the complex human serum. A stable and sensitive glucose electrochemical sensor having zwitterion form of poly(sulfobetaine-3,4-ethylenedioxythiophene)

(PSBEDOT) was developed, where the glucose oxidase (GOx) was trapped directly and then undergoes immobilization on the surface of the substrate by the method of electropolymerization of PSBEDOT via using the aqueous solution of GOx. Therefore, the zwitterion form of PSBEDOT was utilized as a 3D matrix in a hydrated state for encapsulation of GOx, which enhances the lifetime and stability of GOx. Hence, the CP functions as a transducer for transmitting signals formed during the process of oxidation of glucose. The zwitterion form of PSBEDOT-GOx electrode offers better stability in the plasma of human blood than PEDOT-GOx electrode without functionalization. The carbon fiber microelectrode (CFE) on coating with phosphorylcholine-functionalized PEDOT-PC in zwitterion form can be successfully used dopamine monitoring *in vivo*. The PEDOT-PC coated with CFE could resist the adsorption of protein along with retaining the time response and stability for *in vivo* regulating dopamine more effectively than PEDOT coating with CFE or hydroxyl-functionalized PEDOT-OH on coating with CFE. The application of zwitterionic groups of antifouling PEG is to enhance the performance of electrochemical sensing in a biological system and monitoring electrochemical reactions *in vivo*. The property of antifouling in the case of CPs is vital for enhancing the function of electrochemical sensing *in vivo* with emphasizing lifetime and sensitivity. However, some other parameters which influence the performance of sensing include electrostatic forces of interaction with analytes and electrodes of the active surface area [46, 47].

22.6.2 Dual-Functional Antibacterial Surface

Now new advanced polymeric biomaterials although developed, the infection due to the implantation of polymeric materials in the human body is a challenge in the medical sector or specifically in clinical settings. Normally, two types of surfaces named antimicrobial and antifouling are suitable to prevent or reduce such kind of infection. The surface of antifouling submissively prevents the microbe adhesion by using zwitterionic or PEG substances as coating material. The surfaces having antimicrobial properties can vigorously kill the forthcoming microorganisms by releasing antibacterial agents or through the immobilization of killing moieties on its contact. Nowadays, smart antibacterial surfaces (dual-functional surface) is designed and developed, which functions both the releasing and killing. The PEDOT and its corresponding composite materials facilitate antibacterial properties because of their inherent positive charge and capability of carrying ions. The nanocomposite poly(vinylpyrrolidone) sulfobetaines on coating with the nanoparticles of FeO can able to kill the germs or bacteria effectively due to absorption of NIR light from PEDOT. A smart antibacterial surface was developed from sulfobetaine-functionalized PEDOT (PSBEDOT), which can be exchanged in between the zwitterionic antifouling state (PSBEDOT reduction) and cationic antimicrobial state (PSBEDOT oxidation) and exhibits outstanding resistance toward the attachment of *Escherichia coli* K12 bacteria [48].

22.6.3 Controlled Release and Cell Capture

It was observed that the antifouling CP has been established itself as a suitable candidate for controlled release of drugs and ions, which are accredited to the tunable charged state of the backbone of the polymeric material. The redox states of the CPs can be modified by the application of varying surface potentials and the ions produced can move out or into the CPs to establish their electrostatic neutrality. Nowadays, antifouling zwitterionic form of PEDOT-based material and hydroquinone-functionalized EDOT (EDOT-HQ) is developed, which is highly effective for biomedical applications. The hydroquinone group is responsible for the immobilization of aminooxy-terminated molecules due to the bonding of oxime ligation, onto it, which can be obtained through oxidizing hydroquinone and forms benzoquinone. The oxime ligation bonding is highly stable under different physiological conditions, which is again cleaned by the application of specific reduction potential. The zwitterionic EDOT-PC and EDOT-HQ were initially co-electropolymerized and form films of poly (EDOT-PC-*co*-EDOT-HQ). The controlled and regulated way of release and attachment of NIH3T3 cells on the film of this copolymer was obtained. Before coupling with poly(EDOT-HQ-*co*-EDOT-PC) on the RGD peptide, the film of the polymer exhibits appropriate antifouling properties and prevented attachment of NIH3T3. By the application of adequate oxidation potential on the polymeric material film, hydroquinone is converted into benzoquinone. The RGD peptides along with the terminal amino-oxy group have been successfully conjugated on the film (polymer) with subsequent formation of a strong bond with NIH3T3 cells on the surface of the polymeric material. By application of reduction potential, the NIH3T3 cells were successfully from the films of poly(EDOT-HQ-*co*-EDOT-PC), due to the breaking of the RGD peptide bond. The phenylboronic acid-functionalized EDOT (EDOT-PBA) in combination with antifouling EDOT-EG3 forms a new novel glycan-stimulated PEDOT-based nanomaterial, which can be utilized for CTCs purification from our blood samples. The 3D Nano Velcro chip of PEDOT offers a more capture efficiency and permits the mild release of CTC cells and that release can able for effective CTC purification for analysis from the purified form of CTCs. Hence, the existence of antifouling moieties and their density is the vitals cause for the smooth and effective release of cells without any damage or injury. The elements of antifouling may not be required for the release of tiny molecules from the CPs [49, 50].

22.7   CONCLUSION

The conducting polymeric materials are effectively applied in different sectors of the biomedical field such as drug delivery, tissue engineering, bioactuators, cardiovascular, neural electrodes, biosensors, and neuroprosthetic devices. The CPs are found to be extremely higher in biocompatibility as compared with traditional semiconducting materials, metals, and ceramics. The mechanism of conduction in the case of CP is due to the transmission of bipolarons and polarons. However, more research work is required to improve the essential properties of CPs. The antifouling CPs are now considered as one of the most viable areas of advanced biomedical applications, owing to their essential distinct properties. The antifouling CPs are now designed and developed, which can be successfully used in various advanced biomedical applications such as implanted bioelectronics, controlled cell capture, dual-function antibacterial coating, and implanted bioelectronics. But, there are many challenges have still to be needed for the commercialization of the products of antifouling CPs. For the application of antifouling CPs, we have to overcome the basic challenge of adequate stability and durability. The new advanced CPs now offer amazing results like the coupling of antifouling moieties with the fouling release parts. The bound water structure neighboring to the zwitterionic groups and variation of ionic strength is the cause of change in antifouling properties, but still, it does not fully understand. Another significant application is the techniques of surface analysis, which is highly advantageous to know about the mechanism of the properties of antifouling.

REFERENCES

1. Jagur-Grodzinski J (2012) Biomedical applications of electrically conductive polymeric systems. *e-Polymers*, 12(1): 1–19.
2. Naveen M H, Gurudatt N G, Shim Y-B (2017) Applications of conducting polymer composites to electrochemical sensors: a review. *Applied Materials Today*, 9: 419–433.
3. Han R, Wang G, Xu Z, Zhang L, Li Q, Ha Y, Luo X (2020). Designed antifouling peptides planted in conducting polymers through controlled partial doping for electrochemical detection of biomarkers in human serum. *Biosensors and Bioelectronic*, 164: 1–36.
4. Yang L, Wang H, Lü H, Hui N (2020) Phytic acid functionalized antifouling conducting polymer hydrogel for electrochemical detection of microRNA. *Analytica Chimica Acta*, 1124: 104–112.
5. Nezakati T, Seifalian A, Tan A, Seifalian AM (2018) Conductive polymers: opportunities and challenges in biomedical applications. *Chemical Reviews*, 118(14): 6766–6843.
6. Dalsin J L, Messersmith PB (2005) Bioinspired antifouling polymers. *Materials Today*, 8(9): 38–46.

7. Chundawat NS, Chauhan NPS (2019) Conducting polymers with antimicrobial activity. *Biocidal Polymers*, 7: 147–170.

8. Kenry Liu B (2018) Recent advances in biodegradable conducting polymers and their biomedical applications. *Biomacromolecules*, 19(6): 1783–1803.

9. Huang Y, Kormakov S, He X, Gao X, Zheng X, Liu Y, Wu D (2019) Conductive polymer composites from renewable resources: an overview of preparation, properties, and applications. *Polymers*, 11(2): 187–219.

10. Namsheer K, Rout C S (2021) Conducting polymers: a comprehensive review on recent advances in synthesis, properties and applications. *RSC Advance*, 11: 5659–5697.

11. Dan S, Feng D, Ji X, Che J (2014) Application of conducting polymers in controlled drug delivery system. *Progress in Chemistry –Beijing*, 26(12): 1962–1976.

12. Puiggalí-Jou A, del Valle L J, Alemán C (2019) Drug delivery systems based on intrinsically conducting polymers. *Journal of Controlled Release*. 309: 244–264.

13. Maziza A, Özgür E, Bergauda C, Uzunc L (2021) Progress in conducting polymers for biointerfacing and biorecognition applications. *Sensors and Actuators Reports*, 3: 1–17.

14. Bansal M, Dravid A, Aqrawe Z, Montgomery J, Wu Z, Svirskis D (2020) Conducting polymer hydrogels for electrically responsive drug delivery. *Journal of Controlled Release*, 328: 192–209.

15. George P M, Lyckman A W, LaVan D A, Hegde A, Leung Y, Avasare, R, Sur M (2005) Fabrication and biocompatibility of polypyrrole implants suitable for neural prosthetics. *Biomaterials*, 26(17): 3511–3519.

16. Vallet-Regí M, Colilla M, Izquierdo-Barba I, Manzano M (2017) Mesoporous silica nanoparticles for drug delivery: current insights. *Molecules*, 23(1): 47–63.

17. Hu F, Xue Y, Xu J and Lu B (2019) PEDOT-based conducting polymer actuators. *Frontier in Robotics and AI*, 6: 114–130.

18. Bay L, Jacobsen T, Skaarup S, West K (2001) Mechanism of actuation in conducting polymers: osmotic expansion. *The Journal of Physical Chemistry B*, 105(36): 8492–8497.

19. Kongahage D, Foroughi J (2019) Actuator materials: review on recent advances and future outlook for smart textiles. *Fibers*, 7(3): 21–39.

20. Peng C, Zhang S, Jewell D, Chen G Z (2008) Carbon nanotube and conducting polymer composites for supercapacitors. *Progress in Natural Science*, 18(7): 777–788.

21. Guo B, Ma P X (2018) Conducting polymers for tissue engineering. *Biomacromolecules*, 19(6): 1764–1782.

22. Abidian M R, Corey J M, Kipke D R, Martin D C (2010) Conducting-polymer nanotubes improve electrical properties, mechanical adhesion, neural attachment, and neurite outgrowth of neural electrodes. *Small*, 6(3): 421–429.

23. Chen M-C, Sun Y-C, Chen Y.-H (2013) Electrically conductive nanofibers with highly oriented structures and their potential application in skeletal muscle tissue engineering. *Acta Biomaterialia*, 9(3): 5562–5572.

24. Gu K, Kim M S, Kang C M, Kim J-I, Park S J, Kim C-H (2014) Fabrication of conductive polymer-based nanofiber scaffolds for tissue engineering applications. *Journal of Nanoscience and Nanotechnology*, 14(10): 7621–7626.

25. Nemati S, Ki S, Shin Y M, Shin H (2019) Current progress in application of polymeric nanofibers to tissue engineering. *Nano Convergence*, 6(1): 36–51.

26. Jiang L, Wang Y, Liu Z, Ma C, Yan H, Xu N, Su X (2019) 3D printing and injectable conductive hydrogels for tissue engineering application. *Tissue Engineering Part B: Reviews*, 25 (5): 1–37.

27. Mawad D, Artzy-Schnirman A, Tonkin J, Ramos J, Inal S, Mahat M, Stevens M M (2016) Electroconductive hydrogel based on functional poly(ethylenedioxy thiophene). *Chemistry of Materials*, 28(17): 6080–6088.

28. Xu J, Tsai Y-L, Hsu S (2020) Design strategies of conductive hydrogel for biomedical applications. *Molecules*, 25(22): 5296–5309.

29. Giri T K, Thakur A, Alexander A, Ajazuddin Badwaik H, Tripathi D K (2012) Modified chitosan hydrogels as drug delivery and tissue engineering systems: present status and applications. *Acta Pharmaceutica Sinica B*, 2(5): 439–449.

30. Zhao X, Guo B, Ma P X (2015) Single component thermo-gelling electroactive hydrogels from poly(caprolactone)–poly(ethylene glycol)–poly(caprolactone)-graft-aniline tetramer amphiphilic copolymers. *Journal of Materials Chemistry B*, 3(43): 8459–8468.

31. Turnbull G, Clarke J, Picard F, Riches P, Jia L, Han F, Shu W (2018) 3D bioactive composite scaffolds for bone tissue engineering. *Bioactive Materials*, 3(3): 278–314.

32. Athukorala SS, Tran TS, Balu R, Truong VK, Chapman J, Dutta NK Roy Choudhury N (2021) 3D printable electrically conductive hydrogel scaffolds for biomedical applications: a review. *Polymers*, 13: 474–497.

33. Green R, Abidian M R (2015) Conducting polymers for neural prosthetic and neural interface applications. *Advanced Materials*, 27(46): 7620–7637.

34. Harris A R, Morgan S J, Chen J, Kapsa R M I., Wallace G G, Paolini A G (2012) Conducting polymer coated neural recording electrodes. *Journal of Neural Engineering*, 10(1): 1–17.

35. Ravichandran R, Sundarrajan S, Venugopal J R, Mukherjee S, Ramakrishna S (2010) Applications of conducting polymers and their issues in biomedical engineering. *Journal of The Royal Society Interface*, 7(Suppl_5): S559–S579.

36. Cui Z, Yang B, Li R-K (2016) Application of biomaterials in cardiac repair and regeneration. *Engineering*, 2(1): 141–148.

37. Cristallini C, Vitale E, Giachino C, Rastaldo R (2020) Nanoengineering in cardiac regeneration: looking back and going forward. *Nanomaterials*, 10(8): 1587.

38. Wu J-G, Chen J-H, Liu K-T, Luo S-C (2019) Engineering antifouling conducting polymers for modern biomedical applications. *ACS Applied Materials & Interfaces*, 11(24), 21294–21307.

39. Baldissera A F, Miranda K L, de Bressy C, Martin C, Margaillan A, Ferreira C A (2015). Using conducting polymers as active agents for marine antifouling paints. *Materials Research*, 18(6): 1129–1139.

40. Wang W, Cui M, Song Z, Luo X (2016) An antifouling electrochemical immunosensor for carcinoembryonic antigen based on hyaluronic acid doped conducting polymer PEDOT. *RSC Advances*, 6(91): 88411–88416.

41. Wylie M P, Bell S E J, Nockemann P, Bell R, McCoy C P (2020) Phosphonium ionic liquid-infused poly(vinyl chloride) surfaces possessing potent antifouling properties. *ACS Omega*, 5(14): 7771–7781.

42. Hsu C-C, Cheng Y-W, Liu C-C, Peng X-Y, Yung M-C, Liu T-Y (2020) Anti-bacterial and anti-fouling capabilities of poly(3,4-ethylenedioxythiophene) derivative nanohybrid coatings on SUS316L stainless steel by electrochemical polymerization. *Polymers*, 12(7): 1467–1484.

43. Gu L, Xie M-Y, Jin Y, He M, Xing X-Y, Yu Y, Wu Q-Y (2019) Construction of antifouling membrane surfaces through layer-by-layer self-assembly of lignosulfonate and polyethyleneimine. *Polymers*, 11(11): 1782–1792.

44. Goda T, Miyahara Y (2018) Electrodeposition of zwitterionic PEDOT films for conducting and antifouling surfaces. *Langmuir*, 35(5): 1126–1133.

45. Schlenoff J B (2014) Zwitteration: coating surfaces with zwitterionic functionality to reduce nonspecific adsorption. *Langmuir*, 30(32): 9625–9636.

46. Chen Y, Luo S-C (2018) Synergistic effects of ions and surface potentials on antifouling poly(3,4-ethylenedioxythiophene): comparison of oligo(ethylene glycol) and phosphorylcholine. *Langmuir*, 35(5): 1199–1210.

47. Zang C, Wang G, Hein R, Lui N, Luo X, Davis J J (2020) Antifouling strategies for selective in vitro and in vivo sensing. *Chemical Reviews*, 120(8): 1–38.

48. Cao B, Lee C-J, Zeng Z, Cheng Cong H, Cheng G (2016) Electroactive poly(sulfobetaine-3,4-ethylenedioxythiophene) (PSBEDOT) with controllable antifouling and antimicrobial properties. *Chemical Science*, 7(3): 1976–1981.

49. Hsiao Y-S, Ho B-C., Yan H-X, Kuo C-W, Chueh D-Y, Yu H, Chen P (2015) Integrated 3D conducting polymer-based bioelectronics for capture and release of circulating tumor cells. *Journal of Materials Chemistry B*, 3(25): 5103–5110.

50. Hong W Y, Jeon S H, Lee E S, Cho Y (2014) An integrated multifunctional platform based on biotin-doped conducting polymer nanowires for cell capture, release, and electrochemical sensing. *Biomaterials*, 35(36): 9573–9580.

Index

For Product Safety Concerns and Information please contact our EU
representative GPSR@taylorandfrancis.com
Taylor & Francis Verlag GmbH, Kaufingerstraße 24, 80331 München, Germany

www.ingramcontent.com/pod-product-compliance
Lightning Source LLC
Chambersburg PA
CBHW080906220326
41598CB00034B/5489